RESEARCH HANDBOOK ON INTELLECTUAL PROPERTY AND GEOGRAPHICAL INDICATIONS

RESEARCH HANDBOOKS IN INTELLECTUAL PROPERTY

Series Editor: Jeremy Phillips, *Intellectual Property Consultant, Olswang, Research Director, Intellectual Property Institute and co-founder, IPKat weblog*

Under the general editorship and direction of Jeremy Phillips comes this important new *Handbook* series of high quality, original reference works that cover the broad pillars of intellectual property law: trademark law, patent law and copyright law – as well as less developed areas, such as geographical indications, and the increasing intersection of intellectual property with other fields. Taking an international and comparative approach, these *Handbooks*, each edited by leading scholars in the respective field, will comprise specially commissioned contributions from a select cast of authors, bringing together renowned figures with up-and-coming younger authors. Each will offer a wide-ranging examination of current issues in intellectual property that is unrivalled in its blend of critical, innovative thinking and substantive analysis, and in its synthesis of contemporary research.

Each *Handbook* will stand alone as an invaluable source of reference for all scholars of intellectual property, as well as for practising lawyers who wish to engage with the discussion of ideas within the field. Whether used as an information resource on key topics, or as a platform for advanced study, these *Handbooks* will become definitive scholarly reference works in intellectual property law.

Titles in the series include:

The Law and Theory of Trade Secrecy
A Handbook of Contemporary Research
Edited by Rochelle C. Dreyfuss and Katherine J. Strandburg

Research Handbook on Intellectual Property Licensing
Edited by Jacques de Werra

Criminal Enforcement of Intellectual Property
A Handbook of Contemporary Research
Edited by Christophe Geiger

Research Handbook on Cross-border Enforcement of Intellectual Property
Edited by Paul Torremans

Research Handbook on Human Rights and Intellectual Property
Edited by Christophe Geiger

International Intellectual Property
A Handbook of Contemporary Research
Edited by Daniel J. Gervais

Indigenous Intellectual Property
A Handbook of Contemporary Research
Edited by Matthew Rimmer

Research Handbook on Intellectual Property and Geographical Indications
Edited by Dev S. Gangjee

Research Handbook on Intellectual Property and Geographical Indications

Edited by

Dev S. Gangjee

Associate Professor, Law Faculty, University of Oxford; Research Fellow, Oxford Intellectual Property Research Centre, UK

RESEARCH HANDBOOKS IN INTELLECTUAL PROPERTY

Cheltenham, UK • Northampton, MA, USA

© The Editor and Contributors Severally 2016

All rights reserved. No part of this publication may be reproduced, stored in a retrieval system or transmitted in any form or by any means, electronic, mechanical or photocopying, recording, or otherwise without the prior permission of the publisher.

Published by
Edward Elgar Publishing Limited
The Lypiatts
15 Lansdown Road
Cheltenham
Glos GL50 2JA
UK

Edward Elgar Publishing, Inc.
William Pratt House
9 Dewey Court
Northampton
Massachusetts 01060
USA

A catalogue record for this book
is available from the British Library

Library of Congress Control Number: 2015950288

This book is available electronically in the Elgaronline
Law subject collection
DOI 10.4337/9781784719470

ISBN 978 1 84720 130 0 (cased)
ISBN 978 1 78471 947 0 (eBook)

Typeset by Servis Filmsetting Ltd, Stockport, Cheshire
Printed and bound by CPI Group (UK) Ltd, Croydon, CR0 4YY

Contents

List of figures and tables vii
List of contributors viii
Acknowledgements x
List of abbreviations xi
Permissions xiii

1 Introduction: timeless signs or signs of the times? 1
   Dev S. Gangjee

PART I HISTORY AND CONCEPTS

2 French collective wine branding in the nineteenth–twentieth centuries 13
   Alessandro Stanziani

3 'Translating *terroir*' revisited: the global challenge of French AOC labeling 46
   Elizabeth Barham

4 *Terroir* and the sense of place 72
   Laurence Bérard

PART II INTERNATIONAL PROTECTION

5 Geographical Indications under WIPO-administered treaties 95
   Matthijs Geuze

6 Geographical Indications under TRIPS 123
   Daniel Gervais

7 Rethinking GI extension 146
   Michael Handler

8 International protection of Geographical Indications: the WTO multilateral register negotiations 183
   José Manuel Cortés Martín

9 Thinking locally, acting globally: how trade negotiations over
   Geographical Indications improvise 'fair trade' rules   202
   Antony Taubman

PART III   DOMESTIC PROTECTION MODELS

10 A history of Australia's wine Geographical Indications
    legislation   245
    Stephen Stern

11 A comparative analysis of GIs for handicrafts: the link to
    origin in culture as well as nature?   292
    Delphine Marie-Vivien

12 Geographical Indication protection in China   327
    Haiyan Zheng

PART IV   CRITICAL ISSUES

13 Learning to love my PET – the long road to resolving conflicts
    between trade marks and Geographical Indications   361
    Burkhart Goebel and Manuela Groeschl

14 The Budweiser cases: Geographical Indications v trade marks   396
    Christopher Heath

15 Geographical Indications and protected designations of origin:
    intellectual property tools for rural development objectives   440
    Dominique Barjolle

16 Social gains from the GI for Feni: will market size or
    concentration dominate outcomes?   463
    Dwijen Rangnekar and Pranab Mukhopadhyay

17 From *terroir* to pangkarra: Geographical Indications of origin
    and Indigenous knowledge   484
    Brad Sherman and Leanne Wiseman

18 Genericide: the death of a Geographical Indication?   508
    Dev S. Gangjee

Index   549

Figures and tables

FIGURES

15.1	Conceptual framework of the interactions within the system of market-organisation-territory	460
16.1	Proportion of distillers and bottlers of Feni, and their alcoholic concentration	470

TABLES

2.1	Gironde, average price (francs) of a hectolitre of wine	24
11.1	Subject matter categories and types of links	317
11.2	Scope of protection depending upon the nature of the link	320
15.1	The 21 concerned PDO-PGI Product Supply Chains	449
15.2	Coordination and cooperation at the collective management level	450
15.3	Creation of commercial value for the nine PDO cheeses studied	452
15.4	Qualitative measurement of the support to agricultural income effect	454
15.5	Indirect effects of some PDO cheeses	457
16.1	Production of Feni, IMFL and beer	465
16.2	Official description of Feni	468
16.3	Area under cashew plantation	474
16.4	Cashew production in India (2004–5)	475
16.5	Price of cashew liquor and price index	476

Contributors

Elizabeth Barham, Center for Advanced Spatial Technologies, University of Arkansas; Executive Director of the American Origin Products Research Foundation.

Dominique Barjolle, Senior Researcher (Sustainable Agroecosystems, Institute for Agricultural Sciences, Department of Environmental Systems Science), ETH (Federal Institute of Technology), Zurich.

Laurence Bérard, Lead Researcher, Project on Ressources des Terroirs – Cultures, Usages, Sociétés; Joint Research Unit of Eco-anthropology and Ethnobiology, Centre National de la Recherche Scientifique and MNHN, France.

Dev S. Gangjee, Associate Professor in Intellectual Property, Law Faculty, University of Oxford; Research Fellow, Oxford Intellectual Property Research Centre.

Daniel Gervais, Professor of Law; Director of the Vanderbilt Intellectual Property Program, Vanderbilt University.

Matthijs Geuze, Head of the Lisbon Registry in the Brands and Designs Sector, World Intellectual Property Organization, Geneva.

Burkhart Goebel, Regional Managing Partner for Continental Europe, Hogan Lovells; Former Chair, Geographical Indications Committee of INTA.

Manuela Groeschl, Senior Counsel in the legal department of the German Federal Ministry of Finance.

Michael Handler, Associate Professor, Faculty of Law, University of New South Wales.

Christopher Heath, Member of the Boards of Appeal, European Patent Office; former Head of Asian Department, Max Planck Institute, Munich.

Delphine Marie-Vivien, Researcher at UMR Innovation, Cirad, France.

José Manuel Cortés Martín, Professor of Public International Law and International Relations at the Pablo de Olavide University, School of Law, Seville; *Jean Monnet* Professor of European Union Law.

Pranab Mukhopadhyay, Professor of Economics, Goa University.

Dwijen Rangnekar, School of Law, University of Warwick.

Brad Sherman, Professor of Law, University of Queensland.

Alessandro Stanziani, Professor at the Ecole des Hautes Etudes en Sciences Sociales (EHESS); Research Director and Senior Researcher, Centre National des Recherches Scientifiques (CNRS), Paris.

Stephen Stern, Partner, Corrs Chambers Westgarth, Melbourne; International President of the Association Internationale des Juristes du Droit de la Vigne et du Vin 1993–1996; International Vice-President since 1996.

Antony Taubman, Director, Intellectual Property Division, World Trade Organization.

Leanne Wiseman, Associate Professor of Law, Griffith University; Associate Director of the Australian Centre for IP in Agriculture (ACIPA).

Haiyan Zheng, Director of Examination Division I, Trade Mark Office, State Administration of Industry and Commerce (SAIC), P.R. China.

Acknowledgements

Perhaps overly influenced by its frequently controversial subject matter, this book has had a long and difficult gestation period. I am profoundly grateful to all those involved for staying committed to this project.

First, our publisher Edward Elgar, for their supportiveness, professionalism and sheer decency. I cannot praise them enough and acknowledge my indebtedness to them. I am particularly grateful to Luke Adams, Aisha Bushby, Rosemary Campbell, Rebecca Stowell and our series editor, Jeremy Philips.

Second, the contributors to this volume, for their exceptional patience and willingness to update submissions as the months became years.

Third, my partner Nikita, for helping me finish what I had begun. I could not have done this without her support.

As this book neared completion, we were deeply saddened to learn of the death of our colleague Dwijen Rangnekar (17 April 1965–30 October 2015). Dwijen was a globally respected GI scholar as well as a good friend to many of us. His work drew on approaches from sociology, anthropology, evolutionary economics and critical legal studies to produce pioneering and inspiring research. Its influence extended well beyond academia into the fields of international policy making as well as GI implementation in practice. Dwijen was also a generous colleague and wonderful company – foods, wines and spirits were not merely an academic interest. He will be greatly missed.

Abbreviations

ALR	Australian Law Reports
AO	Appellation d'Origine/ Appellation of Origin
AOC	Appellation d'Origine Contrôlée/ Controlled Appellation of Origin
ATMO	Australian Trade Mark Office Decision
BIRPI	Bureaux Internationaux Réunis pour la Protection de la Propriété Intellectuelle/ United International Bureau of Industrial, Literary and Artistic Property
CFI	Court of First Instance of the European Communities
CFR	Code of Federal Regulations (US)
Ch	Chancery (Court in the UK)
CIVC	Comité Interprofessionnel du Vin de Champagne
CJEU	Court of Justice of the European Union (formerly the European Court of Justice or ECJ)
CLR	Commonwealth Law Reports
CTMR	Community Trade Mark Regulation
EIPR	European Intellectual Property Review
ETMR	European Trade Mark Reports
FSR	Fleet Street Reports
FTA	Free Trade Agreement
GC	General Court (European Union) (formerly, the CFI)
GI	Geographical Indication
GIC	Geographical Indications Committee (Australia)
GRUR	Gewerblicher Rechtsschutz und Urheberrecht
HCA	High Court of Australia
ICTSD	International Centre for Trade and Sustainable Development
INAO	Institut National des Appellations d'Origine (previously); Institut National des Appellations d'Origine et la Qualité (at present)
INTA	International Trade Mark Association
IP	Intellectual Property
IPQ	Intellectual Property Quarterly
IPR	Intellectual Property Right
IS	Indication of Source
JIPLP	Journal of Intellectual Property Law and Practice
JO	Journal Officiel (France)

JWIP	Journal of World Intellectual Property
Lisbon	Lisbon Agreement for the Protection of Appellations of Origin and their International Registration of 1958
Madrid	Madrid Agreement for the Repression of False or Deceptive Indications of Source on Goods of 1891
NZLR	New Zealand Law Reports
OHIM	Office of Harmonization for the Internal Market (European Union)
OJ	Official Journal of the European Union
oriGIn	Organization for an International Geographical Indications Network
Paris	Paris Convention for the Protection of Industrial Property of 1883
PDO	Protected Designation of Origin
PGI	Protected Geographical Indication
PIBD	Propriété Industrielle Bulletin Documentaire (France)
RPC	Reports of Patent Cases
SCT	Standing Committee on the Law of Trademarks, Industrial Designs and Geographical Indications (WIPO)
TMD	Trade Marks Directive
TMEP	Trade mark Manual of Examining Procedure (USPTO)
TRIPS	Agreement on Trade Related Aspects of Intellectual Property Rights
UNTS	United Nations Treaty Series
USPTO	United States Patent and Trade Mark Office

Permissions

Grateful acknowledgment is made for permission to reprint the following material:

Excerpts from: Alessandro Stanziani, 'Wine Reputation and Quality Controls: The Origin of the AOCs in 19th Century France' (2004) 18 *European Journal of Law and Economics* 149–167. Reprinted by permission of Springer Verlag.

Excerpts from: Elizabeth Barham, 'Translating *Terroir*: The Global Challenge of French AOC Labeling' (2003) 19(1) *Journal of Rural Studies* 127–138. Reprinted by permission of Elsevier.

Excerpts translated from: Laurence Bérard, 'Du terroir au sens des lieux' (2011) in Claire Delfosse (ed.), *La mode du terroir et les produits alimentaires* 41–55. Reprinted with permission from Paris, Les Indes savants.

Antony Taubman, 'Thinking Locally, Acting Globally: How Trade Negotiations over Geographical Indications Improvise "Fair Trade" Rules' (2008) 3 *IPQ* 231–254. Reprinted by permission of Sweet & Maxwell.

Excerpts from: Christopher Heath, 'The Budweiser Cases: A Brewing Conflict' (2011) in Christopher Heath and Anselm Kamperman Sanders (eds), *Landmark Intellectual Property Cases and Their Legacy* 181–244. Reprinted by permission of Kluwer Law International.

Chapter 16 is based on a study funded by the Economic and Social Research Council (ESRC), UK (RES-061-23-0119) (www.warwick.ac.uk/go/feni). The authors are grateful to Lizette D'Çosta, Rohita Deshprabhu, Santosh Maurya and Suryabhan Mourya who assisted with data collection and processing. Inputs of various commentators especially Rucha Ghate, Jorge Larson Guerra, T.C. James, Latha Nair, among others at Stakeholder meetings held in Goa in 2008 and 2009 are gratefully acknowledged. G.D Bhakta, Mac Vaz and S. Parab were generous with their time and insights of the Feni industry.

1. Introduction: timeless signs or signs of the times?
Dev S. Gangjee

After existing at the margins for over a century, Geographical Indications (GI) scholarship has come of age. Recognising and celebrating its maturity, this edited collection has two objectives: (1) to gather together the most insightful and interesting research, as a convenient and enduring point of reference; and (2) to facilitate an interdisciplinary conversation, allowing future avenues of enquiry to emerge from the themes and insights which follow. In looking back across several decades of deliberations, disagreements and experiments, the contributions in this volume offer up productive ways of looking forwards.

In an increasingly globalised world, place and provenance matter as never before. Regimes regulating the use of GIs set out the conditions under which these signs signalling the provenance of products can be formally recognised and protected, the criteria to be satisfied when collectively using such signs and the extent to which these signs are protected against 'outsiders'. The law relating to GIs therefore protects products such as Rioja, Darjeeling and Café de Colombia. It is certainly true that GI protection is directed towards ensuring accurate information provision about geographical origin and quality in the marketplace, while simultaneously protecting commercially valuable reputations. Such designations are channels of communication and – like trade marks – incorporate this origin-signalling function. Yet it is also claimed that GIs are justifiably different and this assertion rests upon the unique or distinctive link between such regional products and their places of origin. The most influential articulation of this link is the French notion of *terroir*, originally associated with viticulture and periodically reinvented not only in France but subsequently across the EU, then in New World wine-producing countries and most recently across the global South.

The centrality of this link to place opens up an additional dimension of regulation in *sui generis* GI protection regimes. While they are designed to regulate the use of signs in the marketplace, they are additionally – via the formal product specification or *cahiers des charges* – implicated in defining 'authentic' or historically stabilised products as well as collectively generated production techniques. Thus both *signs* and *the products they represent* are the objects of regulation. Moving further back along the chain of production, GIs are seen as potential bulwarks against commoditisation because they do not merely designate what the product is (its appearance or physical and organoleptic qualities) but also where, by whom and how – very specifically – it was made. This anchorage to place and inter-generational, collective investment has led to additional policy implications and justificatory accounts for GI regimes, ranging from enabling endogenous territorial development to biodiversity preservation and even cultural heritage recognition. These justifications extend well beyond the informational efficiency paradigm (ensuring uncluttered signalling in the marketplace), which otherwise accounts for the legal protection of signs against misrepresentation. In précis, these additional justificatory layers are put to work to explain the broader scope of protection as well as enforcement mechanisms for GIs. Consequently, when contrasted with trade marks *sui generis* GI regimes are seen as more prescriptive or demanding. Since the link to place is what sets GIs apart, the verification of this link and causally related product qualities calls for greater scrutiny and public oversight during the registration process, as well as subsequently during commercial use.

As a distinct regime within the categories of intellectual property law, GIs are therefore both conceptually and normatively intriguing. However, there remains the persistently practical as well as controversial question of how best to protect them. Debates occur both at the national level (for instance, involving the choice between *sui generis* or trade mark law regimes) as well as internationally (exploring the unfinished business left behind by the WTO's TRIPS Agreement). Unresolved issues remain in relation to the nature of the link between product and place; the design and implementation of *sui generis* GI protection; whether trade mark law is a viable alternative; the interaction between trade marks and GIs; and ongoing bilateral as well as multilateral negotiations to improve upon the scope – as well as limits – of international GI protection.

These questions have attracted not only specialists in intellectual property law but increasingly, researchers across disciplines in the social sciences and beyond. The greatest strength of this volume is that historians, geographers, sociologists, economists and anthropologists have joined legal scholars, legal practitioners and senior IP bureaucrats in unpacking this link between products and places, as well as the legal claims resting upon this link. This volume is therefore indispensable for both new entrants to this complex yet compelling field and established researchers, who should enjoy the interdisciplinary provocation from experienced contributors. A second virtue is that the contributors include both friends and critics (or in some cases, friendly critics) of GIs.

You will find chapters following divergent paths in their interpretations of WTO panel rulings, the benefits of the Lisbon Agreement, the extent to which GIs can deliver on endogenous development goals or can be responsive to custodians of indigenous knowledge. There is productive disagreement within these pages. The third distinctive feature of this volume is that its coverage is intended to reflect each of the principal institutional layers as well as major debates in this area, in so far as word limits will permit. Readers will find the international treaties and institutional architecture established by WIPO, the WTO and certain bilateral initiatives analysed in detail by expert commentators. This is complemented by in-depth coverage of the origins and operation of the influential French appellation system, distinct new world regimes such as Australia's wine GI system and Asian experiments with GIs, including coverage of China and India. One of the most exciting recent developments – the redrafting of the Lisbon Agreement in 2015 – is analysed in Chapters 5, 6 and 14.

Consequently, one emergent question addressed in different ways is the extent to which a regime arising within a specific European historical and institutional context can be adapted and successfully transplanted elsewhere. A second is the extent to which we may be witnessing the beginnings of a truly global approach to GIs, arising from the acceleration of bilateral and plurilateral agreements and driven by the prospect of international registration. A third line of analysis explores the extent to which GIs could (in theory) and are (in practice) delivering on goals like heritage preservation or embedded development.

The value of an interdisciplinary approach to GIs is evident in Part I, which covers these historical, institutional and conceptual moorings. Chapter 1 opens with the meticulous archival analysis of Alessandro Stanziani, revealing that the French wine appellation system originated in attempts to stabilise and co-ordinate completely new vine-growing and wine-making practices in the post-phylloxera era. As opposed to merely recognising historical or 'authentic' products, these laws were directed towards reconciling innovation with tradition, allowing growers and producers to negotiate the content of acceptable wine production techniques, thereby achieving workable compromises. Disputes usually related to geographical boundaries of regions or the extent to which a practice qualified as traditional or 'loyal'. Accurately informing consumers about the provenance of wine was a secondary or incidental objective. As opposed to their external or consumer-facing signalling functions, Stanziani therefore emphasises the internal dynamics shaping the definitions of both regions and products. He suggests that these early legislative episodes prioritised regulating the fairness of producers' and wine growers' conduct *inter se*, as opposed to consumer protection. This internal aspect of GIs is further developed by the remaining two chapters in Part I.

In Chapter 2, Elizabeth Barham takes up the challenge of 'translating *terroir*', exploring the conceptual origins, normative goals and political economy potential of this controversial link to place. She draws on conventions theory – an institutionalist approach to studying economic co-ordination – to outline the ways in which GIs challenge neo-liberal conceptualisations of industrial-scale agricultural production, by explicitly referencing place or territory. This reconnects people, production and place in significant ways. Barham specifically investigates the French Appellation d'Origine Contrôlée (AOC) regime as a practical operationalisation of the *terroir* concept. Formal recognition as an AOC product requires satisfactory progress through several bureaucratic stages, each of which calls for moments of judgement. These decisions require choices to be made between natural factors, human factors or techniques and the history of the product, which combine to produce the 'typicity' of the product. While reinforcing Stanziani's insight about the negotiated nature of AOCs, Barham is more optimistic and views the drafting of an AOC specification as a combination of research-driven conclusions and negotiated outcomes. This permits market logic to co-exist with other values and ways of viewing the world. Also approaching *terroir* through the lens of social theory, Laurence Bérard – a respected ethnographer of GI products and communities – offers up a genealogy of this concept in Chapter 4. She describes its historical origins as a political resource, within the dual contexts of environmental determinism as well as the politics of retaining regional identities in the face of centralising pressures. Notably, she charts the rise of human or cultural factors and argues that this intergenerational, historical dimension to all such products, linked to collective memory, is what gives place depth. GIs therefore are both constituted by and help to make 'place'.

With a more nuanced understanding of the conceptual foundations in place, Part II proceeds to consider international GI protection. This sequence recommends itself because for many countries, GI protection obligations have been imposed by international treaties, as opposed to emerging first within domestic law. A pair of experienced commentators set out the present framework of international protection, commencing with Chapter 5 by Matthijs Geuze. He reviews WIPO's longstanding commitments in this area, outlining the contents of the three major agreements under its auspices – the Paris Convention of 1883, the Madrid Agreement for the Repression of False or Deceptive Indications of Source of 1891 and the Lisbon Agreement of 1958. These agreements continue to be influential because they have provided much of the raw material and shaped the parameters of the present international regime in TRIPS. Geuze is particularly well qualified to guide readers through the process culminating in the Geneva Act of 2015, which significantly revises the Lisbon Agreement. Key issues considered include the broadening of definitions, the basis for objecting to a registration, conflicts between trade marks and GIs, unprotected status for generic expressions and homonymous GIs (such as Pisco, claimed by both Peru and Chile). Daniel Gervais devotes Chapter 6 to the analysis of the GI provisions in TRIPS, noting that the resurgence of global interest in GIs is owed, in part, to the involvement of developing countries with large catalogues of regional speciality products. He retraces the origins of the GI provisions contained in Arts 22–24 of TRIPS while emphasising that these provisions directly allude to much unfinished business via a 'built-in agenda' for negotiations. A significant element of this is the commitment to establish an international register for wines and spirits. Gervais innovatively proposes that the (potentially upgraded) international register under the Lisbon Agreement may be a receptive home for giving effect to the wines and spirits register and to this extent, his chapter complements the analysis of the Lisbon regime in Geuze's.

The subsequent pair of chapters in Part II tackle the two most controversial aspects of the ongoing GI negotiations at the WTO. In Chapter 7 Michael Handler critically evaluates the case for extending the enhanced protection available to wines and spirits in Art 23 of TRIPS to all products. In what must surely be the definitive review of this topic, Handler convincingly argues that the EU – the principal proponent of extension – has failed to articulate a convincing argument for extension. While other justifications for extension might exist, the arguments presented by proponents are too simplistic. He concludes by labelling the Art 23 extension debate as something of a 'sideshow' when the real action seems to be taking place bilaterally, in the context of preferential trade agreements. The (re)turn to biliateralism is a significant development and is also noted by other contributors such as Taubman, Stern and myself. In Chapter 8, José Manuel Cortés Martín considers the options for establishing a multilateral register for wines and spirits, reviewing the three different visions of registration contained in EU-led, US-led and Hong Kong, China proposals. These have recently been amalgamated into a consolidated draft text, where two prominent and divisive issues are the legal effects of registration (is there a presumption of validity upon registration and should there be an obligation to consult the register?) and participation in the registration system (does it mandatorily cover the entire WTO membership or only those who opt in to this system?).

In the final contribution to Part II, Antony Taubman considers what it means for GI protection in TRIPS to itself be situated within the context of a broader multilateral trading system. This chapter identifies the central difficulty faced by proponents of international GI protection – the notion of territoriality usually limits the protection that is available within national jurisdictional boundaries, while meaning is fluid and varies with context. Since GI products are themselves traded across borders, the meaning of Champagne or Feta (origin-specific or generic?) will inevitably vary across markets. In these circumstances can negotiated outcomes creating specific lists of protected terms produce a trading system that is fairer and enhances overall public welfare? Taubman develops this line of analysis by identifying two methodologies or approaches to international trade negotiations. A 'fix-rules' approach – exemplified by TRIPS – settles on the objective rules expected to apply in the domestic regulatory contexts, so that national authorities interpret these internationally binding rules and apply them consistently. The second, a variant of the classic 'managed trade' or 'fix-outcomes' approach, would settle in advance what the outcomes should be. Bilateral agreements with annexes of protected terms or the Lisbon Register of protected appellations represent this approach, which determines in advance which designations would qualify as GIs and when they would cease to have generic significance. Therefore, what appears questionable to some IP practitioners and scholars – the artificial 'fixing' of semantic content – can be better explained when viewed from this multilateral trade perspective.

In Part III, the perspective shifts to national experiences and experiments with GI protection models. Stephen Stern draws on decades of experience as a GI litigator in Chapter 10, providing us with a detailed and engaging biography of Australian wine GI protection regimes across six distinct phases. Along the way there is coverage of the EU-Australia Wine Agreement (clearly qualifying as one of Taubman's 'fix-outcomes') as well as insights revealing the manner in which national regimes are influenced by international negotiations. Of particular interest is Stern's insider account of the geographical-boundary-delimiting difficulties over the course of the controversial *Coonawarra* wine region litigation. In Chapter 11, Delphine Marie-Vivien also draws on considerable reservoirs of fieldwork as well as academic experience when comparing the GI regimes of France, the EU and India. She asks whether handicrafts can be accommodated within a regime that has historically prioritised the link to origin emphasising natural (that is physical geography) features based on the archetypal subject matter categories of wines and agricultural products or foodstuffs. Marie-Vivien challenges the conventional distinction between agricultural products, wines and spirits on the one hand and crafts on the other, arguing that current registration practice and existing interpretations of the link requirement for agricultural products have already accommodated human or cultural factors to such an extent that excluding handicrafts is no longer justifiable. However, this should be counterbalanced by providing for two different levels of protection, depending on the strength of the link to the region of origin. Finally, Chapter 12 rounds off Part III with a fascinating and detailed account of the trio of concurrently applicable GI protection options in China. Drawing on her operational experience with GI registration, Haiyan Zheng considers the pros and cons of GI protection via trade mark law (as certification or collective marks), via a *sui generis* regime covering all products or via a *sui generis* regime specific to agricultural products. Each regime is administered by a different government agency or department, which gives rise to the potential for both systemic conflicts and overlaps. Different authorities may confer the same GI rights on different right holders, identify incompatible geographical boundaries for the same region and enforce different quality standards, so she favours the trade mark solution as the most effective one.

Part IV completes the volume with a survey of the most enduringly controversial or nascent issues presently confronting GI regimes. Burkhart Goebel (a rare breed of IP litigator with considerable GI experience) and Manuela Groeschl tackle the most controversial of these issues – conflicts between trade marks and geographical indications – in Chapter 13. Drawing on both international law (primarily TRIPS) and European doctrinal materials, they plausibly suggest that disputes between these two categories of signs should be resolved in accordance with the well settled principles of priority, exclusivity and territoriality (or PET). Their preference is for the general trade mark approach to resolving conflicting rights in accordance with the 'first in time, first in right' principle (with room for exceptions), as opposed to *sui generis* approaches which would always prioritise GIs in such conflicts, because of their public values. In Chapter 14 Christopher Heath approaches the same disputed territory but adopts a distinctive method for doing so. In this exceptionally well researched contribution he compiles and analyses a vast swathe of litigated *Budweiser* disputes, between the Czech brewery Budvar and the American company Anheuser-Busch, spanning five continents and four decades. Heath fuses them into a lens to evaluate the effectiveness of international GI protection. He classifies them into cases involving a clash of trade marks, cases involving the Czech appellation 'Budejovicky Budvar/Budweiser Budvar' or its variants being protected under the Lisbon Agreement and cases involving appellation protection under bilateral agreements. Among the most original contributions of this chapter is the detailed analysis of divergent judicial interpretations of the Lisbon Agreement (Art 6 which prevents generic use is particularly controversial). His contribution acts not only as a counterfoil to some of the arguments in the previous chapter but also usefully supplements the analysis of the Lisbon Agreement undertaken in Part II.

The next two chapters approach the topic of GIs being facilitators of regional or territorial development at both theoretical and empirical levels. In Chapter 15 Dominique Barjolle draws on a combination of research expertise and consulting experience to analyse GIs within the context of markets while identifying the institutional factors contributing to successful GIs. Having identified the precise manner in which GIs add value (economic value by signalling product differentiation and guaranteeing quality), she emphasises the significance of well-designed organisations (the representative producer associations) as well as codes of practice in retaining that value. The chapter concludes by considering the extent to which local regions or territories can be the beneficiaries of this value creation and the potential impact of GIs on sustainable development. In both this chapter and the one that follows, the blend of theoretical framing and empirical detail greatly strengthens the arguments. Dwijen Rangnekar and Pranab Mukhopadhyay also consider the social gains and potential for regionally specific value creation in Chapter 16, through a detailed case study of the cashew-based spirit Feni from Goa. They explore the impacts that registration of Feni as a GI could have on stakeholders, concluding that legal protection is insufficient to sustain traditional small-scale producers, who are the principal intended beneficiaries. The central puzzle explored in this chapter is the apparent decline in both production and consumption of Feni within the state of Goa. Interestingly, changes in consumer tastes, linked to the entry of alternative spirit drinks into the Goan market could have eroded Feni's market share, which in turn suggests that enhanced legal protection in isolation will not be effective in reviving the fortunes of small-scale Feni producers.

The penultimate chapter considers the relatively recent argument that GI protection regimes can provide effective protection for indigenous or traditional knowledge. In Chapter 17, Brad Sherman and Leanne Wiseman approach this issue by teasing out the epistemic assumptions underpinning the link between product and place in international GI protection debates. Given the wine-specific origins of GI protection, they identify a scientifically verifiable causal connection between product and place (in the form of local environmental conditions affecting the ripening and flavour of grapes) as the archetype of the link requirement. However, this is increasingly problematic when it comes to artefacts with a greater human dimension, or when it comes to the needs of indigenous peoples with distinct epistemic frameworks and value systems. Sherman and Wiseman explore two approaches – one relating to defining unique or distinctive products by identifying parallels with the naming conventions used in identifying new botanical varieties; the other drawing upon the manner in which indigenous knowledge and links to place have been identified on their own terms (in accordance with the worldviews of indigenous communities) in the context of native title claims to land by Aboriginal communities in Australia. For instance links to place could be proved by allowing songs, dances, stories and paintings as valid categories of evidence. The chapter's value lies in the question it raises: viewed from the perspective of indigenous communities, is it realistic to expect that GIs can be used to protect Indigenous knowledge?

In the final chapter, I have (appropriately enough) considered the death of GIs, by analysing the legal test for genericide. Generic status is reached when a product no longer indicates geographical origin but merely indicates a class of products with certain qualities, such as cheddar cheese. While TRIPS provides that such terms are unprotected, the test for determining generic status is remarkably thin. This chapter draws comparatively on decisions across several jurisdictions to develop a detailed menu of options – both threshold issues and evidentiary categories – which need to be considered when applying the test for generic use in practice. The chapter emphasises that the overarching enquiry should be to assess consumer understanding of the disputed term, as opposed to strategic manoeuvres by producer interests on both sides of this debate.

Having hopefully whetted your appetite all that is left is to recommend this volume to you and to hope that you find it interesting. It has been both an education and a pleasure to gather together some of the most insightful and original research in this fascinating area. In an ideal world, without the constraints of space, the list of contributors would have been even longer. Other legal commentators whose research could have illuminated the pages of this volume include – in no particular order – Justin Hughes, Latha Nair, Michael Blakeney, Roland Knaak, Christine Haight Farley, Irene Calboli, Caroline le Goffic and Jacques Audier. Researchers from across the social sciences exploring the factual claims and policy underpinnings of GIs include Bertil Sylvander, Erik Thévenod-Mottet, Sarah Bowen, Philippe Marchenay, Giovanni Belletti, Cerkia Bramley and Kolleen Guy. My only consolation is that their works are referenced in the chapters which follow. What this volume does showcase is that a conversation has begun between the various research communities. It is one which will greatly enhance our understanding of Geographical Indications in the future.

PART I

HISTORY AND CONCEPTS

2. French collective wine branding in the nineteenth–twentieth centuries
Alessandro Stanziani

1. INTRODUCTION

The set of rules adopted since 1935 defining wine collective labels in France has been a formidable institutional tool to regulate the economic activity of a group of producers. For decades, the winegrowers and merchants of Bordeaux and Champagne were to be protected not only from foreign counterfeiting but also from the temptations of some among them to make unilateral changes in production techniques or simply to cheat. This explains the desire of producers from other regions and of other products to benefit from the same advantages, which led to the multiplication of *Appellation d'Origine Contrôlée* (AOC) labels during the second half of the twentieth century.[1] But if collective designation protection has such strong positive effects, then why did French producers endure discussions on this topic over almost a century and a half (since the revolution up through the mid-twentieth century)? And why are these labels not more widely adopted outside Europe? Furthermore, AOCs are said to express consumers' desires for 'traditions' in winemaking; as such, collective designations will preserve local know-how against globalisation. If this is so, why were AOCs defined well before the current wave of globalisation and, above all, why did local producers disagree for so long over the meaning of 'traditions'?

Our argument will be that the introduction of the AOC label required not unanimous agreement on the part of the economic actors, which was impossible to attain, but a consensus of the majority on playing by rules that were acceptable. However, these rules did not necessarily correspond to what each actor considered optimal. The delimitation of the territories entitled to *appellations d'origine*, the list of accepted production techniques

---

[1] Unlike its predecessor, the *Appellation d'Origine*, in addition to guaranteeing geographical origin the AOC also required that a specified method of production or specific ingredients be used. For contemporary legal definitions, see generally, C Le Goffic, *La protection des indications géographiques en France, en Europe et aux Etats-Unis* (IRPI, Litec 2011).

and the grape varieties identified were the fruit of a painful but necessary agreement among certain actors at the expense of others, who until then had been allowed to produce Bordeaux or Champagne wines. Contrary to standard theories, we argue that collective wine designations did not represent a long-standing tradition, but, quite the contrary, they sought to stabilise completely new vine-growing and winemaking practices in the post-phylloxera era,[2] when organic chemistry entered the agro-food sector. Thus, collective designations were used primarily to provide the producer with a guaranteed income while offering a stable normative framework for negotiating innovation, rather than to inform consumers.

This French scheme enjoyed great success in other European countries, especially in Spain and Italy, where, for reasons similar to those in France, wine production played an important role in market equilibrium and in the introduction of quality norms. We explore the impact of these institutional settings on market dynamics over the long run. The French and subsequently European solution worked relatively well for the better part of the twentieth century, but today it is having a hard time adapting to the new context of a global economy. Outside Europe, the emphasis on *terroir* and traditions is an effective marketing tool for some elites, but it is not enough to offset the development of wines made from a single grape variety or marketed around highly reputed individual trade marks. Especially for new consumers, it is easier to find one's bearings in relation to a particular grape variety (for example Chardonnay) than finding one's way amongst a myriad of AOC labels. The increasing number of these labels has tended to make them commonplace and blur the message that AOC signs were supposed to give to the consumer. On the contrary, Chilean, African, American and Australian wines can be distinguished by easy recognition of their main grape variety as well as a standardised taste that ultimately corresponds quite well to the one imposed by wine guides and marketing campaigns. This is true not only for ordinary wines but very good wines as well. Since information transmission to consumers is otherwise suggested as one of the primary reasons for protecting designations of origin,[3] this chapter sets out to correct this imbalance and offers

---

[2] French vineyards were devastated by a root louse, phylloxera, that was imported in infected vines from America in the latter part of the nineteenth century. See I Stevenson, 'The Diffusion of Disaster: The Phylloxera Outbreak in the *Département* of the Hérault, 1862–1880' (1980) 6 Journal of Historical Geography 47.

[3] See e.g. OECD, 'Appellations of Origin and Geographical Indications in OECD Member Countries: Economic and Legal Implications' December 2000 (COM/AGR/APM/TD/WP(2000)15/FINAL) 7–8, 31–34; C Bramley and

an alternative, historically grounded explanation for the emergence of appellation systems.

2. INTELLECTUAL PROPERTY AND WINE IN NINETEENTH-CENTURY FRANCE

The objective of intellectual property protection is to create incentives that maximise the difference between the value of the intellectual property that is created and used and the social cost of its creation, including the cost of administrating the system.[4] In particular, the protection of trade marks has been considered as a form of indirect protection of the consumer. A primary benefit of trade mark protection is to lower consumer search cost, while eliminating the risk that competitors will free-ride upon investments.[5] Positive effects on investments and innovation have equally been the objects of studies. Producers are discouraged if other producers can easily emulate their product.[6] Trade mark protection has a positive impact on quality: the owner of a valuable trade mark would be reluctant to lower the quality because they would suffer from a capital loss on their investment in the trade mark.[7] To what extent can this analysis be enlarged to collective geographical designations (for example Marseille soap, Bordeaux wine, Cognac) and labels?

In industrial Europe, collective marks were abolished as the heir of corporative systems or guilds – nevertheless, two related legal categories had been progressively developed: generic names and collective trade marks. If an individual trade mark becomes a generic name, trade mark protection ceases.[8] The original producers' aim will thus be to avoid such an outcome. If this resistance is successful then exclusive rights are retained

---

JF Kirsten, 'Exploring the Economic Rationale for Protecting Geographical Indicators in Agriculture' (2007) 46 Agrekon 69.

[4] S Besen and L Raskind, 'An Introduction to the Law and Economics of Intellectual Property' (1991) 5(1) The Journal of Economic Perspectives 3.

[5] WM Landes and RA Posner, 'Trade Mark Law: An Economic Perspective' (1987) 30 Journal of Law and Economics 265; G Stigler, 'The Economics of Information' (1961) 69 Journal of Political Economy 213.

[6] K Arrow, 'Economic Welfare and the Allocation of Resources for Invention', in National Bureau of Economic Research, *The Rate and Direction of Inventive Activity: Economic and Social Factors* (Princeton University Press, 1962).

[7] C Shapiro, 'Premiums for High Quality Products as Returns to Reputations' (1983) 98 Quarterly Journal of Economics 659.

[8] For further details on the process of genericide, see Chapter 18 by Gangjee in this volume.

over what would otherwise have become a generic name. This will reduce the potential competition and increase net profits, while also reducing the cost of information for consumers. In other words, labelling is a way by which a seller can commit to product attributes that are difficult for third parties, such as courts, to verify.[9] Geographical designations are particularly vulnerable to generic use and because of this risk, one would expect that producers of a given area will support the adoption of a law protecting a collective designation, if the costs of negotiating and enforcing this law do not exceed the expected benefit. According to some studies, in a strongly asymmetrical information context, a public law or state regulated quality sign (label or AOC) ensures a better product quality than a private sign, in particular on the wine market.[10] Conversely, other authors have supported the opposite conclusion; public quality signs create a rent to the producer without increasing the consumer information and the quality of the produce.[11] In the case of agro-food products, will local producers willing to prevent the genericide of their product, be encouraged to apply for a collective designation rather than an individual trade mark? Or will they rely on the individual trade mark, while conceding the generic use of the collective sign?

In order to answer these questions, we will study in detail the evolution of collective designations in France during the 'long nineteenth century' (that is, from the revolution of 1789 up through 1914) and its main outcome in the later twentieth century. We will focus in particular on the wine market in so far as French rules and practices on trade marks and collective designations were deeply linked to this market and its main actors and lobbies.

Historically, the question of generic terms in the wine market was raised immediately after the French revolution and the suppression of guilds. Winemakers and traders wished to deal with counterfeiting, especially abroad. The constant argument evoked was that a generic name is conceivable only for standardised or 'manufactured' products, whose

---

[9] Landes and Posner (n 5); B Klein and K Leffer, 'The Role of the Market Forces in Assuring Contractual Performance' (1981) 89 Journal of Political Economy 615.

[10] J-M Bourgeon and B Coestier, 'Private Versus Public Product Labelling' (1996) Cahiers du Théma, n. 9507; D Smallwood and J Conlinsk, 'Product Quality in Markets Where Consumers are Imperfectly Informed' (1979) 93(1) Quarterly Journal of Economics 1; I Png and D Reitman, 'Why are Some Products Branded and Others Not?' (1995) 38(1) Journal of Law and Economics 207.

[11] L Linnemer and A Perrot, 'Une Analyse Economique des Signes de Qualité' (2000) 51(6) Revue Economique 1397.

identification necessarily relies on its name (for example: eau de Cologne). Wine, as a 'natural product', could thus not be admitted to this practice and people located in Germany or in the USA could not use the word Bordeaux or Champagne for their products.[12] The producers' main effort was to avoid a valuable geographical designation becoming a generic name. A number of decided cases attest to attempts made by French producers of Bordeaux, Armagnac, Champagne, and so on to defend their names. While in theory the question was relatively easy to resolve, in practice it was difficult for a French court to prosecute a foreign trader settled abroad. In such situations, a mutual arrangement between the countries involved was the necessary but often insufficient condition to proceed.

It was much easier for courts to stop any attempt made in France to make Bordeaux, Cognac or Champagne a generic name; French producers making this attempt incurred legal penalties.[13] Accordingly, it was held that Art 1 of the law of July 1824:[14]

> totally prohibits and punishes the use on an industrial product of the name of a place other than that of winemaking, or its appearance following any alteration whatsoever; this provision applies not only to the winemaker, but also to any merchant, forwarding agent or publican whomsoever who knowingly exhibits for sale or puts into circulation objects bearing fraudulent or altered names.

This law did not specifically protect the tradesman's mark, nor the collective mark. Litigation on these aspects was thus extremely widespread, despite the uncertainty of the issue. When the name belonged to a city or a region (for example Gironde, Champagne), the false name could still be pursued. The Supreme Court repeatedly declared that one had to interpret the name of a place as the place of manufacture for industrial products, and as that of the harvest for 'natural' produce.[15] Following this last decision, the owner could mark his wines with the name of a different place than that where cellars or buildings are located, but on the condition that the name indicated the place of the vintage.

On 8 June 1847, the chamber of appeals of the Court of Cassation[16]

---

[12] L Lacour, *Des Fausses Indications de Provenance: Contribution a L'étude de la Propriété Industrielle en Droit Français* (Rousseau, 1904) 15–17.

[13] AN F 12 6844, tribunal Grenoble 31/12/1852; A Rendu, *Des Marques de Fabrique et de Commerce et de la Concurrence Déloyale* (Paris, Imprimerie et librairie generale de jurisprudence, Cosse et Marchal imprimeurs, 1857) 12–15.

[14] *Loi du 28 juillet 1824 Relative aux Altérations ou Suppositions de Noms dans les Produits Fabriques* (1825) 7 Bulletin des Lois no 19, 65.

[15] Cour de Cassation, 16 June 1847 and 12 July 1845 (in the context of protecting Champagne), Sirey 47.1.521, 45.1.842.

[16] Cour de Cassation, 8 June 1847, Dalloz 47-I.327.

ruled that estate owners and winegrowers had the right to have the wines from their harvest protected, and deemed that all estate owners of *grand crus* were entitled to stamp the vessels containing their wines with a picture representing the wine, that is, the place where the grapes were harvested, even though the cellars and vats were not located there. This decision naturally opened up a breach in any attempts to legally regulate origin marking; indeed, it allowed wine merchants or winegrowers located, for example, outside the Bordeaux region, to buy products from that region, locate them elsewhere and sell them under the name of Bordeaux.

Thus we can see that existing rules were in place but they did contain gaps and sometimes proved controversial. Another set of circumstances also led to pressure for legal change. The economic and legal system introduced along with the revolution experienced its first setbacks between the years 1835 and 1845 when technological progress, changes in economic networks (colonies, internationalisation of the economy) and monetary as well as speculative fluctuations in France and Great Britain generated a high degree of uncertainty in the markets. As a result, several economic actors and their institutional representatives brought up the need to change the rules of the game. Lively debates took place within trade associations and the Chamber of Peers concerning rampant fraud, particularly in wine and textiles,[17] as well as commercial companies and their effects on trade. In the first case, they pertained to new fabrics and new winemaking techniques; in the second, an attempt was made to protect unfortunate debtors from their creditors, but there was also concern about stock exchange speculation (which motivated the drafting of the Bill of 1838 on the prohibition of limited partnerships with shares). The actors' strategies were questioned and stricter rules to control markets were discussed. The issue was how the principle of contractual freedom could be maintained without encouraging fraud. It was precisely 'innovation' that caused problems and led to a desire for 'standardisation' among merchants and producers who 'would like the government to be able to require a single manufacturing method', although they are aware 'that in the current state of things, such a measure would not have a satisfactory effect'.[18]

Thus, several new rules were introduced to control trade and competition: the law on weights and measures (4 July 1837), the law on bankruptcy (28 May1838) and finally, the law on food fraud of 1851. According to this law, fraud constituted a violation of private property. Protecting the quality of

---

[17] AN F 12 7452, 'Lettre du consul de France à Valparaiso au ministre des Affaires Etrangères, 1 mai 1835'.
[18] AN F 12 7452, 'Rapport du consul de France à Beyrouth du 20 mars 1838'.

foodstuffs encouraged the growth of labour productivity and preserved the reputation of national products. The law introduced a distinction between fraud and adulteration. The first category referred to the behaviour of the actor and the act of deception, whereas the second focused on the product and its characteristics. This solution confirms that, for over half of the nineteenth century, the consumer did not exist from the standpoint of the law, which limited itself to talking about the general 'purchaser', in keeping with the idea that the final consumer, like any professional buying products, should be protected by ordinary contractual rules. There were no special provisions regarding consumption and trade marks.[19]

This lacuna was all the more problematic since from the middle of the 1840s, several estate owners in Gironde, including producers of first vintages (premiers crus), had agreed to sell their wines 'by subscription' to a consortium of traders. According to this contract, the price was settled at such a level that the winegrower could cover current costs and capital expenditure while keeping the quality stable. In exchange, the merchant paid the price well before the harvest. However, after 1847 and more specifically from the beginning of the year 1850, two major producers, Lafite and Haut-Brion, encountered serious problems and sought to terminate these contracts. At the same time, three out of four 'premiers crus' in Bordeaux were not accepted by traders as being 'reference prices' as before. In turn, this setback led to the collapse of the whole hierarchy of wines and prices in Bordeaux. At the very moment when prices as indicated in the subscription contracts of Margaux and Latour collapsed, second rank wines saw their prices climbing. New important changes in ownership followed; Rothschild and other important capitalists took control of various major wine producers.[20] A new stable and well accepted hierarchy of wines was thus required after all these changes. The international exhibition of 1855 was the occasion to settle this new hierarchy. Traders, owners and *courtiers* (brokers) reached an agreement on the classification of all main wines of the area. The issue was the very famous classification of Bordeaux wines of 1855 which was to be a constant reference in quality and price setting over the ensuing hundred years.[21] This process

---

[19] Raymond Théodore Troplong, *De la Vente* (Hingray, 1834); Jean-Marie Pardessus, *Cours de Droit Commercial* (Garnery, 1814).
[20] Dewey Markham, *1855. Histoire d'un Classement* (Editions Féret, 1997); M Lachiver, *Vins, Vignes et Vignerons* (Fayard 1988) 364; P Roudié, *Vignobles et Vignerons du Bordelais, 1850–1980* (CNRS, Paris 1988) 139–140.
[21] Lachiver (n 20) 364; P Roudié, 'Crus classés et châteaux viticoles en bordelais: deux nouvelles notions qualitatives au XIXe siècle?', in G Garrier et al., *Le Vin des Historiens* (Université du vin, Suze-la-Rousse, 1990) 189.

was completed by a new law on trade marks and brands adopted in 1857. It protected the trader's mark while, according to the rapporteur, no protection was granted to collective marks.[22] This left open the question of whether a regional name was protected or not.

Following the decisions of the Court of Cassation in 1845 and 1847 on Champagne, several judicial decisions[23] confirmed that the name of a locality belonged to all the inhabitants who had an interest in exploiting it to make the location of their establishment known, as well as an interest in the place of origin or of manufacturing of their products.[24] On 4 March, 1870, the Angers Court declared that proceedings initiated by the owners of the vineyards of Ay, Bouzy and Sillery and the merchants who bought their products, against the merchants of another district who were selling sparkling wines under these names in Anjou were well founded.[25] On 19 July 1887, the same Court asserted that the Champagne appellation would henceforth be exclusively reserved to wines grown and made in the Champagne region.[26] Because of this, several decisions were obliged to specify what was understood by 'place of fabrication'. The jurisprudence of the appeal courts and the Cour de Cassation acknowledged as the place of manufacture a private domain,[27] a city,[28] a region,[29] or even a whole country.[30]

To sum up, during the nineteenth century judicial interpretations guaranteed the protection of individual trade marks and brands but they refused to take into consideration collective marks and generic names. At the same time, this position was relatively uncertain in so far as the *terroir* appeared among the arguments assuring the protection of individual marks.[31] Moreover, product characteristics were hardly mentioned.

---

[22] Rapport fait au nom de la commission chargée d'examiner le projet de loi relatif aux marques de fabrique et de commerce par Busson, député au corps législatif, reprinted in Rendu (n 13) 399–403, 418–443, 589.

[23] Aix, 27 May 1862; Cassation, 15 July 1863; Grenoble, 11 February 1870, Dalloz 71.2.120; Bordeaux, 1 June 1887, Dalloz 89.2.27; Cassation, 1 May 1889, Dalloz 90.1.470.

[24] Lacour (n 12) 24.

[25] Angers, 4 March 1870, Ann. 1870-231.

[26] AN F 12 6969; Sirey 88-2-209.

[27] Paris, 5 February 1870; Cour de Cassation, 26 April 1872, Dalloz, 74.1.1847.

[28] Court of the Seine, 3 July 1863; Lyons, 4 June 1910.

[29] For the region of Cognac: Court of Bordeaux, 11 August 1886; Cour de Cassation, 2 July 1888, Dalloz, 89,1,111; for Saint-Emilion: Court of Bordeaux, 16 December 1897.

[30] Cour de Cassation, 27 February 1880; R Hodez, *Du Droit à l'Appellation Champagne* (PUF, Paris 1923) 27.

[31] For a detailed survey of the *terroir* concept, see Chapter 3 by Barham in this volume.

Guaranteeing the quality of wine was not synonymous with the protection of the brand. The qualities of the soil, the climate, and so on – factors which were related to the geographical place of origin – would continue to have priority over technical production processes. This was so because, at the very moment when winemaking techniques greatly evolved, 'traditional' techniques were appealed to, in order to certify the truth of the appellation. Otherwise, one could use chemical means to offset product inadequacies linked to climate or natural conditions.[32]

In other words, uncertain legal definitions of the 'origin' of wine caused litigation to increase while encouraging free-riding. However the problem was much more complicated when it came to fixing the precise origin of wine, since the interests of different French agents within the same supply chain were concerned. Two related problems were raised: (i) on the one hand, the relationship between the place where a producer was legally settled and/or registered and that of her product; (ii) on the other hand, the practice of 'mélanges' (blending). The question was whether not only a producer but also a trader could benefit from the use of a territorially identified trade name such as Bordeaux. A trader could be registered in Bordeaux but he could source his grapes from outside the Gironde; conversely, he could work in Montpellier while buying his grapes in Gironde. These issues played out differently in the different wine growing regions. In Champagne, since the beginning of the nineteenth century, négociants[33] dominated the market: they manufactured the products and mixed them. In Gironde, it was much more complicated; if in the eighteenth and in the early nineteenth centuries traders dominated the market, in the middle of the century the strength of the winegrowers increased. Châteaux and individual brands were increasingly successful. But this did not prevent traders from exerting increasing control over the producers, in particular at the end of the nineteenth century because of the economic crisis.[34]

At the same time, winegrowers while being opposed to merchants were themselves divided about winemaking. Thus they established multiple associations in order to pursue their strategies. In Gironde, the unions and the professional associations increased during the last quarter of the nineteenth century and in the early twentieth century. At an institutional level, this phenomenon depended on the law of 1884 that allowed for freedom of association. At the same time, the economic crisis and the uncertainty

---

[32] For further details, see A Stanziani, 'Information, Quality and Legal Rules: Wine Adulteration in Nineteenth Century France' (2009) 51 Business History 268.
[33] Wine merchants and traders who have historically acted as intermediaries between growers and producers on the one hand and buyers on the other.
[34] Roudié (n 20); Lachiver (n 20).

22 *Research handbook on intellectual property and GIs*

dominating the markets encouraged the birth of these associations; but they in turn strongly opposed each other. For example, the chamber of commerce of Bordeaux was in favour of imports, to the mixtures of wines and to the use of new vines and techniques.[35] By contrast, the Chambre syndicale of Bordeaux winegrowers was interested in export, while also preserving wine quality.[36] These conflicts emerged as a result of the fact that several property transfers had been realised. New owners, often former tradesmen, adopted strategies unacceptable to the old owners: some did not hesitate to follow the traders in selling low-quality mixed wines. The divisions within the unions of Saint-Émilion confirm this trend.

These conflicts remained unresolved and even increased with the passing of the years. On the one hand, high-quality wine producers started from the assumption that regional reputation matters in the consumer's choice. The regional label had thus contributed to evidence of 'tradition'. On the other hand, those who relied on quantities gave much less priority to the label than to a quality agreement between the producer and the trader. According to this approach, the consumer's protection would be implicitly guaranteed as well. The persistence of these opposing attitudes, as evidenced in the reports of professional associations,[37] largely explains the protagonists' inability to reach an agreement. In turn, this issue mostly explains the litigation explosion in disputes about the origin of wine during this period.

We may now turn to the impact of judicial interpretations on economic behaviours. We may imagine that if a commonly used regional name is not protected as a mark, the supply will increase and no monopolistic rent will be earned. We can consider the evolution of vineyard surface in Champagne and in Gironde during the period in which neither the law nor the courts protected collective geographical designations.[38] In Aude, the growth was limited during the first half of the nineteenth century, then relatively more sustained after this. In Gironde, the trend was more or less stable; the growth of production met some obstacles when the grapevine became subject to disease (between 1850 and 1880) and, in part, due to the unfavourable conjuncture of economic conditions during the last quarter of the nineteenth century. This means that the supply increased precisely

---

[35] AN BB 18 6024, Chambre de commerce de Bordeaux au ministre de la Justice, 23 mars 1881.
[36] AN BB 18 6024, Rapport de Ambaud, Brouardel, Dubrisay, Gallard, rapporteur au Comité consultatif d'Hygiène, séance du 22 novembre 1880, reproduced in: *Comité consultatif d'hygiène publique, 1881.*
[37] AN F 12 6872, AN F 12 7003, AN BB 18 6024, 6030.
[38] Lachiver (n 20) 594–600.

when the law and the judges did not protect collectively used geographical names.

The impact of these practices on production has been estimated. For example, Gironde produced between 2 and 3 million hectolitres of wine, yet around 6 million hectolitres of Bordeaux wines were put up for sale at this time.[39] Indeed, the Gironde négociants bought grapes and wines in the South of France (Midi), but subsequently, especially with the advent of phylloxera, they did not hesitate to mix their wines with wines imported from Italy, Spain and Algeria. In a similar manner, Champagne traders had traditionally stocked up in the Aube, which assured production close to that of the Marne. With the passing of the years, the Seine-et-Marne was also included in their networks. Finally, the traders of the Marne also bought produce in Touraine and in Anjou. As a result, the supply continuously climbed, and so did the conflicts among the agents.

A further feature of interest is evident among wines from Bordeaux, Burgundy and Champagne: top-ranking wines performed more poorly than the cheaper varieties, which were strongly sustained by the increasing demand of the urban middle classes for 'good but not too expensive wines'.[40] This changing economic environment affected production and vineyard ownership. At the beginning of the twentieth century, French vineyards had been reduced by about 30 per cent in comparison to 1875. The only major exception was the Midi, where the vine-growing area had almost completely recovered. In Gironde, the hectares devoted to wine production diminished by about 15–20 per cent between 1880 and 1900.[41] In fact, phylloxera led to increasing production costs while prices lagged behind, especially from 1893 onwards. At that time, high-quality wines were sold at about 700–850 francs per barrel; 'ordinary Bordeaux' wines were even cheaper and only the very top wines were sold at 1800–6000 francs to the barrel. This means that most of the Bordeaux producers were compelled to either raise their productivity or reduce their costs, or both.

Compared with the problems of winegrowers, the situation of wine traders seems to have been much more favourable: farm prices increased less than wholesale and retail prices (Table 2.1). As a result, in Gironde and in France as a whole, the number of wholesale wine traders and retailers continued to grow. This was not without effects on product

---

[39] Roudié (n 20) 208.
[40] Roudié (n 20) 52; KM Guy, '"Oiling the Wheels of Social Life": Myths and Marketing in Champagne During the Belle Epoque' (1999) 22(2) French Historical Studies 211; G Duby and A Wallon, *Histoire de la France Rurale, t. III: Apogée et Crise de la Civilisation Paysanne (1789E1914)* (Seuil, 1976) 368–369.
[41] Lachiver (n 20) 184.

24 *Research handbook on intellectual property and GIs*

Table 2.1 Gironde, average price (francs) of a hectolitre of wine

Year	Farm prices	Wholesale prices	Retail prices
1870	28	31	49.65
1875	21	31	54.17
1880	43	54	74.34
1885	39.87		75.67
1890	36.10		74.49
1895	34.80		70.66
1900	17.96		64.01

Source: Privat, 1898, table III, appendix I; Ministère des Finances, various years.

identification. Wholesale traders encouraged product standardisation and stabilisation through two main processes: blending wines of different origins and the use of preservatives, antiseptics and artificial colouring. Their business correspondence reveals the pressure these traders put on producers, arguing that the market and the customers wanted to receive stable products (especially wines) from one year to the next.[42] For food industry products, this phenomenon had existed at least since the eighteenth century in Gironde and Champagne where merchants adopted the practice of 'stabilising' products by blending different wines. However, this attitude was strengthened by the expansion of international markets in the 1850s and even more by the development of organic chemistry and the possibility of resorting to preservatives and stabilisers.[43] These orientations prevailed even more when, beginning in 1907, a commission was set up to determine, in compliance with the general law of 1905 a series of decrees defining the accepted characteristics and techniques for the principal products.

In the ensuing discussions and projects, tensions mounted among the various pressure groups: manufacturers, farmers, wholesale traders (each category was often undecided about the criteria to be adopted, depending on the industry and the region concerned), hygienists and administrators (who were also divided). Each group wanted to impose a definition of the product involved and tried to get certain techniques approved at the expense of others. Generally speaking, agricultural producers encouraged detailed product specification, while wholesale traders and their

[42] Revue vinicole, years 1880–1890; ACS, fondo maic, direz. gen. agric. vers. VI, busta 251; AN F 12 6353.
[43] AN F 12 7452; AN F 12 9183; AN F 12 6353.

associations sought wider institutional definitions as well as techniques to stabilise the products in question. The result was usually a compromise between these differing orientations. For example, wine was defined in opposition to 'industrial wines'. The regulations provided a list of lawful manipulations and therefore spread the trends already manifest in the special laws that had prohibited certain practices (dilution, vinage (adding alcohol), sweetening); they summarised the observations of the hygienists (for example the prohibition of sulphuric acid), and of the winemakers of southern France (the prohibition of sweetening) and the Gironde (the prohibition of certain blends).[44] Because of these competing interests, there were a growing number of voices amongst the winegrowers who sought to invoke a state-controlled regulation of the quality problems.

2.1 Judicial Construction of the Wine Market, 1908–1914

An initial solution to these problems would have been to let the judges decide. However, judicial decisions taken on the basis of the laws of 1824 and 1857 had expressed different attitudes concerning collective names. Because of this, actors had difficulties in forming expectations on the outcome of the litigation concerning wine designations and brands. In turn, this encouraged risk-oriented strategies, inciting speculation and free-riding. The vagueness of the law increased the anticipatory costs of the agents.

At the beginning of the century, considering the decrease of exports and the difficulties faced in the home market, a majority of agents was therefore in favour of modifying the legal regime in order to lower the uncertainty of litigation.[45] Parliamentary proceedings put in evidence that the representatives of large Gironde winegrowers and large Champagne (Marne) companies wished to lower the legal uncertainty on this topic. They were therefore in favour of modifying the existing norms. But how was this best achieved? In their campaign, regional appellation supporters stressed the defence of tradition in the interest of consumers. This argument is still crucial in contemporary rhetoric favouring AOCs. Unfortunately, archival records offer very limited support for these claims. Local producers strongly disagreed amongst themselves on the definition of 'local traditions' in vine-growing and winemaking. Let us now consider how attempts to incorporate 'traditional' winemaking techniques played out in practice.

---

[44] AN BB 18 6055.
[45] AN C*I/478.

Art 1 of the 1905 law on fraud and falsification condemned

> anyone who attempts to deceive the contracting party as to the nature, substantial qualities, composition and content of productive principles of any goods, either with regard to their variety or their origin, when, according to conventions or customs, the designation of a falsely attributed variety or origin is to be considered as the main cause for the sale.

Since the law prevented false labelling, a blueprint for the true or genuine product and region of origin was required as the benchmark. The law established the setting up of local commissions in order to fix the regional boundaries of appellation regions. However, these commissions had a merely consultative power and aimed just to lower the political tensions related to the appellations question.[46] These commissions, composed of elected officials and local economic actors, did not manage to overcome the conflicts of the previous years. In the Gironde region, within the commission, the prefect of the Gironde region was surrounded by regional representatives, mayors of towns and *département* councillors, as well as winemakers and wine merchants. However, the work of the commission immediately ran up against two problems: how should the Bordeaux wine region be delimited and, within it, how should the estates and the wines be classified? As the number of winegrowers' associations increased, a majority of the commission finally chose a strictly administrative solution: the Bordeaux region was thus assimilated to the formal boundaries of the Gironde department.[47] This solution set off protests by winegrowers from the communities that were excluded.[48] Similar commissions were established in other concerned areas (Armagnac, Champagne, Cognac). In these cases too, no agreement was reached.[49] In order to overcome this impasse, two solutions were available: either to adopt a centralised decision-making procedure or to revert back to the judges within the existing laws.

Because of the impasse reached by local commissions, a new law was

---

[46] AN C* I/478; Chamber of deputies, 14 June 1907, JO of 17 June 1907; Chamber, 9 July 1907, JO of 10 July 1907, p. 1807; Senate, 7 July 1908, JO of 8 July 1908, p. 912; Senate, 9 July, JO of 10 July, p. 932; 4 July 1907, JO of 10 July 1907, p. 1798; Hodez (n 30) 78.

[47] France is presently divided into administrative units known as regions, which are further subdivided into departments.

[48] AN F 12 6969.

[49] Roudié (n 20); Lachiver (n 20); L Coquet, *Les indicateurs d'origine et la concurrence d'eloyale* (Paris, 1913) 46; Bulletin de l'association 1906; AN C 7392; AN F 12 6969; AN BB 18 6031.

voted into effect on 5 August 1908.[50] Art 1 stated that 'the regions which have exclusive claim to the appellation of product origin ... shall be delimited on the basis of "established local customs"'.[51] Art 2 specified that the associations of winegrowers, wine merchants, and so on would be able to sue for damages in all legal proceedings. New commissions were also set up; unlike the previous commissions, the representatives of several trade associations and agricultural associations were excluded; they were replaced by technicians, agronomists and agronomy professors. The hope was that they would be less influenced by the local lobbies and that they would adopt a consensual solution. Yet this was not necessarily an indication of change. In the Bordeaux region, the new commission confirmed the solution already adopted by the previous commission: the Bordeaux appellation was to be based on the administrative territorial division.[52]

The order of 17 December 1908 delimited the territories to which the Champagne appellation was to be reserved; that of 1 May 1909 performed the same operation with regard to Cognac. Next, the order of 25 June 1909 delimited the regions of Armagnac, Bas-Armagnac, Haut-Armagnac and Tenarèze; that of 21 April 1910 determined the regions authorised to produce *clairette de dieu* and, finally, on 18 February 1911, the Bordeaux appellation was delimited. These provisions were not all inspired by the same principles. Whereas the Bordeaux appellation was reserved solely to communities in the Gironde department, the Cognac and Armagnac appellations were also granted to products made outside of the administrative territories in question.[53] As a consequence, social unrest developed in all the concerned Departments. For those who were excluded, waging war was judged to be less costly than being excluded from a delimited region. Indeed, in their view, the producers that were included enjoyed a monopoly. Conversely, no winegrower was prepared to seek a multilateral agreement with the others and thereby escape conflicts. Such an agreement would require, in fact, circulating information relating to the product, its supply and distribution channels. Furthermore, during

---

[50] Loi du 5 août 1908 Modification de l'Article 11 de la Loi du 1 août 1905 et Completant Cette Loi par un Article Additionnel (11 août 1908) Journal Officiel 5637.

[51] According to the Law of 1908, preventing misleading use required: 'la définition et la dénomination des boissons, denrées et produits conformément aux usages commerciaux... [et] la délimitation des régions pouvant prétendre exclusivement aux appellations de provenance des produits. Cette délimitation sera faite en prenant pour base les usages locaux constants.'

[52] Roudié (n 20) 222–229.

[53] AN F 12 6969.

this period market segmentation operated precisely on the basis of product differentiation excluding any circulation of information on the product or the sales networks.

In the face of increasing social demonstrations and growing dissatisfaction among a large portion of the parties in question, several voices spoke out in the French Parliament in support of suppressing the law concerning delimited regions and transferring exclusive protection of *appellations d'origine* to the courts. On 30 March 1911 a bill was introduced to this end in the Parliament. 'No doubt', asserted the rapporteur, 'it will be necessary to prosecute counterfeiters, professional cheaters, all those who deceive with regard to the type of goods, who thus sell ordinary wine as Bordeaux or Burgundy. Fraud is not new, it stems from the predilection for certain wines.' What was new was the fact that for some time, this type of fraud had reaped the benefits of progress in chemistry; hence, the need for stricter legislation.

Conversely, it was not enough for a vine to be planted in a particular region to yield good wine. With geographical delimitation, on the contrary, the geographical designation would fall into a slump, for people would plant vines everywhere in the region concerned just to take advantage of the regional name. The report concluded that it was necessary to go back to the previous law. It was the job of the courts, and not of the administration, to punish fraud. Trade associations should be granted greater legal authority to represent their members' interests in legal proceedings.[54] In other words, the return to the judicial procedures aimed not only to deactivate the political tensions but also to reach a definition of the quality suitable enough to fight against counterfeiting and free-riding at the same time.

However, the decision of going back to a judicial solution would have been unsatisfactory without specifying on what basis the judges would decide. On 30 June 1911, Pams, the new Minister of Agriculture, proposed a bill 'to set up a method of delimitation by the courts' and no longer by administrative authorities.[55] The first article of the bill stated that judges should take into account the origin of the product, its nature, composition and quality. The reference to quality was the real innovation in the bill. The problem arose, however, of how to define product quality. In the ensuing months, agrarian specialists, political representatives, winegrowers and traders unions gave different definitions of the quality of wine. Thus, on 3 June 1911, the Corbeil Chamber of Commerce sent a

---

[54] AN F 12 6969, Chambre des députés, séance du 30 mars 1910.
[55] Roudié (n 20) 231.

report to the Ministry of Trade in the name of Mr Simon. The author stated from the beginning that he was not in favour 'of delimitations, considering that they are naturally produced by the customer's taste and the price of the goods proposed'. Next, he explained that there are two sorts of Champagne houses in the Marne: the great houses and the long-standing houses on the one hand; and the new houses selling cheap wine, on the other. The first are above any suspicion of fraud; they have always been on good terms with the winegrowers whom they help to bear the ill fortune of mediocre harvests during hard years. On the other hand, in the good years, the new houses buy the inferior wines of the Marne, the Aube and others and make cheap champagne sold at 1.5–3 francs a bottle (the leading brands cost 7–9 francs). 'Yet, in the last few years, the winegrowers of the Marne have not had a harvest; the great houses were able to live off of their reserves, while the other houses went in search of wine, even outside of Champagne. Hence: the problems of the winegrowers of the Marne.'

Next he recalled that the law introduced regions with controlled appellations, setting off protests from the Aube winegrowers who wished to be included in the Champagne region. He stated that he was against all outside constraint, as it was necessary to give winegrowers from the true regions of Champagne the freedom to specify the place of origin; consumers would make the final decision when reading Reims rather than Aube on the label.[56] This speech confirms that the producers of quality wines insisted on *ex-post* delimitation of the regions and appellations; they considered that consumers would choose quality once they were informed. For high-quality products, substantial bounded rationality would impose itself. According to this view, preferences are givens and, in a sense, exist prior to any steps taken by the winegrowers. This attitude was not shared by the merchants and winegrowers who wagered on quantity more than on quality, even if they worked in traditional winegrowing regions. Thus, in a resolution adopted on 14 June 1911, the Bordeaux Chamber of Commerce declared itself to be

> in favour of the punishment of fraud. Regional delimitations contribute positively to this end, but it must not be interpreted in an overly-restrictive sense. In particular, it must not deprive trade of the facilities indispensable to it. Indeed, in the Gironde region, some wines and the wines of some harvests have to be mixed with wine from other regions because they are too weak. The Chamber therefore places its trust in an agreement between wine merchants and wine-growing estates.

---

[56] AN F 12 6969.

Finally, a third attitude was expressed by the merchants of Loir-et-Cher: the resolution adopted on 25 July, 1911 by the Chamber of Commerce of Blois and Loir-et-Cher affirms that:

> At first sight, delimiting regions enables protection of the consumer, at least initially. At the same time, it constitutes an unprecedented privilege: a monopoly in favour of one category of winegrower. A few capitalists are going to monopolise the entire harvest and small and medium-sized merchants will be squeezed out. In the end, this would also hurt the consumer . . . [for] . . . little does he care whether the product sold is actually produced in the region whose name it bears; his concern is that the quality truly correspond to the price he has paid and that it be consistent, regardless of the year. Hence, the need to mix or blend in order to maintain unvarying quality.[57]

According to this approach, quality is defined by taste and not by the characteristics of the product. Standardisation in agricultural matters is not achieved by freezing techniques, but on the contrary, by allowing blends enabling the product to have consistently the same taste. To summarise, different, often incompatible, strategies developed around the issue of regional *appellations*, based on perceptions and categories (quality, consumer, rationality) that also differed. The importance of the beliefs of agents was further reinforced by a highly unstable economic environment and limited circulation of economic information.

2.2 The Attitude of State Authorities in this Context

In July 1911, Fernand David, in the name of the agriculture commission in charge of examining the bill on appellations of origin and regional delimitations, declared: 'We have removed, after the words "their type" the words "their composition and their substantial qualities" which appeared in the bill.' The reason for eliminating these words was that

> it appeared that, by directing attention to the product composition and substantial qualities that it was supposed to guarantee, we would clearly be raising the question of the quality of the product. The legislation to be drafted does not, however, seek to guarantee the uniform quality of the products it protects. It simply wishes, by establishing their origin, to prevent the appellation used to designate them from being fraudulently usurped.

In other words, regional appellations were not supposed to take into consideration the characteristics of products. Winemakers who were depending on quality immediately protested against this approach: how could one

---

[57] Ibid.

defend the name of a wine without first defining the characteristics of the product? The commission's reply was quite clear: mediocre wines could be sold with a Bordeaux appellation merely because they were produced from the grapes of the region. Now that the decision rested with the courts, this approach could hardly be reconsidered. The commission justified this by evoking the difference between manufactured and agricultural goods:

> When we recognised the property rights of wine merchants or winemakers to their marks, the personal rights of writers and artists over their works, the exclusive right of the inventor to dispose of his patented invention, we never meant to assert that the mark, the work or the invention would be of good or bad quality. Their freedom to produce remains full and complete, at their own risk and it is up to the public to assess its value.[58]

Given the considerable influence of atmospheric conditions on wine production and its quality,

> who would even dream of creating a uniform type, or even an average type for each of the innumerable wines of France? ... the *appellation d'origine*, by itself, is meaningless. Indeed, when one speaks of Bordeaux wines, the name of Bordeaux does not designate anything in itself, since the city of Bordeaux has never contained vineyards. It takes on its full meaning, however, when it is a question of products grown on the same soil, under the same climatic conditions, from the same grapes that have always been used to make the type of wine called Bordeaux. The name of origin in this case is a generic name, the appellation of which is determined by the nature of the product.[59]

Consequently, the protection of the consumer could no longer be separated from that of the mark owner: 'The law of 1905 considers the damage caused to consumers who are deceived about the origin; we think that the winegrower, who is the legitimate mark owner, should also be compensated.' This was an attempt to mediate between the conflicting interests of winegrowers, merchants and consumers. Unfortunately, during the following months, a number of petitions from traders' federations, Chambers of Commerce, winegrowers' associations, and others were addressed to the Ministry of Trade and to Parliament.[60] In particular, high-quality wine producers backed ex-post procedures because they start from the assumption that regional reputation matters in the consumer's choice. The label had thus put into evidence 'tradition'. On the contrary, those who wagered on quantities gave much less priority to the label than

---

[58] AN F 12 6971, *Chambre des députés*, n. 1136, 7 July 1911.
[59] Ibid.
[60] AN F 12 6969, *Chambre des députés*, n. 2564, 27 February 1913.

to a quality agreement between the producer and the trader. The consumer's protection would follow from this. There is a third possibility to consider, according to which the information given to the consumer on the label is not very relevant, in so far as the choice is made on the basis of the taste rather than the origin of the wine. The information is only hedonistic and legislation in accordance with this approach would leave little room for monopolies (except in the case of State intervention in their favour).

These conflicts were reproduced in the parliamentary debates which continued throughout the spring and summer of 1912.[61] The problem raised at the time was twofold: on the one hand, it was necessary to take into account all of the merchants who bottled wines at locations outside of the typical regions, but using wines made from grapes grown in the soil of the appellation: in these cases, should an *appellation d'origine* be applied? On the other hand, in the event of fraud, how should liability be meted out between, let us say, the winemaker and the merchant or the publican? In July 1912, the commission in charge of this question proposed the following wording:

> [W]hosoever shall affix, or display, by adding, removing or altering in any way whatsoever on manufactured or packed goods, the name of a winemaker or merchant other than the one who is the author, or the name of the winemaking company or the trade firm other than the ones where the aforementioned goods were made or packed, or finally, the name of a place other than the one where it was made in the case of a product of origin, shall be punished by the penalties provided for by article of the law of 1 August, 1905.
>
> Whosoever knowingly uses a geographical denomination to designate products that are different by their origin or their nature from those to which this denomination applies, and, if they are wines, differ by their composition and their substantial qualities, drawn mainly from the grape varieties, cultivation methods and soil, in accordance with fair, established, local practices shall also be punished by the same penalty.
>
> Any merchant, forwarding agent or publican whosoever who knowingly displays for sale or puts into circulation goods or products thus fraudulently named will also be subject to the same penalty.

In relation to the existing law and the bill formerly discussed, the commission thus suggested generalising the application of Art 1 and extending it to packed and manufactured goods as well as to winegrowers and merchants. At the same time, the bill back-pedalled on the relationship between the *appellation d'origine* and the definition of quality. The words 'their composition and their substantial qualities', which had appeared in the governmental bill, were reintroduced. Indeed, the commission deemed

---

[61] AN F 12 6969.

that it was necessary to protect the quality of the product; this protection was, by the way, requested by the vintners' associations and most of the merchants' trade associations.

> One obviously would never dream of imposing a special, uniform type of wine for each *cru*. It would seem, however, that despite the external conditions under which a wine is produced, it will retain its distinctive qualities if it has been obtained from the same soil, with the same grape varieties that have made it renowned.[62]

This proposal raised a storm of protests in the country and in the Parliament, particularly on the part of representatives of merchants and winegrowers less concerned about the quality of the product. These agents insisted on having the appellation of origin essentially linked to the geographical location of the vineyards. It was not until the month of February that a new bill was proposed by the agricultural commission.[63]

The new report specified that there 'has been no infringement of the principle of the law: delimitation by legal means has replaced administrative delimitation'. Opinions in business circles were divided. For winegrowers, who had made great sacrifices to reconstruct their vineyards, the mere fact of being harvested in a region should earn the benefit of the appellation for the product, regardless of the grape varieties used. 'The merchants do not share this opinion: in their view, a wine should have the right to the *appellation d'origine* only in so far as it meets the full definition of the product.' 'Finally, there is the point of view of the consumer, which must also be taken into consideration. The consumer wants a wine which, in accordance with a certain name, will have certain specific characteristics.' How were these various points of view to be reconciled? The records note that the Minister of Justice was asked

> if the notion of quality was in conflict with the law of 1824. The reply, dated 19 December 1912 is as follows: 'The enumeration, in Art 1 of the law of 1824, of the means used with a view to deceiving the consumer about the identity of an object sold would be usefully supplemented and specified by adding, with regard to wines, the idea of substantial quality and that, consequently, the new law should not contain any rule contrary to the principles in force thus far.'

This tendency had been confirmed in a contemporary judgement of 24 January 1912, issued by the civil court of Fontainebleau, within the scope

---

[62] AN F 12 6969.

[63] AN F 12 6969 French Parliament session no. 2564 of February 27, 1913: supplementary report of the agricultural commission in charge of examining the bill on protection of *appellations d'origine* and regional delimitations.

of a lawsuit for usurpation of a name of origin of a natural product. The judgement, as quoted by the commission rapporteur, declared:

> given that X was perfectly aware of the fact that what makes the reputation of a product is the special flavour that it has, the particular taste given to it by circumstances independent of the will of the person who grows it, whether it results from the nature of the soil, its richness in one chemical or natural product or another, its exposure to sunlight, temperature, climate, in short, everything that makes up the environment which contributes to giving it what connoisseurs designate by the name of 'taste of the terroir' has, by his way of acting, sought to unite all of the conditions most likely to deceive the buyer.

It is obvious, the rapporteur comments, that the essential factor remains the origin and that a product made using fair practices and whose origin is beyond doubt will have the right to the appellation. In other words, the origin is of prime importance. The notion of quality plays a secondary role, but it is taken into account. However, the rapporteur again notes:

> winegrowers have raised objections to this project: the courts would have problems and if the controlled origin was not sufficiently protected, foreign wines could have the right to the same appellations if they met the conditions provided for the grape varieties, the method of cultivation and the land ... Even if these fears seem exaggerated, the commission has thus modified the text: anyone who knowingly uses the geographical name to designate products differing from those to which local, fair, established practices have attributed this name by virtue of their origin, their nature and their substantial qualities will be punished by the same penalty.

Moreover, the commission decided that it was necessary to ensure protection of *appellations d'origine* not only at the national but also at the international level. It was therefore necessary to request this protection from the other powers, on a reciprocal basis. To achieve this, they would have to rely on the International Institute of Agriculture in Rome.[64]

This was an attempt to mediate between the conflicting interests of winegrowers, merchants and consumers. The attempt was all the more difficult in that, in an international market, the question of controlled appellations can only be solved on the basis of international agreements. At the same time, the commission was aware that these interests were difficult to reconcile, which is what drove it to take the role of trade associations into account. Unlike earlier versions of the bill, this time trade associations were granted the legal right to act as representatives in legal proceedings. However, this solution was based on the firm assumption that allowing

---

[64] AN F 12 6969.

the associations to take part would reduce conflict among economic agents. The economic and social upheavals that rocked the environment of winegrowers and wine merchants during those years did not necessarily ensure such an outcome. On the contrary, the forces of opposition seemed once again to prevail. During the following months, a number of petitions from employers' federations, Chambers of Commerce, winegrowers' associations, and others were addressed to the Ministry of Trade and to Parliament.[65] As a general rule, the Chambers of Commerce required the protection of established quality, defined above all by taste, whereas the winemakers' associations insisted rather on the grape variety or the location.[66]

It was in this context that an agreement was finally signed between the winegrowers and wine merchants of the Gironde region, who until then had failed to agree on the definition of quality or the controlled appellation, and therefore on the issue of regional delimitation. This agreement concerned both the delimitation of the Bordeaux region and the characteristics of the products that would be entitled to the appellation. Reaching a consensus was the direct consequence of the non-cooperative solutions of the previous years when tensions between wine merchants and winegrowers in the Bordeaux region had weakened the impact of the Bordeaux reputation in maintaining stable relations and thus on the markets. Instability had been increased by the transfer of land ownership[67] as well as the arrival of new tradesmen, merchants, and so on in the market. In the absence of a strong legislative solution, however, their individualistic and often free-riding behaviour was incapable of ensuring advantageous economic solutions, once the initial successes began to wear off. This was all the more true as the reputation of the product, and even of the region, had dropped sharply. These problems, such as arriving at a clearer definition of the market segments, both from the standpoint of supply as well as of demand, were the factors motivating the agreement of 1913. However, this outcome did not lead the way to a more far-reaching agreement at the national level. This was so because of two main reasons: first, agents relying on quantity rather than on quality were less concerned by regional reputation; secondly, even among Bordeaux agents, this agreement did not include a perfect circulation of information on products, namely on the cépage composition.

---

[65] Ibid.
[66] See also discussions in the press: *le Moniteur vinicole*, 3/6/1913, p. 1: 'Le Projet Pams amendé par le syndicat national'; *Feuille vinicole de la Gironde*, 9/1/1913.
[67] G Postel-Vinay, *La Terre et l'Argent* (Albin Michel, 1997).

In the wake of these events, the Parliamentary commission modified its bill and a new version was submitted to the Parliament in November 1913.[68] Compared to the previous bills, the definition of the product in this bill no longer included wine characteristics, that is, the grape variety, the soil, and so on; instead, emphasis was placed on the trade mark and the geographical denomination. Thus, the winegrowers had prevailed over the wine merchants. There were a number of reasons for this outcome. First, from a legal standpoint, a clear distinction was now made between product falsification (its adulteration) and fraudulent use of trade marks and trade names. Second, the strategies of the economic agents changed. Not only winemakers and wine merchants but also winegrowers began using processes to maintain unvarying taste despite climatic variations from year to year. Blending grape varieties was done precisely for this purpose. This solution was accompanied by the increasing role played by reputation in the strategies of the various agents. That is why the bill mentioned legal protection not only for the winegrower but also for the merchant and the winemaker.

The converging interests of winegrowers and wine merchants did not, however, include those of merchants and distillers who would be penalised by the new bill, either due to the location of the vineyards, or the product characteristics, or yet again the technical solutions they adopted (which were not necessarily compatible with 'established local practices'). The bone of contention was once again the circulation of economic information. Indeed, in order to be applied, these provisions concerning the *appellation d'origine* and the reputation of the agents required an increased circulation of information about the product, among the economic agents themselves on the one hand, and between them and government authorities on the other. In particular, Art 4 of the bill asserts that 'any person engaged in the wholesale trade of wine, sparkling wine, natural sweet wine, liqueur wine, any person who has a wholesale account with an excise office will be required to keep a special account of all sales and purchases of products with controlled appellations'. Consequently, as a memorandum from the Ministry of Trade dated 14 January 1914 noted, 'a number of merchants dealing in "liqueur" wines have sent the Ministry a protest against the provisions of Art 4 of the bill. . . . The central association of French distillers has raised the same problem.' Both groups demanded the removal of the article.

---

[68] French Parliament session no. 3187 on 7 November 1913, Report of the Agricultural Commission in Charge of Examining the Bill on the Protection of *appellations d'origine* and Regional Delimitations.

The Minister of Trade then referred the question to the Ministry of Finance, about whose reply there could be little doubt:

> Contrary to the belief of the petitioners, regarding the account of wines purchased and sold, there is no need to record the grape varieties, which the interested parties may continue to process, as they have in the past, in complete freedom, without being required to provide any declaration or conduct any procedure. Thus, this account will not contain any information about winemaking processes. Consequently, the civil servants in charge of inspections will examine only these accounts, which do not disclose any information about processes.[69]

This reply confirmed the previous trends. At least at the administrative and normative levels, protection of reputation and regional delimitations did not go so far as to define product characteristics. At the same time, the door was left open to anyone who wished to make a complaint to the courts. While this reply satisfied the winegrowers of the Gironde, it could only annoy the wine merchants who were interested in delivering wines in large quantities with unvarying tastes.[70] The First World War forced any solution to the problem into abeyance.

2.3 AOCs in the Interwar Years and its Aftermath

This debate started up again after the war; the law of 6 May 1919 reiterated the principle adopted in 1911, which left the decision about whether or not a wine was entitled to an *appellation d'origine* to the judges. The decision was supposed to be based on fair, unchanging, local customs. However, a review of the case law from 1919 to 1928 reveals that no attempt was made to define these terms more precisely. As a result, disputes multiplied through the years, almost always over what was meant by a fair, unchanging, local custom.

Tensions sharpened with the crisis in 1929: domestic and international demand strongly fell, while wines from the New World increasingly took shares of the market from French producers. In July 1931, the government decided to support a bill proposed by winemakers in the south of France to raise taxes on output of more than 100 hectolitres per hectare, while prohibiting the planting of new vineyards by winegrowers that already had more than 10 hectares or produced more than 500 hectolitres.

However, these measures soon proved inadequate to curb the crisis, which was linked to falling income, reduced consumption and increased

---

[69] AN F 12 6969.
[70] Ibid.

product imports from the United States. In 1934, a new law gave bonuses to winegrowers who uprooted vine plants and stopped growing hybrids. Although the collapse of prices subsequently ceased, the problem of appellation marking remained. Indeed, many producers reacted to the new rules by selling their surplus production under false names and labels. The winegrowers' associations of the Bordeaux region succeeded in passing a decree proposal at the national conference of winegrowers' associations that was sent on to the government and quickly transformed into the Decree of 30 July 1935.[71] This was the famous law that brought into being *Appelations d'Origine Contrôlée* (AOC) as we know them today. The decree not only limited the surface area of origin, but also called for a list of accepted techniques and grape varieties for each region. The regulations were concerned with the process rather than the product per se, which continues to be the essential difference today between AOC and other labels. The AOC label controls production techniques without referring to the quality of the final product, whereas other labels focus on the quality aspect. Definition of territory is still central to the French wine appellation. The unity of political identity and production is apparent in all the regions making wine that are eligible for an appellation, where the name of the commune is often synonymous with the name of the wine. This association is particularly marked in Burgundy. The identification of place names with territory begins at a more detailed scale than the commune. Throughout France fields are often named after attributes of their sites. Only in Burgundy, more than 1000 lieux-dits (that is small geographical areas bearing a traditional name) have been identified. Most of them take their name from the physical characteristics of the site, while others take their name from human endeavours to raise plants and animals.

Within the AOCs category, a clear hierarchy has been established. Less than 5 per cent of French vineyards with a right to an appellation are classified as grands crus and each level lower in the hierarchy has a larger area than that above it. However, as in previous periods, purely technical and environmental features would have been unable to reach this institutional outcome without the action of lobbies. Each commune has its own syndicat of vignerons without whose cooperation INAO can do very little.[72] The appellation system is underpinned by a complicated network

---

[71] *Décret-loi du 30 juillet 1935 Relatif à la Défense du Marché des Vins et au Régime Economique de l'Alcool* (31 Juillet 1935) Journal Officiel 8314.
[72] For an updated account of INAO's more limited role, see Delphine Marie-Vivien, 'The Role of the State in the Protection of Geographical Indications: From Disengagement in France/Europe to Significant Involvement in India' (2010) 13 JWIP 121.

of organisations which represent different groups in the industry. The industry is supported by a highly professional bureaucracy. These groups and people have to identify and negotiate the main aspects of production subjected to control, that is, the varieties of grapes to be grown, cultural practices in the vineyard and yields of grapes or wine. In most regional appellations, the number of authorized varieties is fewer than five, and individual varieties are restricted as to the appellation in which they may be used. Specifying the varieties that the viticulturalist is permitted to grow in any appellation means that innovation in varieties is prohibited in many vineyard regions of France, notably the most successful ones.

3. COLLECTIVE APPELLATIONS: INVENTING TRADITION TO NEGOTIATE INNOVATION

We can draw a few conclusions regarding the issue of collective brands or designations, which was raised immediately after the revolution and continued to generate political debate and legal disputes throughout the nineteenth and twentieth centuries. Those disputes pertained to three interrelated aspects: the recognition of collective appellations; the protection of a distinctive sign for each producer or for the merchant; and the definition of wine as a natural or a manufactured product. These three aspects referred to one and the same problem, namely the process of agricultural industrialisation and the hierarchies between trade and production. It is untrue that France continued to protect collective appellations as a legacy of the guilds. During the course of the nineteenth century the reverse was true: justices, legal scholars and an overwhelming majority of political representatives rejected the idea of 'appellations régionales' precisely because they were considered as a resurgence of guilds, an obstacle to the market and, above all, because they were estimated as impossible to identify in so far as productive economic spaces were exposed to constant changes.

Towards the middle of the century, the recognition of individual appellations encouraged the development of vineyards, which fuelled rather than impeded the uncertainties relating to market dynamics and the outcome of disputes. During the last quarter of the nineteenth century, however, the system experienced a major crisis coinciding with market internationalisation and the sudden appearance of synthetic chemicals in agriculture. At that point, the uncertainties became more radical and new playing rules were required. Traders seemed to take control of the wine market, and this was even more difficult for winegrowers, since phylloxera had increased their dependence on credit and the international market was quickly changing. Industrial property succeeded in imposing the recognition of collective brands (regional appellations); however, opinions were divided regarding the geographical limits and technical or other criteria to be taken into consideration when defining the authentic product. The reason was that winegrowers of a given area were themselves divided about the extension of the concerned region and the techniques and practices to be considered as 'traditional' and 'loyal'. The introduction of the AOC label thus required – not unanimous agreement on the part of the economic actors, which was impossible to attain – but a consensus of the majority on playing rules that were acceptable but did not necessarily correspond to what each group considered optimal. The delimitation of the territories entitled to *appellations d'origine* and the list of accepted techniques and grape varieties were the fruit of a painful but necessary agreement among certain actors at the expense of others who until then had been allowed to produce Bordeaux or Champagne wines. In addition to their local, unchanging character defined by the appellation, AOC wines were also to be the result of 'fair' customs, in which fairness referred not only to spontaneous or emergent trade practices but also to the way in which they were defined by a group of actors and institutionalised as such. This is a far cry from the AOC as the expression of a *terroir* and its traditions. The set of rules, justified on behalf of consumers and their interests, was aimed in fact at ensuring guaranteed income to a group of producers and an institutional framework in which any innovation would be negotiated. This solution was to become a formidable institutional tool for the economic activity of a group of producers. For decades, the winegrowers and merchants of Bordeaux and Champagne were to be protected not only from foreign counterfeiting but also from the temptations of some among them to make unilateral changes in production techniques or simply to 'cheat'. Hence the desire of producers from other regions and of other products to benefit from the same advantages, which led to the multiplication of AOC labels during the second half of the twentieth century. Contrary to standard theories, appellations and collective trademarks were used primarily to provide the producer with a guaranteed income while offering a stable normative framework for negotiating innovation, rather than to inform consumers. From this standpoint, AOCs have three complementary goals: first, to ensure fair practices within a group of producers; second, on this ground, to better protect winegrowers against traders; and, third, to protect them from international competition. Collective appellations protected not only winegrowers against traders, but also helped small winegrowers. Still, between 1990 and 2003, while the total area of vines in the EU fell by 0.7 per cent per year, the average (wine) farm size only rose from

7.6 to 9.2 hectares.[73] Most European wine grapes therefore continue to be produced by thousands of relatively small holdings, each employing only a few salaried workers.[74]

French AOCs rules enjoyed great success in other European countries, especially in Spain and Italy where, for reasons similar to those in France, wine production played an important role in market equilibrium and in the introduction of quality norms. French lobbies and criteria (mainly from the Bordeaux and Champagne areas) have strongly influenced the contemporary European legislation on geographical indications protection. The regulations for quality wine in the EC involve much more than identification of geographic source. They clearly come within the definition of *appellations d'origine*, because they include demarcation of the area of production, vine varieties to be grown, cultivation methods, winemaking methods, minimum natural alcoholic strength by volume, and yield by hectare. These rules have been strongly supported by French representatives and their list and the eligible criteria are continuously re-negotiated (in particular between French and Italian representatives). In France, collective appellations served to control a group of producers faced with the emergence of innovations, on the one hand, and with the impact of trade on the other. The invention of traditions aimed precisely at closing the ranks of winegrowers and producers.

4. CONTEMPORARY TRENDS: GLOBAL WINES OR CONVERGING RULES?

The French and later European solution worked relatively well for a good part of the twentieth century, but today it is having a hard time adapting to the new context of a global economy. Within this scope, and particularly outside Europe, the emphasis on *terroir* and traditions is an effective marketing tool for some elites, but it is not enough to offset the development of wines made from a single grape variety or highly reputed individual trade marks. Nowadays, the total area in vineyards with no right to an appellation is about 50 per cent of the total areas under vines in France (24 per cent in 1981). This change has occurred as much of the viticulture areas of the Rhone valley and the Midi have planted better

---

[73] Andy Smith, 'Globalization within the European Wine Industry: Commercial Challenges but Producer Domination', paper presented at UACES annual conference, session 9, 1–3 September 2008.
[74] AMJO Coelho and J-L Rastoin, 'Globalisation du marché du vin et stratégies d'entreprise' (2001) 264(1) Economie Rurale 16.

varieties and applied to the INAO for the status of VDQS (*vin délimité de qualité supérieure*), often a step on the way towards AOC. From the point of view of average consumers, it is indeed easier to get one's bearings in relation to a particular grape variety (for example Chardonnay) than among a myriad of AOC labels. The increasing number of these labels has tended to make them commonplace and blur the message that AOC signs were supposed to give to the consumer. By contrast, Chilean, South African, American and Australian wines can be distinguished by easy recognition of their main grape variety as well as a standardised taste that ultimately corresponds quite well to the one imposed by wine guides and marketing campaigns, not only for ordinary wines but even for very good wines.

However, important changes have occurred in the US wine market and its regulation. Regardless of how advanced standardised agriculture may have become, it concerned only certain aspects of production and product characteristics. Thus, the criteria of wine denomination and presentation continued to evolve due to the introduction of new techniques; the composition and so-called 'minority' percentages of grape varieties generated constant conflict between winemakers, and then between winemakers and consumers, as well as the use of a given stabiliser.[75] Together with the increasing differentiation in wine quality, this opened the way to most recent trends in lawmaking and quality identification. Indeed, during last decades, not only did domestic production of wine greatly increase, but, importantly, vertical integration and concentration went along with increasing product quality differentiation. Some areas made their own reputation while some producers have sought to capitalize on and protect the goodwill they have developed in association with their wine regions. The federal wine-labelling regulatory scheme promulgated in 1978 by the US Bureau of Alcohol, Tobacco and Firearms (ATF) in many ways can be analogised to a comprehensive municipal zoning scheme. The ATF's wine-labelling regime limits the use of certain terms on wine labels to specific products in much the same way as zoning plans limit certain types of property uses to specific areas. The 'properties' regulated by the ATF

---

[75] M Maher, 'On Vino Veritas? Clarifying the Use of Geographic References on American Wine Labels' (2001) 89 California Law Review 1881; J Morilla Critz, A Olmstead and P Rhode, 'Horn of Plenty: The Globalization of Mediterranean Horticulture and the Economic Development of Southern Europe, 1880–1930' (1999) 59(2) Journal of Economic History 316; CF Nuckton, *Demand Relationships for California Tree Fruits, Grapes and Nuts: A Review of Past Studies* (Giannini foundation special publication 3247, Berkeley 1978); M Tracy, *Agriculture in Western Europe: Challenge and Response, 1880–1980* (2nd edn Granada, 1982).

include ownerless descriptive terms (for example, geographic place names, grape variety names and 'estate bottled') and geographic brand names, often registered as trade marks by the wine producers who use them. These are geographic indications rather than *appellations d'origine*. Only names of producing areas are registered.[76]

The US regulation finds its rationale in the fact that previous rules about the use of the regional indicators were vague and difficult to enforce prior to the establishment of the viticulture areas. Among these rules was the requirement that 75 per cent of the grapes must originate from the locality named on the label. In the regulation of 1978 the requirement became 85 per cent for viticultural areas and 75 per cent for single counties or single states. In order to enforce these rules, it was necessary to define the viticultural areas. Although these criteria were influenced by the geographic determinism permeating commentaries on the French industry, other aspects of the French system – such as its hierarchical nature and emphasis on quality – were not included. US criteria include nothing about the wine itself. Thus, like most zoning schemes, ATF's wine-labelling scheme encompasses a number of 'nonconforming uses'. Unlike most zoning schemes, however, ATF's current scheme has neither provisions for limiting the expansion of nonconforming uses, nor provisions for the eventual termination of such uses. Geographic terms have appeared as wine type designations, as part of grape variety names, and as a 'generic' or 'semi-generic' wine type. A geographic term may appear as part of a grape variety name. Grape variety names may be used as wine type designations, provided the variety name appears on a list of approved names. The ATF gives 16 examples of semi-generic terms that describe a wine type for the purposes of American labelling. Fifteen of these terms are European in origin; among them, some of Europe's most famous wine regions, including Champagne, Chablis, Burgundy, Sauterne, and Chianti. These terms may be used on wines derived from American grapes provided a true appellation of origin appears in conjunction with the semi-generic term.[77] The US considers names such as chablis, burgundy, sherry and port as generic, so that names such as 'California Chablis' are permitted. They are considered by the ATF to describe the type of wine rather than its origin. However, in recent years, trade mark law has expanded to grant trade mark protection beyond that provided by the likelihood of confusion test. This additional protection is

---

[76] For a detailed review of similar zoning experiments in Australia, see Chapter 10 by Stern in this volume.

[77] Maher (n 75).

against 'dilution' of famous marks so that the use of a mark by another will gradually diminish the distinctive value of a famous mark, even in the absence of consumer confusion.[78] The complaint of French wine producers is that such use may dilute their appellations.

From this standpoint, one may observe that in recent years, the opposition between continental Europe (with collective appellations) and 'the rest of the world' (without these signs) is increasingly weakening in so far as one single grape (*monocépage*) wines are quickly developing in France, while origin identification is increasingly becoming the object of legal and economic interest in the USA, the Americas, Australia and South Africa. At the same time, pressures are emerging in the US to improve the labelling system. As we have shown, the entire US system of viticultural areas has been conceived as a tool for discouraging labelling that might mislead the consumer. However, as recent debates and litigation testify, it seems doubtful that consumers are aware of the location, let alone the boundaries, of the more obscure viticultural areas. And even for well-known names such as Napa Valley, few consumers are likely to be aware of these areas' precise boundaries. As in France, the US definitions have created regional identity crises: Napa Valley has never ceased to grow and expand. This converging trend finds a confirmation in the increasing dispersion of vines over the French territory (previously concentrated) and its opposite, a concentration of main vines in some California areas.

These trends help to explain widely discussed phenomena such as quite similar processes of vine growing and winemaking leading to similar tastes in some Californian, Chilean, Italian or French areas. Uniform taste is no more a matter of low quality and cheap wines but it can equally be important for high-quality wines. From this standpoint, still persistent (but narrowing) differences in legal institutions (collective wine appellations in particular) seem to have a diminishing impact on wine quality. Nowadays pressures to change the institutional definition of wine (and the AOCs) in Europe proceed along this same direction. Still, producers and representatives have to choose between two strategies: they may either relax AOC conditions in order to better enter the mass world market or restrict them. In the first case, this will lead to a solution close to the American one (origin identifications, but not as appellation 'brands'); in the latter case, a different strategy, closer to that of the 1930s will be at work. The first solution runs the risk of erasing any specificity

---

[78] For a discussion of dilution in the context of GIs, see Chapter 7 by Handler in this volume.

of French and European wines; the second one would require a much stricter evaluation of the AOCs wines by the INAO; this, in turn, will imply several exclusions from the label and, for those who will benefit from it, a strong internal cohesion. Both conditions nowadays seem difficulty to satisfy.

3. 'Translating *terroir*' revisited: the global challenge of French AOC labeling
Elizabeth Barham

This chapter returns after a decade has passed to an article originally published as 'Translating Terroir: The Global Challenge of French AOC Labeling'.[1] I would like to thank the editor of this volume for the opportunity of reopening the arguments made in that article, after what has been a fairly tumultuous decade for origin and 'local' food projects of all kinds, as well as for international trade negotiations related to them. The substance of the article has held up well over the time, but a number of endnotes and references have been added pointing to publications and resources that have become available since the article was originally published. In particular, the reader may wish to refer to a collection of chapters and case studies focused primarily on the European experience with origin product labeling, with some insights on the US situation and that of developing countries, published in 2011.[2]

1. INTRODUCTION

Since the signing of the WTO Agreement in 1994, a longstanding struggle between the European Union and the United States has been intensifying. Other countries from around the world have aligned themselves with either the EU or the US in this debate.[3] The dispute is little known to those outside of international law and trade circles, and yet the outcome could have important repercussions for rural development globally. It concerns the portion of the WTO Agreement – TRIPS – which deals with intellectual property and addresses geographical indications (GIs) in Arts 22–24.[4]

---

[1] E Barham, 'Translating Terroir: The Global Challenge of French AOC Labeling' (2003) 19 Journal of Rural Studies 127 (hereafter, Barham (2003)).

[2] E Barham and B Sylvander (eds), *Labels of Origin for Food: Local Development, Global Recognition* (CABI, 2011).

[3] For background, see, in this volume, Chapter 13 by Goebel and Groeschl as well as Chapter 9 by Taubman.

[4] For details, see Chapter 6 by Gervais in this volume.

The agreement defines these as 'indications which identify a good as originating in the territory of a [m]ember [country], or a region or locality in that territory, where a given quality, reputation or other characteristic of the good is essentially attributable to its geographical origin'.

GIs are members of the larger family of marks known as labels of origin, of which they are one type. They have been used extensively in association with wine and spirits, but can also be applied to cheeses, meat products and other foods. In Europe, 'traditional' or 'typical' products often carry labels of origin. Recent agro-food literature contains many references to the growing consumer demand for these products, often as evidence of the emergence of a new rural development paradigm.[5] Most authors identify origin labeled products as important manifestations of 'local', 'quality' or 'endogenous' food systems.[6] They are seen as contributing to the 'consumer turn' which may portend major shifts in the conventional agricultural model.[7]

---

[5] T Marsden, J Banks and G Bristow, 'Food Supply Chain Approaches: Exploring their Role in Rural Development' (2000) 40(4) Sociologia Ruralis 224; J Murdoch, T Marsden and J Banks, 'Quality, Nature, and Embeddedness: Some Theoretical Considerations in the Context of the Food Sector' (2000) 76(2) Economic Geography 107; JD van der Ploeg, H Renting, G Brunori, K Knickel, J Mannion, T Marsden, K de Roest, E Sevilla-Guzmán and F Ventura, 'Rural Development: From Practices and Policies Towards Theory' (2000) 40(4) Sociologia Ruralis 391. Since 2003, several publications have examined the linkage between traditional, typical and origin products and actors' strategies for rural development in a globalized marketplace, primarily in the European context. For examples, see Barham and Sylvander (n 2); L Bérard, and P Marchenay, *From Localized Products to Geographical Indications: Awareness and Action* (CNRS 2008); Claire Delfosse (ed), *La Mode du Terroir et les Produits Alimentaires* (Les Indes Savantes 2011); M Fonte and AG Papadopoulos (eds), *Naming Food After Places* (Ashgate 2010); INAO, *Local Flavor: A Tour of French Origin-Linked Products* (Hachette Pratique 2006) and the articles in a special issue: (2010) 13(2) JWIP. For the *terroir* concept as applied to the US context, typically emphasizing positive aspects from the perspectives of gastronomy, culture and history, see Rowan Jacobsen, *American Terroir: Savoring the Flavors of Our Woods, Waters, and Fields* (Bloomsbury 2010) and Amy Trubek, *The Taste of Place: A Cultural Journey into Terroir* (University of California Press 2008). However, several articles by American researcher Sarah Bowen and colleagues have critically assessed the rural development claims associated with origin products. See the works cited in S Bowen and MS Gaytán, 'The Paradox of Protection: National Identity, Global Commodity Chains, and the Tequila Industry' (2012) 59(1) Social Problems 70 (pointing to evidence of the failure of governance structures to adequately take into account vulnerable socioeconomic groups and the environment).

[6] Jan Douwe van der Ploeg and A Long (eds), *Born from Within: Practice and Perspectives of Endogenous Rural Development* (Van Gorcum 1994).

[7] Ben Fine and E Leopold, *The World of Consumption* (Routledge 1993); M FitzSimmons and D Goodman, 'Incorporating Nature: Environmental

Gilg, in fact, estimates that as global agricultural production differentiates into a bipolar system of high volume 'day-to-day' foods produced and distributed by multinational corporations and lower volume niche or specialty products such as those produced under labels of origin, the latter category could come to account for as much as 30 percent of overall food sales due to their higher value.[8]

Labels of origin, including GIs as defined in TRIPS, warrant closer attention from rural development scholars concerned with local and quality food products for two reasons. First, ongoing negotiations within the WTO will inevitably determine the future legal interpretation and global scope of GIs.[9] The European Union is leading the demand for a global registry that would protect place names covered by GIs as unique for the purposes of labeling agro-food products (that is, Roquefort cheese may only be labeled such if it originates in Roquefort, France). The US opposes such a move. Should the strong position of GIs supported by the EU be undermined as a result of these negotiations, a potentially important rural development tool could be lost or severely constrained. Because there may be more at stake for sustainable rural development than initially meets the eye, analyses are urgently needed from scholars and rural development practitioners on both sides of the debate to inform the policy-making process.[10]

Narratives and the Reproduction of Food' in B Braun and N Castree (eds), *Remaking Reality: Nature at the Millennium* (Routledge 1998) 194; J Urry, *Consuming Places* (Routledge 1995).

[8] A Gilg, *Countryside Planning: The First Half Century*, 2nd edn (Routledge 1996) 71. Recent forays of agri-food giants such as Walmart into 'local food' merchandizing may begin to muddy the distinction in American consumers' minds between 'local' products and the more structured regimes of origin products that are the focus in this chapter. For more discussion of 'local food' and GIs in the US context, see D Giovannucci, E Barham and R Pirog, 'Defining and Marketing "Local" Foods: Geographical Indications for US Products' (2010) 13 JWIP 94.

[9] Since the original 2003 publication of this writing, the WTO Doha Round has become mired in disputes and conflicting agendas, with progress on GIs coming to a practical standstill. For further details on WTO GI negotiations, see Chapter 7 by Handler and Chapter 8 by Martín in this volume.

[10] Notwithstanding the work already cited, the entire topic of origin products and GIs is still relatively little known, and even less understood, in the United States, the primary locus of opposition to a global regime recognizing GIs. In fact, in 2010, a decision by the US Department of Agriculture to cancel funding for a research project that would have created a national listing of American GIs reflects resistance to the organization of even a US-based system of recognition for its own origin products. Information about the intent of the original project can be found in E Barham (ed), *American Origin Products: Protecting a Legacy* (oriGIn, Geneva 2010).

Secondly, recent agro-food literature does not always adequately take into account that there are technically several types of labels of origin which vary according to how they are implemented within different nation states and within trading blocs such as the EU. For example, distinctions are not always made between labels that promote a product by indicating its origin but are not protected by the state, and labels that enjoy both state protection and administrative support, as in the case of GIs. Labels with differing degrees of state financial and institutional support can be expected to have respectively different impacts on rural development. Without a clear understanding of how different labels of origin function in practice, analysts run the risk of confounding situations that are empirically quite distinct on the ground. Resulting research categories and typologies may be misleading and analyses based on them incomplete or inaccurate.[11] Research in this area would also be of benefit to developing countries, many of whom have begun the process of implementing national GI regimes.[12]

This chapter takes a closer look at the importance of GIs by considering how they reconnect people, production and place in the context of rural development. Section 2 points out some key underlying causes for the differing national positions on GIs taken at the WTO level. The influence of special interests plays a key role in US opposition to a global registry of protected place names, and this aspect is addressed. But the argument is also made that a more fundamental feature of the debate derives from the

---

[11] The plethora of labeling schemes can also leave Europeans confused at times. For more detail on distinguishing particular types of EU labels of origin, see generally Barham and Sylvander (n 2). This work grew out of two major European research projects directed by Bertil Sylvander: Development of Origin Labelled Products: Humanity, Innovation and Sustainability (DOLPHINS) and the *Syner-GI* project. For outputs from both projects, see www.origin-food.org

[12] See Agence Française de Développement & Fonds Français pour l'Environnement Mondial, *Indications Géographiques: Qualité des Produits, Environnement et Cultures* (Savoirs communs no. 9, AFD and FFEM 2010); S Bowen, 'Embedding Local Places in Global Spaces: Geographical Indications as a Territorial Development Strategy' (2010) 75(2) Rural Sociology 209; P van de Kop, D Sautier and A Gerz, *Origin-Based Products: Lessons for Pro-Poor Market Development* (Bulletin/Development Policy & Practice (372), 2006). Training and advice to developing countries in this regard is provided by: (1) the Organisation for an International Geographical Indications Network (oriGIn), at: http://www.origin-gi.com/; (2) the Swiss Association for the Development of Agriculture and Rural Areas (Agridea), at: http://www.agridea.ch/. See also M Bagal and M Vittori, *Practical Manual on Geographical Indications for ACP Countries* (oriGIn 2011). For India, see Chapter 16 by Rangnekar and Mukhopadhyay; for China, see Chapter 12 by Zheng, in this volume.

status of GIs as a special kind of intellectual property. Drawing on recent theories advanced in the agro-food literature, particularly embeddedness perspectives and conventions theory, this section explores how GIs challenge conventional agricultural practice due to their explicit reference to place or territory.

Section 3 explores aspects of this challenge in more depth by examining a prominent type of geographical indication, the French *Appellation d'Origine Côntrolée* (AOC). Products covered by AOC labels are controlled by the state to assure both their territorial origin and their conformity to precise rules for production and processing that guarantee their 'typicity', or distinctive character. The AOC system is the oldest of the European label of origin systems and is widely regarded as the most strict and thoroughgoing of its kind. It is, in this sense, a model of reference for origin labeled products. The system is guided by the concept of *terroir*, a French word without a suitable English translation. Section 3 discusses the meaning of this term, then examines how it is institutionally expressed in the AOC labeling process. This process is supportive of rural development related to agricultural products in many respects, particularly in marginalized areas. But there are also possibilities for slippage in the system in terms of meeting its self-stated goals. This section addresses both sides of this issue.

The conclusion returns to a broader consideration of the potential of GIs for supporting the emergence of a new paradigm for agriculture with some necessarily brief reflections on how they may fare when applied in different parts of the world.

2. WHAT IS AT STAKE IN THE GI DEBATE?

2.1 Special Interests

The effect of adopting a system of geographical indications at the global level is immediately troubling for many corporations based in the United States, and explains much of the US opposition to GIs in WTO negotiations. Problems arise for US producers who took pre-existing European place names for their products – a frequent phenomenon in other countries that experienced heavy European immigration as well, including Canada, Australia and many countries in Latin America. Immigrant business owners of European extraction were familiar with geographical names from their home countries that were associated with quality products. They used those same names to promote their own products, riding on the coat tails of the original product's reputation (or, depending on one's point of view, using them allusively, to indicate general product qualities). In the US and some other countries, such place names have been treated as generic names for certain types of products, to the chagrin of countries where the actual regions are located (for example, Champagne and Chablis in France).[13] Thus, the establishment of a global registry reserving GIs for the country of origin would clearly have important impacts on a number of labels currently in use, and on the trade position of these products.[14]

US resistance to a global registry preserves the status quo in 'generic' names, as well as the investments American corporations have made in promoting brand names that derive from certain geographical names. One notable example of this is American Budweiser beer, produced from 1876 in St Louis by the Anheuser-Busch Company. The Budweiser Budvar brewery in the Czech Republic laid claim to being the 'original' Budweiser beer producer, and decades of increasingly bitter intellectual property law suits ensued around the world.[15] Note that US Budweiser is the number one selling beer in the world, reflecting years of adroit advertising strategy as well as significant expenditure. It would have been naive to assume that Anheuser-Busch would relinquish the European name without a struggle, and the same holds true for many other corporations worldwide that find themselves in similar situations.[16] Such conflicting interests make compromise extremely difficult, but perhaps not impossible. Much

---

[13] The Wine Agreement between the US and the EU moved the impasse on wine names forward to some extent, although ambiguities remain. See B Rose, 'No More Whining about Geographical Indications: Assessing the 2005 Agreement between the United States and the European Community on the Trade in Wine' (2007) 29 Houston Journal of International Law 731. For related developments in Australia, see Chapter 10 by Stern in this volume.

[14] For recent developments on the international registry, see Chapter 8 by Martín in this volume.

[15] For a comprehensive review of this litigation, see Chapter 14 by Heath in this volume.

[16] While the lawsuits continue for now, a majority of the shares of Anheuser-Busch were purchased in 2008 by the Belgian-Brazilian company InBev, becoming AB InBev (see http://www.ab-inbev.com/). In January 2012, the company was then able to purchase the Budejovicky Mestansky Pivovar brewery in Czech town of Ceska Budejovice, one of the two breweries laying claim to the Budweiser name in international courts. The remaining Czech Republic brewery, Budejovicky Budvar, is larger, but may itself be a target for purchase by AB InBev. See http://www.bendbulletin.com/article/20120115/NEWS0107/201150333/. This merger could have repercussions for both litigants. More importantly, perhaps, it could remove a major opponent to a US system of GI recognition from the playing field, one that has spent considerable sums in the past on lobbying in the US to maintain the national anti-GI stance.

depends on the time horizon (short- or long-term) taken into account, and the vicissitudes involved in building a world trade system capable of recognizing and protecting GIs.

2.2 Geographical Indications and the Conventions of Place

On a deeper level, geographical indications as a form of intellectual property challenge the law, culture and economic logic of American business, oriented as it is towards liberal economic theory based on individual ownership. The United States is familiar and comfortable with trade marks as a way of protecting the intellectual property associated with a business name. Trade marks belong to individuals or corporations (the latter being treated as individuals under US law),[17] and can be bought and sold as a business asset. If they are infringed, it is up to the individual or corporation to defend their rights to the name before a court. The US has taken the position that the current system of international trade mark protection can be used to protect origin labeled products, and that it provides sufficient legal tools to protect American GIs.[18]

EU countries disagree. They point out that labels of origin 'belong' to the region itself and are only administered by state governments, the latter preventing consumer fraud by overseeing certification systems and other controls. Individual producers within territories covered by geographical indications cannot buy, sell or inherit the rights to the name of the territory, as they can with trade marks. Nor can they move their production out of the region and retain the region's name, as a corporation might when using a trade mark.[19] If producers are located in a territory that is protected by a geographical indication, they are not obliged to use the name in their product labeling. In fact, they are only allowed to do so if they follow the requirements for certification. Finally, in the case of usurpation of the name it is the state that potentially intervenes – a consideration of crucial

---

[17] The emergence of the 'Occupy' movement in the US and now globally has perhaps begun to gradually shift the discourse concerning the status of corporations as persons before US law (see http://en.wikipedia.org/wiki/Occupy_Wall_Street, for an overview of the emergence of the Occupy movement with Occupy Wall Street). More specifically, one of the central goals emerging from the movement is the pursuit of a legislative overturn of the legal recognition of corporations as persons (see http://www.shutdownthecorporations.org/).

[18] For the position of the US Patent and Trade Marks Office (USPTO), see: www.uspto.gov/ip/global/geographical/index.jsp

[19] Precisely such a conundrum faced a UK GI producer ('Newcastle Brown Ale') which sought to relocate its production facilities. It was compelled to give up the GI. See Cancellation Request for Newcastle Brown Ale [2006] OJ C 280/13.

importance for small-scale producers who are not able to afford costly legal battles, particularly at the international level. It can be argued that because GIs are the only form of intellectual property specifically related to place or territory, they represent a type of collective property.[20] While recognizing their existence as a distinct regime, the World Intellectual Property Organization (WIPO) has set up a special committee dedicated to developing a better understanding of how they function as intellectual property.[21]

Concepts of neo-liberal economic theory that lie behind the economic push towards globalization posit a frictionless economy where neither space nor time impedes the free flow of goods, labor and capital. However, as conceived in the European context, as a form of collective property anchored to specific places, GIs challenge this picture in significant ways. Traditional, 'typical' products are enmeshed in both the place and history of their area of production.[22] When they are awarded GI status by the body authorized to conduct product reviews, the presence of the GI name on the label carries specific messages to the consumer not only about the place of production, but also about the *process* of production. This is distinctly different from ordinary 'brand' labels that carry information on the inherent qualities of the product alone (ingredients, and so on) as it stands before the consumer, but do not bring to mind the 'story' of the product's history.

In previous work I had identified the focus on *process* as a key distinguishing feature of product labels that attempt to connect to non-market values held by consumers, such as eco-labels (that is, organic, fair trade, rain-forest friendly, and so on).[23] But while eco-labels may promote laudable goals, they travel with the product, so to speak, more akin to a simple brand, informing the consumer of *how* the product was produced to some extent, but not necessarily *where*. Labels of origin, on the other hand, hold the potential of re-linking production to the social, cultural and environmental aspects of particular places precisely because they must

---

[20] In legal discourse it is unclear as to what type of property, if any, a GI is – public, private, common or some variant/hybrid of these three categories – and TRIPS does not settle the issue. See D Gangjee, *Relocating the Law of Geographical Indications* (CUP 2012) 202–206.

[21] The Standing Committee on the Law of Trade Marks, Industrial Designs and Geographical Indications. Documents from its meetings are available at: http://www.wipo.int/

[22] D Barjolle, S Boisseau and M Dufour, 'Le Lien au Terroir: Bilan de Travaux de Recherche' (Institut d'Économie Rurale 1998).

[23] E Barham, 'Towards a Theory of Values-Based Labeling' (2002) 19(4) Agriculture and Human Values 349.

specify the 'where' of a thing. Origin products are thus distinguished from anonymous mass produced goods, opening the possibility of increased responsibility to place.

Theorizing in the agro-food literature has often identified connections with place and tradition as important aspects of both the production and consumption of 'quality' foods. Buttel outlines a brief history and characterization of this branch of the sociology of agriculture, which he identifies as 'by far the most heterogeneous'.[24] He traces influences of the French regulationist school, as later modified by the critiques of Goodman and Watts[25] and the work of French scholars on conventions theory,[26] as making major contributions to the conceptualization of 'quality'. These authors, and others concerned with the related theme of 'local' food,[27] are indebted to the work of Karl Polanyi[28] for the concept of 'embeddedness', which postulates that free market capitalism must be subject to social and environmental constraints if it is not to destroy the basis of the economy itself.

Conventions theory can be viewed as one way to interpret how social constraints are placed on the market to re-embed it in non-market concerns.[29] It posits the process of 'justification' or legitimation as a fundamental cognitive act that serves to produce lasting agreements among social actors and ultimately maintain order. Moments when a shared choice of action or decision is called for, or 'situations' in the theory, require actors to make use of their assumptions, or world views, to 'qualify' things and

---

[24] FH Buttel, 'Some Reflections on Late Twentieth Century Agrarian Political Economy' (2001) 41(2) Sociologia Ruralis 165, 174.

[25] D Goodman and M Watts, 'Reconfiguring the Rural or Fording the Divide? Capitalist Restructuring and the Global Agro-Food System' (1994) 22(1) The Journal of Peasant Studies 1.

[26] B Sylvander, 'Conventions de Qualité, Marchés et Institutions: le Cas des Produits de Qualité Spécifique' in G Allaire & R Boyer, *La Grande Transformation de l'Agriculture: Lectures Conventionnalistes et Régulationnistes* (INRA 1995) 167 (see also other contributions in the same edited volume); L Boltanski and L Thévenot, *De la Justification: Les Économies de la Grandeur* (Gallimard 1991). The book is also now available in English (L Boltanski and L Thévenot, *On Justification: Economies of Worth* (Princeton University Press 2006)) but I have relied on the original French version.

[27] C Clare Hinrichs and T Lyson (eds), *Remaking The North American Food System* (University of Nebraska Press 2008); C Clare Hinrichs, 'Embeddedness and Local Food Systems: Notes on Two Types of Direct Agriculture Market' (2000) 16 Journal of Rural Studies 295.

[28] Karl Polanyi, *The Great Transformation* (Rinehart [1944] 1957).

[29] J Wilkinson, 'A New Paradigm for Economic Analysis?' (1997) 26 Economy and Society 305.

people. This allows them to decide which object, configuration of objects, or path of action is preferable. Because situations tend to call upon more than one world view, compromises are required among the world views involved. Compromise itself is not considered pejorative within the theory, but rather necessary. The compromise reached, in fact, represents the agreed upon path of action that allows things to move forward on a stable basis until internal or external change again requires a moment of shared justification.[30]

Boltanski and Thévenot[31] developed six sets of world views (or 'conventions'), based on major works of Western political philosophy. Only one of these worlds – the *market* world – corresponds closely to current neo-liberal economic thinking. The other five counter or mitigate it in various ways.[32] These world views or 'logics' have been invoked by a number of authors researching origin labeled products.[33] Marescotti, for example, analyzes coordination mechanisms used by actors in marketing channels linked to a 'typical' agro-food product as evidence of the importance of the *domestic* and *civic* worlds.[34] The domestic convention emphasizes the importance of the home, family and community. It stresses the role of face-to-face relationships, loyalty and trust in maintaining a functioning society. Civic logic asserts the need for adherence to shared principles for the common good, within the expectation that personal interest will be placed below the interest of the collectivity. Marescotti concludes that because *market* logic does not account very well for how consumers evaluate the quality of typical products, 'the most appropriate quality convention for typical products seems to be the result of a compromise between *domestic* and *civic* logic' (emphasis added).[35]

---

[30] The theory does not address situations in which violence is used to impose one point of view, and situations in which individuals refuse to participate in reaching social agreement (i.e., solitary or detached action).

[31] Boltanski and Thévenot (1991) (n 26).

[32] The six worlds are: the world of inspiration, the domestic world, the world of opinion, the civic world, the market world and the industrial world. Boltanski and Thévenot take the general position that the market form of justification is currently overstepping its bounds in significant ways, and that critical attention is therefore needed to the social process of judging or legitimating our actions.

[33] See, in particular, the collection of essays in B Sylvander, D Barjolle and F Arfini (eds), *The Socio-Economics of Origin Labelled Products in Agri-Food Supply Chains: Spatial, Institutional and Co-ordination Aspects*, 2 Vols, Proceedings of 67th EAAE Conference, Le Mans, France, 28–30 October 1999, (INRA 2000).

[34] A Marescotti, 'Marketing Channels, Quality Hallmarks and the Theory of Conventions' in Sylvander et al. Vol. 2 (n 33) 103.

[35] Ibid 116.

Label of origin systems are particularly useful examples of both embeddedness and conventions theories at work. By insisting upon a strong link in production to the ecology and culture of specific places, they re-embed a product in the natural processes and social context of its territory. But because nature does not speak for itself, but rather through the 'translation' that is the production process, conventions theory is helpful in analyzing how social systems of coordination and organization legitimate and perform a given interpretation of the product. The legitimation process, to be effective, must be carried out not only within the territory of production but nested within multiple levels of coordination from the local to the global. In addition, the consumer's acceptance of the product must also be legitimated, which gives the word 'quality' its full meaning.

Aligning these various levels of coordination and making the needed connections to consumers can be seen as an operation of closing the loop on production in terms of environmental and social accounting. Capital is no longer operating in a 'virtual' world without time or place.[36] Instead, a product's 'story' becomes the responsibility of the producers. A label of origin reconnects the product with a specific place and opens up the possibility that producers, as well as consumers, can be held accountable for their actions *in that place*. The dynamic nature of label of origin systems, furthermore, makes them potential candidates for acting as 'platforms for resource use negotiation', identified as crucial to the regulation of collective forms of property.[37] Here again, conventions theory can be a useful framework for analyzing the workings of such platforms, which incorporate actors at different levels, with quite different interests. Examining specific labeling cases and systems helps us situate processes of legitimation within the complex webs of relationship, both horizontal and vertical, that actually make such systems work.

The next section explores one such example, the French AOC label of origin. It begins with a discussion of the concept of *terroir*, which the AOC system draws upon in establishing its regulation of each product to be labeled. In the process of regulation, a number of situations arise that call for legitimation, or qualification, of both the product and the producers, creating potential platforms for judgment. Conventions theory can then be useful for analyzing the act of passing judgment, bringing embeddedness

---

[36] JG Carrier & D Miller, *Virtualism: A New Political Economy* (Berg 1998).
[37] Although GIs have not been studied extensively as examples of common property regimes, there are many fruitful parallels to be drawn. See, for example, the collection of articles in the special issue (1999) 16(3) Agriculture and Human Values, devoted to multiple-use commons, collective action and platforms for resource use negotiation.

concepts and the logics or worlds of conventions theory into play in the same framework. The purpose here is not to examine any particular AOC label in detail, but rather to draw on multiple examples to show how the overall system reproduces the cultural concept of *terroir*, at the same time connecting it to more macro systems of regulation through the AOC's status as a GI. The AOC is particularly interesting to consider as a GI because it influenced the development of the European Union Protected Designations of Origin (PDO), to the point that once an AOC is awarded in France there is very little questioning of its legitimacy at the level of the EU.[38] Its influence is also being extended to other parts of the world as countries increasingly request assistance from the French government in adapting the system to their particular situation.[39] An understanding of the AOC process can therefore be useful for making comparisons to label of origin systems outside of France.

3. AOC LABELING AND THE REGULATION OF TERROIR

3.1 Translating *Terroir*

Historically, *terroir* refers to an area or terrain, usually rather small, whose soil and micro-climate impart distinctive qualities to food products. The word is particularly closely associated with the production of wine. A *terroir* can be identified, for example, as one that produces a *grand cru*, or a particularly excellent wine. It can also be said that a certain wine

---

[38] This is based on an interview conducted in Brussels in 2001, with EU Commission employees responsible for reviewing PDO applications. Admittedly within the EU there are disputes over whether certain names should be considered generic (e.g. Feta cheese; see Chapter 18 by Gangjee in this volume). However the focus here is on demonstrating that a very culturally defined and locally grounded concept such as *terroir* is connected to more macro levels of decision making via a label of origin system, creating *potential* openings for local voice at these higher levels. Details on EU quality policy labeling are available at: http://ec.europa.eu/agriculture/quality/

[39] INAO *Rapport d'Activités* (annual report, 2010) 15 (indicating that the agency received 16 international delegations, including representatives of China, Canada, Macedonia, Morocco and Paraguay, and sent its own representatives out on request to eight different countries, including Bosnia, Russia and Turkey). The case of Quebec, Canada, is also illustrative in this regard. See E Barham, 'The Lamb that Roared: Origin Labeled Products as Place Making Strategy in Charlevoix, Quebec' in Hinrichs and Lyson (n 27) 277.

has a *goût*, or taste, of its particular *terroir*. Figuratively, *terroir* can also designate a rural or provincial region that is considered to have a marked influence on its inhabitants. It is said in French, for example, that certain customs or idioms are rooted in their *terroir*, or that a person strongly conveys a sense of the *terroir* of their birth and upbringing.[40] This concept of *terroir* relates to a time of much less spatial mobility, when change occurred at a slower pace. *Terroir* products, in this interpretation, resulted from long occupation of the same area and represented the interplay of human ingenuity and curiosity with the natural givens of place.

The concept of *terroir* reaches its most elaborate expression in the case of wine, and so it is no accident that one encounters it most frequently in texts on this subject. James E Wilson describes it in the following way:[41]

> Terroir has become a buzz word in English language wine literature. This lighthearted use disregards reverence for the land which is a critical, invisible element of the term. The true concept is not easily grasped but includes physical elements of the vineyard habitat – the vine, subsoil, siting, drainage, and microclimate. Beyond the measurable ecosystem, there is an additional dimension – the spiritual aspect that recognizes the joys, the heartbreaks, the pride, the sweat, and the frustrations of its history.

Wilson goes on to explain that there is an assumption among winegrowers that each terrain 'should be allowed to be itself and produce the wine for which nature endowed it'.[42] In other words, winemakers are striving to produce a wine that is special in the sense that it bears the 'signature' of their style of vinification while not interfering with the 'natural' taste that wines produced from that terrain should display. The historical *terroir* concept viewed wine production as a complex dance with nature, with the goal of interpreting or translating the local ecology, displaying its qualities to best advantage. A great deal of knowledge about the local terrain is needed for success, as well as respect for local natural conditions that can be expressed through the wine.

The notion of *terroir* has met with renewed interest in recent years, however, and is refocusing discussions of how the old is made new in the ongoing process of history. Its interpretation today reflects the involvement of powerful social actors, such as farmers unions (syndicats), whose

---

[40] The definition and examples given here are drawn from the *Petit Robert*, a well-known standard French dictionary. I thank one of the anonymous reviewers for pointing out the importance of this influence.

[41] JE Wilson, *Terroir: The Role of Geology, Climate and Culture in the Making of French Wines* (Mitchell Beazley 1998) 55.

[42] Ibid.

influence helped shape its meaning as labels of origin for wine began to emerge in France in the early nineteenth century.[43] It is also part of the recent surge of interest in all things to do with history, heritage or 'patrimoine'.[44] Bérard and Marchenay have made multiple contributions to this literature,[45] and the online journal *Anthropology of Food* has devoted a special issue to 'Patrimoines alimentaires'.[46] Well-known French historian Jacques Revel refers to this as a sort of national obsession with what he terms *patrimonialization*, which began more or less around the year 1980. Declared to be the Year of Patrimoine by president Valéry Giscard d'Estaing, patrimonialization refers to the effort to trace, record and commemorate – in conjunction with museums and monuments – all sorts of events, both majestic and mundane, related to French history. Natural places, landscapes and traditional foods have also been swept up in this frenzy. Revel states: 'It is as though, little by little, the French acquired the habit of considering the ensemble of the infinitely diverse traces of their collective experience as a treasure that urgently needed to be conserved and protected, a base that grounded them.'[47]

Part of this fascination with the past can be attributed to the enormous amount of rapid change that France has witnessed over the latter half of

---

[43] RC Ulin, *Vintages and Traditions: An Ethnohistory of Southwest French Wine Cooperatives* (Smithsonian Institution Press 1996); Delfosse (n 5); M Demoissier, *Hommes et Vins: Une Antrhopologie du Vignoble Bourguignon* (Éditions Universitaires de Dijon 1999). See also Chapter 2 by Stanziani in this volume.

[44] S Blowen, M Demossier and J Picard (eds) *Recollections of France: Memories, Identities and Heritage in Contemporary France* (Berghahn Books 2000); M Lamont and L Thévenot, *Rethinking Comparative Cultural Sociology: Repertoires of Evaluation in France and the United States* (CUP and Éditions de la Maison des Sciences de l'Homme 2000); M Rautenberg, A Micoud, L Bérard and P Marchenay, *Campagnes de Tous Nos Désirs: Patrimoines et Nouveaux Usages Sociaux* (Maison des Sciences de l'Homme 2000).

[45] Many of the publications by Laurence Bérard and Philippe Marchenay concerning the construction, documentation and evolution of French *terroir* products as part of French 'patrimoine' are available at 'Ressources des Terroirs': http://www.ethno-terroirs.cnrs.fr/. See also Chapter 4 by Bérard in this volume; L. Bérard, and P. Marchenay, *From Localized Products to Geographical Indications: Awareness and Action* (CNRS 2008).

[46] Available at: http://aof.revues.org/. See especially: 'Patrimoine alimentaires' (2011) Vol. 8; 'From Local Food to Localized Food' (2007) Vol. S2 March; and 'Local Foods' (2005) Vol. 4.

[47] J Revel, 'Histoire vs Mémoire en France Aujourd'hui' (2000) 18(1) French Politics, Culture & Society 1, 2. An anonymous reviewer notes that French rural *heritage* has been revived at earlier periods in its history as well. See S Peer (1998), *France on Display* (New York University Press, NY 1998).

the twentieth century, particularly since the 1960s. In this sense, seeking out traditional products can be seen as one reflection of the 'malaise' of the French concerning modernization and globalization,[48] as well as their concern over recent health and safety threats that have arisen in the food sector. The taste for history in the form of *'produits de terroir'*, therefore, reflects in part the ongoing construction of a collective representation of the past through food that is perhaps largely unconscious for consumers. At the same time, *terroir* also reflects a conscious and active social construction of the present by various groups concerned with rural areas in France (social and economic organizations, state agency personnel, academics), who jostle for position in their efforts to recover and revalorize elements of the rural past to be used in asserting a new vision of the rural future.[49] As such, the concept of *terroir* is being hotly debated in France and has become a highly politicized notion for some. Anthropologist Muriel Faure recounts the process of patrimonialization of rural cuisine in the case of Beaufort cheese, a product from the Northern Alps.[50] She traces the 'refinement' of this agricultural product into a cultural object, assisted along the way by a host of local actors not directly implicated in production of the cheese, but very concerned with the tourist and development value of claiming such an 'authentic' element of national heritage in their region. One result of this tendency in rural areas across France, and indeed in many parts of Europe, is a concern that the countryside and the customs defined as consonant with it will undergo a process of 'Disneyfication', becoming living museums for visitors from the city, a kind of 'rurality under glass' for the consumption of privileged consumers.

A different misgiving is expressed by academics and intellectuals in

---

[48] E Barham, 'Social Movements for Sustainable Agriculture in France: A Polanyian Perspective' (1997) 10(3) Society & Natural Resources 239.

[49] On the social construction of *terroir* as a tool for rural development, see Barham and Sylvander (n 2); S Bowen, 'Embedding Local Places in Global Spaces: Geographical Indications as a Territorial Development Strategy' (2010) 75(2) Rural Sociology 209; M Demossier, 'Culinary Heritage and *Produits de Terroir* in France: Food for Thought' in Blowen et al. (eds), *Recollections of France* (n 44) 141. Also significant in this regard are the experiences of the European Commission's Leader+ program, which identified and supported local initiatives for rural development, linking groups of local actors located in different European countries with shared objectives. Many such projects centered around developing and promoting 'local' products, which were often origin labeled/GI products. See: http://ec.europa.eu/agriculture/rur/leaderplus/index_en.htm (one search category is 'Adding value to local products').

[50] M Faure, 'Un Produit Agricole "Affiné" en Objet Culturel: Le Fromage Beaufort dans les Alpes du Nord' (1999) 33 Terrain 81.

France, who watch with some disquiet the public's apparent need to recover a sense of the 'authentic' and the 'true' – without realizing the contested nature of these terms – by returning to the foods associated with their rural roots.[51] Bérard and Marchenay note similarities in wording and tone between recent advertisements praising *terroir* products and the rhetoric and reactionary ideology of the Pétain era in France.[52] However, over and against an inward looking, xenophobic interpretation of *terroir*, they argue for acceptance of history as a point of reference, a kind of rootedness that participates in and supports universalistic values such as liberty and tolerance. In this view:

> [Products of the *terroir*] are synonyms for cultural diversity, reflections of the evolution of a society, of its attachment to certain habits of consumption, and not the guardians of a culture that is fixed and turned in on itself. Locality participates in the construction of identity, it doesn't suffocate it.[53]

Thus *terroir* is undergoing a process of cultural re-evaluation whose outcome is still uncertain, but which potentially points towards a future that includes a valued past without becoming either rigid or exclusionary.

Because of the strong interaction between natural and social factors that is reflected in the concept of *terroir*, it could be considered representative in some respects of the 'shared corporeality' of agro-food practice described by FitzSimmons and Goodman. They insist that 'ecology and social relations, the production and reproduction of nature and society, be located within a unified analytical frame'.[54] The current debate over the meaning of *terroir* provides one such frame, as well as reconnecting us to the embeddedness approaches to agro-food studies discussed earlier. However, because it is a fluid, cultural construction, it can be difficult to apply the insights of conventions theory to its evolution. We now turn to a consideration of the AOC label of origin process as a practical application of the concept of *terroir*. The discussion emphasizes how the system is organized to make the transition from *produit de terroir* as a concept to the 'qualified' agro-food entity that becomes an AOC labeled product. This system of regulation brings into focus a number of 'situations', or

---

[51] C Ray, 'Culture, Intellectual Property and Territorial Rural Development' (1998) 38(1) Sociologia Ruralis 3.
[52] L Bérard and P Marchenay, 'Terroirs, Produits et Enracinement' (1998) 43 Association Rhône-Alpes d'Anthropologie 16 (Special Issue: 'Pour une anthropologie impliquée. Argumentations face aux extrémismes'). See also Chapter 4 by Bérard in this volume.
[53] Ibid 3.
[54] FitzSimmons and Goodman (n 7) 195.

moments of judgment, in which we can witness how the non-market worlds of conventions theory are called upon to find a legitimating compromise.

3.2 The French AOC Administrative Process

The AOC regulatory system has been evolving in recent years along with EU recognition of labels of origin.[55] It has had to expand to encompass more products than wine and spirits (cheeses, meats, and so on), and in step with the overall cultural reconsideration of *patrimoine*, it has begun to emphasize more strongly the cultural and collective aspect of the link between the product and its *terroir*. Critics might object that the system is heavily bureaucratic, and as a result is rather costly.[56] Initial requirements for obtaining an AOC label are complex, as are ongoing requirements for retaining it. There are opportunities for slippage and contradictions in the system, several of which are pointed out below. Nonetheless, it has been remarkably successful in developing the French wine industry and, more recently, in supporting the development of a myriad of other products.

The presence of an AOC label on a product reflects the completion of a multi-level process of negotiation from the local to the state level. When this negotiation is successful, it results in a product that is strongly embedded in the natural, social, cultural and political dimensions of its territory. In this instance, AOCs provide support for a view of GIs as potential sites for the emergence of a new system of agriculture that would reaffirm and support unique values associated with local places. From the producer's point of view, gaining approval for a new AOC is a long and arduous process.[57] Producers must organize themselves into a union that will later uphold the rules of production at the local level. Individual producers or private enterprises cannot request an appellation in France; it must be a collective process, and these take time.[58] The producers' union prepares a

---

[55] For recent developments, see Delphine Marie-Vivien, 'The Role of the State in the Protection of Geographical Indications: From Disengagement in France/Europe to Significant Involvement in India' (2010) 13 JWIP 121.

[56] In the case of AOCs, as opposed to some other labeling schemes (e.g. Label Rouge poultry) this cost was previously subsidized by the French government, reflecting the importance they place on this type of production, which is linked to France's reputation in the area of food quality.

[57] See INAO *GI Applicants' Guide* (07/10/2005). The guide, along with copies of completed applications approved by INAO, are available for download at www.inao.gouv.fr/

[58] The requirement for collective producer application has not always been the case in France, but is increasingly prominent. Presently, Art 49(1) of EU Regulation 1151/2012 states: 'Applications for registration of names under the

detailed dossier that is submitted to the Institut National des Appellations d'Origine et la Qualité (INAO).[59] The dossier typically takes a year or more to prepare, but it can take much longer in complicated cases. In the dossier, producers must:

- explain precisely the reasons motivating their request for an AOC;
- provide evidence that the name (appellation) they are requesting has a historical reputation with consumers and is known to them, and that the name is associated by the consumers with the product; this usually requires assembling written evidence of use of the name, as well as interviews;
- establish the product's close tie to the *terroir* of origin based on natural factors, and on human factors or '*savoir faire*', which together should produce the product's 'typicity', or special, traditional character;
- furnish evidence that will allow an evaluation of how well the product distinguishes itself from other similar products which exist on the market;
- describe the area of production and the exact procedures involved in cultivation (wine) or production (cheese, meat products) of inputs and in processing;
- carry out an economic study of the product, including existing and potential markets, prices, distribution channels, value-added of this product relative to other products which are similar, and so forth.

The dossier is forwarded by INAO through expert committees formed to review products by specialty (wine, dairy product, meat, and so on). These committees are composed of representatives from the production, processing and distribution sectors affected by the request, assisted by INAO professionals.[60] After the dossier has been thoroughly reviewed,

---

quality schemes ... may only be submitted by groups who work with the products with the name to be registered.' In France, this collective aspect has been increasingly emphasized in recent years and applications that did not meet this criterion have been rejected (M Claude Béranger, INAO, personal communication, 2002).

[59] In 2006, INAO became responsible for all products under official 'quality' signs, including the AOC, GI, Label Rouge, STG (Guaranteed Traditional Speciality) and organic labels. While the agency's acronym did not change, their name was changed to the *Institut National de l'Origine et de la Qualité* to reflect their new responsibilities.

[60] See INAO, *The Official Quality and Origin Signs* (2012) available at www.inao.gouv.fr/. For products such as cheeses and meats, for example, this includes attention to how the animals are raised, what they are fed, whether they are to be

if accepted it is forwarded, at which point it is sent to the appropriate National Committee for review.

National Committees are made up of professionals selected from among appropriate Regional Committee members, accompanied by national level experts in export and distribution and consumer representatives. They are also assisted by INAO representatives. The National Committee appoints a special Review Commission for each request, including National Committee members along with producers and other professionals chosen from outside the requesting region. Their work can take another year or more to complete, and can include hearings. They report their recommendation to the National Committee (accept, postpone or refuse) and establish final production rules. If the request is accepted, a commission of experts is formed to fix the definitive boundaries for the AOC. The project is then sent to the Minister of Agriculture with draft legal text to establish and protect the appellation. The Minister of Agriculture signs the text into law (no changes are allowed at this point) and forwards it through proper channels for publication in the Official Journal, at which point it becomes part of the French legal system. The new appellation is now protected as the collective property of the producers, as well as part of the agricultural, gastronomic and cultural heritage of France.

After an AOC is awarded, it is monitored more locally by one of a number of recognized entities with expertise relative to the particular type of product.[61] They respond to producer requests for assistance, but they also carry out unannounced farm visits to assure that producers are following the production techniques required for their label. Because of the wide scope of expertise required in considering new AOC requests, INAO commonly retains technical consultants in any number of fields, such as geologists, soil scientists, plant scientists, anthropologists, sociologists and historians, to carry out its work. As a private-public entity, it holds final jurisdiction over all appellations, and has historically established and enforced regulations concerning almost every aspect of growing, making and marketing AOC products down to a very fine level of detail. However,

---

grazed only, etc. – in other words, detail on every step of the process that precedes the actual processing.

[61] A list of the control entities, their expertise and contact information, is available on the INAO website. They are trained with specialized knowledge of the products they control as well as knowledge of the local area of production. There are some similarities here between INAO control mechanisms and the implementation of organic standards at the federal level, a US labeling scheme which also requires certification procedures and on-farm visits which are carried out by certifying entities recognized by the US Department of Agriculture.

it has recently begun to transfer some of these functions to the collective producer organizations.[62]

The complexity of this process reflects the seriousness with which it is addressed by various levels of producers and professional experts. Each stage in the process presents a possible opportunity for discussion, debate and reflection, and ensuing negotiations confront participants with the necessity of justifying their positions. The history of care taken with the process has often – but not always – been rewarded by consumers in the form of increased product sales and higher commanding prices. INAO administrators attending a 2001 conference in Die, France,[63] praised the success of AOC production as a value added strategy within French agricultural policy, as well as its contribution to encouraging the preservation of rural *heritage* and the maintenance of traditional landscapes. They reported that export sales for AOC products were continuing to rise and were highly profitable. In 1999, new export revenues for wine from France exceeded those of grains – quite an accomplishment considering that France is the second largest grain exporter in the world.[64] The fact that France is considered today to be the top, or among the top, world producers of quality wine is a reflection to some extent of the care taken with its AOC system.

Because AOCs have at times helped maintain agricultural profitability in zones that are considered difficult or marginal, they are seen as an important tool for rural development by the French government. And they have helped many traditional, historic products remain in production and competitive on the market that might otherwise have disappeared. They have helped create rural employment, both directly and indirectly, in associated industries such as tourism.[65] Because France is the most visited country in the world,[66] this last contribution is not negligible in terms of

---

[62] See Delphine Marie-Vivien, 'The Role of the State in the Protection of Geographical Indications: From Disengagement in France/Europe to Significant Involvement in India' (2010) 13 JWIP 121.

[63] 'Protéger les produits, protéger les paysages, le role des labels et des appellations d'origine dans la construction du patrimoine culturel' 14–17 May 2001, Die, France, organized by the École Nationale du Patrimoine.

[64] A Valadier, 'AOC: l'Instrument de Développement Economique. Terroir, Tradition et Modernité' (March 2000) Lettre Trimestrielle d'INAO 4.

[65] J Bessière, 'Local Development and Heritage: Traditional Food and Cuisine as Tourist Attractions in Rural Areas' (1998) 38(1) Sociologia Ruralis 21; E Croce & G Perri, *Food and Wine Tourism: Integrating Food, Travel and Territory* (CABI Tourism Texts, Wallington, Oxfordshire 2010).

[66] UNWTO Tourism Highlights (United Nations World Tourism Organization, 2013) 6; Available at http://publications.unwto.org/

the national economy, and has helped stabilize population in some rural areas previously considered in decline.

The decision to award an AOC designation to a product, and by association to its region of origin, is based on the strength of the link between the two. Evaluation of this link depends directly on the concept of *terroir*. In an important study of how INAO agents use various terms when examining a request for an AOC, the concept of *terroir* was selected as the most important notion out of 27 different concepts used by the 112 agents included in the study.[67] Experts are called upon to make this judgment in three main categories: natural factors (tie to the local environment or ecological niche), human factors (*savoir faire*, or particular techniques and know-how confined to that area), and history (public knowledge of a product as originating in that area, recognition of the association between product and place that is consistent and widespread). Each factor is investigated individually as well as in terms of how they combine to determine the 'typicity' of the product. To obtain an AOC a product must incorporate all three aspects of its *terroir* and carry them forward to the consumer, but the natural and human factors are decisive.

3.3 Natural Factors

For those most knowledgeable about the AOC system, natural factors are the most important determinants of how well a product represents its *terroir*. In the study of INAO agents just mentioned, approximately half of those interviewed felt that *terroir* referred primarily to natural qualities of a geographic area (soil, micro-climate, slope, exposure, and so on), and the other half felt that *terroir* meant a blend of these natural factors with human factors. Thus, of the three factors to be considered, the tie to nature figures most prominently in determining a product's tie to its *terroir*.[68] Perhaps because of this, it is often thought that the relationship

---

[67] S Scheffer and F Roncin, 'Qualification des Produits et de Terroirs dans la Reconnaissance des Produits en Appellation d'Origine Contrôlée' in L Lagrange (ed), *Signes Officials de Qualité de Développement Agricole* (Lavoisier Clermont-Ferrand 1999) 37; S Scheffer and B Sylvander 'The Effects of Institutional Changes on Qualification Processes: A Survey at the French Institute for Denomination of Origins (INAO)' in F Arfini and C Mora Zanetti (eds), *Typical and Traditional Products: Rural Effects and Agro-Industrial Problems*, Proceedings of 52nd EAAE Seminar, Parma, Italy: Università di Parma (1998) 463.

[68] In November 2005, a conference was held in Paris to take stock of the 20 years of research on origin products which had been funded by the French government ('Produits Agricoles et Alimentaires d'Origine: Enjeux et Acquis Scientifiques'). Sponsored by INRA (the national French agricultural research

works in the other direction, as well. That is, many people assume that AOC products have a beneficial effect on the environment, or that they are produced in a more sustainable fashion. Is this necessarily the case? According to Bérard and Marchenay,[69] the state does subject AOC areas to a higher level of scrutiny in terms of compliance with environmental regulation. As a result of the standard-setting negotiations that go on during the application phase of an AOC, environmental restrictions on AOC producers may be higher than for other types of production, but this is not a given. It is sometimes claimed that AOCs contribute to the preservation of biodiversity, but the situation can be complicated.[70] For example, in the case of an AOC requested for the châtaignes d'Ardèche (chestnuts of the Ardèche region), only 19 of the more than 60 varieties developed by peasant farmers over several centuries were retained as eligible for the AOC, leading the person in charge of assisting the producers in the formulation of their request to question whether an AOC can actually be detrimental to biodiversity.[71] On the other hand, without the protection of the AOC, it is highly likely that most of the 19 varieties actually retained would have been lost from production over time.

In some cases, the preservation of traditional methods of production does result in a clear environmental gain. Saint Joseph wine is one example. Over the centuries, Saint Joseph came primarily from steep hillsides that did not lend themselves to cultivation by mechanized methods. Vines, particularly near the tops of hills, occupied terraces built of stone at a time when tree removal was done by hand. However, over the last century the boundaries set by the AOC encompassed more than just the steep hills. These boundaries were recently redrawn by INAO to realign

---

institute) and INAO, the conference included presentations from a wide range of disciplines. At the conference, a clear schism emerged between researchers emphasizing the importance of nature versus those emphasizing the human or social factors in AOC systems. The dividing line, unsurprisingly, fell between disciplines focused on the natural sciences versus those versed in the social sciences. This contention cannot be resolved here, but it is worth noting that the possibility of a genetic trace, or marker, that can be used to identify an origin product, particularly for the process of boundary making for an AOC area, is always welcome and viewed as indisputable evidence of uniqueness.

[69] Personal communication, 2002 (on file with the author).

[70] P Marchenay and L Bérard, 'Les Variétés Locales dans le Contexte Agro-Alimentaire Actuel: Mythes et Réalités' in N. Eizner (ed), *Voyage en Alimentation*, Conference Proceedings of the Association des Ruralistes Français, 12–13 October 1989 (ARF Editions, Paris 1995) 263.

[71] S Sabot, 'Une AOC Nuit-Elle à la Biodiversité?' in *Avenir Agricole de l'Ardèche* (Privas 2000).

the production area more closely with its historical boundaries.[72] As a result, the terraces, which had been allowed to deteriorate, were rebuilt and the skills needed to maintain them relearned. The change fulfilled more than an historical and aesthetic mandate, however. Without the terraces, erosion was increasing on the hillsides, which could eventually result in mudslides, removing all topsoil and vegetation and leaving nothing but bare rock. The story of this AOC is not unusual in its blending of human history and environmental protection, to the point where it is sometimes difficult to disentangle the two. Finally, the delineation of an AOC area presents a challenging task of ecological analysis that is a crucial aspect of the overall system. This analysis confronts all of the challenges that are well known to ecologists who attempt to define ecological regions elsewhere,[73] except that in the case of a *terroir* the regions are usually at a smaller scale than those used by ecologists.[74]

INAO continues to review AOC delineations as the need arises to make certain they are environmentally and historically coherent. This is in part a response to trade negotiations within the WTO which have brought pressure to bear across the EU for more harmonized standards. The application of science required to arrive at these new delineations can, on the one hand, be seen as a process of rationalization that will eventually undermine the more cultural and historical interpretations of *terroir*. On the other hand, some view this process as a re-legitimation of science itself, as it is being called to the service of more than market objectives. The key to this difference is the emphasis of the *terroir* concept on interpreting what is there in nature to be known, rather than viewing nature as an obstacle to be overcome or controlled for production.

3.4 Human Factors, or *'Savoir Faire'*

Terroir delineation must also take into account factors that typically do not concern ecologists, such as the current and historical geographic distribution of the human know-how or *savoir faire* associated with the product. While there may be a clear cradle of historical production, the extent of

---

[72] INAO Report, *A l'Origine, il y Avait les Terrasses: Cas du Saint Joseph* (INAO, Paris 2001).
[73] R Bailey, *Ecosystem Geography* (Springer 1996).
[74] According to Robert Bailey, a widely respected expert on eco-regional geography, the only US state that has delineated eco-regions down to the scales comparable to AOC delineation is Missouri (personal communication with Mr Bailey). See TA Nigh and WA Schroeder, *Atlas of Missouri Ecoregions* (Missouri Department of Conservation, 2002).

currently existing knowledge about the product still has to be determined. Laurence Bérard, an anthropologist who has often been called in as an expert to work on this aspect of evaluation for AOC requests, states that knowing how to take these human factors into account is *the* problem for awarding a new AOC.[75] She begins by working closely with producers to define key technical aspects of production. Researchers document these techniques in detail, as well as the producers' ideas of the spatial distribution of their skill. Taking the producers' delimitation as a starting point, researchers then look *within* the area for problems, defined as situations in which techniques diverge too widely from the established norm. They then gradually push outwards towards the edges, looking for the point at which the typicity of the product is no longer present.

Defining the exact boundaries and definition of an AOC can be controversial among producers. Some who participated in the AOC request might find themselves outside of the final boundaries. 'Free riders' may appear, demanding to be included. Two neighbors who follow slightly different processing methods may find that one of them is included while the other is not. Producers who want to use the AOC may balk at the production requirements. In the case of wine, for example, production ceilings are often imposed as a means of supply-side control.[76]

Finding a link between a product and a *terroir* also means that the effect should, in principle, be scientifically demonstrable. There are a number of research teams in Europe investigating how a *terroir* marks the taste of its products in terms of chemical composition and other factors. However, for many products, this determination is made on the basis of tasting panels. Because the products under consideration are also historical artifacts that generally predate the industrialization of production methods, organizing tasting panels raises the question of choosing competent judges.[77] They are usually selected from the area of production, and include producers, connoisseurs (retired producers or consumers with many years of experience

---

[75] See L Bérard, O Beucherie, M Fauvet, P Marchenay and C Monticelli, 'Indication Géographique Protégée. Critères de Zonage' (2000) 38 Le point sur. . . 53.

[76] I am indebted to an anonymous reviewer for a reminder of this fact. However, it should also be noted that many producers eligible for an AOC devote only a percentage of their production to the AOC, selling the remainder under a different label.

[77] F Casabianca and C de Sainte Marie, 'Typical Food Products and Sensory Assessment: Designing and Implementing Typicality Trials' in Sylvander et al., Vol. 2 (n 33) 269 (considering how sensory evaluation enters into establishing parameters of acceptability for a product whose 'authenticity' is linked to collective experience and tasters' memories).

with the product), 'mediators' such as chefs and specialized vendors and distributors, and government agency personnel assigned to follow and document the request process. Producers' voices remain important in this phase, but they must weigh in among the voices of other experts who may not be from their region, or may be representing the interests of consumers (for example, specialty retailers). The compromises reached further embed acceptance of the final result in broader networks, horizontally at the local level but also vertically through production and distribution chains encompassing the different levels of expertise involved in regulation and control.

The result of all of the research and negotiation involved in establishing an AOC is the *cahier de charges*, or certification requirements. These remain open to change, however, as producers can and do return to them from time to time to make adjustments.[78] This raises the issue concerning which actors in a territory can take part in the organization and potential future evolution of the production methods. In the past, only producers organized AOC production in interaction with the state. Increasingly, other actors are getting involved who are pursuing broader rural development objectives linked to the cachet and consumer draw of specialty products in their area. These actors can include regional tourism bureaus, agents from park areas in or near the AOC area, regional economic development specialists, representatives of local businesses such as hotels and restaurants, and producers of other regional agro-food products or handicrafts.

AOCs are now clearly recognized as important contributors to the economic and agricultural structure of a region, as well as to its shared identity. And so while producers must still cooperate among themselves to prepare the actual request for INAO, they now find they are involved with larger networks of players in the process. It is in analyzing these discursive contexts that scholars are finding fertile ground for the application of conventions theory. The AOC process, and label of origin systems more broadly, provide a trace on shifts taking place in the rural agro-food arena, and can contribute to reflection on societal changes that reach well beyond the rural.

---

[78] L Bérard and P Marchenay, 'Le Vivant, le Culturel et le Marchand: Les Produits de Terroir' (2000) 194 Autrement 91.

4. CONCLUSION

This chapter has made the argument that the current expansion of label of origin systems globally represents an opportunity to examine new forms of local–global connections in the making. Embeddedness perspectives combined with the conventions theory framework can inform analyses of these systems as they evolve. Together, they help open up the discussion of whether we are moving towards further industrialization of the agro-food system, or perhaps instead witnessing a moment of legitimation crisis within that system that will lead to further development of alternatives. While industrialization of the food system has certainly continued apace since the original version of this writing was published in 2003, the emergence of social movements such as Slow Food International[79] and the burgeoning 'local food' movement in the United States would seem to indicate that the legitimation crisis of the modern industrial food system is far from over.

By emphasizing the nature of negotiation and compromise, conventions theory does not rule out the importance of market logic that drives the industrial system. But it does help us explore how market logic co-exists with other ways of viewing the world that constrain it within social, historical and ecological limits. It allows us to treat the multi-level negotiations of origin labeled production as instances of what David Goodman has termed 'an international bio-politics of agriculture and food', one that raises the critical questions of 'Where, how, and by whom is food to be produced, processed, marketed, and consumed?'[80] And it helps us better understand and interpret the reasons behind social movements of resistance to the industrialized food system.

Like the discursive field of organic production, the discourse around labels of origin brings in issues of 'the incorporation of nature (both symbolic and biophysical), social movements, consumers and food scares, regulatory politics, contest over corporate involvement, and issues of standards and meanings'.[81] But labels of origin tie all of these questions to specific places, arguably intensifying and clarifying discussions of who will ultimately benefit, and on what basis of legitimation.

---

[79] See http://www.slowfood.com/

[80] D Goodman, 'Organic and Conventional Agriculture: Materializing Discourse and Agro-Ecological Managerialism' (2000) 17(3) Agriculture and Human Values 215, 217.

[81] H Campbell and R Liepins 'Naming Organics: Understanding Organic Standards in New Zealand as a Discursive Field' (2001) 44(1) Sociologia Ruralis 21, 23.

4. *Terroir* and the sense of place
Laurence Bérard*

1. INTRODUCTION

The word *terroir* is used in a variety of contexts and remains controversial. *Terroir* stands in opposition to globalisation and displacement. It is a concept used at every level of communication, generally employed as a sounding board for the issues of the moment. French in origin, the concept of *terroir* is woven into the political and cultural history of France and is still hotly debated in the research community. Today it is more relevant than ever, linked to the French-championed principle of protecting the geographical origins of agricultural products and foodstuffs. For better or worse, that principle is steadily gaining global acceptance, which raises a number of issues, not least in relation to the link between product and territory. It turns out that what is feasible albeit challenging in France is often unworkable in developing countries. The *terroir* approach has certain inherent weaknesses that tend to give undue prominence to its French origins. But it can also represent a valuable tool for local development. First though, more thought must be given to the nature and extent of the link with place, and the issues at stake in international negotiations. For that, we must take a closer look at France, exploring the reasons behind its current devotion to *terroir* – what the word means, how it is used and why what works on paper does not always work in the real world. *Terroir* must be viewed in a global context. Not as a French passion that France hopes to export, but as a key factor in understanding how societies across the globe are linked to the particular places they occupy. All the hype surrounding *terroir* is symptomatic of our image-obsessed culture. Here we look beyond origin protection legislation, in the form of GI or AO protection, and think afresh about how a sense of place is fostered in different communities around the world.

In this chapter we explore the history and development of the *terroir* concept, along with the implementation issues it raises. Later, we focus on

---

* This chapter elaborates on a previous publication: L Bérard, 'Du terroir au sens des lieux' in Claire Delfosse (ed), *La mode du terroir et les produits alimentaires* (Les Indes Savantes, Paris 2011) 41.

the nature of the relationship with place, and also look more generally at how people understand and use the space around them.

2. *TERROIR* IN THE FRENCH CONTEXT

The role of *terroir* in France is closely linked to the French nation-building project. The concept itself emerged largely as a result of human geography influences and its precise definition remains debatable.

2.1 The Weight of History

Anthropologists tend to avoid the word *terroir* because of its tainted associations with regionalism, and Vichy regionalism in particular. Until recently ethnographers were reluctant to conduct research in this area, mindful of anthropology's contribution to the 'folk museum' ('Musée des arts et traditions populaires') founded in 1936 at the time of the Front Populaire.[1] But *terroir* cannot be reduced to the narrow meaning it acquired under the Vichy regime, when the idea was employed to support a new French state grounded in a glorified agrarian past, folklore and traditionalism.[2] In line with its reactionary, conservative ideology, the Vichy regime sought refuge in anachronism by reinstating, albeit temporarily, a rural social order that has become synonymous with one of the darkest periods in French history. Hiding behind that measure, however, is a national–local dichotomy that has persisted for more than 200 years. France since the 19th century has been shuttling to and fro between the general and the particular, shifting from one extreme to the other as it considers the thorny question of recognition or negation of the weight of local memories.[3] Of particular relevance here is the process of post-revolution nation-building. France after 1789 was a unified and uniform entity that relied on a merging of differences: 'local' was seen as the opposite of 'national', a contradiction that could not be glossed over and had to be eliminated. This was what later became known as the Jacobin view, and it stood at the core of the administrative reorganisation that took place between 1799 and 1802.

---

[1] See the seminal contribution by Daniel Fabre, 'L'ethnologie Française à la Croisée des Engagements (1940–1945)' in Jean-Yves Boursier (ed), *Résistances et Résistants* (L'Harmattan 1997) 319, 319–400.

[2] Daniel Fabre, 'Le Manuel de Folklore Français d'Arnold Van Gennep' in Pierre Nora (ed), *Les Lieux de Mémoire*, Tome III (Gallimard, Paris 1992) 641, 671.

[3] Thierry Gasnier, 'Le Local: Une et Divisible' in Nora (n 2) 463, 465.

It was the way in which the 1789 revolutionaries, overwhelmingly drawn from the urban upper-middle classes, tackled the question of French diversity in light of national unity that really gave birth to the regional issue. They created the problem of regional cultures, you could say, by their insistence on the importance and immutability of local specificity . . . The will for political unity, in other words, turned the spotlight on local realities that related to essentially rural places elsewhere.[4]

A few years later, French historian Jules Michelet revisited the question of sacrificing local specificity for the sake of national unity. In his major work, *Histoire de France*, Michelet writes:

In this manner has been formed the general, the universal spirit of the country; the local has disappeared daily . . . Local *fatalités* have been overcome and man has escaped from the tyranny of material circumstances . . . In this marvellous transformation, the spirit has triumphed over matter, the general over the particular, and the ideal over the real.[5]

The 24 provinces listed in Michelet's *Tableau de la France* are presented as insular, self-contained unities, each one rendered sterile by its particular relationship with the soil, the climate and the kind of person they produce.[6] The 19th century saw a concerted political drive to get rid of those particularities, marginalising local specificity for the sake of nation-building. This Jacobin-style vision of a new-look France, reorganised into administrative *départements*, coincided with a proliferation of learned societies that were based on the new administrative divisions and run by local dignitaries, 'dignitaries' perhaps, rather than 'worthies' – a coincidence that clearly illustrates the huge ambivalence felt about local places and local knowledge resources.

2.2 The Role of Human Geography and Pedology

The relationship between the 'national' and the 'local' in representations of French territory underwent significant transformation in the latter half of the 19th century. The French defeat of 1870 redrew the map of France, at a time when French geographer Paul Vidal de la Blache was usher-

---

[4] Mireille Meyer, 'Vers la Notion de Cultures Régionales (1789–1871)' (2003) 33(3) Ethnologie Française 409, 410 [Special Issue: Cultures Régionales. Singularités et Revendications].

[5] 'Tableau de la France', Chapter 1 Book III, L'Histoire de France, cited by Gasnier (n 3) 227 (translation from: Stephen Kippur, *Jules Michelet: A Study of Mind and Sensibility* (SUNY 1981)).

[6] Gasnier (n 3) 473.

ing in a new approach to geography. Considered the father of human geography, la Blache was 'pivotal in the daunting task of redefining the French nation on the basis of local environments'.[7] Vidalian geography focuses on identifying natural regions (or *pays*) and the variety of landscapes and lifestyles (*genres de vie*) encompassed within those areas and resulting from the interaction between human groups and their natural milieus. Vidal's approach is exemplified in his *Tableau de la Géographie de la France* (1903), which is prefaced to Ernest Lavisse's *Histoire de la France*. It is worth noting that these ideas were hotly debated by the social scientists of the time. Durkheimians like François Simiand (co-founder of the French School of Sociology) argued for research based on abstract places that glossed over any distinction between one place and another and transcended any preconceived geographical frame. French historian Lucien Febvre and his followers meanwhile sided with geographers, endorsing the rationale behind the regional monographs of the Alpine and Rhodanian culture produced by Vidal's students in the early 20th century.[8] Costume, language and food emerged as distinguishing marks of locality, part of a broad regionalist tendency that ranged from the exploration of new territorial unities to the promotion of the infinite variety found within France.

The early 1930s marked the arrival of pedology, a new scientific discipline that moved the goalposts somewhat by focusing on the role of the soil. Pedology originated in Russia in 1879 and was introduced to France in 1934 by French soil scientist Albert Demolon.

> Pedology was developed to provide a systematic approach to soil science, using standard criteria to assess the physical properties of soils. . . . Soils with designated names have a genetic history, functional behaviour and composition that serve as indicators of soil aptitude. Add to that a particular set of ecological, edaphic [i.e. relating to the soil] and weather conditions, and what you get in the end is 'terroir'. The products coming out of that area share certain characteristics related to quality and quantity – cue the concept of the 'cru'.[9]

As the discipline of geography evolved, that approach was increasingly compelling for French geographers like Roger Brunet, who

---

[7] Ibid 504.

[8] Roger Chartier, 'Les Sciences Sociales et la Région: Un Regard Rétrospectif' (1981) 1 Le Monde Alpin et Rhodanien 81.

[9] Stéphane Hénin, 'Utilisation du Sol et Maintien de la Fertilité: Systèmes de Production et Systèmes de Culture' in *Deux siècles de progrès pour l'agriculture, 1789–1989* (Académie d'Agriculture de France; Lavoisier 1990) 124.

describes *terroir* as a place that is defined by its particular physical characteristics.[10]

The protection and subsequent use of the term *terroir* must be seen in this two-fold historical context: on the one hand, environmental determinism; on the other hand, a preoccupation with locality that persisted from the 19th to the early 20th century, but in a country with a long history of centralisation. The precepts of the Vidalian School were particularly influential in this regard, providing the impetus to regional studies based on small, coherent entities with shared lifestyles, histories and agrarian systems. It was in this context that the first appellations of origin emerged and were defined in France[11] – within the framework of a burgeoning field of geography whose practitioners were already well versed in the defence of locally relevant products. Geographers like Lucien Gallois, Paul Vidal de la Blache and Albert Demangeon[12] laid the groundwork for the French law of 1919 relating to the protection of product names – names that were associated with given places and coherent environments, ones defined by local customs and time-honoured practice. The actual term *terroir* is not mentioned anywhere in the laws of 1919, 1927 or 1966.[13] The first two refer only to a 'production area', and the third to 'a geographical milieu, including natural and human factors'. But the notion of the 'petit pays' – meaning a region, district or locality – is implicit, alongside a more deterministic interpretation that confines itself to physical factors.

2.3 An Endlessly Debated Concept

Though the term '*terroir*' often appeared in publications, its meaning remained largely confined to physical factors, focusing on the role of the soil and climate. The *terroir* was a given factor; it had always existed, pre-dating man, who only served to reveal its potential. That definition still stands in some (chiefly viticultural)[14] contexts, even if it is regarded as

---

[10] R Brunet, *Le mots de la géographie* (Reclus, La documentation française, 3ème édition 1994) 482, cited by Claire Delfosse, 'Noms de Pays et Produits du Terroir: Enjeux des Dénominations Géographiques' (1997) 26(3) Espace Géographique 222, 229.

[11] For their legal emergence, see Chapter 2 by Stanziani in this volume.

[12] Claire Delfosse, 'Le Pays et ses Produits: Défense et Illustration d'Une Identité' (2004) 139–140 Etudes Sociales 117, 121–123.

[13] See respectively: Loi du 6 Mai 1919 Relative à la Protection des Appellations d'Origine; Loi du 22 Juillet 1927 Modifie La Loi Du 6 Mai 1919; Loi n° 66-482 du 6 juillet 1966 Modifie La Loi Du 6 Mai 1919.

[14] Cf. R Morlat, G Barbeau and C Asselin, 'Facteurs naturels et humains des terroirs viticoles français: méthode d'étude et valorisation' in Pascale Moity-Maïzi

rather too simplistic. Taking wine-growing *terroirs* as an example, French historian Roger Dion observed that the quality of a wine is the expression of a particular social milieu and what he called 'human will'. Writing in 1952, Dion is quoted as follows: 'The role of the land in the making of a grand cru scarcely goes beyond that of the material used in the making of a work of art.'[15] Most of the works relating to *terroir* take into account the human factor. To quote George Bertrand, 'there is a subtle dialectic between the ecological complex and the historical-economic complex' that comes from the constant interaction between the agrisystem, production and human effort.[16] Other interpretations followed, mostly by geographers who questioned the relationship between land and human effort. The development of that ambiguous concept was the focus of an article by Philippe Roudié, who stressed how difficult it was for geographers to agree on a definition.[17] Pierre George wrote in the 1970s that '*terroir*' was a catch-all term that could have several meanings in agricultural geography.[18] Thirty years later, Diry[19] likewise regarded *terroir* as a vague concept that was far from receiving unanimous support in geographical circles. This commonplace usage is recent – and does seem to epitomise the problems entailed in defining the ways communities take ownership of their local environments. The Africanist approach, for instance, involved a multidisciplinary research programme lasting several years, bringing together geographers, sociologists and ethnographers for the purpose of establishing an atlas of African *terroirs* based on '*terroir* monographs'. The authors defined *terroir* as 'a portion of territory appropriated, improved and utilized by the group residing there and for which it provides a means of livelihood'.[20] There was obviously a problem with that definition to judge

---

et al. (eds), *Systèmes agroalimentaires localisés: terroirs, savoir-faire, innovations* (INRA Département de Recherches sur les Systèmes Agraires et le Développement 2001) 111.

[15] Roger Dion, *Le Paysage et la Vigne. Essais de Géographie Historique* (Payot 1990) 226.

[16] Georges Bertrand, 'Pour une Histoire Ecologique de la France Rurale' in G Duby and A Wallon (eds), *Histoire de la France Rurale* (Le Seuil, Paris 1975) 37, 74.

[17] Philippe Roudié, 'Vous Avez dit 'Terroir'? Essai sur l'évolution d'un Concept Ambigu' (2001) Journal International des Sciences de la Vigne et du Vin 7 (Special Issue).

[18] P George, *Dictionnaire de la géographie* (Paris, P.U.F., 1970) 416, cited in Roudié (n 17) 8.

[19] Ibid.

[20] Gilles Sautter and Paul Pélisser, 'Pour un Atlas des Terroirs Africains: Structure-Type d'Une Etude de Terroir' (1964) 4(1) L'Homme 56, 57.

from the frequent necessity to envisage alternative solutions – substitute the de facto *terroir* for the *terroir* under study, and drawing on the physical setting or, on the contrary, the purely social setting, in the absence of any autonomous socio-spatial organisation.[21] Reviewing their research in a special issue of *Etudes Rurales* (1970) devoted to African and Madagascan terroirs, the authors argue that there is no substitute for the word *terroir*. They make a particular distinction between *terroir* and '*finage*' ('community territory' in French-speaking West Africa); likewise between '*terroir foncier*' (the notion of ownership contained within '*finage*') and '*terroir d'utilisation*' (*terroir* defined by use). Clarifying their comments, Pélisser and Sautter point out that 'terroir studies, far from being confined to a narrow field of enquiry, range across every factor that is somehow pivotal to a group of people living in the same space or with common social norms, on land that they farm and manage on a collective basis'.[22] The concept of variable geometry as used in France seemed concerned here with agrarian structures and types of land occupation. It represented a tool for development by collecting data on relevant variables that included the biocultural heritage linking a community to a place. These data were the basis of the local land register, providing the research input for a major cartographic survey carried out by the French authorities some 30 years later. The official use of those data, however, particularly as reflected by the mapping survey, showed scant regard for the complexity of land use in Africa and sparked serious conflict over access to resources and their control. Several authors[23] have expressed reservations about using the concept of *terroir* in the context of emerging and developing economies.

The question continues to provoke discussion among French researchers. Marc Dedeire develops the theme of *terroir*-based agriculture.[24] Jacques Maby presents *terroir* as a 'knowledge organisation tool'.[25] *Terroir* is increasingly seen as a tool for local development and spatial

---

[21] Ibid 58–59.
[22] Pierre Pélisser and Gilles Sautter, 'Bilan et Perspectives d'Une Recherche sur les Terroirs Africains et Malgaches (1962–1969)' (1970) 37–39 Etudes Rurales 7, 24.
[23] See e.g. Thomas J Bassett, Chantal Blanc-Pamard and Jean Boutrais, 'Constructing Locality: The Terroir Approach in West Africa' (2007) 77(1) Africa 104.
[24] Marc Dedeire, *Le Concept d'Agriculture de Terroir* (PhD dissertation, Economic Sciences, University of Montpellier I, 1997).
[25] Jacques Maby, *Campagnes de Recherche: Approche Systémique de l'Espace Rural* (Habilitation à diriger des recherches, UMR Espace, University of Avignon, 2002) 28, cited by Jean-Claude Hinnewinkel, *Les Terroirs Viticoles: Origines et Devenir* (Féret, 2004) 5–6.

planning, generating what agronomist Jean-Pierre Deffontaines describes as projects in the broader human interest. He defines *terroir* as encompassing 'players, with their histories and ways of life, plus activities, especially agricultural practices – which are defined as specific, concrete modalities adopted by farmers to implement farming techniques – and a particular stretch of land with its ecological profile, layout, area, neighbourhood and landscape'.[26]

This kind of dynamic understanding of *terroir*, projecting it into the future on the basis of projects that harness collective wisdom, addresses a very contemporary issue. The concept of *terroir* is highly adaptable due to its polysemous meaning and broad scope, and it is therefore easy to see why it continues to occupy such a prominent place. The hierarchy of French wine as we know it today is the outcome of those local ways of life generated by an aspect of *terroir* that lies at the heart of France's greatest vineyards. Wines are ranked according to a vibrant and constantly evolving system that straddles time and protects local interests. Senior anthropologist Marion Demossier explores this process in her recognition of the Burgundy *climates*.[27] Soil, aspect and climate are certainly important, but most important of all is the ability to make a difference and to make it known.

The French AOC (*Appellation d'Origine Contrôlée*) system is so inextricably linked with *terroir* that it makes no sense without it, especially where wine is concerned. To quote the INAO (Institut National des Appellations d'Origin – guardian of the AOC system): 'With AOC wines, everything starts with the particular relationship between the grape variety and the terroir: this is what defines the wine's identity, which is then expressed through local know-how.'[28] As the INAO gradually extended its remit to include the agri-food sector as a whole, it also broadened its approach in consultation with INRA, France's centre for agricultural research, thereby arriving at the definition of *terroir* we find today, published in the INAO guide to applicants seeking registration as an AOC/PDO:[29]

> A geographical area with defined boundaries where a human community generates and accumulates across its history a collectively developed knowledge of production based on a system of interactions between bio-physical and human factors. The combination of techniques involved in production reveals

---

[26] Jean-Pierre Deffontaines, 'Commentaires à l'article de Lucette Laurens' (1997) 5(2) Natures, Sciences, Sociétés 60, 60.

[27] Marion Demossier, 'Beyond Terroir: Territorial Construction, Hegemonic Discourses and French Wine Culture' (2011) 17 Journal of the Royal Anthropological Institute (N.S.) 685.

[28] INAO, *Rapport de politique générale* (1992) 26.

[29] INAO GI Applicants' Guide (07/10/2005); available at: www.inao.gouv.fr

originality, confers 'typicity',[30] and leads to a reputation for a product originating from this geographical area.

Other organisations share this preoccupation with *terroir* – most notably 'Terroir et Cultures', which campaigns for the 'recognition, promotion and value enhancement of "terroirs" and the goods and services originating from specific geographical areas around the world'.[31] The adoption in 2001 of the UNESCO Universal Declaration of Cultural Diversity raised cultural diversity to the level of 'the common heritage of humanity'. Following that Declaration, 'Terroirs et Cultures' was the impetus behind 'Planète Terroirs':[32] a charter and international definition of *terroir*, signed at UNESCO headquarters in 2005, recognising the intrinsic cultural diversity of *terroir* and its potential as an alternative to standardisation and uniformity. An accompanying mission statement uses openly militant language to drive home a six-point agenda that revolves around two major foci: 'the strength of terroir'; and the conviction that 'the future needs terroirs', especially for the sake of sustainable development, food safety and helping the local economy. The charter concludes with an inset calling for a clear, international definition of *terroir* – even though the word itself is untranslatable – and attaches a revised and extended version of the INAO definition of *terroir*. On-going projects include maintaining of a worldwide network of *terroirs*, co-sponsored by UNESCO, through the development of an international approach to *terroir*. Three forums have been held to date: two in France and a third, in 2010, in Chefchaouen, Morocco, on 'diversity, sustainability, terroir and development'.

3. THE ROOT OF THE MISUNDERSTANDING

It is one thing to say that *terroir* is an easy-to-use tool, and quite another to prove it – in practice, there are complex social factors involved. The word

---

[30] The concept of typicity, uniquely French, is quite recent. It is derived from the idea of a typical, or representative product whose characteristics, particularly those pertaining to its taste, are strongly linked to its place of origin. See M-T Letablier and F Nicolas, 'Genèse de la "typicité"' (1994) 14 Sciences des aliments 541; F Casabianca, B Sylvander, Y Noël, C Béranger, JB Coulon, F Roncin, G Flutet and G Giraud, 'Terroir et typicité: un enjeu de terminologie pour les Indications Géographiques' in Claire Delfosse (ed), *La mode du terroir et les produits alimentaires* (Les Indes savantes, Paris 2011) 101–117.
[31] The 'Terroirs et Cultures' association no longer exists.
[32] The 'Planète Terroirs' association no longer exists.

is now such common currency that it has become a catch-all, obscuring the many different ways in which people relate to places, in France and elsewhere.

3.1　*Terroir* – a 'Long Quiet River' by No Means

'Terroirs et Cultures' is a case in point, symbolising the importance that the French attach to the *terroir* and the issues it raises. *Terroir* is a 'tool for sustainable development'; it plays a role in the 'safeguarding of diversity and sustainability'; it must be 'internationalised' and 'its potential unlocked': the association's platform reads like a *terroir* manifesto, founded on an entity that is not up for discussion since it has already been defined. But scientists remain divided over the issue of *terroir*. The concept is certainly interesting and can sometimes produce an ideal synergy of the possible and the humanly do-able. But today more than ever *terroir* stands at the crossroads of multiple converging interests, raising a host of thorny issues that demand closer scrutiny. The challenge lies in distinguishing between the fact and fiction of *terroir*.

While it is true that *terroir* implies specific collective skills that confer 'typicity' (another term that is not easy to define), plus a shared history and cultural profile, none of this is necessarily set in stone. In regions with harsh environmental conditions, agriculture is largely consistent with traditional practice – French mountain cheese is a good example. In more clement regions, it can be quite a different story. Take Bresse poultry for instance.[33] Though production is linked to a defined geographical area, the methods are at odds with local advances in agriculture that have made Bresse famous as an agricultural centre. Bresse poultry production is labour-intensive, time-consuming and almost entirely manual. It appears to have little common heritage with regional agriculture as a whole, but for the French it is a national institution. All the ingredients are there – a specific method of production, shared history, collective skills and professional identity linked to a defined geographical area. But how is one to conceive of terroir as a developmental tool in this context? Social organisation – particularly farming and land ownership schemes can be an obstacle to the promotion of *terroir*-linked products. Take the case of Dombes carp,[34] which are farmed in ponds mainly owned by city folk who

---

[33] Also registered as an EU PDO Volailles de Bresse, Dossier No. FR/PDO/0117/0145. All Dossiers can be accessed via the EU DOORS Database, at: http://ec.europa.eu/agriculture/quality/door/list.html

[34] See e.g. L Bérard and P Marchenay, 'Les étangs de la Dombes' in *Agricultures singuières* (Montpellier, IRD Editions 2008) 111; L Bérard and

reserve them for fishing and the pursuit of leisure and privilege. Very few of the local farmers have a strong say in the matter, however proficient they are as fishermen themselves.

The size of the French supermarket sector gives it a huge impact on the food-processing industry, reinforcing the latter's hefty market share even though relations between the two are not always cordial. Localised products that remain in high demand are no exception to this rule, particularly cured meats and dairy products. France's largest industrial dairy company, dairy cooperative Lactalis, which is also Europe's leading producer of raw-milk cheese, is represented in the vast majority of French AOCs. As a result, milk or cheese produced in those defined geographical areas is increasingly sold to dairy plants falling outside the remit of the AOC, operating at national or international levels. This contradicts the first principle of AOC regulation, seriously undermining localised production and with it local development, to say nothing of the repercussions for local knowledge systems and product culture in general. Consider the case of AOC Normandy Camembert,[35] made with unpasteurised milk in keeping with a traditional recipe specified by the INAO. Would the great 'Camembert war', which very nearly spelt the end of the road for unpasteurised milk, have been quite as fierce if there had not been such major commercial interests at stake? Every *terroir* is a place of dynamic interaction between local forces and balances of power, between complex mechanisms that dictate the uses of heritage and a revisiting of tradition that is often controversial. All of these factors deserve close attention if we are to understand the role of *terroir* in local development.

3.2 *Terroir* – Concealing Origin

The importance that the French attach to *terroir*, a word almost exclusively used in the context of local production, distracts from a clear explanation of the link with the physical environment as a factor in the protection of origin. No one would deny that the term is closely associated with territory, meaning a piece of land with a particular soil, aspect and other physical factors. Yet two types of regulation are envisaged by the measures in force at European level, most especially by the legal framework that makes it possible to establish and protect the relationship

---

P Marchenay, *Les produits de terroir. Entre cultures et règlements* (CNRS Editions 2004) 120–122.

[35] Florence Boulanger, 'Camembert, les enjeux du débat' (Mar–Apr 2007) 26 Profession fromager 10–11.

between a product and a place by reserving the use of a particular name. One is Protected Designation of Origin (PDO), which is closely modelled on the French AOC system (and whose name is now AOP, or *Appellation d'Origine Protégée*), and which takes into account the effect of natural factors. It refers to a product 'whose quality or other characteristics are essentially or exclusively due to a particular geographical environment with its inherent natural and/or human components and which is produced, processed and developed within the delimited geographical area'.[36] The other is Protected Geographical Indication (PGI), which makes no mention of natural factors. It designates a product whose geographical origin is defined by 'a specific quality, reputation or other characteristics ... and the production and/or processing and/or preparation of which take place in the defined geographical area'.[37] Natural factors actually play little part in some localised production systems, particularly in industrialised countries – but France refuses to get the message.[38] It persists in a dogmatic view of protection of origin that comes across loud and clear in some of the product specifications laid down for PGI applicants. Small wonder then, that the Commission's approval procedures are fraught with misunderstandings.

The protection of geographical indications is in accordance with internationally recognised principles set out in the provisions of the TRIPS Agreement, which refer to geographical indications as 'indications which identify a good as originating in the territory of a Member, or a region or locality in that territory, where a given quality, reputation or other characteristic of the good is essentially attributable to its geographical origin'.[39] This closely resembles the definition of a PGI. In the lead up to Regulation 1151/2012 the European Commission at one point intended to move towards a single system of regulation for both the PGI and the PDO, in line with a Communication adopted on 28 May 2009 laying down strategic orientations to improve the EU's agricultural policy related to product quality. Some might argue that this was partly France's

---

[36] See Art 5(1) of Regulation (EU) No. 1151/2012 on Quality Schemes for Agricultural Products and Foodstuffs [2012] OJ L343/1. Its predecessors were: Council Regulation (EC) No 510/2006 of 20 March 2006 on the Protection of Geographical Indications and Designations of Origin for Agricultural Products and Foodstuffs [2006] OJ L93/12; Council Regulation (EEC) 2081/92 [1992] OJ L208/1.
[37] Regulation 1151/2012, Art 5(2).
[38] For the legal history of the compromise between France and Germany which resulted in the PDO/PGI distinction see D Gangjee, 'Melton Mowbray and the GI Pie in the Sky' (2006) 3 IPQ 291.
[39] TRIPS, Art 22.1.

fault – a consequence of formulating applications for PGIs as if this were a 'lesser form' of PDO. If the two systems had been brought under a single set of regulations, we can be assured that the wording of the legislation would have been closer to the definition of a geographical indication than that of a PDO. This would have been regrettable, because the current emphasis on natural and human factors makes for a protection system that is ideally situated to identify and assess the impact of biodiversity and cultural differences on localised production systems. Emerging markets should be aware of this when drawing up their own frameworks for protection. EU Regulation 1151/2012 took this ambiguity into account by confirming the existence of the two forms of protection. Let us hope that this distinction remains clear well into the future.

3.3 Is '*Terroir*' Untranslatable? Here's Proof

The French say, not without a certain pride, that the term '*terroir*' is impossible to translate. There is no equivalent term in any other language. It is uniquely French, the mark of France's distinctive character and much-vaunted geographical variety. France is certainly an amazingly varied country, but couldn't the same be said of Italy or Portugal? More than likely, says historian Thierry Gasnier, but that has nothing to do with it.[40] France's attachment to local diversity is partly linked to its history. *Terroir*, what's more, is closely related to the protection of geographical origin and to French winegrowing, which is where origin protection was first introduced and where it remained confined for many years. Based on the concept of the AOC, those early regulations were revised and reinforced through successive crises, tightening up loopholes for the sake of better enforcement.[41] The law of 1927, for instance, established a de facto link between origin and quality, by making quality dependent on the existence of *terroir* and particular grape varieties. Passed in the wake of phylloxera, fraud, over-production and administrative delimitation, the new law was an attempt to bring order out of chaos – a joint undertaking between producers in different winegrowing regions, administrative authorities and wine merchants. Economic factors were very much a part of the equation. Everything revolved around wine, though the law also covered, albeit loosely, a selection of other products – mainly cheeses plus certain fruits and vegetables. It would be more than 60 years before the

---

[40] Gasnier (n 3) 463 ('that comparable diversity exists in neighbouring countries is irrelevant').
[41] For historical background, see Chapter 2 by Stanziani in this volume.

AOC structure and rule system was officially extended to the entire agri-food sector, under the law of 2 July 1990.[42]

The INAO was the state labelling body for all French food products until the new agricultural orientation law of 2006, and did not find it easy applying a wine-business mindset and approach to other sectors, principally because the link with *terroir* is more difficult to demonstrate. Do charcuterie and particularly everyday cheeses really exhibit those effects of *terroir* so often celebrated in wines? The relative importance of climate, pedology and human environment is still up for debate within INAO.[43] Already controversial enough for the French, the concept proves even more difficult to translate to other countries, including those of southern Europe that are otherwise culturally and linguistically close to France – never mind the vast majority of emerging economies that don't have a wine culture. New World countries on the other hand have often welcomed a stripped down version of the concept, usually reduced to the role of natural factors, and capitalised on its potential to add value to vineyards now operated on a commercial footing.[44]

'*Terroir*' also only partially addresses the complex issues raised by some of the production systems typical of emerging and developing economies. Coffee production in Ethiopia, for instance, a country whose economy revolves around coffee-growing, is based on a classification and grading system directly linked to quality in the cup (and setting aside any mention the influence of particular ethnic groups).[45] Nevertheless, to a certain extent, preaching the powers of *terroir* has eventually paid off, since the term is now part of the international vocabulary. However in the absence of any specific geographical and historical frame of reference, *terroir*

---

[42] For further details, see Delphine Marie-Vivien, 'The Role of the State in the Protection of Geographical Indications: From Disengagement in France/Europe to Significant Involvement in India' (2010) 13 JWIP 121.

[43] See Scheffer and Roncin's study of the different concepts used by INAO agents: Sandrine Scheffer and François Roncin, 'Qualification des Produits et des Terroirs dans la Reconnaissance des Produits en Appellation d'Origine Controlee' in Louis Lagrange (ed), *Signes officiels de qualité et développement agricole: Aspects techniques et économiques*, Actes de Colloque/SFER Société Française d'Economie Rurale SFER, 14–15 avril 1999 à Clermont-Ferrand (Technique et Documentation, Paris 1999) 37.

[44] For a critical overview of this in the Australian context, see Chapter 10 by Stern in this volume.

[45] Cindy Adolphe and Valérie Boisvert, 'Nommer et Contrôler: Les Appellations de Café en Ethiopie' (Paper presented at UNESCO International Symposium: Localizing Products: A Sustainable Approach for Natural and Cultural Diversity in the South? 2009).

can sometimes change meaning – a point made by Yveline Poncet in her Chilean case study.[46] She comments on the 'snob value' of the French word *'terroir'*, whether used by Chile's major companies to 'gain market visibility in the global arena', or by small and medium-size farmers in Coquimbo (the region under study) to illustrate the localised nature of production.

4. AN ANTHROPOLOGICAL INTERPRETATION OF PLACE

Given the limitations of the *terroir* approach, which as we have seen oversimplifies the relationship between place and product, we need to restate the problem of *terroir* in the larger context of globalisation and what it implies for the way people live their spaces.

4.1 A Sense of Place

People construct their spatial spheres of action by setting boundaries, by occupation and by transformation, designating and distinguishing a given place 'by considering it in all of its forms and from every aspect, and generally stamping it with the unmistakable mark of their identity'.[47] Localised agri-food production is a part of that process, helping to foster a sense of place through a wealth of specialised products that engage with local society in all sorts of ways and on all sorts of production levels. Polymorphism apart, all local food production systems maintain a particular relationship with their surroundings; their place within a given area involves historical precedent and shared know-how, a common thread that links them in place and time. This collective historical dimension is what defines their origin, allowing us to think of them as a family of products that make sense at a local level. The time-span in question may range from several centuries to just a few decades, but there is a definite sense of historical precedence. That historical dimension, linked to the

---

[46] Yveline Poncet, Tchansia Koné and Héctor Fabián Reyes, 'La Diversité des Produits de Terroir en Pays Emergent: Un Chemin Malaise' in Marie-Christine Cormier-Salem and Bernard Roussel (eds), Les Produits de Terroir au Service de la Diversité Biologique et Culturelle (Revue de sciences sociales, N° 50, IRD éditions 2009) 17, 19–20.

[47] Gérard Lenclud, 'L'Ethnologie et le Paysage. Questions sans Réponses' in Claudie Voisenat (ed), *Paysage au Pluriel: Pour une Approche Ethnologique des Paysages* (Editions de la Maison des Sciences de l'Homme, Paris 1995) 3, 11.

collective memory that has been handed down through generations, is what gives a place depth. Thus, in order to understand the diversity of localised products we need to look at the cultural criteria that link a place with a particular history and social group. The collective dimension of a product is what makes it a part of local culture and helps to distinguish origin (meaning 'to be from a place') from provenance (meaning 'to issue from a place but with no particular connection with that place').

Some localised production systems maintain powerful links with the physical environment, based on complex interactions between bio-physical and human factors that are consistent with AOC philosophy and reflected in the definition of *terroir*. For the sake of simplicity, all localised products involving historical precedent and collective practices are usually lumped together under the catch-all label *'produits de terroir'*. Though widely used in France, this term makes no distinction between origin and provenance, creating a huge temptation to pass provenance off as something it is not and doing nothing to address consumer demands for clarity. We suggest therefore that *'produits de terroir'* should be exclusively reserved for products that show a discernible link to *terroir*. Outside France, two good alternative terms are 'origin-based product' or simply 'localised product'. Both are easy to translate and used to best advantage by authors Petra Van de Kop, Denis Sautier and Astrid Gerz in their book *Origin-Based Products*.[48]

4.2 Place in Effervescence

In France there is sustained debate about the respective significance of *savoir-faire*, history and 'milieu' in justifying the relationship with a given territory and usage of a geographical name. It all started centuries ago with the reorganisation and subsequent development of French territory. What drives the success, or revival, of local production systems is a process of innovation and transformation that borrows from other sources, but revolves around an already established element, or set of elements, pertaining to the product that are developed to a greater or lesser extent. The protection of geographical indications is a principle that has been adopted at the international level, encouraging emerging and developing economies to seek recognition for products whose quality is linked to geographical origin[49] – with varying degrees of success, depending on

---

[48] P Van de Kop, D Sautier and A Gerz (eds), *Origin-Based Products: Lessons for Pro-Poor Market Development* (KIT, Amsterdam 2006).
[49] See E Barham and B Sylvander (eds), *Labels of Origin for Food – Local Development, Global Recognition* (CABI, Wallingford 2011); D Giovannucci, T Josling, W Kerr, B O'Connor and M Yeung, *Guide to Geographical Indications:*

the country. New World countries, especially the USA and regions like Quebec, have barely begun to tap the potential in this area. Numerous initiatives are underway, at every level of activity, aimed at exploring the full implications of 'local' and ways to take ample advantage of the concept in local agriculture and food production.[50] Among the questions already raised are how to motivate, manage and steer local production, and what local sources or studies can be cited to support the link with place of a particular product.[51] All of these questions encourage us to rethink that link for the purposes of cases where, unlike France, there is no requirement to reinforce and optimise what already exists, presenting an exciting opportunity to view this issue from a broader perspective. A new understanding of territory is emerging, based on an alternative approach to localised production, because the producers' own view of it also matters, as became clear in the 'Taste of Place' conference held in Vermont in 2008 and organised by Amy Trubek.[52] Some products are already rooted in place, like apples, apple juice and ice cider. Maple syrup is a particularly good example of the close ties between place, natural raw materials and specific *savoir-faire*, and clearly eligible for AOC registration, albeit such recognition may need to take into account that the natural boundaries of maple territory straddle the frontier between the US and Canada. Other products, such as cheeses for instance, have no tradition to speak of or commonality of typicity, but are the focus of strenuous development efforts at the individual producer level. For Amy Trubek the focus of the Taste of Place initiative, which involves producers, researchers and policymakers in the state, remains the consideration of a PGI for maple syrup; however artisan cheese and cider and ice cider[53] are also part of discussions and projects. A series of collaborative research projects have considered a broad definition of *terroir*, one that involves natural and human contexts as intrinsic to economic development, emergent in ideas of sensory quality, and crucial to cultural identity. Some

---

Linking Products and their Origins (International Trade Centre, Geneva 2009); FAO and SINER-GI, *Linking People, Places and Products: A Guide for Promoting Quality Linked to Geographical Origin and Sustainable Geographical Indications*, 2nd edn (FAO, Rome 2009–2010).

[50] For concrete illustration see the 'Taste the Local Difference' programme in the NW part of Michigan, in the counties surrounding Traverse City. See: http://www.mlui.org/food-farming/projects/

[51] See Chapter 15 by Barjolle in this volume.

[52] See also Amy B Trubek, *The Taste of Place: A Cultural Journey into Terroir* (University of California Press, Berkeley and Los Angeles 2008).

[53] 'Apples (Old, New and Blue) Grow in Cornwall', available at: http://goo.gl/K33qgx

results are a sensory tool for maple syrup, analysis of *terroir* in the supply chain for Vermont artisan cheese, and studies of consumer perception and commitment in the greater New England region.

In Missouri, an applied research programme mounted by the Department of Rural Sociology of the University of Missouri-Columbia was created to promote local produce, especially wine, as part of a drive to develop agri-tourism in the area.[54] The programme was originally run in partnership with the Missouri Department of Agriculture and the Missouri Division of Tourism, and drew upon the region's winegrowing history while also highlighting the present-day activities of local farmers and restaurants capable of providing a food pairing experience to go with the local wines. The programme director, Dr Elizabeth Baham, who later moved to the University of Arkansas, was asked in 2010 by OriGin, the Organization for an International Geographical Indications Network,[55] to consider the adjustments required to the current legal framework for the protection of 'American Origin Products' or 'AOP' – a nod no doubt to the European acronym 'AOP'. Her findings are published in a manual, co-authored with other American experts, setting out the current situation and suggesting where to go from here.[56] One direction this work has taken is the organisation of the American Origin Products Association, and more recently the American Origin Products Research Foundation.[57] This is part of a more general interest in place. Increasing numbers of American researchers are rethinking the concept of place – what it means and why people feel attached to the places where they live.[58]

'Farmers markets' for instance are a growing trend, alongside the interest in local products of known provenance. In France, geographical proximity matters like never before, championed by the peasant farmers'

---

[54] See: E Barham, D Lind and L Jett, 'The Missouri Regional Cuisines Project: Connecting to Place in the Restaurant' in Peggy F Barlett (ed), *Urban Place: Reconnecting with the Natural World* (MIT Press, Cambridge, MA 2005) 141; E Barham, 'The Missouri Regional Cuisines Project: Geographical Indications as a Rural Development Opportunity' (2009) 20(4) Rural Research Report 8.

[55] The Organization for an International Geographical Indications Network (OriGin) is the first international network of Geographical Indications producers, representing some 150 associations (upwards of two million people) from around 40 countries. See: www.origin-gi.com

[56] E Barham (ed), *American Origin Products (AOPs): Protecting a Legacy* (Origin, Geneva 2010).

[57] See http://www.aopcentral.us/

[58] See e.g. Laura B DeLind and Jim Bingen, 'Place and Civic Culture: Rethinking the Context for Local Agriculture' (2006) 21(2) Journal of Agricultural and Environmental Ethics 127–151.

organisation AMAP (Association for the Preservation of Peasant Agriculture) through weekly market outlets where consumers can buy fresh market-garden produce ordered direct from the grower. Initiatives to connect producers and consumers are proliferating, in particular with peri-urban agriculture and various 'buy direct' efforts linking consumers to producers nearby. However, as the local has taken on more importance in France it has been reduced in meaning to proximity, to the detriment of a sense of product origin.

4.3 The Great 'Globalisation Versus Localisation' Debate

Globalisation, which is centred on the accelerated circulation of capital, people, goods and ideas,[59] disrupts conventional wisdom by forcing us to think differently about our relationship to place. The idea of local roots has no apparent place in a techno-scientific world wedded to free trade and progress, where growth, rationality and productivity inevitably correlate with the emergence of a rootless society that has lost all ties to the land.[60] This is a point stressed by the authors of *L'Equivoque Ecologique*,[61] who insist on the need to give contemporary currency to place and a sense of attachment to one's home in geographical terms. Freedom, social belonging, universal awareness – it all starts with a sense of place. What chance is there of that today?

The result of this broad circulation of people and goods is a shortening of time and a shrinking of space. Both of these effects, says French anthropologist Marc Augé, are closely linked to excess:

> an over-abundance of events and space that makes modern time increasingly difficult to understand, space less and less intelligible, leading to featureless suburban landscapes and a proliferation of non-places that conflict with the sociological concept of place, one which for Mauss and a whole generation of French ethnologists is supported by cultural legacies located in time and space.[62]

---

[59] Marc Abélès, *Anthropologie de la Globalisation* (Payot, Paris 2008).
[60] Pierre Alphandéry, Pierre Bitoun and Yves Dupont, *L'équivoque écologique* (La Découverte, Paris 1992)., between growth, rationality and productivity and the loss of everything supporting an attachment to the land. The authors refer, specifically, to the 'essential correlation, unique to trading and techno-scientific societies roots and a sedentary existence'.
[61] Ibid.
[62] Marc Augé, *Non-lieux: Introduction à Une Anthropologie de la Surmodernité* (le Seuil, Paris 1992) 48; also translated: *Non-places: Introduction to an Anthropology of Supermodernity* (Verso Books, London 1995).

This is precisely what some people find wrong with anthropology – this obscure relationship with the land, verging on a fetishistic attachment to place and the 'micro-level', that favours cultures largely impervious to change and divorced from the context of space and time.[63] Then again, 'what does this reference to land mean today, in a context marked by a very general process of "deterritorialisation" where people and products circulate on a global scale, in a procedure that repeatedly defines and redefines cultural distinctions?'[64] In the same vein, Akhil Gupta criticises anthropologists for considering place as a given factor without bothering to understand how it is perceived, constructed and experienced.[65] Arjun Appadurai also takes an interest in the concept of 'locality' – meaning to come from a place – as a consequence of 'localised production'. He too questions the false evidence on which this so often relies, by examining local production techniques and the ways in which people become attached to the place they live in and transform it materially. All of this, says Appadurai, relates to the perpetuation of local lifestyles, which depends on 'the uninterrupted interaction of localized space and time with local people who possess the knowledge to reproduce locality'.[66]

Globalisation moves the goalposts and raises endless questions. The discipline of ethnology may well be based on the study of self-contained communities, but ethnologists have long ago shed the belief that the society they study functions as a closed circuit divorced from the noise of the world. It does, all the same, remain true that people still live and breathe in identifiable places, where they are sheltered to varying degrees from those global forces that seek to interact with and profoundly modify their particular codes of living. Geographical indications take their place today among those global forces by opening a space for thinking differently about local development.

---

[63] Abélès (n 59).
[64] Ibid 88–89.
[65] Akhil Gupta and James Ferguson (eds), *Culture, Power, Place: Explorations in Critical Anthropology* (Duke University Press, Durham NC 1997).
[66] Arjun Appadurai, *Après le Colonialisme: Les Conséquences Culturelles de la Globalisation* (Payot, Paris 2005) 261; see also Michel Agier, 'Quel Temps Aujourd'hui en ces Lieux Incertains?' (2008) 185–186 L'Homme 105 (Special Issue: L'anthropologue et le contemporain. Review of Marc Augé).

PART II

INTERNATIONAL PROTECTION

5. Geographical Indications under WIPO-administered treaties
Matthijs Geuze*

1. INTRODUCTION

Whether geographical names or other denominations with a geographical connotation can be Geographical Indications (GIs) or Appellations of Origin (AOs) may be assessed differently from country to country. For example, the fact that some of these indications are protected as intellectual property rights in some jurisdictions, whereas in other jurisdictions they are considered as generic indications for a certain kind of product, has generated extensive debate for many years. This introductory section will outline WIPO's longstanding engagement with these debates over several decades.

More than a century ago, discussions between governments concerning the protection of GIs at the international level resulted in the incorporation of a number of provisions on the subject into the Paris Convention for the Protection of Industrial Property[1] and in the conclusion of the Madrid Agreement for the Repression of False or Deceptive Indications of Source on Goods.[2] Such discussions have continued to take place ever since, resulting in modifications of these provisions, the conclusion of many bilateral or plurilateral agreements, the adoption of the Lisbon Agreement for the Protection of Appellations of Origin and their International

---

* While the author is Head of the Lisbon Registry in the Brands and Designs Sector of the World Intellectual Property Organization (WIPO), the views expressed in this chapter are personal, rather than those of the Organization.

[1] The Paris Convention for the Protection of Industrial Property, 20 March 1883, as revised at Stockholm on 14 July 1967, 828 UNTS 305 (1972) (hereafter, the Paris Convention). While the Convention was adopted in 1883, it was revised several times, most recently at Stockholm in 1967, with further amendments concerning administrative provisions in 1979. All WIPO treaties are available at: www.wipo.int/treaties/en/

[2] Madrid Agreement for the Repression of False or Deceptive Indications of Source on Goods, 14 April 1891, 828 UNTS 389 (1972) (hereafter, the Madrid Agreement). The Agreement was revised several times, most recently in 1958, in Lisbon. An Additional Act concerning administrative provisions was adopted in 1967, in Stockholm.

Registration[3] and most recently, the adoption of the Geneva Act of the Lisbon Agreement in 2015.[4]

Discussions initiated by WIPO took place in 1974–1975, relating to proposals for a New Multilateral Treaty on the Protection of Geographical Indications (Committee of Experts on the International Protection of Appellations of Origin and other Indications of Source); on proposals for the revision of the Paris Convention between 1978 and 1980 (Working Group on Conflict Between an Appellation of Origin and a Trade Mark); on proposals for the revision of the Paris Convention between 1982 and 1984 (Working Group of Main Committee I established to consider Article 10*quater*); and on proposals concerning the possible conclusion of a new treaty for the international protection of Appellations of Origin and indications of source or the possible revision of the Lisbon Agreement for the Protection of Appellations of Origin and their International Registration (Committee of Experts on the International Protection of Geographical Indications, 1990).[5]

Issues concerning the protection of GIs were also discussed in the Uruguay Round of Multilateral Trade Negotiations between 1986 and 1994. As a result, the TRIPS Agreement[6] contains a number of provisions concerning the protection of GIs, in particular in Section 3 of Part II (Arts 22 to 24). Built into these provisions is an agenda for further work on certain issues by WTO Member governments. Work on these built-in agenda items of the TRIPS Agreement is on-going, either in the WTO TRIPS Council, since 1996, or in the context of the Doha Development Agenda, since 2001.[7]

Between 2000 and 2003, WIPO Member States also discussed various issues concerning GIs in the Standing Committee on the Law of Trade Marks, Industrial Designs and Geographical Indications (SCT).[8]

---

[3] Lisbon Agreement for the Protection of Appellations of Origin and their International Registration, 31 October 1958, 923 UNTS 205 (1974) (hereafter, the Lisbon Agreement). The Agreement was revised in 1967, in Stockholm, and amended in 1979, both times as regards its administrative provisions.

[4] Geneva Act of the Lisbon Agreement on Appellations of Origin and Geographical Indications, 20 May 2015 (LI/DC/19) (hereafter, Geneva Act of the Lisbon Agreement).

[5] For these deliberations, see D Gangjee, *Relocating the Law of Geographical Indications* (CUP, Cambridge 2012) 195–199.

[6] Agreement on Trade-Related Aspects of Intellectual Property Rights, 15 April 1994, in the Marrakesh Agreement Establishing the World Trade Organization, Annex 1C (1994) 33 ILM 1125, 1197 (hereafter, TRIPS).

[7] For details, see Chapter 6 by Gervais in this volume.

[8] Documents from these meetings are available at: http://www.wipo.int/meetings/en/topic.jsp?group_id=14

Documents were prepared by WIPO and discussed in the SCT on the nature of the right and its historical background, the existing systems of protection, obtaining protection abroad,[9] conflicts between trade marks and GIs or between homonymous GIs,[10] definitional aspects[11] and the application of the territoriality principle in relation to GIs.[12] Also, in 2000 and 2001, a Working Group, convened by the Director General of WIPO, met twice to review the Regulations under the Lisbon Agreement.[13] As a result, revised Regulations were adopted by the Lisbon Union Assembly in 2001 and entered into force in 2002.

In September 2008, the Lisbon Union Assembly once again established a Working Group, to explore possible improvements of the Lisbon system. This Working Group on the Development of the Lisbon System (Appellations of Origin) met ten times between 2009 and 2014, ultimately recommending the convening of a Diplomatic Conference for adopting a Revised Lisbon Agreement, which will be considered in more detail later in this chapter. Membership of the Working Group was open to representatives of Lisbon Member States as well as other WIPO Member States, intergovernmental organizations and non-governmental organizations.

As a result of these deliberations, at the first session of this Working Group,[14] WIPO conducted a survey aimed at identifying how the system might become more attractive for users and prospective new members of the Lisbon Agreement while preserving the principles and objectives of the

---

[9] WIPO Secretariat, 'Document SCT/6/3 Rev. on Geographical Indications: Historical Background, Nature of Rights, Existing Systems for Protection and Obtaining Protection in Other Countries' 2 April 2002 (SCT/8/4); with Addendum 2 April 2002 (SCT/8/5).

[10] WIPO Secretariat, 'Possible Solutions for Conflicts between Trade Marks and Geographical Indications and for Conflicts between Homonymous Geographical Indications' 8 June 2000 (SCT/5/3).

[11] WIPO Secretariat, 'The Definition of Geographical Indications' 1 October 2002 (SCT/9/4) and WIPO Secretariat, 'Geographical Indications' 25 March 2003 (SCT/10/4).

[12] WIPO Secretariat, 'Geographical Indications and the Territoriality Principle' 1 October 2002 (SCT/9/5).

[13] For the issues addressed, see the following WIPO documents: 'Questions to be Examined With a View to the Modification of the Regulations under the Lisbon Agreement' 10 May 2000 (LI/GT/1/2); 'Proposals for Modification of the Regulations under the Lisbon Agreement' 19 January 2001 (LI/GT/2/2); and 'Notes Concerning the Proposals for Modification of the Regulations under the Lisbon Agreement' 19 January 2001 (LI/GT/2/3).

[14] WIPO, 'Summary by the Chair' 20 March 2009 (LI/WG/DEV/1/3) and 'Report' 3 September 2010 (LI/WG/DEV/1/4).

98   *Research handbook on intellectual property and GIs*

Agreement.[15] WIPO also conducted a study on the relationship between regional systems for the protection of GIs and the Lisbon system as well as the conditions for accession to the Lisbon Agreement by intergovernmental organizations.[16] Following the survey and study, the Working Group entered into treaty drafting mode. Initially WIPO was requested to prepare treaty texts on a number of specific topics, notably: definitions (AOs, GIs, traditional non-geographical indications, products from transborder areas), means of protection, scope of protection, application and registration procedures, refusal and invalidation procedures, prior users, and accession by intergovernmental organizations.[17] However, since its fourth session, the Working Group has worked on a new draft instrument concerning the international registration of GIs and AOs as well as draft regulations under such an instrument. This drafting work was completed at the Group's tenth session in October 2014, after which the Preparatory Committee for the Diplomatic Conference for the Adoption of a Revised Lisbon Agreement on Appellations of Origin and Geographical Indications transformed the two texts into the Basic Proposal for this Conference, subsequently hosted by WIPO in May 2015.[18]

Since 1988 WIPO has also organized, in a biennial rhythm and in cooperation with individual Member States, international symposia dedicated to the subject of Geographical Indications (GIs). Open to the general public, those symposia constitute a platform for the exchange of information and views for a wide spectrum of participants, governmental and non–governmental, institutional and private, on various issues relating to the protection of GIs.[19]

---

[15] WIPO, 'Results of the Survey on the Lisbon System' 18 June 2010 (LI/WG/DEV/2/2).

[16] WIPO Secretariat, 'Study on the Relationship between Regional Systems for the Protection of Geographical Indications and the Lisbon System and the Conditions for, and Possibility of, Future Accession to the Lisbon Agreement by Competent Intergovernmental Organizations' 6 August 2010 (LI/WG/DEV/2/3).

[17] WIPO, 'Draft Provisions on Certain Matters Addressed by the Working Group in the Context of the Review of the Lisbon System' 16 March 2011 (LI/WG/DEV/3/2).

[18] The Basic Proposal consists of the draft *New Act of the Lisbon Agreement on Appellations of Origin and Geographical Indications* and the draft *Regulations under the New Act* (LI/DC/3 and 4).

[19] Symposia documents are available at: http://www.wipo.int/meetings/en/topic.jsp?group_id=14

2. DEFINITIONS

2.1 Indications of Source and Appellations of Origin under the Paris Convention; Geographical Indications under TRIPS

Art 1(2) of the Paris Convention defines the notion of 'industrial property' as having as its object, among others, Indications of Source (IS) and Appellations of Origin (AO). The provision does not require Paris Convention countries to provide protection to all objects of industrial property mentioned,[20] but the importance of Art 1(2) lies in Art 2 of the Paris Convention, which refers to 'industrial property' when stipulating that nationals of Paris Convention countries shall enjoy in any other Paris Convention country the same protection as the nationals of that country, that is, national treatment. Thus, for example, if a country provides protection in respect of an IS or AO, it would be obliged to provide the same protection to nationals from other Paris Convention countries and not be entitled to require reciprocity of protection for the same object. Such a country can, however, require that the nationals of other Paris Convention countries comply with the conditions and formalities that the country in question imposes upon its own nationals in respect of the protection concerned.

Art 1.2 of TRIPS defines the term 'intellectual property' as referring to all categories of intellectual property that are the subject of Sections 1 through 7 of Part II of the Agreement. One of these sections deals with GIs, but no explicit reference can be found in Sections 1 to 7 of Part II of TRIPS to the IS or AO. However, by virtue of Art 2.1 of TRIPS, the provisions of Arts 1 to 12 and 19 of the Paris Convention are incorporated by reference into Parts II, III and IV of the Agreement. Thus, it can be argued that the IS and AO, as covered by the notion of 'industrial property' in Art 1(2) of the Paris Convention are also categories of intellectual property that are the subject of Sections 1 through 7 of TRIPS.

2.2 Indications of Source

The term 'indication of source' is used in Arts 1(2) and 10 of the Paris Convention. As mentioned above, by virtue of Art 2.1 of TRIPS, these provisions also apply among Members of the WTO, whether or not they are Member States of the Paris Convention. The term is also used in the

---

[20] GHC Bodenhausen, *Guide to the Application of the Paris Convention for the Protection of Industrial Property* (Bureau de l'Union, Geneva 1969) 24.

Madrid Agreement on Indications of Source. There is no definition of the IS in any of the abovementioned treaties, but Art 1(1) of the Madrid Agreement on Indications of Source contains language which clarifies what is meant by the term.[21] It reads as follows:

> All goods bearing a false or deceptive indication by which one of the countries to which this Agreement applies, or a place situated therein, is directly or indirectly indicated as being the country or place of origin shall be seized on importation into any of the said countries.

Consequently, an IS can be defined as an indication referring to a country, or to a place in that country, as being the country or place of origin of a product. It is important that the IS relates to the geographical origin of a product and not to another kind of origin, for example, an enterprise that manufactures the product. This definition neither implies that the product on which an indication of source is used should have a specific quality or one or more specific characteristic(s), nor that use of the indication should relate to a specific product. Examples of indications of source are the mention or illustration, on a product, of the name or picture of a country, or indications such as 'made in . . . . .'. In terms of the scope of protection, the prohibition against false indications would apply where goods do not originate from the place indicated, while the bar on deceptive indications would cover literally true, yet misleading uses (for example placing 'Made in Paris' on fashionable garments, without specifying Paris, Texas).

2.3 Appellations of Origin

The term 'appellation of origin' is used in Art 1(2) of the Paris Convention which, as mentioned above, also applies among Members of the WTO. The Guide to the Paris Convention states that AOs are considered to be a species of the genus IS, characterized by their relationship with quality or characteristics derived from the source.[22] As a consequence, AOs would also fall under the obligations of Art 10 of the Paris Convention, albeit only to the extent that its provisions require Paris Convention countries and WTO Members to protect an IS against misleading uses. By the same token, it could be argued that WTO Members are under an obligation to protect AOs, to the extent that the TRIPS Agreement requires them to protect GIs. Of course, if higher standards of protection apply in any of these countries, the national treatment provisions of the Paris Convention

---

[21] WIPO (SCT/8/4) (n 9), para 4.
[22] Bodenhausen (n 20) 23, at (j).

and/or the TRIPS Agreement would require the application of those higher standards, as mentioned under Section 2.1 above.

There is no definition of the term 'appellation of origin' in any of the abovementioned treaties, but such a definition is contained in the 1958 Lisbon Agreement. The Lisbon Agreement establishes an international registration-based system for its Member States to facilitate protection abroad for AOs already recognized and protected in their country of origin. Protection is subject to the international registration of that AO with WIPO. Art 2(1) of the Lisbon Agreement defines an AO as follows:

> 'Appellation of origin' means the geographical denomination of a country, region, or locality, which serves to designate a product originating therein, the quality or characteristics of which are due exclusively or essentially to the geographical environment, including natural and human factors.

Under this definition, the product for which an AO is used must have a quality or characteristics which are due exclusively or essentially to a geographical environment existing in the area indicated by the appellation of origin in question – a geographical environment that can be determined by natural and human factors. In addition, the product must have reputation, as required by Art 2(2) of the Lisbon Agreement:

> The country of origin is the country whose name, or in which is situated the region or locality whose name, constitutes the appellation of origin which has given the product its reputation.

Furthermore, the provisions of Art 1(2) of the Lisbon Agreement lay down that an AO must be 'recognised and protected' in the country of origin. This condition means that the AO must be constituted by a geographical denomination that is protected in the country of origin as the denomination of a geographical area (country, region or locality) recognized as serving to designate a product that originates therein and meets certain qualifications. Such recognition of the denomination must be based on the reputation of the product, and protection of the appellation of origin must have been formalized by means of legislative provisions, administrative provisions, a judicial decision or any form of registration.[23] The Records of the Diplomatic Conference that adopted the Lisbon Agreement in 1958 explicitly mention that the words 'recognised and' were added before 'protected as such', in order to bring the provision in harmony with the principle that an appellation of origin

---

[23] Lisbon Regulations, Rule 5(2)(a)(vi).

always applies to a product enjoying a certain reputation.[24] The manner in which the recognition and protection, required to justify its international registration, take place is determined by the domestic legislation of the country of origin. Examples of protected appellations of origin are 'Bordeaux' for wine, 'Szeged' for ground paprika, 'Tequila' for spirit drinks, 'Parmigiano-Reggiano' for cheese, or 'Chulucanas' for ceramics.[25]

Following the review of the Lisbon system (see Section 5, below), the Geneva Act of the Lisbon Agreement defines an AO in Arts 2(1)(i) and 2(2) as follows:

> Any denomination protected in the Contracting Party of Origin consisting of or containing the name of a geographical area, or another denomination known as referring to such area, which serves to designate a good as originating in that geographical area, where the quality or characteristics of the good are due exclusively or essentially to the geographical environment, including natural and human factors, and which has given the good its reputation;
>
> A geographical area of origin as described in paragraph (1) may consist of the entire territory of the Contracting Party of Origin or a region, locality or place in the Contracting Party of Origin. This does not exclude the application of this Act in respect of a geographical area of origin, as described in paragraph (1), consisting of a trans-border geographical area, or a part thereof.

2.4 Geographical Indications

The TRIPS Agreement contains a Section dealing with the protection of GIs (Part II, Section 3). Art 22.1 of TRIPS provides the following definition:

> Geographical indications are, for the purposes of this Agreement, indications which identify a good as originating in the territory of a Member [of the World Trade Organization], or a region or locality in that territory, where a given quality, reputation or other characteristic of the good is essentially attributable to its geographical origin.

---

[24] *Actes de la Conférence des Lisbonne* (Bureau International de l'Union, Geneva 1963) 859 (unofficial translation from the official French text); see also WIPO, 'Possible Improvements of the Procedures under the Lisbon Agreement' 10 February 2009 (LI/WG/DEV/1/2 Rev), Annex II, FN 2.

[25] Information concerning all registered AOs under Lisbon are available on the 'Lisbon Express' database, at: www.wipo.int/ipdl/en/search/lisbon/search-struct.jsp. The online resources include an interactive World Map allowing for a structured search on registered AOs, including information on the product to which an appellation of origin applies, its area of production, the holders of the right to use the appellation of origin, any refusals or invalidations notified by member countries, etc.

This definition resembles to some extent the definition for appellations of origin in Art 2 of the Lisbon Agreement. However, Lisbon requires that the quality or characteristics of the product in question be due exclusively, or essentially, to the geographical environment, including natural and human factors, and must have resulted in a certain reputation of the product in its country of origin. The TRIPS Agreement covers goods which have a given quality, reputation or other characteristic – that is, in the singular – that is essentially attributable to their geographical origin.[26]

GIs thus defined in accordance with TRIPS are not necessarily disqualified from international registration under the Lisbon Agreement, provided, of course, that their country of origin is party to the Agreement. The Fourth Commission of the Diplomatic Conference that adopted the Lisbon Agreement in Lisbon in 1958 confirmed the view that, where the laws of numerous countries did not distinguish between AOs and IS, these countries could only adhere to the Agreement if the Agreement would give a clear indication as to what was an AO meeting the requirements of the Agreement. 'By introducing a definition for AOs into the Agreement itself, such definition could be invoked for the purposes of registration, *without prejudicing a national definition, whether broader or more precise in scope*' (emphasis added).[27] Such a definition could be invoked by authorities of countries refusing protection in their territories and would also serve as a yardstick for national courts to assess whether any given geographical denomination, even when registered as an appellation of origin, did actually fall under the terms of the Lisbon Agreement.[28]

In this regard, it should be noted that, in recent years, a number of registrations effected under the Lisbon system are based on protection of the denomination in question in the country of origin as a GI. Other Lisbon Member States are of course free to refuse protection, if the GIs in question, in their view, do not meet the AO definition of Lisbon. To date, none have done so, although a court could still invalidate the effects of the international registration in their territory. However, perhaps, there is reason to accept international registrations of GIs under Lisbon, as Lisbon countries may well provide protection to both AOs and GIs under their national law and because Article 5(3) of the Lisbon Agreement stipulates, with reference to Article 4 of the Agreement, that refusal declarations are

---

[26] For further comparisons between the AO and GI, see Chapter 6 by Gervais in this volume.
[27] *Actes des Lisbonne* (n 24) 859.
[28] WIPO (LI/WG/DEV/2/3) (n 16) para 35.

104 *Research handbook on intellectual property and GIs*

not detrimental to the other forms of protection the owner may be entitled to claim in the country concerned.

Following the review of the Lisbon system (see Section 5, below), the Geneva Act of the Lisbon Agreement in Arts 2(1)(ii) and 2(2) defines GI as follows:

> Any indication protected in the Contracting Party of Origin consisting of or containing the name of a geographical area, or another indication known as referring to such area, which identifies a good as originating in that geographical area, where a given quality, reputation or other characteristic of the good is essentially attributable to its geographical origin;
>
> A geographical area of origin as described in paragraph (1) may consist of the entire territory of the Contracting Party of Origin or a region, locality or place in the Contracting Party of Origin. This does not exclude the application of this Act in respect of a geographical area of origin, as described in paragraph (1), consisting of a trans-border geographical area, or a part thereof.

3. DIFFERENCES IN MEANS OF PROTECTION

Protection of GIs and AOs at the national and regional levels is characterized by the existence of a wide variety of different legal concepts.[29] Those concepts were developed in accordance with different national legal traditions and within a framework of specific historical and economic conditions. These differences have a direct bearing on important questions such as conditions for protection, entitlement to use and scope of protection. As regards the various means of protection at the national level, three main systems can be distinguished, namely:

(i) legislation establishing a *sui generis* system, under which special titles of protection or rights are granted, exclusively applicable in respect of GIs or AOs and independent from any other intellectual property right;
(ii) trade mark legislation allowing GIs to be protected through the registration of collective marks or certification marks (or guarantee marks);
(iii) other legislation, such as administrative schemes; legislation providing protection against unfair competition, consumer protection legislation, or labelling requirements for the sale of products.

---

[29] A useful document in this regard is WTO, 'Review under Article 24.2 of the Application of the Provisions of the Section of the TRIPS Agreement on Geographical Indications' 24 November2003 (IP/C/W/253/Rev.1).

Apart from the three main approaches, there are also other means of protection, which may be country-specific. These different approaches do not necessarily apply on an exclusive basis and in practice there is cumulative protection (for example a designation being registered both under *sui generis* law and trade mark law). Moreover, some national or regional systems provide different means of protection for different categories of products, which accentuates cumulative protection.

Following the review of the Lisbon system (see Section 5, below), the Geneva Act of the Lisbon Agreement confirms that 'Each Contracting Party shall be free to choose the type of legislation under which it establishes the protection stipulated in this Act, provided that such legislation meets the substantive requirements of this Act'.[30] 'Contracting Parties that do not distinguish in their national or regional legislation as between AOs and GIs shall not be required to introduce such a distinction into their national or regional legislation.'[31] These Contracting Parties will protect AOs as being covered by the definition for GIs.

3.1 *Sui Generis* Protection Systems

About 100 years ago, in France,[32] for the first time a law was enacted recognizing the existence of AOs and laying down conditions for their protection. The system for the protection of AOs was developed out of a perceived need to provide a remedy against fraudulent commercial practices involving the origin of agricultural products and, in particular, products of the vine.[33] Although such practices are probably as old as trade itself, they occurred increasingly during periods when there was a shortage of supply of certain products, such as was the case with certain wines during the phylloxera crises in some European vineyards in the nineteenth century.[34] The origin of this law lay partly in the fact that the legal means to protect indications of source against false or deceptive use were insufficient and partly in the emerging need to encourage local, traditional

---

[30] Art 10(1).
[31] Art 9.
[32] *Loi du 6 Mai 1919 Relative à la Protection des Appellations d'Origine* 8 Mai 1919 Journal Officiel 4726; [1919] Propriété Industrielle 61. For a comprehensive historical review of French regulation, see Chapter 2 by Stanziani in this volume.
[33] INAO, *L'appellation d'Origine Contrôlée*, 11.
[34] J-M Girardeau, 'The Use of Geographical Indications in a Collective Marketing Strategy: The Example of Cognac' in WIPO, *Symposium on the International Protection of Geographical Indications, Somerset West, Cape Province, South Africa, September 1 & 2 1999* (WIPO Pub. No. 764) 70. See also Chapter 3 in Gangjee, *Relocating the Law of Geographical Indications* (n 5).

methods of production. As a result, a special title of industrial property was introduced for a special category of IS, namely AOs. However, it was not until 1958, in the Lisbon Agreement, that AOs were defined as meaning the denomination of a country, region or locality that serves to designate a product originating therein, the quality or characteristics of which are due to the geographical environment, including both natural and human factors. Under a *sui generis* system, only such products which originate from a specific area and which owe their specific quality to their place of origin are protected under the special title. The protection is based on a legislative or administrative act, or can be the result of a judicial or administrative decision, or of registration, in which important parameters, such as the production standards and the demarcation of the geographical area, are defined. Such systems may also comprise a mechanism of quality controls. *Sui generis* systems exist in respect of AOs and/or GIs in various jurisdictions.

As mentioned above, *sui generis* systems were created because of the insufficiency of the protection against false or deceptive use, as available in respect of IS. Consequently, these systems specify protection against unfair competition more broadly, that is, not only in respect of use of the GI or AO as such, but also in respect of use thereof in combination with delocalizing qualifiers, such as 'kind', 'type', 'style', 'imitation', or the like. There are also advantages when it comes to the enforcement of rights. Proceedings are usually instigated by public law (that is state) bodies, such as fair trading bodies, authorities entrusted with the administration of protected AOs or consumer protection bodies. Often, authorities competent for the enforcement of fiscal laws play an important role in prosecuting the fraudulent use of protected AOs. In some jurisdictions, such *sui generis* systems are only available for the protection of GIs identifying specific products, such as, for example, wines, spirits or agricultural products and foodstuffs.

3.2 Registration of Collective Marks or Certification Marks

In some countries, titles of protection for GIs and AOs are available under trade mark law, particularly in the form of collective marks or certification marks (in some countries, guarantee marks).[35] It should be noted that it is difficult to register a geographical term as an individual trade mark, since marks must not be descriptive (describing geographical origin, as

---

[35] WIPO (SCT/8/5) (n 9) paras 20–22; also Chapter C. III(a) (for a helpful summary of such marks).

opposed to specific trade origin) or deceptive (falsely suggesting geographical origin). Geographical signs are exempted from this general principle if they have acquired distinctive character through use (for example Singapore Airlines), or their use is fanciful (for example Amazon. com) and therefore not deceiving as to the origin of the goods on which the trade marks are used. The concepts of collective mark and certification mark differ from country to country. Depending on the applicable national or regional law,[36] a collective mark or certification mark may serve to indicate, *inter alia*, the origin of goods or services, and may therefore to some extent be suitable for the protection of GIs and AOs.

Use of a collective mark is only allowed to the members of a collective body. Such a body can be an association or cooperative of manufacturers, producers or traders. The collective mark is owned by the association which controls use of the mark, but is not allowed to use the mark itself. The association will grant the right to use the mark exclusively to its members and has the right to prohibit its use by persons who are not members of the association. Use of the collective mark is governed by regulations which, normally, have to be submitted to the industrial property office together with the application for registration. The question as to whether a GI or AO can be registered as a collective mark depends entirely on a given national or regional law.

Certification marks and guarantee marks are owned by a certification authority, for example, a local government agency or another body which is not engaged in the production or the trade of the products concerned, but has to ensure that the goods bearing the certification mark possess the certified qualities. A certification mark may be used to certify, *inter alia*, the origin of products or services. The application for the registration of a certification mark has to be accompanied by regulations which govern the use of the certification mark. Regarding the registrability of GIs or AOs as certification marks or guarantee marks, the same applies as mentioned above in respect of collective marks.

Where a GI or AO has been registered as a certification mark or guarantee mark, it may normally be used by anyone whose products comply with the requirements set out in the regulation.[37] The institution which owns the registered certification mark or guarantee mark has the right to prohibit the use of that mark by persons whose products do not comply with

---

[36] A helpful overview is found in WIPO, 'Technical and Procedural Aspects Relating to the Registration of Certification and Collective Marks' 15 February 2010 (SCT/23/3)

[37] Ibid., para 19.

108 *Research handbook on intellectual property and GIs*

the requirements set out in the regulations. The protection of GIs and AOs through collective marks or certification marks is enforced under general trade mark law. In other words, protection is accorded against unauthorized commercial use by third parties, which would result in a likelihood of confusion, which is the usual test for trade mark infringement. Depending on the circumstances, broader protection against 'dilution'[38] for reputed marks may be available in principle, although there do not seem to be any reported court decisions in this regard. Collective, certification and guarantee marks can be registered under the procedures of the Madrid system for the international registration of marks, administered by WIPO.[39] This is explicitly stipulated in the Common Regulations under the Madrid Agreement and Protocol.[40]

3.3 Other Laws – Administrative Schemes; Business Practices

3.3.1 Administrative schemes

Where a GI or AO is used on goods the marketing of which is subject to an administrative approval procedure, this procedure may be applied also for controlling its use in respect of such goods. The example that comes readily to mind is wines and spirits, the sale of which is regulated in many countries.[41]

Under an administrative approval procedure for product labels, the authority entrusted with the administration of that scheme controls whether the product for which marketing authorization is sought complies with relevant legal requirements, including the permitted use of a GI or AO on the label of the product. If the requirements for approval are not met, for example, because the use of a given GI or AO on the product is not permitted, marketing approval is not granted.

Administrative schemes usually form part of the broader framework of laws preventing unfair trading, so their implementation varies across jurisdictions. As far as GIs and AOs are concerned, some of these laws do

---

[38] For a discussion of 'dilution', see Chapter 7 by Handler in this volume.

[39] The Madrid system for international trade mark registration administered by WIPO is adhered to by 93 States and the European Union and OAPI (as of April 2015). It offers applicants the possibility of having a mark protected by filing one application, in one language (English, French or Spanish), with one set of fees, in one currency (Swiss Francs). International protection is based on a prior national or regional (e.g. EU) application or registration and the mark is then protected in each of the designated contracting parties. For details, see: http://www.wipo.int/Madrid/

[40] More specifically, Rule 9(4)(a)(x).

[41] WIPO (SCT/8/4) (n 9) paras 34–36.

not enable the holders of the collective goodwill attached to a GI or AO to take individual action in order to protect that goodwill. Rather, they provide for an administrative mechanism aiming at preventing misleading use of GIs or AOs on products. Where such misleading use occurs despite the administrative procedures to be followed, usually administrative or criminal sanctions will apply.

3.3.2 Laws focusing on business practices, including unfair competition

Apart from the benefits derived from specific industrial property titles of protection, as mentioned above, GIs and AOs are also protected through certain laws which focus on business practices, such as laws relating to the repression of unfair competition, consumer protection laws, or laws concerning the labelling of products. Those laws do not create an individual industrial property right over the GI or the AO, but indirectly afford protection, namely in so far as they prohibit certain acts or conduct which may involve unauthorized use of a GI or AO. The use of a certain GI or AO in respect of goods or services not originating from the respective area may be misleading and may thus deceive consumers. Furthermore, such use may constitute a misappropriation of the goodwill of the person who is entitled to use the GI or AO. Such misleading practices can be the subject of an action for unfair competition.

All countries apply norms and standards providing commercial operators with protection against unfair business practices that might be employed by competitors.[42] A distinction can be made between norms and standards dealing with unfair business practices focusing on the unfairness of such practices vis-à-vis competitors, and norms and standards focusing on the misleading of consumers. Article 10*bis*(3) of the Paris Convention gives the following examples of acts of unfair competition and requires their prohibition by its Member States:

(i) all acts of such a nature as to create confusion by any means whatever with the establishment, the goods, or the industrial or commercial activities, of a competitor;
(ii) false allegations in the course of trade of such a nature as to discredit the establishment, the goods, or the industrial or commercial activities, of a competitor; and
(iii) indications or allegations the use of which in the course of trade is liable to mislead the public as to the nature, the manufacturing process, the characteristics, the suitability for their purpose, or the quantity, of the goods.

---

[42] Bodenhausen (n 20) 145–146; WTO (IP/C/W/253 Rev.1) (n 29), Table I, footnote P.

It is evident that while Art 10*bis*(3)(ii) is applicable to competitive relationships (that is between competitors), (iii) is more broadly drafted with consumer protection interests in mind. It applies in situations where a trader makes misleading allegations about its own goods to the public. Another point to note is that (iii) does not apply to misleading allegations concerning the commercial or geographical origin of products. As to such acts, in so far as they are not covered by Art 10 of Paris, national legal systems will determine whether they are acts contrary to honest practices in industrial or commercial matters and for that reason constitute acts of unfair competition.[43] A similar distinction between protecting competitor and consumer interests is found in Art 22.2(a) and (b) of TRIPS.

Although the conditions for a successful action for unfair competition vary from country to country, the following basic principles appear to be generally recognized.[44] In order to be protectable through an action for unfair competition, a given GI or AO must have acquired a certain reputation or goodwill. Such an action further requires that the use of the GI or AO in respect of goods or services not originating from the respective geographical area is misleading, so that consumers are deceived as to the true place of origin of the goods or services. Under some national laws, proof of damages or the likelihood of damages caused by such misleading practices is required. In common law jurisdictions, the specific tort of passing off is the primary means of preventing unauthorized GI or AO use. A plaintiff must establish that goodwill or reputation is attached to the goods on which the GI or AO is regularly used and which are supplied by him; that the defendant misrepresents to the public that the goods offered by him originate from the plaintiff; and that he is likely to suffer damage from such a misrepresentation.

Whereas the principle that misleading use of a GI or AO may give rise to an action for unfair competition is generally recognized, the outcome of such an action is uncertain. In particular, the extent to which the GI or AO in question must have acquired a reputation may vary from country to country. It may be required that the GI must have been used in the course of trade for a certain time and that an association between the GI and the place of origin of the products and services must have been created amongst the relevant trade and consumer circles. Moreover, protection accorded to a GI or AO following a lawsuit based on passing off

---

[43] Bodenhausen (n 20) 146.
[44] For an overview of such general principles, see WIPO, *Protection against Unfair Competition: Analysis of the Present World Situation* (WIPO Publication No 725(E), Geneva 1994).

or unfair competition is only effective between the parties involved in the proceedings, and the entitlement to protection of a given GI or AO must be demonstrated every time enforcement of the protection of that GI is being sought.

4. ELIGIBILITY CRITERIA; RELATIONSHIP WITH THE GEOGRAPHICAL AREA

In previous studies, WIPO has investigated the concepts of quality,[45] reputation[46] and other characteristics,[47] as contained in definitions laying down eligibility criteria for GIs and AOs, and their assessment in jurisdictions employing them. It also provides an explanation of the correlation between one or more elements of such a definition and the demarcated area where the product originates.

In terms of the nature of the link between product and place, Art 2(1) of the Lisbon Agreement requires the quality or characteristics of the product to be 'exclusively or essentially due to the geographical environment, including natural and human factors'. Art 22.1 of the TRIPS Agreement requires a given quality, reputation or other characteristic of the good to be 'essentially attributable to its geographical origin'. Both the terms – 'due to' and 'attributable to' – require evidence that the product's typicality results from its geographical origin. The two provisions differ as to the nature of the evidence. Under Art 2(1) of Lisbon, evidence should be available consisting of elements relating to the geographical environment of the production area. Art 22.1 of the TRIPS Agreement is much less precise in that regard, as it just calls for evidence showing that the typicality of the product is somehow related to its geographical area of origin. However, under both provisions, the evidence is crucial for establishing the boundaries of the production area on the basis of criteria that can be verified objectively. Obviously, the producers of the product are best placed to provide the evidence. They should, however, make sure that the evidence can be verified objectively, so that, if it is questioned that the typicality of the product results from its geographical environment (Lisbon) or its geographical origin (TRIPS), the issue can be dealt with on the basis of the evidence provided. The evidence may also serve as proof in infringement proceedings, in particular evidence allowing for the

---

[45] WIPO (SCT/10/4) (n 11) paras 10–13 and 27–30.
[46] Ibid., paras 23–26.
[47] Ibid., paras 31–36.

product's traceability. Traceability is a concept that refers to the system used to monitor the product from the production to the marketing stage.

As mentioned above, under *sui generis* systems, the protection is based on a legislative or administrative act, or can be the result of a judicial or administrative decision, or of registration, in which important parameters, such as the production standards and the demarcation of the geographical area, are defined. In order to ensure that the products using the GI or AO possess the specified qualities, a *sui generis* system may also comprise a control mechanism under which the competent authorities regularly carry out quality controls. Other protection systems employed in respect of GIs or AOs use other criteria to prove the required link between the geographical area of production and the typicality of the product. Systems exist requiring, for example, just evidence showing that the geographical area concerned differs in geological terms from the surrounding areas. Certain examples from national and regional systems will help illustrate these different approaches.

In the European Union, four different regional *sui generis* systems apply, that is in respect of wines, spirits, aromatized wines, and agricultural products and foodstuffs.[48] In addition, collective, guarantee or certification marks are being used by producers of GIs and AOs from EU Member States as a complementary tool in order to ensure additional protection in respect of elements such as logos, symbols or other figurative elements that, while containing a GI or AO, cannot be registered as GIs or AOs. Moreover, producers use these marks to inform consumers of their affiliation to a specific producer association or consortium.[49] While

---

[48] See respectively Council Regulation (EC) No 479/2008 of 29 April 2008 on the Common Organisation of the Market in Wine [2008] OJ L148/1; Council Regulation (EC) No 110/2008 of 15 January 2008 on the Definition, Description, Presentation, Labelling and the Protection of Geographical Indications of Spirit Drinks [2008] OJ L39/16; Regulation (EU) No 251/2014 of the European Parliament and of the Council of 26 February 2014 on the definition, description, presentation, labelling and the protection of geographical indications of aromatised wine products [2014] OJ L84/14; Regulation No 1151/2012 of the European Parliament and of the Council of 21 November 2012 on Quality Schemes for Agricultural Products and Foodstuffs [2012] OJ L323/1. This replaced Council Regulation (EC) No 510/2006 of 20 March 2006 on the Protection of Geographical Indications and Designations of Origin for Agricultural Products and Foodstuffs [2006] OJ L93/12.

[49] For example, Roquefort cheese benefits from both an EU designation of origin (PDO) and a Community trade mark that includes the designation of origin (OHIM Registration No. 001514124). Likewise, Parmigiano Reggiano benefits from an EU designation of origin (PDO) and has five related Community collective marks (OHIM Registrations Nos. 001126481, 005882394, 005882444, 005882469, 006103899). See generally, WIPO (LI/WG/DEV/2/3) (n 16) paras 80–85.

the Community Trade Mark Regulation[50] provides for the possibility of registering a Community collective mark which may contain signs or indications with a geographical connotation, the Regulation in question does not define GIs or AOs and, consequently, does not provide for titles of protection necessarily indicating a link between quality or characteristics and geographical origin.

Chinese trade mark law provides for the registration of GIs as collective or certification marks while requiring the applicant to present information containing evidence of: (i) the quality, reputation or any other characteristic of the goods indicated by the GI; (ii) the correlation between the given quality, reputation or any other characteristic of the goods and the natural and human factors of the region indicated by the GI; and (iii) the boundary of the region indicated by the GI.[51] Thus, the definition under Art 16 of the Chinese Trade Mark Law has been implemented, which defines a GI as a means to indicate the origin of goods the special qualities, credibility or other characteristics of which are primarily determined by the natural factors or other humanistic factors of the place indicated.[52]

In the United States, an application for the registration of a certification mark must contain a so-called 'certification statement', which may consist of, for example, a particular regional origin of the goods. A concrete example for a certification statement would be the following: 'The mark certifies that the cheese is blue moulded or white cheese produced within the county boundaries of Leicestershire, Derbyshire and Nottinghamshire, England, with no applied pressure, forming its own crust of coat and made in cylindrical form, from full cream milk produced by English dairy herds.'[53]

5. MODERNIZATION OF THE LISBON SYSTEM

5.1 The Lisbon Agreement

Historically, *sui generis* GI protection or administrative schemes have been associated with a registration-based form of protection, which

---

[50] Council Regulation (EC) No. 207/2009 on the Community Trade Mark [2009] OJ L 78/1.
[51] Trade Marks (Collective and Certification Marks) Regulation 2003, Art 7.
[52] For details, see Chapter 12 by Zheng in this volume.
[53] US Certification mark for Stilton cheese, Registration Number 0921358. However, the mark appears to have expired now.

114   *Research handbook on intellectual property and GIs*

delineates entitlements and provides greater assurances to rights-holders.[54] One of the results of the Lisbon Diplomatic Conference of 1958, which had attempted, *inter alia*, to improve the international protection for GIs within the framework of the Paris Convention and the Madrid Agreement on Indications of Source, was the adoption of the Lisbon Agreement for the Protection of Appellations of Origin and their International Registration.[55]

As WIPO Director General Francis Gurry noted in 2008, at the commemoration ceremony of the Lisbon Agreement's 50th Anniversary, its conclusion back in 1958 was a remarkable achievement.[56] True, the Agreement has been heavily criticized, but, even among the traditional AO countries, it had, for many years, not been possible to find common ground on the establishment of an international registration system for AOs. In Lisbon, in 1958, negotiators managed to agree on a definition that could be invoked for the purposes of international registration, without prejudicing a national definition, whether broader or more precise in scope.[57] They also reached agreement on the level of protection that Member States should provide in respect of international registrations and, moreover, they found a way to lay down a large degree of flexibility in the provisions of the Agreement without impinging on the effectiveness of the protection to be accorded to international registrations.

Nevertheless, some of its provisions prevented several countries from acceding to the Lisbon Agreement and, as a result, the Agreement today still has a limited membership. In this regard, in particular, Article 5(6) on prior use and Article 6 shielding AOs against becoming a generic indication could be mentioned. It should also be said, however, that, in the decades after the Agreement came into force, international awareness about the flexibility of its provisions diminished so much that it almost evaporated. For example, the Agreement has become known as being restricted to AOs consisting strictly of geographical names. However, in 1970, its Member States decided, in the Lisbon Union Council, that

---

[54] A summary of existing international notification and registration systems for GIs is contained in: TRIPS Council, 'Overview of Existing International Notification and Registration Systems for Geographical Indications for Wines and Spirits' 17 November 1997 (IP/C/W/85) and 'Overview of Existing International Notification and Registration Systems for Geographical Indications Relating to Products Other Than Wines and Spirits' 2 July 1999 (IP/C/W/85/Add.1).

[55] Currently (June 2015), 28 States are party to the Lisbon Agreement (for a complete list, see http://www.wipo.int).

[56] See the WIPO home page at http://www.wipo.int under 'Director General'; 'speeches'.

[57] See Section 2.3, above.

the Agreement also allowed for the international registration of non-geographical denominations that had acquired a geographical character through use or by virtue of an administrative decision[58] and, indeed, the Lisbon Register contains several of such AOs, for example *Reblochon*, for the well-known cheese from the Savoie and the Haute Savoie, or *Bons Bois*, for one of the spirit drinks from the Cognac region. At the same time, the Lisbon Union Council accepted[59] that the Lisbon Agreement did not rule out the international registration of AOs consisting of a geographical denomination accompanied by the name of the product, for example *Olives de Nyons*, *Prosciutto di Parma* or *Café Chiapas*.

Furthermore, the right of Lisbon Member States to refuse to protect an internationally registered AO in their territory has apparently required clarification. Once the competent authority of a Lisbon Member State has received notice from WIPO of the international registration of an appellation of origin, that competent authority has the right to submit a declaration of refusal, which has to meet two basic requirements:[60]

- a time requirement: the declaration of refusal has to be notified to the International Bureau within a period of one year from the date of receipt by that country of the notice of registration; and
- a requirement regarding content: the declaration of refusal has to specify the grounds on which the refusal is based – for instance, a country may refuse to protect an AO because it considers that the AO has already acquired a generic character in its territory in relation to the product to which it refers; or because it considers that the geographical denomination is not in conformity with the definition of an AO in the Lisbon Agreement; or because the AO would conflict with a trade mark or other right already protected in the country concerned.

Such refusals are not necessarily cast in stone. If a contracting country that has issued a refusal subsequently decides to withdraw the declaration of refusal, procedures are foreseen under the Lisbon system for having such withdrawals recorded in the International Register. On the other hand, however, if no declaration of refusal is submitted but the effects, in a given Lisbon Member State, of an international registration are, subsequently,

---

[58] Lisbon Council, 'Report of the Fifth Session' September 1970 (AO/V/8), at para 19.
[59] Ibid., para 18.
[60] Art 5(3) of the Lisbon Agreement and Rule 9 of its Regulations.

invalidated by a court and the invalidation is no longer subject to appeal, the competent authority of the country concerned is to notify the International Bureau accordingly. Following such a notification, the International Bureau enters the invalidation of the effects of the international registration in that Lisbon Member State in the International Register.

Interestingly, the membership of the Lisbon Agreement started growing again after the conclusion of the TRIPS Agreement. While there had been no new accession to the Agreement between the mid-1970s and the mid-1990s, from the mid-1990s its membership has increased from 16 to 28 States. Also, about 20 per cent of the total number of Lisbon registrations have been effected only since then.

5.2 Review of the Lisbon System

Similar to the Madrid and Hague systems (concerning, respectively, the international registration of trade marks and industrial designs), the Lisbon system facilitates the registration of industrial property rights at the international level on the basis of provisions laying down the procedural rules governing the international registration procedure. However, unlike Madrid and Hague, the Lisbon Agreement also lays down a number of provisions regarding the protection to be accorded, at a minimum, to internationally registered AOs. Thus, Art 3 specifies that the Member States are to protect AOs registered at the International Bureau against any usurpation or imitation of the AO, even if the true origin of the product is stated or if the appellation is used in translated form or accompanied by terms such as 'kind', 'type', 'make', 'imitation' or the like.

Another parallel with the Madrid and Hague systems is that also the Lisbon system has been the subject of attempts to adapt the system in order to allow for a wider membership. In the case of the Madrid system, this eventually led to the conclusion of the Madrid Protocol, in 1989, and in the case of the Hague system to the conclusion of the Geneva Act of the Hague Agreement, in 1999. After unsuccessful attempts between the mid-1970s and the early 1990s to revise the Lisbon system, a new review exercise started after the abovementioned 50th Anniversary of the Lisbon Agreement in 2008. A Working Group was established, in September 2008, by the Lisbon Union Assembly, which held its first session in March 2009. As a result of the recommendations agreed at that session, the Assembly extended the mandate of the Working Group, so as to allow the Working Group to engage in a full-fledged review of the Lisbon system. The objective to be pursued by the Working Group is to review the international registration system of the Lisbon Agreement so as to make the system more attractive for users and prospective new members, including the potential for accession by intergovernmental organizations, while preserving the principles and objectives of the Lisbon Agreement.

In order to assist the Working Group in this review, the International Bureau was requested to prepare two documents:

- a document reflecting the results of a *survey* on the Lisbon system among stakeholders, in the widest possible sense, that is Member State and non-Member State governments, intergovernmental organizations (IGOs), non-governmental organizations (NGOs) and interested circles;[61] and
- a document reflecting a *study* on the relationship between regional systems for the protection of GIs and the Lisbon system and the conditions for the possible accession to the Lisbon Agreement by competent IGOs.[62]

Arising from the Working Group's discussion of these documents in 2010, the International Bureau was requested to prepare, for the third session of the Working Group, draft provisions on a number of topics. These included definitions for GIs and AOs, the scope of protection for GIs and AOs, prior use, applications for GIs and AOs concerning products from trans-border areas, accession criteria for IGOs, and procedures in Contracting Parties prior to the issuance of possible refusals and for challenging refusals issued.[63]

At the closure of its third session, in May 2011, the Working Group requested the International Bureau to prepare a new draft instrument containing these provisions and incorporating not only revisions arising from the discussions but also any further draft provisions required to make the new instrument as complete as possible, while leaving open the question as to the legal means by which it might be formalized. Accordingly, the International Bureau prepared such a Draft New Instrument (DNI) as well as Draft Regulations (DR) under the DNI, along with corresponding explanatory notes,[64] which were discussed at the fourth session of the Working Group, in December 2011.

Also at its fifth session, in June 2012, the Working Group discussed revised versions of the DNI, the DR and the explanatory notes.[65] As

---

[61] WIPO (LI/WG/DEV/2/2) (n 15).
[62] WIPO (LI/WG/DEV/2/3) (n 16).
[63] WIPO (LI/WG/DEV/3/2) (n 17).
[64] WIPO documents LI/WG/DEV/4/2 to 5.
[65] Ibid.

118 *Research handbook on intellectual property and GIs*

instructed by the Working Group, the subsequent focus of the review of the Lisbon system was on developing an international registration system. This concerned not only procedures for the international registration of AOs and GIs, as is the case for the Madrid (trade marks) and Hague (designs) systems, but also provisions relating to the scope of protection for AOs and GIs. While the system would encompass two definitions, one for AOs and the other for GIs, Contracting Parties would not be obliged to provide protection on the basis of two separate definitions. However, any Contracting Party providing protection only on the basis of a GI definition would be required to protect AO registrations as GIs, except where a ground for refusal exists.

At its sixth session, the Working Group agreed on the basic approach for the level of protection to be accorded to AOs and GIs alike, even though some issues remained to be resolved as to the precise wording of the ensuing provisions.[66] The same happened at the seventh session in respect of the provisions laying down safeguards for prior trade marks and certain other rights. The provision addressing conflicts between AOs/GIs and prior trade marks was drafted on the basis of the relevant TRIPS provisions, in light of the WTO Panel Reports in the Australia-EU and US-EU disputes with regard to EC Regulation 2081/92. At the eighth session, in December 2013, the Working Group considered provisions laying down the content of protection in respect of AOs and GIs, in addition to the question of generics and questions concerning conflicts between AOs/GIs and prior trade marks or other rights. This led to a renewed debate between Lisbon Member States and non-Member States, notably Australia and the United States.[67] The debate continued during the ninth and tenth sessions in 2014.[68]

Nevertheless, agreement was reached at the Working Group's seventh session on a recommendation to the Lisbon Union Assembly aimed at the convening of a Diplomatic Conference to conclude the negotiations.[69] The Assembly decided that this Diplomatic Conference should be held in 2015.[70] At the tenth session of the Working Group, the number of issues still on the negotiating table was reduced to 17, as reflected in the Notes on the Basic Proposal for the New Act of the Lisbon Agreement on Appellations of Origin and Geographical Indications.[71] Some of these

---

[66] WIPO documents LI/WG/DEV/6/6 and 7.
[67] WIPO document LI/WG/DEV/8/7.
[68] WIPO documents LIWG/DEV/9/7 and LI/WG/DEV/10/7.
[69] WIPO documents LI/WG/DEV/7/6 and 7.
[70] WIPO document LI/A/29/2.
[71] WIPO document LI/DC/5, at [4].

related to the draft provisions on scope of protection, for example as regards the flexibilities in respect of generic elements of an AO or GI and the safeguards in respect of other rights. Others concerned, for example, certain aspects of the international registration procedures and the applicable fees.

5.3 The Geneva Act of the Lisbon Agreement

The Lisbon system review was concluded at a Diplomatic Conference held from 11 to 21 May 2015 at WIPO headquarters in Geneva. The result, the Geneva Act of the Lisbon Agreement, responds to the objective of the review by making the Lisbon system more attractive for users and prospective new Members, while preserving the principles and objectives of the Lisbon Agreement. The Geneva Act refines and modernizes the legal framework of the Lisbon system and allows for the accession by IGOs. It expands the coverage of the Lisbon system beyond AOs and will provide for an international registration system in respect of AOs as well as other GIs. The definitions for the AO and GI are modelled on those of the Lisbon Agreement and TRIPS, respectively, while containing certain clarifications, in particular: (i) that each definition can accommodate non-geographical terms if such terms are known as referring to a geographical area; and (ii) that a geographical area of origin can also consist of a trans-border geographical area, or – if the Contracting Parties where the area is located cannot find agreement on the establishment of a joint AO or GI – the part of the area situated in their own territory.

In view of the different legal systems that exist around the world for the protection of GIs, and the flexibility that the Geneva Act of the Lisbon Agreement stipulates in this regard (see Section 3, above), both the provisions on the application procedures and those on the scope of protection of AOs and GIs contain important modifications compared to those of the current Lisbon Agreement.

International applications are to be filed by the Competent Authority of the Contracting Party of Origin in the name of: (i) the beneficiaries, that is those having the right to use the AO or GI; or (ii) a natural person or legal entity having legal standing under the law of the Contracting Party of Origin to assert the rights of the beneficiaries or other rights in the AO or GI. Direct filings by those referred to under (i) or (ii) will be possible, but only if the legislation of the Contracting Party of Origin so permits and the Contracting Party of Origin has made a declaration to that effect.[72]

---

[72] Art 5(2) and Rule 5(2)(a)(iii); and Art 5(3).

A Contracting Party may notify the Director General of WIPO that, for the protection of the AO or GI in its territory, the application must be signed by a person as referred to under (ii). A Contracting Party may also notify the Director General that the application must be accompanied by a declaration of intention to use, or that the application must indicate particulars concerning, in the case of an AO, the quality or characteristics of the good, and its connection with the geographical environment of its geographical area of production, and, in the case of a GI, the quality, reputation or other characteristic of the good, and its connection with its geographical origin. If such a declaration-based requirement is not met, the International Bureau will not reject the application, but the AO or GI concerned will not be protected in a Contracting Party that has notified the requirement in question. Protection can, however, be extended to such a Contracting Party subsequently, once the requirement is met. The same will apply in respect of a Contracting Party that has notified the Director General that, for protection of the AO or GI in its territory, a fee has to be paid to cover the cost of substantive examination of the international registration by its competent authority. Such a Contracting Party may also require an administrative fee relating to the use by the beneficiaries of the AO or GI in its territory.[73]

As regards the content of protection, Art 11 of the Geneva Act of the Lisbon Agreement distinguishes between use of the AO or GI in respect of goods of the same kind and use in respect of goods that are not of the same kind, or services. Each Contracting Party has to provide the legal means to prevent use in respect of goods of the same kind not originating in the geographical area of origin or not complying with any other applicable requirements for using the AO or GI. In respect of goods that are not of the same kind, or services, the legal means have to be provided to prevent use of the AO or GI, if such use would indicate or suggest a connection between those goods or services and the beneficiaries of the appellation of origin or the geographical indication, and would be likely to damage their interests, or, where applicable, because of the reputation of the appellation of origin or geographical indication in the Contracting Party concerned, such use would be likely to impair or dilute in an unfair manner, or take unfair advantage of, that reputation. Such protection – whether in respect of goods of the same kind, not of the same kind, or services, or the registration of later trade marks – must also apply to use of the AO or GI amounting to its imitation, even if the true origin of the goods is indicated, or if the appellation of origin or the geographical indication is

---

[73] Art 7(4); Rule 5(3) and (4); Rule 6; and Rule 16(2).

used in translated form or is accompanied by terms such as 'style', 'kind', 'type', 'make', 'imitation', 'method', 'as produced in', 'like', 'similar' or the like. In addition, the legal means have to be provided to prevent any other practice liable to mislead consumers as to the true origin, provenance or nature of the goods.

Article 13 of the Geneva Act of the Lisbon Agreement provides for the necessary safeguards in respect of prior trade marks, personal names in business and rights based on a plant variety or animal breed denomination. Other safeguards are contained in Agreed Statements by the Diplomatic Conference. While Article 12 stipulates that AOs and GIs cannot be considered to have become generic in a Contracting Party, the Diplomatic Conference adopted an Agreed Statement which states that it is understood that Article 12 is without prejudice to the application of the provisions of this Act concerning prior use, as, prior to international registration, the denomination or indication constituting the appellation of origin or geographical indication may already, in whole or in part, be generic in a Contracting Party other than the Contracting Party of Origin, for example, because the denomination or indication, or part of it, is identical with a term customary in common language as the common name of a good or service in such Contracting Party, or is identical with the customary name of a grape variety in such Contracting Party. In addition, another Agreed Statement, concerning Article 11, states that it is understood that, where certain elements of the denomination or indication constituting the appellation of origin or geographical indication have a generic character in the Contracting Party of Origin, their protection shall not be required in the other Contracting Parties. For greater certainty, a refusal or invalidation of a trade mark, or a finding of infringement, in the Contracting Parties under the terms of Article 11 cannot be based on the component that has a generic character.

6. CONCLUSION

The Working Group has consistently worked towards a refinement and modernization of the legal framework of the system established under the Lisbon Agreement, while building on its principles and objectives and in the light of the relevant provisions of the TRIPS Agreement. It has also introduced provisions for the possible accession by intergovernmental organizations administering a regional system for the protection of AOs and/or GIs.

Despite the relatively limited participation by both Lisbon Member States and non-Lisbon Member States in the ten sessions of the Working

Group, the Diplomatic Conference that adopted the Geneva Act of the Lisbon Agreement on 20 May 2015 was attended by all 28 Lisbon Member States and 89 non-Lisbon WIPO Member States plus representatives of several IGOs and NGOs, including, in particular, the European Union and OAPI. As a result, we may possibly witness, in the not too distant future, the coming into existence of a truly international register for AOs/GIs.

6. Geographical Indications under TRIPS
*Daniel Gervais**

1. INTRODUCTION

The TRIPS Agreement signed in April 1994[1] imposed on all Members of the World Trade Organization (WTO) (a) an obligation to protect Geographical Indications (GIs) against deceptive uses, and (b) an obligation to protect GIs used in connection with wines and spirits at a higher level.[2] In addition, TRIPS Art 23.4 foresees the establishment of a multilateral register for GIs on wines (not spirits) and mandates negotiations to that end, a part of the Agreement's 'built-in agenda'.[3] Despite the seemingly arcane character of this debate, it is far from a minor issue. Billions of trade dollars are at stake. Not surprisingly, it has taken centre-stage in the Doha Round.[4] Indeed, the establishment

---

* I wish to express my gratitude to all those who provided me with comments on earlier drafts. Special thanks to Mr Matthijs Geuze, Head of the Lisbon Registry in the Brands and Designs Sector, WIPO; to the editor, Professor Dev Gangjee; and to Amy Cotton (USPTO). Naturally, I take full responsibility for errors and omissions, and the views expressed.

[1] *Agreement on Trade-Related Aspects of Intellectual Property Rights*, 15 April 1994, Marrakesh Agreement establishing the World Trade Organization, Annex 1C, *Results of the Uruguay Round of Multilateral Trade Negotiations: The Legal Texts*, 1869 UNTS 299, 33 ILM 1125, 1197 [hereafter, TRIPS]. As of 26 June 2014, 160 countries and territories are Members.

[2] The latter level resembles the protection against 'usurpation and imitation' contained in the 1958 Lisbon Agreement, which prohibits the use of a protected GI even where there is no actual or potential consumer deception. See Lisbon Agreement for the Protection of Appellations of Origin and their International Registration, 31 October 1958, as revised at Stockholm on 14 July 1967, as amended on 28 September 1979, 923 U.N.T.S. 205 [hereafter, Lisbon Agreement]. A new Act was adopted in Geneva in 2015. See note 30 below.

[3] The 'built-in agenda' is a list of issues on which no consensus could be reached at the end of the Uruguay Round and on which parties agreed to continue their discussions after the entry into force of TRIPS. See Daniel Gervais, *The TRIPS Agreement: Drafting History and Analysis* (3rd edn, Sweet & Maxwell, 2008) 29; and Jayashree Watal, *Intellectual Property Rights in the WTO and Developing Countries* (Springer, 2001) 263–265.

[4] Reference is made here to the WTO Doha Development Round of Multilateral Trade Negotiations which was launched in Doha (Qatar) in 2001.

123

of a multilateral register and/or notification system of GIs as foreseen in TRIPS was portrayed as an essential ingredient of a successful outcome in the (ongoing) Doha Round. In an effort to move the international registration debate forward, a detailed set of proposals by WTO members was consolidated into a composite draft text by the Secretariat in April 2011.[5] In the more recent past, it was one of the obstacles to completing negotiations on the Transpacific Partnership (TPP). It may be even harder to solve in the context of the Transatlantic Trade and Investment Partnership (TTIP) between the European Union and the United States.

Traditionally, those demanding higher levels of protection for GIs have been European negotiators, but they are not – or no longer – alone. Several other WTO Members, many of them developing countries, now insist on an extension of the higher level of protection contained in Art 23 to products other than wines and spirits, a debate usually referred to as the 'extension issue'.[6] They consider the current emphasis on, and higher level of protection of, alcoholic beverages to be both culturally discriminatory and a commercial impediment to their ability to collect higher rents that would be associated with GIs on other products. In fact, several members of that group are rich in products associated with traditional knowledge and see the extension of higher levels of protection (relatively few developing countries are well known as producers and exporters of products of the vine)[7] as a way to repair historical wrongs,[8] and more broadly as a way to 'de-Westernize' intellectual property rules, which some consider systematically discriminatory because they favor Western methods of creation, invention, marketing and production.[9] Several of those countries see GIs as a way of protecting but also of globally marketing rural and traditional

---

[5] See 'Multilateral system of notification and registration of geographical indications for wines and spirits', Report by the Chairman to the Trade Negotiations Committee, document TN/IP/21 of 21 April 2011 [hereinafter the Composite Text]. For detailed analysis, see also Chapter 8 by Martín in this volume.

[6] See Gervais (n 3) 84. Additionally, see Chapter 7 by Handler in this volume.

[7] There are significant exceptions of course. As of 2008, there were four among the top 10 wine producing countries: Argentina, Chile, China and South Africa. These are of course advanced developing economies. See http://www.wineinstitute.org/files/WorldWineProductionbyCountry.pdf

[8] See David Vivas-Eugui, 'Negotiations on Geographical Indications in the TRIPS Council and Their Effect on the WTO Agricultural Negotiations: Implications for Developing Countries and the Case of Venezuela' (2001) 4(5) JWIP 703. See also Gervais (n 3) 104–107.

[9] 'Western' is used here not as a geographical reference but in its usual meaning as a reference to the most industrialized nations.

products at a higher price, which they assert would lead to 'development from within', that is, 'an alternative development strategy that prioritizes local autonomy and broad, community-wide development goals'.[10] As Min-Chiuan Wang noted:

> In sharp contrast to the consistency of European countries' attitude toward geographical indication protection, Asian countries have taken different stands in this divide. Some of them have actively pursued the expansion of 'enhanced' protection to cover products beyond wines and spirits. Countries taking this position include India, Pakistan, Thailand, and Sri Lanka. Their position has converged with that of the European Union: belonging to the group of 'Friends of GIs,' these countries have struck a bargain with the European Union, which wishes to put into practice a compulsory system of registration and notification.[11]

For example, among the denominations currently on the Lisbon register for products other than wines and spirits, one finds crafts and coffee from Mexico, one of the few developing nations to have made more than token use of the Lisbon system.[12] While not (yet) a Lisbon member, India has also indicated a willingness to develop its GI potential, notably for tea and rice.[13] Those types of product denominations are still the exception, however. Of the active entries on the Lisbon register as of 2009, almost 75 percent were for wines, spirits and beer.[14] Another striking aspect of the register as it stands now is the geographical concentration of the entries: 11 countries hold 97.5 percent of all entries, and in fact the top three hold over 78 percent, with one country, France, holding 62.5 percent of the

---

[10] See Sarah Bowen, 'Development from Within? The Potential for Geographical Indications in the Global South' (2009) 13(2) JWIP 231. See also Chapter 15 by Barjolle in this volume.

[11] Min-Chiuan Wang, 'The Asian Consciousness and Interests in Geographical Indications' (2006) 96 TMR 906, 939–940.

[12] For example, Lisbon-protected appellations such as Talavera (No. 833) for 'handcraft objects' and Ambar de Chiapas (No. 842) registered for 'semi-precious stones of vegetal origin, for its use in derivative products, namely, jewellery, art objects and religious objects'. See F Addor and A Grazioli, 'Geographical Indications beyond Wines and Spirits: A Roadmap for a Better Protection for Geographical Indications in the WTO/TRIPS Agreement' (2002) 5(6) JWIP 865; SK Soam, 'Analysis of Prospective Geographical Indications of India' (2004) 8(5) JWIP 679, 684.

[13] See K Das, 'International Protection of India's Geographical Indications with Special Reference to "Darjeeling" Tea' (2006) 9(5) JWIP 459; Soam (n 12); see additionally Chapter 16 by Rangnekar and Mukhopadhyay in this volume.

[14] The Lisbon Express Registry database is at: www.wipo.int/ipdl/en/search/lisbon/search-struct.jsp. See also Chapter 5 by Geuze in this volume.

total (almost 90 percent of which are for wines and spirits). This seems to reinforce the claims about de-Westernization and the flaws of the current system's emphasis on specific products. Indeed, there are no developing countries in the top five holders of Lisbon registrations.[15]

GI protection may impact the future of global food consumption patterns and lead to shifts in agricultural models.[16] As such, beyond their importance as calculated in trade terms, GIs also have environmental significance as a part of agricultural and food policy.[17] The debate has thus captured the imagination of a number of consumer groups, who insist on proper labeling of products to clarify their origin, partly, it seems, so that they may buy more locally produced products and reduce the carbon footprint of their consumption patterns, but also because of the 'quality assurance factor'[18] associated with certain GIs.[19] Whether as cause or effect, GIs are progressing, despite claims that they mostly lead to higher prices without corresponding tangible benefits.[20]

Opponents of GI protection object that the higher price resulting from a GI is demonstrably irrational because no measurable objective quality difference exists between a GI product and its non-GI equivalent. However, this argument can easily be refuted, unless one is prepared to throw trade marks under the bus with GIs. Indeed, trade marks also perform an 'irrational' yet well accepted function in guiding behavior. As McCarthy noted:

> An economist who draws up a set of criteria for market analysis finds that conclusions flow from the criteria set up. If price, quality, and rationality are the only criteria of an economic system, then emotional consumer choices do not fit into this economic model. Advertising investment in promoting such choices are [sic] then regarded as wasteful and non-productive. The problem is that human beings, not economists' symbols, purchase products. Moreover, as noted earlier, modern economic analysis teaches that brand loyalty is not

---

[15] The first developing country (Cuba) is ranked in eighth place with 2.3 percent of entries, mostly for cigars.

[16] See Chapter 3 by Barham in this volume; Bowen (n 10) 2.

[17] See William van Caenegem, 'Registered GIs: Intellectual Property, Agricultural Policy and International Trade' (2004) 26 EIPR 170, 171.

[18] See Alberto Francisco Ribeiro de Almeida, 'Key Differences between Trade Marks and Geographical Indications' (2008) 30 EIPR 406.

[19] See David Goodman, 'Rural Europe Redux? Reflections on Alternative Agro-Food Networks and Paradigm Change' (2004) 44(1) Sociologia Ruralis 3, 5.

[20] Nevertheless, certain commentators believe that GIs might *increase* competition. See Massimo Vittori, 'The International Debate on Geographical Indications (GIs): The Point of View of the Global Coalition of GI Producers' (2010) 13(2) JWIP 304.

irrational consumer behavior. It is a common sense, rational method of reducing shopping or 'search' costs.

Additionally, who can agree on a definition of 'irrationality' when it comes to buying goods? Where is this buyer who only buys goods on the basis of price and quality alone, eschewing all feelings and emotional impulses? He or she sounds like quite a dull person.[21]

There are of course other arguments against GI protection and I discuss them in the rest of the chapter. To explicate the global debate surrounding GIs, I examine, first, the emergence of GIs during the Uruguay Round and the culmination of that debate, Articles 22, 23 and 24 of the TRIPS Agreement. I then consider the work on the issue of GIs in the TRIPS Council and other WTO bodies since the adoption of TRIPS as part of the Uruguay Round package. I conclude by considering the interface between the new TRIPS system and the Lisbon Agreement.

2. GEOGRAPHICAL INDICATIONS IN THE TRIPS AGREEMENT

2.1 The Emergence of GIs in TRIPS

The emergence of the GI section in Part II of the TRIPS Agreement, namely Articles 22 to 24, can be traced to a proposal by the European Communities.[22] That text reflected the substantive level of protection of the Lisbon Agreement (*inter alia*, against any usurpation, imitation or evocation, even where the true origin of the product is indicated or the appellation or designation is used in translation or accompanied by expressions such as 'kind', 'type', 'style', 'imitation' or the like),[23] and also the following provision on registration:

> In order to facilitate the protection of geographical indications including appellations of origin, an international register for protected indications shall be established. In appropriate cases the use of documents certifying the right to use the relevant geographical indication should be provided for.[24]

---

[21] *McCarthy on Trademarks and Unfair Competition* (4th edn, West, looseleaf, 2011) vol. 1, para 2:38.

[22] See GATT Doc., 'Draft Agreement on Trade Related Aspects of Intellectual Property Rights' 29 March 1990 (MTN.GNG/NG11/W/68). During the TRIPS negotiation the European negotiators were representing the European Communities. Since the Maastricht Treaty (1993), the corresponding entity is known as the European Union.

[23] Ibid., Art 20(1).

[24] Ibid., Art 21(3).

An Australian text on geographical indications, tabled a few weeks after the EC proposal,[25] provided for protection by requiring parties to refuse registration or invalidate a trade mark suggesting the territory or part thereof of a party with respect to goods not originating in that territory, *where this could mislead or confuse the public*, and by prohibiting the use of such an indication. However, the draft provisions did not apply:

(a) to the prejudice of holders of rights relating to an indication identical with or similar to a geographical indication or name and used or filed in good faith before the date of entry into force of this [amendment] [Annex] in the contracting party;
(b) with regard to goods for which the geographical indication or name is in the common language the common name of goods in the territory of that contracting party, or is identical with a term customary in common language.

Looking at the two proposals, one sees a reflection of an Old World attachment to the protection of the designation of the geographical origin of certain products as providing rights higher in status than ordinary trade marks, contrasted with a New World view anchored in the reality of commerce and trade mark law, which espouses a 'first-in-time, first-in-right' (FITFIR) approach. The two main issues for negotiation were clearly delineated:

- Should GIs be protected at a *higher level than trade marks* (without the requirement for confusion/deception) and what then would be the status of existing marks?
- Should a *register* for GIs be established, and if so, where and how?

The draft TRIPS text produced under the responsibility of GATT Director General Arthur Dunkel in 1991 – and included in the second Draft Final Act Embodying the Results of the Uruguay Round of Multilateral Trade Negotiations – split the difference on the first issue by providing special protection (under geographical indications) only for wines and spirits.[26] On the second issue, negotiators agreed to disagree and included a provision that would obligate them to negotiate 'concerning the establishment of a multilateral system of notification and registration

---

[25] GATT Doc. 'Draft text on Geographical Indications, Communication from Australia' 13 June 1990 (MTN.GNG/NG11/W/75).
[26] GATT Doc. (MTN.TNC/W/FA). There were few changes between that text and the final (adopted) one.

of geographical indications for wines eligible for protection in those Members participating in the system'.[27]

The solution adopted in the end by the negotiators in TRIPS left unanswered, first, the nature of the difference, if any, between the notion of Geographical Indications, on the one hand, and the pre-existing notion used in the Lisbon of Appellation of Origin (AO), on the other hand. This potential difference may matter because TRIPS was the first multilateral text to use the notion of GI. The change reflects the skepticism of a number of New World countries, including the United States, vis-à-vis the Lisbon Agreement. Let us see the actual differences between the two notions.

2.2 GI or AO: The TRIPS 'Solution'

While the definition of GI contained in Art 22.1 of TRIPS resembles Art 2 of the Lisbon Agreement, it also differs on a number of points: (a) AOs under the Lisbon Agreement are *geographical names* of a country, region, or locality, while GIs under TRIPS are *any indication* pointing to a given country, region or locality, but not necessarily limited to the *name* of a country, region or locality – what matters is the indication that the good originates in the territory of a member or a part thereof;[28] (b) AOs under Lisbon designate a *product*, while a GI under TRIPS identifies a *good*, the term traditionally used in GATT/WTO contexts to differentiate goods from services (that is, the definition does not include services); (c) finally, AOs refer to a *geographical environment*, including natural and human factors,[29] while TRIPS uses a more general concept of *geographical origin*. Those differences are not huge. Indeed, the negotiating history of the Lisbon Agreement demonstrates that the Lisbon definition was understood to provide considerable flexibility to Member States. More importantly, the Geneva Act of the Lisbon Agreement (2015) refers to both AOs and GIs and no longer states that AOs must be protected 'as such', which

---

[27] TRIPS, Art 23.4.
[28] Lisbon is not restricted to 'geographical names'. Although the English translation of '*dénomination*' into 'name' seems to be referring only to names that are *stricto sensu* geographical, denominations that are not geographical (for example 'Reblochon' for cheese) can also be registered under Lisbon. This was made clear by the Lisbon Union Council (the predecessor of the Lisbon Union Assembly) in 1970.
[29] For examples of how far this could be applied, see F Gevers, 'Geographical Names and Signs Used as Trade Marks' (1990) 8 EIPR 285, and by the same author, 'The Future Possibilities of International Protection for Geographical Indications' [1991] Industrial Property 154.

strongly suggests that the terminological divide between AOs and GIs can be bridged.[30]

2.3 Substantive Protection for GIs in TRIPS

The main difference between the approaches of different groups of industrialized countries during the TRIPS negotiations was not semantic. It resided in the appropriate standard for protection. Some believed that the use of an indication must constitute *unfair competition* to give rise to a remedy while for others an element of *deception* (that is, evidence that the public is misled) was necessary.[31] The solution found in TRIPS was to include aspects of both approaches. The general rule (for goods other than wines and spirits)[32] is that the use of indications must be liable to mislead the public.[33] For the higher protection accorded wines and spirits, there is no need to prove this element of deception.

There was also disagreement on whether a sui generis system would be required to satisfy these international obligations. TRIPS Art 22.1(a) directs WTO Members to provide the *legal means* to prevent use of any means in the designation or presentation (packaging, pictorial representations, and so on) of a good that is false ('indicates or suggests that the good in question originates in a geographical area other than the true place of origin') or deceptive ('in a manner which misleads the public as to the geographical origin of the good'). In the same vein, WTO Members must also provide *legal means* to prevent any act of unfair competition according to Art 10*bis* of the Paris Convention, which includes a prohibition of 'indications or allegations the use of which in the course of trade is liable to mislead[34] the public as to the nature, the manufacturing process, the characteristics,

---

[30] Geneva Act of the Lisbon Agreement on Appellations of Origin and Geographical Indications and Regulations under the Geneva Act of the Lisbon Agreement on Appellations of Origin and Geographical Indications, 20 May 2015 (LI/DC/19) [hereinafter the Geneva Act].

[31] There was also of course a debate on 'generic appellations', which found its way into Art 24. See also Chapter 18 by Gangjee in this volume.

[32] Under Art 23, there is no such need to prove that the public might be misled or that the act constitutes unfair competition.

[33] In respect of indications of source, the rule there is also linked to a possible misleading effect (see Art 10*bis*(3)). Art 10*bis* of the Paris Convention is also included under Art 2.1 of TRIPS.

[34] Art 22.2(a) uses 'which misleads', which may be interpreted as requiring evidence that the public was in fact misled, though that can often be presumed from the circumstances when a false indication is used.

the suitability for their purpose, or the quantity, of the goods'.[35] Art 10*bis*(3)(iii) of the Paris Convention extends not just to indications but also to any allegation (but not to presentations) that are misleading. Art 10*bis*(3)(iii) has a partially different scope in that it applies to the 'nature, the manufacturing process, the characteristics, the suitability for their purpose, or the quantity' of the goods. Art 22.2 of TRIPS makes the application of this provision mandatory 'in respect of geographical indications'. However, one must bear in mind that Art 10*bis* of the Paris Convention was also more generally incorporated into TRIPS via Art 2.1.

Art 22.3 extends the protection of GIs squarely into the realm of trade marks. In fact, in certain jurisdictions GIs are protected under that system, not under a sui generis regime.[36] According to Art 22.3, a WTO Member must, either ex officio if its national law so permits, or at the request of an interested party, refuse or invalidate[37] the registration of a trade mark which contains or consists of a geographical indication if (a) the goods do not originate in the territory indicated, and (b) use of the indication in the trade mark for such goods in the territory of the Member concerned is (again) of such a nature as to mislead[38] the public as to the true place of origin.[39]

Art 22.4 applies to homonymous indications, that is where the territory, region or locality of a country is the same or similar to a known territory, region or locality of another country. This may happen in the case of former colonies, for example.[40] When, say, French nationals emigrated to

---

[35] At the 1958 Lisbon Conference, proposals to expressly include 'origin' under Art 10*bis* were refused. See Albrecht Conrad, 'The Protection of Geographical Indications in the TRIPS Agreement' (1996) 86 TMR 11, 36 et seq. It could be said that this gap has now been filled.

[36] For an illustration of the trade mark approach, see D Gervais, 'A Cognac after Spanish Champagne: Geographical Indications as Certification Marks' in RC Dreyfuss and JC Ginsburg (eds), *Intellectual Property at the Edge: The Contested Contours of IP* (Cambridge University Press, 2014) 130.

[37] Compare with 'cancellation' in TRIPS Arts 15 and 19.

[38] As noted above, Art 22(2)(a) uses 'which misleads', Art 10*bis*(3) of the Paris Convention uses 'is liable to mislead' and Art 22(3) uses 'is of such a nature as to mislead'. The latter two tests seem very close indeed. The likelihood that the public will be misled may, as in the case of trade marks, be inferred in appropriate circumstances.

[39] This element could exclude marks having acquired a secondary meaning, i.e. which may contain a formally geographical term such as AMAZON but over time come to be understood by the public as the trade source indicator for an online retailer. Also, the first condition must be read in conjunction with Art 22.1: territory includes a region or locality within the territory.

[40] See TP Stewart (ed.), *The GATT Uruguay Round: A Negotiating History (1986–1992)* (Kluwer Law International, 1993) 2304.

another country and founded a village or town, they may have given it the name of their village or region of origin, which may be famous for a special kind of cheese. In such a case, if the 'second' village also produced cheese under that name it could (depending on the circumstances of each case, of course) falsely represent the origin of the cheese in the minds of the average consumer, thereby free-riding on the reputation acquired by the first users of the indication. Falsehood is the required threshold here. The average consumer of the good in question must believe that the good originated in a country other than the real country of origin.

A higher level of protection for GIs used for wines and spirits was added to TRIPS.[41] Under Art 23.1, using a geographical indication identifying wines or spirits for wines and spirits not originating in the place indicated by the indication is prohibited, even where the true origin of the wines and spirits concerned is indicated and/or a translation is used and/or the indication is accompanied by expressions such as 'kind', 'type', 'style', 'imitation' or the like. There is thus no need here to show that the public might be misled or that the use constitutes an act of unfair competition.

Art 23.3 deals with the case of similar (homonymous) indications for wines and spirits, whose use is not misleading (or deceptive) under Art 22.4.[42] In such cases, both indications may be protected and WTO Members concerned must determine the practical conditions necessary to differentiate wines and spirits from both indications. In doing so, they must ensure that consumers are not misled and that the producers concerned are treated equitably. This may require creative solutions to distinguish the goods, perhaps by limiting linguistic versions where the languages spoken in the territories concerned are not the same or by stipulating labelling requirements. Finally, Art 23.4 requires that negotiations be undertaken in the TRIPS Council to establish an international notification and registration system for geographical indications for wines (not spirits).[43]

Art 23.4 is supplemented by a number of provisions in Art 24 in which WTO Members agreed to enter into negotiations aimed at increasing the protection of individual geographical indications under Art 23. It is understood that Art 24 was the result of difficult negotiations between WTO Members, including many European nations, that wished to protect indications for wines and spirits by not legitimizing 'past sins', while others

---

[41] The addition of spirits occurred at the end of the negotiations. See GATT Doc., 'Progress of Work in Negotiating Groups: Stock-Taking' 7 November 1991 (MTN.TNC/W/89/Add.1) 8.
[42] An example is 'Rioja' wine produced in both Argentina and Spain.
[43] Spirits were added by the 2001 Doha Declaration.

were concerned that it might affect acquired rights and/or uses.[44] The result of the negotiations was only partly satisfactory for both sides, in the sense that, on the one hand, protection was granted but not in a way comparable to EU regulations and without an agreement to establish a registration system comparable to that of the Lisbon Agreement, and, on the other hand, safeguards for 'acquired rights' were neither complete nor permanent. For example, Art 24.4 allows continued use of any indication identifying wines or spirits used in any form in a continuous manner *with regard to the same goods or services*[45] either (a) for ten years preceding the signing of the Marrakesh Agreement on 15 April 1994, or (b) in good faith preceding that date. In these circumstances, it may continue to be used by the same person, who must be a national or domiciliary of the country where the use took place.

Finally, under Art 24.5, a WTO Member is not obliged to protect a geographical indication against a conflicting trade mark, provided that an application for registration of the mark was filed or the mark registered, or the right acquired by use (and the trade mark used in good faith) in the WTO Member concerned either (i) before the TRIPS Agreement became applicable in the Member concerned or (ii) before the indication in question was protected in its country of origin.[46]

3. GIs IN THE DOHA ROUND

One of the three paragraphs in the Ministerial Declaration that launched the Doha Round in 2001 is entirely devoted to GIs. It reads as follows:

> With a view to completing the work started in the Council for [TRIPS] on the implementation of Article 23.4, we agree to negotiate the establishment of a multilateral system of notification and registration of geographical indications for wines and spirits by the Fifth Session of the Ministerial Conference. We note that issues related to the extension of the protection of geographical indications provided for in Article 23 to products other than wines and spirits

---

[44] The concerns conventionally relate to: (a) acquired interests in individual trade marks which have built up goodwill over time and are threatened by GIs; and (b) the freedom to use a formerly geographical term in a generic manner (e.g. cheddar for a style of cheese). For further details, see respectively Chapter 14 by Heath and Chapter 18 by Gangjee in this volume.

[45] The services referenced here would include, for example, advertising and promotion.

[46] See WIPO, 'Possible Solutions for Conflicts between Trademarks and Geographical Indications and for Conflicts between Homonymous Geographical Indications' 8 June 2000 (SCT/5/3) 11–12.

will be addressed in the Council for TRIPS pursuant to paragraph 12 of this declaration.[47]

There are two issues, one vertical and the other horizontal, that have been standing in the way of progress on substantive discussions. The vertical issue is the fact that many GIs for wines and spirits are considered generic or semi-generic in the United States and other New World countries as well. The horizontal issue is the interface between GIs and trade marks (which, to some WTO Members, are the same thing) and specifically whether the first in time, first in right principle would continue to apply. We now take a look at both issues before turning to progress made in the Doha context.

3.1 Semi-generic Denominations

At the heart of the GI debate in TRIPS are dual-purpose denominations, in particular those for wines and spirits such as Chablis or Champagne.[48] Those denominations are considered *semi*-generic in the United States and in some other countries because they are understood by many consumers as being able to refer both to a geographical origin (in the two examples above, specific regions in France) and to a type of product (using the same two examples, a crisp white wine made mostly with Chardonnay grapes and a Chardonnay-blend bubbly white wine, respectively).[49] Where the designation is used by itself (e.g. Champagne), it may still retain connotations of geographical origin but when used in conjunction with a qualifier (e.g. Californian) the term is often considered 'generic', particularly in common law jurisdictions, and is thus interpreted as a signal of certain product qualities instead of its geographic origin. The existence of semi-generic denominations often stems from migratory patterns. They are used by New World producers to associate a product produced with a type of product or location associated with the Old World.[50]

Much of this debate was put to rest as far as EU wines are concerned

---

[47] Ministerial Declaration adopted on 14 November 2001, document WT/MIN(01)/DEC/1 of 20 November 2001, at para. 18.
[48] See EC Creditt, 'Terroir vs. Trade Marks: The Debate over Geographical Indications and Expansions to the TRIPS Agreement' (2009) 11(2) Vanderbilt Journal of Entertainment & Tech. Law 425, 443.
[49] See http://www.terroir-france.com/region/
[50] See Elizabeth Barham, 'Translating Terroir: The Global Challenge of French AOC Labelling' (2003) 19 Journal of Rural Studies 127, 128; PM Brody, '"Semi-Generic" Geographical Wine Designations: Did Congress Trip Over TRIPs?' [1999] TMR 979.

by the 'Wine Pact' with the US.[51] As provided for in Art 6 of the Pact, the United States agreed to limit the use of semi-generic appellations[52] to wines originating in the EC. A grandparent-type exception is provided for wines not originating in the EC on which a term (Appellation) mentioned in Annex 2 to the Pact was used before 13 December 2005 and for which a certificate of label approval (COLA) has already been issued.[53] The Wine Pact is situated at the confluence of two traditions and rocked by shifting consumer preferences. There are three clusters of factors that affect the quality of a wine, namely: (a) grapes (variety); (b) soil and climate; and (c) wine-making ability (know-how and technique). The combination of all three produces a unique product, related to the French concept of *terroir*.[54] This link between a product and the *terroir* can be traced back to the fifteenth century in Europe and is best epitomized by the system of *Appellations d'Origine Contrôlée* (AOC) in France.[55] The AOC system established 'by the Law of the 30th of July of 1935 has created a specific type of French wine: AOC wines. These wines use the notion of *terroir* to distinguish themselves from the other wines. A *terroir* relies on natural and human factors and their specificities'.[56] AOC wines often command

---

[51] *Agreement of 10 March 2006 between the European Community and the United States of America on trade in wine* [2006] OJ L 87/2 (24 March 2006) [Hereinafter, Wine Pact]. See also Agreement in the Form of an Exchange of Letters Between the United States of America and the European Community on Matters Related to Trade in Wine, Document EUUSA/CE/en1 and en2, 23 November 2005, available at: http://www.ttb.gov/wine/wineagreement.pdf?cm_sp=ExternalLink-_-Federal-_-Treasury.

[52] See Wine Pact, Annex 2. The list of semi-generic denominations, which signal both a geographical origin and a type of product, is as follows: Burgundy (but not *Bourgogne*), Chablis, Champagne, Chianti, Claret, Haut Sauterne, Hock, Madeira, Malaga, Marsala, Moselle, Port, Retsina, Rhine, Sauterne, Sherry and Tokay.

[53] COLAs are issued by the Alcohol and Tobacco Tax and Trade Bureau, under 24 CFR 4.30. See http://www.ttb.gov/forms/f510031.pdf. Provisions implementing Art 6 were introduced into US law by s 422 of the (appropriately named?) *Tax Relief and Health Care Act of 2006*, signed by the President on 20 December 2006.

[54] In Spain, local table wines are known as 'vino de la tierra', or wine from certain 'land'. The French concept of 'vin de pays' is a close cousin. There is no good translation for '*terroir*'. Like many others, including WIPO, I thus decided to use the French term. For detailed analysis of *terroir*, see Chapter 3 by Barham and Chapter 4 by Bérard in this volume.

[55] As defined in *Code de la Propriété Intellectuelle* (France), art. L. 721-1. For contemporary developments, see Chapter 11 by Marie-Vivien in this volume.

[56] David Menival, 'The Greatest French AOCs: A Signal of Quality for the Best Wines' working paper (2007), at p. 1, http://www.wineecoreports.com/

a higher price as a result (monopoly rent). Under the AOC system and a number of similar systems all administered by the *Institut national de l'origine et de la qualité* (INAO), a number of products (wines, spirits, but also cheese, candy, and so on) can be identified as having been produced in a certain region not only if the geographic provenance is factually correct but if certain codified guidelines for the production were followed.[57] A system based on a high level of protection for denominations of origin emphasizes the second cluster of factors. A white wine made with, say, Sauvignon grapes, will not be the same even if made by the same person using the same technique in Loudoun county in Virginia, the Loire valley of France or the Marlborough region of New Zealand. The acidity of the soil and the amount of rain and sun exposure will affect the outcome.[58] Indeed, climate variations will lead to significant differences in wine produced in any given region. This does not comport well with a trade-based view of symbols on products as reflecting the rights of the first user.

3.2 'First in Time, First in Right'

The application of the first in time, first in right principle, according to which rights to a particular trade mark belong to the first user, is opposed to the application of GIs to existing marks.[59] The phrase is a shorthand expression for the combined principles of priority and exclusivity. The first party to acquire rights to a sign can claim exclusive use of the sign. Since the United States and a number of other New World countries protect GIs as *collective or certification marks*, their view is that the first users of a 'GI mark' in their territory should have the exclusive right to use it. The European Union and others consider at least some GIs as superior to ordinary trade marks, though they also agree that coexistence is often possible.

TRIPS contains several rules to regulate conflicts between GIs and trade marks which are the same or similar. As mentioned above, under TRIPS Art 22.3,[60] a WTO Member must, either ex officio if its national law so

---

Working_Papers/Abstract/WP_2007/Menival.pdf. For the origins of this system, see Chapter 2 by Stanziani in this volume.

[57] See http://www.inao.gouv.fr/. For wines, see James E Wilson, *Terroir: The Role of Geology, Climate and Culture in the Making of French Wines* (University of California Press, 1999); and Barham (n 50) 131–132.

[58] Among the soil-related factors that are most important are the drainage capacity, salinity and the ability of the soil to retain heat, thus encouraging ripening and the development of stronger roots.

[59] See Chapter 13 by Goebel and Groeschl in this volume.

[60] One of four main provisions dealing with the trade mark/GI interface in TRIPS. See also Arts 23.2, 24.3 and 24.5.

permits, or at the request of an interested party, refuse or invalidate[61] the registration of a trade mark that contains or consists of a geographical indication if (a) the goods do not originate in the territory indicated, and (b) use of the indication in the trade mark for such goods in the territory of the 'Member' concerned is *of such a nature as to mislead* the public as to the true place of origin. Art 23.2 more or less corresponds to Art 22.3, but applies specifically to indications identifying wines and spirits, except of course that deception (misleading the public as to the true place of origin) does not have to be shown.[62]

However, the most important conflict rules are contained in Articles 24.5 and 24.6. Under the former, a *trade mark may trump a GI* (the opposite of the EU position other than perhaps for famous marks), if an application for registration of the mark was filed or the mark registered, or the right acquired by use (and the trade mark was in fact used in good faith)[63] in the WTO Member concerned either before the TRIPS Agreement became applicable in that Member,[64] or before the indication in question was protected in its country of origin.[65]

The WTO panel report in the *EC – Trademarks and Geographical Indications I* dispute explained that the coexistence of a protected indication and a trade mark was a limited exception justified under Art 17.[66] This

---

[61] See text accompanying footnotes 37–39 above.

[62] For US implementation, see s 522 of the *Act to Implement the Results of the Uruguay Round of Multilateral Trade Negotiations* (Public Law 103–465 of 8 December 1994), entitled 'Nonregistrability of Misleading Geographic Indications for Wines and Spirits'.

[63] This test is sometimes difficult to apply, as evidence of good (or bad) faith is not always easy to produce. Showing bad faith based entirely on circumstances is sometimes rendered more difficult in legal systems that presume good faith until the contrary is shown. In applying the test, the fact that an indication is particularly well known and/or used (directly or indirectly) by undertakings located in or near the 'true' place of origin should be taken into account.

[64] According to TRIPS, Art 65.1 the date is 1 January 1996 for the most industrialized nations. Developing countries other than least-developed ones had to apply most substantive provisions of the Agreement as of 1 January 2000.

[65] See WIPO 'Possible Solutions for Conflicts between Trade Marks and Geographical Indications' (n 46) 11–12.

[66] WTO Panel Report, *European Communities – Protection of Trade Marks and Geographical Indications for Agricultural Products and Foodstuffs* 15 March 2005 (WT/DS174/R) (adopted on 20 April 2005); WT/DS290/R (complaint by Australia), 15 March 2005 (adopted on 20 April 2005). The Panel concluded, 'with respect to the coexistence of GIs with prior trade marks, the Regulation is inconsistent with Article 16.1 of the TRIPS Agreement but, on the basis of the evidence presented to the Panel, this is justified by Article 17 of the TRIPS Agreement' (Australia Report, para. 7.686, U.S. Report, para. 7.688).

provision recognizes a 'fair use' exception, stating: 'Members may provide limited exceptions to the rights conferred by a trade mark, such as fair use of descriptive terms, provided that such exceptions take account of the legitimate interests of the owner of the trade mark and of third parties'. The EU had argued that since GIs were descriptive as regards the origin (and potentially the quality) of products – for example, Parmigiano Reggiano describes a cheese made in a certain region of Italy with specific qualities – they fit within the Art 17 exception. The purpose of Art 17, the Panel noted, was to allow a trade mark to be registered (or applied for) and used, even if it is identical with or similar to a GI,[67] provided the trade mark is applied for, registered or the rights acquired through use, either before the WTO Member concerned must apply Art 23 or before the indication is protected in its country of origin. This means that a WTO Member may continue to use first-in-time, first-in-right as the principal conflict resolution norm, but coexistence of prior trade marks and subsequent GIs is a permitted derogation under Art 17.

Certain commentators go a (big) step further and argue that TRIPS actually *mandates* the application of 'first in time, first in right'.[68] Not only would a trade mark survive a challenge by a GI owner, but the trade mark owner could oppose the use of the identical or similar GI in the country where the trade mark is used and/or registered.[69] I find it difficult to support this view. As the WTO panel explained, the coexistence of a protected indication and a trade mark is a *permitted exception* to trade mark rights, not a right.[70] The Panel agreed with the EC that, although its GI Regulation allowed it to register GIs even when they conflicted with a prior trade mark, the Regulation, as written, was sufficiently constrained to qualify as a 'limited exception' to trade mark rights. However, the Panel agreed with the United States and Australia that the TRIPS Agreement did not allow unqualified coexistence of GIs with prior trade marks (for example, a prior well-known mark might not coexist with a GI). Moreover, several countries protect indications not by a sui generis

---

[67] See Lisbon Agreement, Art 5(6) and also Florent Gevers, 'Geographical Names and Signs Used as Trade Marks' [1990] EIPR 285.

[68] Clark W Lackert, 'Geographical Indications: What Does the WTO TRIPS Agreement Require?'(1998) Trade Mark World 22, 24.

[69] See Florent Gevers, 'Topical Issues in the Protection of Geographical Indications', in *WIPO Symposium on the International Protection of Geographical Indications, Eger, Hungary 1997* (WIPO 1999) 155–156; and Henning Harte-Bavendamm, 'Geographical Indications and Trade Marks: Harmony or Conflict?', in *WIPO Symposium on the International Protection of Geographical Indications, Cape Town, 1999* (WIPO 2000) 63–67.

[70] Supra (n 66).

or specific intellectual property right, but as (collective or certification) marks, or even under unfair competition law. In such cases, the conflict is between two trade marks, not between a trade mark and a sui generis GI right. Common law courts have often granted limited remedies under equitable rules to allow the coexistence of two trade marks (usually one of which was acquired through use).[71] Finally, TRIPS specifically provides for the coexistence of indications in Art 23.3 (homonymous GIs for wines). Demonstrating that TRIPS *mandates* 'first in time, first in right' is thus an uphill battle.

Genericide rules are also relevant in this context. Genericide is when an indication is no longer distinctive of a region or territory because it has become generic for a type of product.[72] For example, 'china' (for dishware) no longer indicates a product made exclusively in China. TRIPS Art 24.6 limits GI protection when they have become generic and is different from the Lisbon Agreement in that regard.[73] Art 24.6 provides that WTO Members may decide not to protect a geographical indication used in connection with foreign goods or services for which the relevant indication *is identical with the term customary in common language as the common name for such goods or services in the territory of that Member*. The article also states that Members are not required to protect foreign geographical indications 'with respect to products of the vine for which the relevant indication *is identical with the customary name of a grape variety* existing in the territory of that Member as of the date of entry into force of the WTO Agreement' (emphasis added). Nothing in TRIPS indicates that this determination, effectively a legal ruling that a geographical indication has become generic in a particular jurisdiction, should be performed outside a Member State's courts or by reference to any law other than that of the Member State. Put differently, it is up to each Member to decide what is or is not generic within its borders.[74]

3.3 Progress on the TRIPS Notification and Registration System

The negotiations on the establishment of a TRIPS notification and registration system for GIs have focused on six main issues, namely:

---

[71] See WIPO Secretariat, 'Historical Background, Nature of Rights, Existing Systems for Protection and Obtaining Protection in Other Countries' 2 April 2002 (SCT/8/4) 5–9. This is also apparently the solution in the WIPO analysis. Ibid., 28.
[72] See Chapter 18 by Gangjee in this volume.
[73] Lisbon Agreement, Art 6.
[74] This is precisely what the US did in 27 CFR §§4.24 and 12.31.

1. Notification
2. Registration
3. Legal Effects/Consequences of Registration
4. Fees and Costs
5. Special and Differential Treatment
6. Participation.

While TRIPS only mandates negotiations on a *notification and registration* system for wines, this was extended to spirits by the Doha Declaration.[75] The key issues are: whether all WTO Members must participate in the new system; what standard will be used by the WTO to register notified GIs (that is, does the term imply an *examination* and, if so, by whom); and the legal effect (if any) of the notification/registration. A number of important but more technical issues also arose, including: who notifies on behalf of a Member; what information must be supplied; whether one may oppose (which probably implies publication); how the information is disseminated; resources available at the WTO Secretariat, and the role that WIPO might play (and in that case, any fee-sharing arrangement).

Participation is a key element of course. Most GIs are owned by Old World entities. However, there may not be a fundamental conflict between GI protection and the approach to the sale of at least food and wine products in the New World. Developing countries and others are beginning to see the potential of GI protection in raising prices paid for other commodities such as coffee, cocoa and tea, as well as textiles and various crafts. Their interest in GIs may increase, and they may be able to negotiate a form of special or differential (S&D) treatment. In such a scenario, it would be perilous for New World nations to stay outside a new system designed and operated by others in or for territories in which the non-participants have trade interests. Put simply, American and other trade mark holders would then face GIs in countries party to the new system but there would be little they could do.

On administrative issues, the April 2011 composite draft text released by the WTO is an attempt at finding a compromise.[76] It reflects a relatively modest role for the WTO Secretariat. Basically, all notified GIs would be registered. According to the Hong Kong Proposal reflected in that text, the Secretariat's role would be limited to a 'formality examination

---

[75] For details on the register, see Chapter 8 by Martín in this volume.
[76] The text is available as an Annex to TRIPS Council, 'Report by the Chairman to the Trade Negotiations Committee' 21 April 2011 (TN/IP/21) (hereafter, Composite Text).

of the notifications and ensure that documents submitted are in order'.[77] However, the entire set of proposals on the system, and specifically on which documents and information would be submitted, and the role of the Secretariat (examination, and so on) is square-bracketed, indicating an absence of consensus.

The issue of the legal effect of the registration/notification system is even more problematic. An example from the April 2011 text should suffice to illustrate the level and scope of the disagreement:

> E.1 Each [participating]^{JP,IND,SG,BRA,CUB} WTO Member [commits to ensure]^{JP,BRA} [shall provide]^{EU} that [its procedures include the provision to]^{JP,BRA} [domestic authorities shall]^{EU} consult the [Database]^{JP} [Register and take its information into account]^{EU} when making decisions regarding registration and/or protection of trade marks and geographical indications [for wines and spirits]^{JP,SG,BRA} in accordance with its [laws and regulations]^{JP,BRA,COL} [and]^{COL} [domestic procedures]^{EU,COL}.]^{JP,EU,COL} [78]

There is not even consensus on the imposition of an obligation to *consult* the register. Other more ambitious proposals aim to make registered GIs prima facie valid and to impose on a party invoking generic status the burden of establishing it.

3.4 Extension to All Products

The extension of higher GI protection in accordance with Art 23 of TRIPS to products other than wines and spirits is similarly complex, both politically and legally.[79] Bulgaria, the Czech Republic, Iceland, India, Liechtenstein, Slovenia, Sri Lanka, Switzerland and Turkey, now joined by Egypt, Kenya and Pakistan, have submitted a joint paper proposing the extension of high protection for geographical indications applied to products other than wines and spirits.[80] The proposal drew

---

[77] Ibid at C.2.

[78] The two and three-letter superscripted symbols refer to the Members who made the relevant proposals: Brazil, Colombia, Cuba, the European Union, India, Japan and Singapore.

[79] For detailed analysis, see Chapter 7 by Handler in this volume.

[80] Communication from Bulgaria et al., 'Implementation of Art 24.1' 18 Sep 2000 (IP/C/W/204) and 2 December 2000 (IP/C/W/204/Rev.1). By contrast, the appeal of a revised Lisbon Agreement would be a high level of protection available to a broader membership. See Art 10 of the draft Revised Lisbon Agreement and safeguards in respect of prior trade marks (Art 13(1)). See also the Annex to the Summary by the Chair adopted at the sixth session of the WG in December 2012 (WIPO document LI/WG/DEV/6/7) and the Annex to the Summary by the

its legal basis from the 1996 WTO TRIPS Council Report, which had noted that the possible extension of additional protection was part of the TRIPS built-in agenda.[81] India has also presented two papers on the same matter.[82] Discussions thus far have been driven by two opposing groups – one in favour of the extension, and the other against. The latter group included Argentina, Australia, Brazil, Canada, China, Colombia, Mexico, New Zealand, Paraguay, the Philippines, the United States and Uruguay. They assert that TRIPS does not include an obligation or even indicates a willingness to extend high GI protection (along the lines of Art 23, without requiring deception) to products other than wines and spirits.[83] As of this writing, the issue seems to be making little, if any, progress.

3.5 Locus of the Registration/Notification System

Asking *where* the TRIPS Article 23.4 system will be located, and by whom it will be administered, and then what, if any, the interface with the existing Lisbon register would be, are all valid questions. The WTO, which administers TRIPS, has no experience in establishing or creating a multilateral intellectual property registration system. In fact, under the cooperation agreement entered into between WIPO and the WTO, WIPO administers the registration and notification system contained in Article 6*ter* of the Paris Convention and made applicable to WTO Members who are not party to the Paris Convention.[84] Could WIPO administer the Lisbon register *and* a parallel system established under Article 23.4 of the TRIPS Agreement?

Chair adopted at the seventh session in April/May 2013 (WIPO document LI/WG/DEV/7/7).

[81] WTO 'Report (1996) of Council for TRIPS' 6 November 1996 (IP/C/8) para. 26.

[82] TRIPS Council, 'Communication from India' 12 July 2000 (IP/C/W/195) and (IP/C/W/196).

[83] TRIPS Council, 'Communication from the European Communities' 30 May 2001 (IP/C/W/260).

[84] Admittedly a short list: Cape Verde, Fiji, Kuwait, Maldives, Myanmar, the Solomon Islands and Taipei, as well as Hong Kong and Macau (China) and the European Union. See Gervais, *The TRIPS Agreement* (n 3) 757–759. See also *Agreement Between the World Intellectual Property Organization and the World Trade Organization*, 22 December 1995, Art 3.1 (available at http://www.wto.org/english/tratop_e/trips_e/wtowip_e.htm), which provides that the 'procedures relating to communication of emblems and transmittal of objections under the TRIPS Agreement shall be administered by the International Bureau in accordance with the procedures applicable under Article 6*ter* of the Paris Convention'.

The adoption in May 2015 of the Geneva Act of the Lisbon Agreement may make this less likely, not more.[85] To achieve the aim of becoming a worldwide register *attractive* to both current members (who by and large protect GIs under a sui generis regime) and those who use trade mark law, the revised Agreement needed to contain two elements.[86] First, the new Act had to eliminate or at least limit the scope of the rule according to which a GI could not become generic anywhere in the world unless and until it is generic in its country or origin. Otherwise, this clashes with the territorial nature of trade mark law and notably common law doctrines of acquiescence and abandonment. That did not happen, as the Geneva Act basically replicates the genericide regime (lex originis) of the 1958 Act.[87] Second, the obligations to protect GIs under the new Act had to match TRIPS to avoid substantive renegotiation, either by referring to TRIPS or by incorporating TRIPS. While some TRIPS language is reflected in the new Act, it goes beyond this, as it extends the higher or 'trade mark plus' (that is, without the need to show deception or confusion) level of protection to *all GIs*. In TRIPS it is limited to wines and spirits.[88] Therefore, for countries joining the 2015 Geneva Act of the Lisbon Agreement, this is a de facto answer (that some WTO members may not like) to the ongoing debate in the Doha Round on extension of the higher level of protection to goods other than wines and spirits.

On the administrative side, the Geneva Act requires applicants to designate countries (as, for example, does the Madrid system for trade marks). Applicants must pay a per-country fee if one has been set. However, the new Act does not expressly allow for maintenance or renewal fees, which are typically required in trade mark regimes.

---

[85] For details on the negotiation process, see Chapter 5 by Geuze in this volume.

[86] For comprehensive analysis of this potential, see Daniel Gervais, 'Reinventing Lisbon: The Case for a Protocol to the Lisbon Agreement' (2010) 11(1) Chicago Journal of International Law 67 (including a detailed review of all entries on the Lisbon register as of late 2009); Christophe Geiger, Daniel Gervais, Norbert Olszak and Vincent Ruzek, 'Towards a Flexible International Framework for the Protection of Geographical Indications' (2010) 1(2) WIPO Journal 147; and Daniel Gervais, 'The Misunderstood Potential of the Lisbon Agreement' (2009) 1(1) WIPO Journal 87 (inaugural issue – on invitation).

[87] Geneva Act of the Lisbon Agreement, Arts 8(1) and 12. A fuller analysis of the interaction between the Geneva Act and TRIPS will be available in Daniel Gervais, 'Irreconcilable Differences? The Geneva Act of the Lisbon Agreement and the Common Law' (2015) 53(2) Houston Law Review 339.

[88] Geneva Act of the Lisbon Agreement, Art 11.

Finally, as will become clear once the Acts of the Diplomatic Conference are published, the remarks made by several countries (Argentina, Australia, Chile, Japan, Korea, Panama, United States and Uruguay) at the last plenary session confirm that the May 2015 Diplomatic Conference may have increased the divide instead of building bridges, at least in the short term. Yet, bilateral and regional trade negotiations may 'force' even those very reluctant countries to join the Lisbon system. Put differently, if and when significant overlap between WTO membership and Lisbon membership emerges, it may be possible to use Lisbon as the notification and registration system foreseen in TRIPS.

4. CONCLUSION

Geographical indications and their close cousins, appellations of origin, are near and dear to the heart of many Old World producers of wines, spirits and other products of the *terroir*. Their first function is as symbols of national or regional identity anchored in long-standing traditions and know-how. However, GIs are also strategic commercial tools allowing for the capture of additional rents in domestic as well as international markets. EU negotiators have understandably tended to insist on protection of the former, but any discussion of the subject among interested parties almost inevitably seems to revolve around the second function. Consumers might see both. New World buyers may wish to partake of the traditions, *savoir-faire* and perhaps also the *savoir-vivre* associated with certain GIs, and they apparently are prepared to pay more for GI-backed products.

The EU is not alone in desiring GI protection. Developing countries have realized the commercial and developmental advantages of GIs, but typically in areas other than those favored by the TRIPS Agreement, namely wines and spirits. On the other side of this debate, New World producers, many of whom have been using denominations similar (sometimes intentionally so) to Old World ones do not want to lose the goodwill associated with their trade marks, and argue that, absent consumer deception, no regulatory intervention is required.

The complexity of this multifaceted backdrop presents a challenge to WTO Members trying to fulfill the Art 23.4 mandate to negotiate (and presumably establish) a notification and registration system for GIs. At the time of writing, the list of issues to be settled is too long to hope for a quick ministerial intervention. Several important details on the implementation and legal effect of the system must be ironed out. Unfortunately, contrary to what might have been hoped, the lack of flexibility demonstrated during the Diplomatic Conference means that the Lisbon system, as modernized in May 2015, is not an obvious way forward. That said, EU-led efforts to broaden its geographic reach in the coming years will remain highly visible on the global GI radar.

7. Rethinking GI extension
Michael Handler*

1. INTRODUCTION

It is well accepted that the final form of the GI provisions of the TRIPS Agreement represented a messy, politically expedient compromise between the EU and the US, designed largely so that the Agreement, which was otherwise so mutually beneficial to those parties, could come into being.[1] Nowhere is this compromise more evident than in the dual minimum standards of GI protection set out in Arts 22 and 23.

Under Art 22.2, WTO Members are under a general obligation in respect of other Members' GIs to prevent:

(a) the use of any means in the designation or presentation of a good that indicates or suggests that the good in question originates in a geographical area other than the true place of origin in a manner which misleads the public as to the geographical origin of the good; [and]
(b) any use which constitutes an act of unfair competition within the meaning of Article 10*bis* of the Paris Convention (1967).

Nothing in the TRIPS Agreement stipulates how Members are to afford such protection under their domestic laws.[2] Thus, Members can comply with their Art 22.2 obligations through non-GI specific laws that prevent parties from engaging in trade misrepresentations involving the use of geographical insignia in which other parties have a collective interest or

---

* My thanks go to Robert Burrell, Dev Gangjee and Megan Jones, and to Qi Jiang for his research assistance.

[1] See generally D Gervais, *The TRIPS Agreement: Drafting History and Analysis* (3rd edn, Sweet & Maxwell, 2008), pp. 24, 26, 305. See also P Demaret, 'The Metamorphoses of the GATT: From the Havana Charter to the World Trade Organization' (1995) 34 Columbia Journal of Transnational Law 123, 166; R Okediji, 'Public Welfare and the Role of the WTO: Reconsidering the TRIPS Agreement' (2003) 17 Emory International Law Review 819, 849–850.

[2] On the contrary, Art 1.1 of the TRIPS Agreement provides that 'Members shall be free to determine the appropriate method of implementing the provisions of this Agreement within their own legal system and practice'.

reputation.³ That is, certification trade mark schemes that allow producers or collectives to register their GIs as certification marks and bring infringement actions against unlicensed parties that use such signs are sufficient to ensure that a country meets its Art 22.2 obligations. So too are laws that prohibit third parties from engaging in 'passing off' or related causes of action that turn on whether consumers have been misled by certain deceptive marketing practices. A significant flipside of the misrepresentation-based Art 22.2 standard is that in countries that choose to comply with this obligation by the abovementioned methods – the US, Canada, Australia and New Zealand being noteworthy examples – not all commercial uses of foreign GIs will be proscribed. Such signs can potentially be legitimately used in circumstances where consumers in those countries either do not attribute geographical significance to the sign at all or do not expect there to be any connection between the sign and the place of production (for example, because the sign is used with qualifying material that makes clear the true place of origin). Thus, in non-European countries where 'feta' and 'parmesan' are understood only as types of cheese, or where a term such as 'kanterkaas' for cheese would not be recognised as having any geographical significance, Art 22.2 would not be implicated, even though the use of each of these terms as GIs is tightly restricted under EU law.⁴ Similarly, Art 22.2 does not require a country to proscribe the sale of goods clearly marketed as 'Mortadella Bologna. Product of Australia'.

The Art 22.2 standard is hardly exceptional to those familiar with

---

³ Some commentators have suggested that the reference to 'unfair competition' in Art 22.2(b) suggests that this provision might impose a higher standard than mere misleading conduct under Art 22.2(a): see, e.g., A Taubman, 'Thinking Locally, Acting Globally: How Trade Negotiations Over Geographical Indications Improvise "Fair Trade" Rules' [2008] IPQ 231, 251–252. However, Art 10*bis* of the Paris Convention has never been interpreted to require parties to apply such a higher level of protection, and a strong argument can be made that the three examples of 'unfair competition' in Art 10*bis*(3) that each refer to misleading conduct should be read as defining the core of the action: see, e.g., W Cornish, 'Genevan Bootstraps' [1997] EIPR 336, 337; D Gangjee, *Relocating the Law of Geographical Indications* (CUP, 2012) 52–59. Art 22.2(b) does, however, have additional work to do in catching conduct that is misleading other than as to geographical origin: see the discussion in Section 3.3.

⁴ It should also be noted that the fact that the first two terms are protected at all in the EU is highly controversial: see the ECJ's decisions in Joined Cases C-465/02 and C-466/02, *Federal Republic of Germany and Kingdom of Denmark v Commission* [2005] ECR I-9115 (on 'Feta') and Case C-132/05, *Commission v Federal Republic of Germany* [2008] ECR I-957 (on 'Parmesan').

trade mark or consumer protection laws, and few developed countries needed to amend their domestic laws when the TRIPS Agreement came into force to comply with this obligation. However, Art 23.1 imposes an additional obligation on Members to afford a higher minimum level of protection to a subset of GIs. More particularly, it requires Members:

> [to proscribe the] use of a geographical indication identifying wines for wines not originating in the place indicated by the geographical indication in question or identifying spirits for spirits not originating in the place indicated by the geographical indication in question, even where the true origin of the goods is indicated or the geographical indication is used in translation or accompanied by expressions such as 'kind', 'type', 'style', 'imitation' or the like.

Art 23.1 thus establishes something much closer to a total bar on the use of a GI, irrespective of its reputation or consumer understandings of the term in the country where protection is sought, but only in respect of two product types – wines and spirits. Thus, a New Zealand distiller could not call its locally made product 'New Zealand cognac' or even describe it as 'cognac-style brandy'. Similarly, Art 23.1 would prevent a South African winemaker from adopting the brand name 'Bull', given that this word translated into Spanish is 'Toro', which is a protected wine GI in Spain.[5] The bar on the use of such GIs is not, however, absolute: Art 23.1 (like Art 22.2) is subject to a number of exceptions in Art 24 that permit the use of terms in one Member even if these are recognised as GIs elsewhere. The key exceptions permit the use of certain pre-existing trade marks that are similar to GIs as well as terms that have become generic product descriptors in the country in which protection is sought.[6] It is this second exception that means that nothing in the TRIPS Agreement requires a country to proscribe the use of terms such as 'champagne', 'port' and 'burgundy' if these are understood in that country as generic terms for sparkling, fortified and a type of red wine respectively.

In the GATT Uruguay Round negotiations over TRIPS, the EU had sought the higher level of GI protection to apply to all goods,[7] reflecting the approach long taken under the domestic GI laws of a number of its members (notably France and Italy) and which had been adopted

---

[5] See http://www.dotoro.es/en/index.php.
[6] TRIPS, Arts 24.5 and 24.6.
[7] See GATT Doc MTN.GNG/NG11/W/68 (29 March 1990), Art 20(1); see also Chapter 6 by Gervais in this volume.

at an EU-wide level for agricultural foodstuffs,[8] wines[9] and spirits,[10] as well as under the little-utilised Lisbon Agreement for the Protection of Appellations of Origin and their International Registration 1958. The EU was also not prepared to provide any safeguards for terms that had become generic in other countries, nor was it prepared to accommodate pre-existing trade mark rights in any way.[11] The US, supported by countries such as Australia, considered that parties should only be required to prevent misleading uses of non-generic GIs, and only in respect of wines.[12] Such countries were particularly concerned about the imposition of minimum standards of protection that would effectively allow for the 'repropertisation' of generic product descriptors. The final form of Arts 22–24 of the TRIPS Agreement, which also included a provision deferring negotiations on the establishment of a multilateral register for wine and spirit GIs,[13] was not the result of the parties arriving at a common understanding as to the intrinsic merits of protecting certain GIs in particular ways. Rather, it was the product of hard bargaining in the form of concessions made to the EU in relation to its wine and spirit GIs (which the EU had consistently argued were particularly vulnerable to 'misuse' in foreign markets) in return for the EU agreeing in concurrent trade in agriculture negotiations to reduce its domestic subsidies.[14]

Unsurprisingly, from very soon after TRIPS came into force the EU

---

[8] See currently Regulation No 1151/2012 of the European Parliament and of the Council of 21 November 2012 on Quality Schemes for Agricultural Products and Foodstuffs [2012] OJ L323/1. This replaced Council Regulation (EC) No 510/2006 of 20 March 2006 on the Protection of Geographical Indications and Designations of Origin for Agricultural Products and Foodstuffs [2006] OJ L93/12, itself replacing Council Regulation (EC) No 2081/92 [1992] OJ L208/1.

[9] See currently Council Regulation (EC) No 479/2008 of 29 April 2008 on the Common Organisation of the Market in Wine [2008] OJ L148/1 (replacing Council Regulations (EEC) No 2392/86 [1986] OJ L208/1 and (EC) No 1493/1999 [1999] OJ L179/1).

[10] See currently Council Regulation (EC) No 110/2008 of 15 January 2008 on the Definition, Description, Presentation, Labelling and the Protection of Geographical Indications of Spirit Drinks [2008] OJ L39/16 (replacing Council Regulation (EEC) No 1576/89 [1989] OJ L160/1).

[11] GATT Doc MTN.GNG/NG11/W/68 (29 March 1990), Arts 20–21.

[12] GATT Doc MTN.GNG/NG11/W/70 (11 May 1990), Art 19.

[13] See TRIPS Agreement, Art 23.4.

[14] See JH Reichman, 'Compliance with the TRIPS Agreement: Introduction to a Scholarly Debate' (1996) 29 Vanderbilt Journal of Transnational Law 363, 387; T Josling, 'The War on *Terroir*: Geographical Indications as a Transatlantic Trade Conflict' (2006) 57 Journal of Agricultural Economics 337, 351. See also WTO Docs IP/C/M/29 (6 March 2001), para 93; IP/C/W/289 (29 June 2001), para 9.

started pushing for the Agreement to be amended so that the Art 23 level of protection applies to GIs for all goods, not merely wines and spirits.[15] This position has been supported by a group of mainly developing countries, most of which can point to one or a small number of domestic GIs that already have a significant international export market.[16] This push has been coupled with a proposal to have a binding multilateral GI register, the effect of which is likely to create certain presumptions in favour of protecting a country's GIs in other countries.[17] This agenda has, however, been staunchly opposed by a group of agricultural exporters such as the US, Australia, New Zealand, Argentina, Chile and Canada, who claim, in short, that expanded protection is unnecessary and unjustified. This disagreement, which has played out primarily in the Council for TRIPS at the WTO, has become known as the 'GI extension debate', the key feature of which has been the intransigence of the two camps, and the sameness, indeed triteness, of the arguments put forward in favour of extension. In over ten years there has been no meaningful movement towards resolution of the issue.[18]

---

[15] See WTO Doc TN/IP/W/11 (14 June 2005). On the problematic vagueness of the drafting of the EU's proposed amendment to Art 23.1, see J Hughes, 'Champagne, Feta and Bourbon: The Spirited Debate about Geographical Indications' (2006) 58 Hastings Law Journal 299, 384–385. The EU is also proposing that Art 23.2 (dealing with the registration of trade marks containing wine or spirit GIs) and Art 23.3 (dealing with homonymous wine GIs) be amended so that they apply to GIs for all goods: see generally WTO Docs IP/C/W/353 (24 June 2002), paras 14–29; TN/IP/W/11 (14 June 2005).

[16] Key developing countries that have consistently supported the EU's agenda (and which have valuable domestic GIs) include Cuba, India, Jamaica, Kenya, Pakistan and Sri Lanka. Only more recently has China supported the EU's agenda: see WTO Doc TN/C/W/52 (9 July 2008); WTO Doc TN/C/W/60 (19 April 2011).

[17] See currently WTO Doc TN/C/W/52 (9 July 2008), Annex, paras 1–2; WTO Doc TN/IP/21 (21 April 2011). The EU even went as far as presenting, as part of the WTO agriculture negotiations in 2003, a list of 41 of the most famous European GIs (including contentious terms such as 'Feta', 'Parmigiano Reggiano' and 'Prosciutto di Parma') and demanding that all other countries cease using these terms other than to identify European products, irrespective of any TRIPS flexibilities (see European Commission, 'WTO Talks: EU Steps Up Bid for Better Protection of Regional Quality Products', Press Release IP/03/1178, Brussels (28 August 2003), http://europa.eu/rapid/press-release_IP-03-1178_en.htm). Although the EU has since seemed to have retreated from demanding 'clawback' protection for these particular GIs, its initial demand gives a strong indication of its tactics in attempting to export its GI standards throughout the world.

[18] As noted by the Director-General: WTO Doc WT/GC/W/633 (21 April 2011), para 17. In fact, the two camps cannot even agree on whether they have

This chapter analyses the GI extension debate, focusing in particular on how and why the EU, as the main advocate of stronger global standards of protection, has failed to articulate a cogent rhetorical framework that would ease its policy agenda to general, worldwide acceptance.[19] It criticises some of the simplistic reasoning that has been used to seek to justify GI extension and attempts to explore whether other justifications for extension that have not been clearly articulated by the EU and its supporters might exist. This task in turn involves trying to unpack some of the underlying tensions between the advocates and opponents of extension that underpin the entire debate. The chapter concludes by arguing that the GI extension debate is ultimately something of a sideshow, and that increased attention needs to be paid to how GIs are being treated in preferential trade agreements. It is here that such terms are dealt with and indeed 'traded' more openly as instruments of agricultural and viticultural policy in the context of attempts to secure greater market access for wines, spirits and foodstuffs.

---

agreed to negotiate on extension. Advocates of extension argue that Art 24.1 of the TRIPS Agreement contains a built-in agenda to negotiate on this issue. This Article provides in full:

> Members agree to enter into negotiations aimed at increasing the protection of individual geographical indications under Article 23. The provisions of [Art 24.4 to 24.8] shall not be used by a Member to refuse to conduct negotiations or to conclude bilateral or multilateral agreements. In the context of such negotiations, Members shall be willing to consider the continued applicability of these provisions to individual geographical indications whose use was the subject of such negotiations.

However, Art 24.1 can really only be read as providing that Members have agreed to enter negotiations aimed at extending protection for *specific* wine and spirits GIs (e.g., by agreeing that the genericism exception in Art 24.6 will not apply), particularly in the context of separate bilateral or multilateral negotiations. This interpretation is supported by most commentators on the issue: see, e.g., R Knaak, 'The Protection of Geographical Indications According to the TRIPS Agreement' in F Beier and G Schricker (eds), *From GATT to TRIPS – The Agreement on Trade-Related Aspects of Intellectual Property Rights* (Weinheim: VCH, 1996) 138–139; P Brody, 'Protection of Geographical Indications in the Wake of TRIPS: Existing United States Laws and the Administration's Proposed Legislation' (1994) 84 Trademark Reporter 520, 535; Josling (n 14) 352, 355. Further, while there is growing support for the idea that negotiations on GI extension form part of the Doha Work Programme (see WTO Doc TN/C/W/52 (19 July 2008)), these arguments are hollow in that nothing in the Doha Declaration committed parties to negotiate on GI extension due to the absence of any in-built agenda in the TRIPS Agreement.

[19] Cf. NW Netanel, 'Copyright and a Democratic Civil Society' (1996) 106 Yale Law Journal 283, 306.

2. CONTEXTUALISING THE EXTENSION DEBATE

Before assessing the merits of the arguments for and against GI extension, something needs to be said about the interests and motivations of the parties on either side of the debate. This is because the disagreement is about much more than intellectual property policy. Rather, the disagreement also needs to be seen as the product of competing understandings about the appropriate relationship between governments, agricultural producers and the land and, at its core, is about attempts to secure economic advantages for producers of particular types of agricultural and viticultural goods in international markets.

To appreciate the final point above, some political and historical context is needed. While most countries might be able to point to locally-made products sold by reference to their geographical origin, it is not the case that valuable GIs are equally distributed throughout the world. Rather, the vast majority of the world's established GIs are located within a number of European countries that have long sought in their rural policies to control and privilege a particular type of localised production of foods and alcoholic beverages. France provides the clearest example. As far back as the 19th century the French state sought to encourage traditional agricultural and viticultural practices in various regions, to manage production levels and to subsidise rural producers. It did so by developing laws that established organisations to delimit the boundaries of regions in which goods were produced, to administer the registration of the names of such regions and, most importantly, to define the qualities, characteristics and methods of production of goods from such regions whose producers were entitled to use the registered name. Because it was thought that a registered name indicated unique qualities and characteristics of goods as well as source, these laws also gave entitled producers the right to prevent non-entitled traders from using the registered name outright or from describing the broadly similar qualities of their own goods – even where the true origin of their goods was indicated.[20] These laws thus had the effect of entrenching certain agricultural and viticultural production methods, while also allowing certain regional producers to

---

[20] For more detailed consideration of the French system of GI protection, see, e.g., W Moran, 'The Wine Appellation as Territory in France and California' (1993) 83 Annals of the Association of American Geographers 694; W van Caenegem, 'Registered Geographical Indications: Between Intellectual Property and Rural Policy – Part II' (2003) 6 JWIP 861; Chapter 2 by Stanziani in this volume; Gangjee, *Relocating the Law of Geographical Indications* (n 3) ch 3.

charge premiums for their goods based on their perceived uniqueness and state-sanctioned quality.[21]

More importantly for present purposes, it was the above model of rural policy and GI regulation that proved to be highly influential at a European level in the latter part of the twentieth century, since it meshed with the EU's common agricultural policy goals of improving the incomes of farmers and safeguarding the rural economy. Despite some degree of internal opposition, EU-wide notification and registration schemes for Member States' GIs for wines, spirits and agricultural foodstuffs were set up in the late 1980s and early 1990s, with the stated aim of fostering the production of quality regional produce that was bound up with the uniqueness of 'place'.[22] The most noteworthy feature of these schemes is the high level of protection afforded to GI owners: the 'misuse, imitation or evocation' of protected GIs by third parties is prohibited, even where this does not result in consumer deception, and it is stipulated (somewhat disturbingly to any scholar of historical linguistics) that protected GIs cannot become generic.[23] Given its domestic policies and its large number of locally-protected GIs, the EU thus has a clear interest in seeking to ensure that its GIs are protected at much the same levels in foreign export markets – an interest that was to some extent thwarted in the GATT Uruguay Round negotiations.

Not all countries, however, share the EU's rural policies or its conception of GIs. Agricultural exporters such as the US and Australia, for example, have not traditionally sought to manage agricultural and viticultural production by prioritising small-scale, artisanal and localised production through their rural policies.[24] Instead, agricultural production in these countries tends to be dictated more by market considerations such as the cost of labour and obtaining raw materials than by fixed geographical location. While it would be a major oversimplification to

---

[21] See, e.g., J Chen, 'A Sober Second Look at Appellations of Origin: How the United States Will Crash France's Wine and Cheese Party' (1996) 5 Minnesota Journal of Global Trade 29, 35 (the French system 'segments the production market and shields it from outside competitors, thus helping to prop up farming and related industries as significant sources of jobs').

[22] See European Commission, *Protection of Geographical Indications of Origin, Designations of Origin and Certificates of Special Character for Agricultural Products and Foodstuffs: Guide to Community Regulations* (2nd edn, 2004), pp. 4–5, at: http://ec.europa.eu/agriculture/publi/gi/broch_en.pdf.

[23] See currently Regulation 1151/2012, Arts 6(1), 13(1); Regulation 479/2008, Art 45; Regulation 110/2008, Arts 15–16.

[24] W van Caenegem, 'Registered GIs: Intellectual Property, Agricultural Policy and International Trade' [2004] EIPR 170, 173.

characterise the rural policies of such countries as non-interventionist and entirely free-market driven, the overarching agricultural policy framework in these countries remains different from that in place in Europe.[25] Even when the state intervenes in the US or Australia, it tends not do so in ways that chime with the logics of GI protection: in terms of agricultural production, state intervention is normally only focused on ensuring that minimum, usually health-driven, standards are met.[26] Thus, historically, these countries have not seen the need to establish registration schemes for regional names specifically to benefit rural, traditional producers. This affords at least a partial explanation for why such countries have relatively few globally-recognised GIs, and why such countries have been content to rely on traditional trade mark and consumer protection laws in order to safeguard against the misleading use of geographical insignia in the marketing of goods.

A further complication as to how GIs have been conceptualised in such 'new world' countries is that over the 19th and 20th centuries a significant number of famous European GIs for alcoholic beverages and foodstuffs were adopted as generic product descriptions in those countries. One explanation for this is that early agricultural and viticultural producers in such countries wished to emulate the products of the 'old world', and adopted European geographical insignia to try to give the best possible indication of the characteristics of their newly produced goods for their local consumers. For example, in Australia the 'Victorian Champagne Company' was established in 1881 with a view to producing local sparkling wine[27] and this company developed a local 'sparkling burgundy' shortly afterwards.[28] Sparkling wine made in accordance with the *méthode champenoise* (but with different grapes) was first made in rural Victoria in 1891 and sold as 'champagne'.[29] Both types of wine were produced by Australian entrepreneurs who employed French winemakers to advise

---

[25] See generally, J Dibden, C Potter and C Cocklin, 'Contesting the Neoliberal Project for Agriculture: Productivist and Multifunctional Trajectories in the European Union and Australia' (2009) 25 Journal of Rural Studies 299.

[26] For more detailed discussion, see M Handler and R Burrell, 'GI Blues: The Global Disagreement over Geographical Indications' in K Bowrey, M Handler and D Nicol (eds), *Emerging Challenges in Intellectual Property* (OUP, 2011) 134–135.

[27] *The Argus*, Melbourne, 26 September 1881, p. 6.

[28] M Clarkin, 'Red Fizz' (2000) 74(10) Law Institute Journal (Victoria) 46.

[29] Museum of Victoria, 'Hans William Henry Irvine, Vigneron & Politician (1856–1922)', http://museumvictoria.com.au/collections/themes/2690/hans-william-henry-irvine-vigneron-politician-1856-1922

them on how best to imitate French oenological practices in rural Australia.[30] A related reason for why European terms became generic in the new world relates to the fact that such countries have substantial immigrant communities, and locally-made products using European names were either produced by such people or produced to appeal to them. An example of the latter is an Australian 'feta' produced in the mid-20th century that was marketed to Greek post-war immigrants.[31] It is worth noting that the production of the Australian versions of 'burgundy', 'champagne' and 'feta' described above all predated these terms being formally recognised as GIs in France and Greece respectively.[32]

Given the different perspectives on GIs outlined in the previous paragraphs, it is not surprising that countries such as Australia and the US have viewed with scepticism the EU's domestic GI regime and the TRIPS extension campaign it is spearheading. To these and similar countries the EU's plans tend to appear to be little more than a crude attempt to secure monopolies for European producers by shielding certain valuable product names from competition, coming at a time when the EU is being required to phase out other forms of export subsidies for its agricultural

---

[30] See ibid and Clarkin, 'Red Fizz' (n 28). On the experience of German immigrants to the wine-producing Barossa Valley region of South Australia, see M de Zwart, 'Geographical Indications: Europe's Strange Chimera or Developing Countries' Champion?' in A Kenyon, WL Ng-Loy and M Richardson (eds), *The Law of Reputation and Brands in the Asia Pacific* (CUP, 2012).

[31] K Farrer, *To Feed a Nation: A History of Australian Food Science and Technology* (CSIRO Publishing, 2005) 176–177. See also Case C-317/95, *Canadane Cheese Trading AMBA and Adelfi G Kouri Anonymos Emoriki Kai Viomichaniki Etaireia v Hellenic Republic* [1997] ECR I-4681, para 17 (Ruiz-Jarabo Colomer AG) (noting the production of generic versions of feta in Europe in the post-war period 'to satisfy demand from communities of Greek immigrants in non-member countries'). For a recent illustration of how the term 'fetta' is understood in Australia, see the Federal Court of Australia's decision in *Yarra Valley Dairy Pty Ltd v Lemnos Foods Pty Ltd* [2010] FCA 1367 (involving a dispute over Yarra Valley Dairy's claim to trade mark rights in Australian-made 'Persian Fetta'). Lemnos Foods also produces a fine Australian-made haloumi, marketed as 'Cyprus style cheese': cf. the Cypriot application for EU GI registration of 'Halloumi' in 2009, since withdrawn (Dossier No. CY/PDO/0005/01243), at: http://ec.europa.eu/agriculture/quality/door/list.html, and the unsuccessful attempt to register HALLOUMI and ΧΑΛΛΟΥΜΙ as Community trade marks: Joined Cases T-292/14 and T-293/14, *Cyprus v OHIM* (7 October 2015).

[32] See D Gangjee, '(Re)Locating Geographical Indications: A Response to Bronwyn Parry' in L Bently, J Davis and J Ginsburg (eds), *Trade Marks and Brands: An Interdisciplinary Critique* (CUP, 2008) 389–390 (on the regulation of 'Champagne' in France between 1908 and 1927); *Canadane Cheese Trading* (n 31), para 20 (on the regulation of 'Feta' in Greece in the late 1980s).

producers.[33] What is equally unsurprising is that the EU has not sought to present its case for GI extension by focusing on the trade benefits that would accrue to its producers. Rather, it has attempted to make the case in more neutral terms. In countless communications to various WTO bodies it has focused on the perceived inadequacies of the present TRIPS regime and the benefits of GI extension for consumers and for producers generally, especially those from developing countries. It is to these arguments we now turn. However, when assessing these purported justifications for extension, the significant economic advantages that would accrue to European producers in relation to their already established GIs if the TRIPS Agreement were to be amended must always be kept firmly in mind. If none of these justifications are convincing, this raises real questions about whether expanded GI protection is in fact designed to do little else but set up protectionist measures aimed at propping up a European agricultural sector that is struggling to remain competitive in global markets.

3. ASSESSING THE PURPORTED JUSTIFICATIONS FOR EXTENDED GI PROTECTION

3.1 What the Imbalance between Arts 22 and 23 Does and Does Not Tell Us

The arguments that the EU and its supporters have deployed in favour of GI extension take a consistent form, whether raised in WTO communications or by commentators sympathetic to extension. The first argument that tends to be raised focuses on the lack of any justification for the different levels of GI protection contained in Arts 22 and 23 of the TRIPS Agreement. It is said that because the different minimum standards represented a political compromise made at the time of the GATT Uruguay Round negotiations and because there is nothing about wine and spirit GIs that mean that they deserve a higher level of protection than other GIs, then, as a consequence, the Art 23 level of protection should apply to all products.[34] Of all the arguments raised by advocates of extension, this

---

[33] See, e.g., C Lister, 'A Sad Story Told Sadly: The Prospects for US-EU Food Trade Wars' (1996) 51 Food & Drug Law Journal 303.

[34] See, e.g., WTO Docs IP/C/W/353 (24 June 2002), paras 4, 12 (referring to WTO Docs IP/C/W/204/Rev.1 (2 October 2000), paras 6–7; IP/C/W/247/Rev.1 (17 May 2001), paras 7–8, 15–16); and TN/C/W/14 (9 July 2003), paras 3, 9. This chain of reasoning has also been employed by various commentators in favour of exten-

is by far the weakest. The obvious problem with this argument is that it treats as axiomatic the very point that needs to be independently proved: that the Art 23 level of protection is an *appropriate* global minimum standard. There is, in fact, little disagreement between advocates and opponents of extension over the fact that the imbalance between Arts 22 and 23 is theoretically unsound, but this could just as easily support the conclusion that the additional protection for wines and spirits should be removed. Only those arguments that go to the issue of the appropriateness or otherwise of the Art 23 standard should be taken seriously.

What is in fact more interesting about the above argument is the attempt to frame the extension debate as something that can and should be considered entirely separately from other international trade negotiations. The clear suggestion is that whatever concessions on GIs were made to bring the TRIPS Agreement into being, these are irrelevant to the issue of whether the Agreement should be amended. Yet such a suggestion is hardly likely to be convincing to those countries that (as was described in the Introduction) made those concessions in the first place after having argued that the higher level of GI protection, even for wine and spirit GIs, was unnecessary.[35] As will be explored in more detail in the Conclusion, one of the major difficulties with the TRIPS GI extension debate is that extension is being presented as something that should be implemented as a matter of right, rather than as part of a broader set of interrelated trade negotiations.[36] Having said this, it is still important to see whether there are in fact stand-alone justifications for expanded GI protection that might make opponents of extension, who would otherwise be reluctant to accept any change to the TRIPS Agreement without receiving trade concessions in return, rethink their position.

3.2 Stand-Alone Arguments for Extension: An Overview

Proponents of GI extension tend to rely on a number of arguments in seeking to explain why the Art 23 level of protection should apply to all

---

sion: see, e.g., S Balganesh 'Systems of Protection for Geographical Indications of Origin: A Review of the Indian Regulatory Framework' (2003) 6 JWIP 191, 203; A Lang, 'On the Need to Expand Article 23 of the TRIPS Agreement' (2006) 16 Duke Journal of Comparative & International Law 487, 494–497.

[35] See, e.g., P Fowler and A Zalik, 'A US Government Perspective Concerning the Agreement on the Trade-Related Aspects of Intellectual Property: Past, Present and Near Future' (2003) 17 St John's Journal of Legal Commentary 401, 407.

[36] See, e.g., WTO Docs IP/C/W/360 (26 July 2002), para 3; IP/C/W/386 (8 November 2002), para 3.

products. These can be grouped into two categories. The first, essentially negative, set of arguments focuses on the alleged inadequacies of the Art 22 standard. More particularly, it is said that Art 22 generates uncertainty of outcome in different countries in relation to litigation over the same GI, that it creates significant costs for GI owners seeking to enforce their rights in foreign markets, and that it does nothing to prevent GIs from becoming generic, or to prevent 'free-riding'. The second set of arguments addresses the costs and benefits of extension. Broadly speaking, it is said that extension would improve consumer choice and information, that it would better reward the investment of producers in maintaining the quality of goods marketed under GIs and that it would help developing country producers in particular to secure access to foreign markets. It is also said that whatever the costs that would be incurred by countries having to implement the higher level of protection in their domestic laws, these would be more than offset by the abovementioned benefits of extension.[37]

There is something of a normative hollowness to the above arguments. As presented by advocates of extension, these arguments are very much an accretion of theoretical and pragmatic concerns, where no particular issue is given any particular weight, and which ultimately skirt around the question of what, exactly, justifies the imposition of a global standard that takes the form of Art 23.1 of the TRIPS Agreement. Thus, rather than addressing each of the above arguments in turn, a more fruitful approach is to go back to first principles and explore the justifications for protecting GIs and the normative consequences of such justifications. This approach, which focuses more on the harms that are or may be caused by third party use of GIs, helps to flesh out some of the more important potential arguments in favour of extension that tend not to have been articulated in careful detail by the EU and its supporters. It also helps to expose the fundamental weaknesses of such arguments.

---

[37] For the most convenient summary of the EU's and its supporters' arguments on extension, see WTO Docs IP/C/W/353 (24 June 2002) and TN/C/W/14 (9 July 2003). For arguments by commentators in favour of extension that raise some or all of these concerns, see in particular F Addor and A Grazioli, 'Geographical Indications Beyond Wine and Spirits: A Roadmap for a Better Protection for Geographical Indications in the WTO TRIPS Agreement' (2002) 5 JWIP 865; Lang (n 34); M Vittori, 'The International Debate on Geographical Indications (GIs): The Point of View of the Global Coalition of GI Producers – oriGIn' (2010) 13 JWIP 304.

3.3 Justifying GI Protection: Traditional Economic Arguments

Given that a GI is intended to convey information about the origin of a collective's goods as well as information about 'reputation' or 'quality' attributable to that origin, the most obvious way of seeking to justify protecting GIs would be to look to the economic rationale for protecting trade marks, which operate as badges of origin (albeit usually of a single trader's goods).[38] The dominant justification for protecting trade marks is that they improve the efficiency of the market by lowering consumers' search costs. That is, trade marks allow consumers to quickly identify goods and services they liked previously, thereby enabling them to make faster and more efficient purchasing decisions. In this way, trade marks also provide traders with incentives to maintain and guarantee the quality of their goods and services so that consumers can be reasonably confident that a branded product has much the same qualities as similarly-branded goods or services they have acquired before.[39] Broadly similar economic arguments can be made about the information communicated by GIs and the incentives they afford to traders to maintain the quality of goods marketed under GIs.[40]

What is key, however, is that the above justifications only support a level of legal protection that prevents third parties from engaging in conduct that disrupts the above-mentioned origin and/or quality guarantee functions. It is for this reason that, for the most part, the core legal protection afforded to trade marks in most jurisdictions is closely tied to a standard that looks to whether consumers are confused or misled by a third party's branding practices, the harm being the increase in

---

[38] See also K Raustiala and S Munzer, 'The Global Struggle Over Geographic Indications' (2007) 18 European Journal of International Law 337, 354–359 (assessing other justifications for protecting GIs normally used to explain other forms of IP, such as labour and desert, 'firstness', moral rights and incentives to create, and finding none of these to support the Art 23 standard).

[39] For well-known articulations of these arguments, see N Economides, 'The Economics of Trademarks' (1988) 78 Trademark Reporter 523; W Landes and R Posner, *The Economic Structure of Intellectual Property Law* (Cambridge, MA: The Belknap Press of Harvard University Press, 2003), ch 7.

[40] See, e.g., OECD Working Party on Agricultural Policies and Markets of the Committee for Agriculture, *Appellations of Origin and Geographical Indications in OECD Member Countries: Economic and Legal Implications* (December 2000), Annex 1, http://www.origin-food.org/pdf/meet0901/oecd.pdf; C Bramley and J Kirsten, 'Exploring the Economic Rationale for Protecting Geographical Indicators in Agriculture' (2007) 46 Agrekon 69, 74–77.

consumer search costs and the diversion of trade to that third party.[41] This in turn suggests that a misrepresentation-based standard for GI protection – that is, the Art 22.2 standard – is easily justified. At this point, it is worth emphasising the extent of protection for GI owners that is potentially provided by this standard. Art 22.2 requires Members to do more than ensure the prevention of conduct that causes confusion as to the provenance of goods. Given that Art 22.2(b) is not tied to misrepresentations as to geographical origin, this standard offers a strong degree of protection to GIs whose reputations have crossed over to foreign markets, where misrepresentations are made as to goods having certain qualities or characteristics associated with such GIs. The 'extended passing off' action that developed in a line of British Commonwealth cases from the 1960s is an excellent illustration of this. In these cases, European GI owners were able to restrain the sale in the UK of products marketed as 'Spanish Champagne',[42] 'Sherry' from South Africa, Cyprus and Australia,[43] 'Scotch Whisky' from Ecuador,[44] and British-made 'Elderflower Champagne',[45] and to prevent the sale in New Zealand of 'Australian Champagne'.[46] Importantly, in none of these cases were the sellers of these products found to have engaged in misleading conduct as to the geographical origin of their goods.[47] Rather, in each case the misrepresentation in question was that the goods had certain qualities and the cachet that was known to be exclusively associated with goods produced by the owners of the GI in question. While there are some ongoing difficulties with determining the precise contours of the extended passing

---

[41] It must, however, be noted that although the prevention of consumer confusion has always been understood as justifying trade mark rights, the core action for infringement of a registered trade mark in some jurisdictions has never turned on showing that a misrepresentation took place or even that confusion was likely. See further R Burrell and M Handler, *Australian Trade Mark Law* (OUP, 2010), chs 1 and 10.

[42] *Bollinger v Costa Brava Wine Company Ltd* [1960] Ch 262 (Ch).

[43] *Vine Products Ltd v MacKenzie & Co Ltd* [1967] FSR 402 (Ch). See further (n 70) below.

[44] *John Walker & Sons Ltd v Henry Ost & Co Ltd* [1970] RPC 489 (Ch).

[45] *Taittinger SA v Allbev Ltd* [1993] FSR 641 (CA).

[46] *Wineworths Group Ltd v Comité Interprofessionel du Vin de Champagne* [1992] 2 NZLR 327 (New Zealand Court of Appeal).

[47] See also *Chocosuisse Union des Fabricants Suisses de Chocolat v Cadbury Ltd* [1998] RPC 117 (Ch), where UK chocolate maker Cadbury was prevented from selling a locally-made 'Swiss Chalet' chocolate, with Laddie J finding that a substantial number of consumers would think that Cadbury's goods were not only Swiss but that they would have the particular qualities and characteristics of Swiss chocolate. This reasoning was not disturbed on appeal: [1999] RPC 826 (CA).

off action,[48] the existence and scope of such an action casts serious doubt on claims made by advocates of extension that anything less than the Art 23 standard affords inadequate or even no protection for GIs.[49]

More important for present purposes, however, is that the economic justifications set out above based on analogies with trade mark law *cannot* explain the higher level of GI protection under Art 23 that does not turn on consumer confusion. Conduct that neither misrepresents the geographical origin of goods nor misrepresents their qualities does not impact on consumer search costs and does not impair the origin or quality guarantee functions of the GIs in question. To return to an example raised in the Introduction, the term 'kanterkaas' may well be an EU-registered GI for cheese from the Friesland and Westerkwartier regions of the Netherlands,[50] but it would be fair to say that it is a sign that has little or no extant reputation in Australia. As such, Australian consumers seeing cheese marketed as 'Australian kanterkaas' or under the brand name 'Kanter Cheese' are highly unlikely to be misled, either as to the origin or qualities of the goods. There is no justification for the law intervening to restrain such conduct on the basis of the economic rationales described above.[51]

Indeed, it is worth noting there would seem to be two ways in which consumer search costs would in fact be *increased* if the EU's extension agenda were to be implemented. First, some qualifiers such as 'imitation'

---

[48] See Burrell and Handler (n 41) 455–456.

[49] For examples of such claims, D Vivas-Eugui, 'Negotiations on Geographical Indications in the TRIPS Council and their Effect on the WTO Agricultural Negotiations – Implications for Developing Countries and the Case of Venezuela' (2001) 4 JWIP 703, 712 (arguing that extension is needed so that 'misleading identification of products' can be prevented); B Babcock, 'Geographical Indications, Property Rights and Value Added Agriculture' (2003) 9(4) Iowa Ag Review 1, 3, at: http://www.card.iastate.edu/iowa_ag_review/fall_03/IAR.pdf.

[50] See Dossier No. NL/PDO/0005/0059, at: http://ec.europa.eu/agriculture/quality/door/list.html

[51] Even in the longer term, the idea that consumer search costs will be lowered if only one party is allowed to use the sign in question rests on an untested assumption about how consumers will respond to a particular type of terminology. It cannot be taken for granted that consumers will latch on to GIs as providing a second-order description of source that will allow them to make more informed purchasing decisions, a point that is to some extent supported by empirical work on consumer responses to GIs: see, e.g., C Bonnet and M Simioni, 'Assessing Consumer Response to Protected Designation of Origin Labelling: A Mixed Multinomial Logit Approach' (2001) 28 European Review of Agricultural Economics 433.

or 'like' when used alongside a GI in the marketing of a product[52] will often provide useful information to consumers as to the qualities or characteristics of that comparable product. An outright prohibition on such qualifiers would make it harder for the competing trader to describe its comparable product, particularly where there is no straightforward generic term that can be used as a substitute for the GI, and consequently harder for consumers to ascertain the qualities or characteristics of the third party's product.[53] Secondly, although the EU's plans for the multilateral GI register have become more clear in recent years, one interpretation of the EU's latest proposed model is that it will be difficult for countries to resist protecting foreign GIs that are recognised in some contexts as generic terms but in others as carrying geographical significance.[54] If the US, for example, were required to recognise 'Feta' as a GI, consumers would have to become accustomed to the fact that this term would denote only sheep's milk cheese in brine produced in particular regions in Greece, and that soft white cheese in brine produced in countries such as Australia, New Zealand, Denmark or Bulgaria formerly sold under that name would henceforth be known by a new generic descriptor. Not only will consumer search costs increase,[55] but the costs of developing and educating consumers as to these new generic names would likely be borne

---

[52] This, of course, assumes that such qualifiers are presented in such a way so that the overall appearance or marketing of the goods is not misleading.

[53] See Hughes (n 15) 381; I Calboli, 'Expanding the Protection of Geographical Indications of Origin under TRIPS: "Old" Debate or "New" Opportunity?' (2006) 10 Marquette Intellectual Property Law Review 181, 202.

[54] The EU's current proposal contemplates that Members' GIs will be entered on a multilateral register, which 'in the absence of proof to the contrary ... shall be considered as ... prima facie evidence' in a protecting country that the registered term meets the TRIPS definition of a GI. It is further stated that authorities in the protecting country 'shall consider assertions on the genericness exception laid down in TRIPS Article 24.6 only if these are substantiated': WTO Doc TN/C/W/52 (19 July 2008), Annex, para 2. The meaning of this is unclear, and it could be argued that the effect of registration is to put the onus on a party challenging the GI owner's rights in the protecting country to show that the sign is in fact generic. Recent litigation over the registration of 'Feta' as a GI in the EU shows the difficulties that parties using such terms generically might face in attempting to make out their case against recognition of a term as a GI. For consideration, see D Gangjee, 'Say Cheese: A Sharper Image of Generic Use through the Lens of *Feta*' [2007] EIPR 172.

[55] See WTO Docs IP/C/W/289 (29 June 2001), para 25; IP/C/W/360 (26 July 2002), para 26; IP/C/W/386 (8 November 2002), para 26. See also B Goebel, 'Geographical Indications and Trademarks – The Road from Doha' (2003) 93 Trademark Reporter 964, 988–989; see, further, Chapter 13 by Goebel and Groeschl in this volume.

by the product manufacturers and in turn passed on to consumers. That GI extension has the potential to increase consumer search costs in such ways calls into question the EU's assertions that GI extension would only benefit consumers by providing them with greater information about their purchases.

3.4 Preventing Dilution? Preventing Misappropriation?

If extended GI protection cannot be justified by reference to its positive impact on consumer search costs, the EU and its supporters face the more daunting task of finding a justification divorced from that which has been employed to explain the traditional contours of trade mark protection. Nevertheless, alternative arguments do exist, particularly for the type of additional protection offered by legal prohibitions on the 'dilution' of famous trade marks, which, like Art 23 of the TRIPS Agreement, do not turn on consumer confusion. Given that anti-dilution provisions are a well-established feature of US trade mark law, it is perhaps surprising that the EU and its supporters have not explicitly sought to draw analogies with such provisions in seeking to justify their extension agenda. However, when the theoretical basis of these laws is unpacked, it can be seen that they offer only a weak justification for Art 23 levels of protection for GIs.

The harms against which anti-dilution laws are said to protect are notoriously elusive. Given that such laws do not depend on consumer confusion, it is rare to see anti-dilution laws justified by reference to any benefit or protection that they might provide to consumers. It has been suggested that anti-dilution protection lowers 'imagination costs' for consumers by preventing them from having to think harder to recall a famous brand when seeing the same sign used by another party on its goods, even if such consumers are not confused as to the origin or other qualities of that other party's goods.[56] However, this 'imagination costs' argument has been the subject of scathing academic criticism[57] and enjoys no general acceptance, especially amongst the US judiciary.[58] The more widely accepted explanation for anti-dilution protection is that it serves to protect the brand owner's investment by preserving the 'attractive

---

[56] *Ty, Inc v Perryman*, 306 F 3d 509, 511 (7th Cir, 2002).

[57] See, most notably, R Tushnet, 'Gone in 60 Milliseconds: Trademark Law and Cognitive Science' (2008) 86 Texas Law Review 507; G Austin, 'Tolerating Confusion about Confusion: Trademark Policies and Fair Use' (2008) 50 Arizona Law Review 137.

[58] See *Moseley v V Secret Catalogue, Inc*, 537 US 418, 429 (2003) (anti-dilution laws 'are not motivated by an interest in protecting consumers').

force' of its famous mark or marks. Most relevantly for present purposes it is said that an owner of a well-known brand needs to be protected against dilution by 'blurring', this being the atrophying of the mark by its use across a range of (usually dissimilar) goods by parties other than the brand owner, which is said to impair the mark's source distinctiveness. An argument could be made that GI owners face similar problems. That is, it could be argued that the unchecked use of GIs by third parties in foreign countries will impair the distinctiveness of the GI and its ability to communicate unambiguous information about origin and quality, and that such a harm can only be prevented by an outright prohibition on the third party use of such terms that does not turn on the demonstration of any negative impact on consumers.

There are, however, a number of difficulties with relying on anti-dilution laws to justify the need for GI extension. Putting to one side the serious competition and speech implications for other traders in being prevented from making non-confusing use of certain signs in the marketing of their goods,[59] the key problem is that the harm that is said to be caused by dilution is something that needs to be taken as an article of faith.[60] As an empirical matter, it has never been convincingly demonstrated that a single, non-confusing use of a famous sign has any impact on that sign's source distinctiveness,[61] and much the same would apply to GIs. If it cannot be shown that an individual use of a famous mark or a foreign GI causes any particular harm, why then should the law intervene? The counter-argument to this is that 'blurring' is a type of progressive harm: its effects might not be perceptible in individual cases, but if nothing is

---

[59] For detailed arguments that anti-dilution laws pose an anticompetitive threat to market efficiency, see R Klieger, 'Trade Mark Dilution: The Whittling Away of the Rational Basis for Trademark Protection' (1997) 58 University of Pittsburgh Law Review 789.

[60] This point has been made by a number of recent commentators: see, e.g., C Haight Farley, 'Why We Are Confused about the Trade Mark Dilution Law' (2006) 16 Fordham Intellectual Property, Media & Entertainment Law Journal 1175, 1184 ('[t]he main problem with dilution law is that it provides a remedy without a supportable theorization of the harm' and querying whether the harm of dilution in fact exists). See also C Long, 'Dilution' (2006) 106 Columbia Law Review 1029, 1037; M LaFrance, 'No Reason to Live: Dilution Laws as Unconstitutional Restrictions on Commercial Speech' (2007) 58 South Carolina Law Review 709, 716–717.

[61] Attempts have recently been made to use cognitive science to demonstrate that 'blurring' does in fact exist in individual cases in an attempt to justify anti-dilution laws: see, e.g., M Morrin and J Jacoby, 'Trade Mark Dilution: Empirical Measures for an Elusive Concept' (2000) 19 Journal of Public Policy and Marketing 265. For trenchant criticism of these efforts, see Tushnet (n 57).

done about these individual cases the source distinctiveness of the famous mark will inevitably be eroded.[62] However, this argument still fails to articulate the harm that would result in the absence of legal protection: it assumes that 'death by a thousand cuts' will occur but without explaining what the 'death' in fact involves and why it is inevitable. It might be argued that allowing the potential whittling away of a mark's source distinctiveness would result in suboptimal investment in the development of new brands.[63] But this argument is unconvincing. Apart from the fact that attempts to justify trade mark type regimes by reference to the need to preserve incentives for 'brand entrepreneurs' have been subjected to considerable criticism,[64] the problem in the case of GIs is that incentive-based arguments can never justify the *ex post* creation of monopolies over existing subject matter, leaving the EU's core demands for protection of its established GIs unmet. More generally, it is incumbent on advocates of extension to demonstrate that the absence of the Art 23-level protection for GIs for goods other than wine and spirits in particular countries has led to an underinvestment in the production and marketing of such goods under GIs. This is something that they have failed to do.

A potentially stronger argument that might be made by advocates of GI extension that draws on trade mark dilution theory is that the progressive harm that is sought to be protected by the Art 23 standard is that of 'genericide'. Even though the focus of anti-dilution laws tends to be on the prevention of the use of a famous sign on dissimilar goods to ensure that the sign communicates unambiguous information about a single source, it is sometimes said that one potential type of dilution is where the distinctiveness of the sign in question becomes so undermined by third party use that the sign ends up becoming a generic product descriptor, such that the 'owner' of the sign loses any exclusive value in it.[65] This is clearly only likely to happen when the third party use is on the same goods as those

---

[62] For an early articulation of this argument, see F Schechter, 'The Rational Basis for Trademark Protection' (1927) 40 Harvard Law Review 813, 830.

[63] On trade mark laws being justified on the basis that they provide incentives for the creation of new brands, see, e.g., V Chiappetta, 'Trade Marks: More Than Meets the Eye' [2003] University of Illinois Journal of Law, Technology and Policy 35; M Richardson, 'Trade Marks and Language' (2004) 26 Sydney Law Review 193.

[64] See D Gangjee and R Burrell, 'Because You're Worth It: *L'Oréal* and the Prohibition on Free Riding' (2010) 73 Modern Law Review 282, 290 and the sources cited therein.

[65] See T Martino, *Trademark Dilution* (Oxford: Clarendon, 1996), ch 9. See also H Carty, 'Dilution and Passing Off: Cause for Concern' (1996) 112 Law Quarterly Review 632, 645.

provided by the 'owner'. This dilution-based argument would seem to map on to the concerns raised by advocates of extension that the Art 23 level of protection is needed to prevent GIs from becoming generic,[66] concerns that would seem to be given weight by the number of famous European GI owners aggrieved by the fact that their terms have become generic descriptors in other countries or by the expenditure they have had to incur to prevent their GIs becoming generic.[67]

An obvious problem with the above argument is that much of the conduct sought to be prevented under the Art 23 standard, such as use of a GI in a comparative manner or use in translation, is highly unlikely ever to make the GI generic.[68] A more fundamental problem, however, is that it assumes that the likelihood of a GI becoming generic turns entirely on the level of legal protection afforded to such GIs, when the far more significant reason relates to the conduct of the GI owner in enforcing what rights are in fact available to it. Whatever difficulties GI owners might have faced historically in seeking to protect their interests abroad – and it should be noted that some of these difficulties have been overstated[69] – if a GI is to become generic now, this would most likely be due to the inactivity of the GI owner in failing to take action under existing laws to protect its interests. If a GI owner wishes to sell its products in certain export markets, and those countries rely only on the Art 22 standard of protection, there are steps that the owner can take to ensure that its GI does not come to be adopted as the name of the goods in question products. The most obvious of these is to register the GI as a certification trade mark, which can in most cases be secured even before the owner maintains a sufficient reputation to bring a passing off or similar action to protect its interests. Advocates of extension have not been able to demonstrate convincingly that GIs have or are

---

[66] See, e.g., WTO Doc IP/C/W/353 (24 June 2002), para 13.

[67] See, e.g., the statements in *Consorzio del Prosciutto di Parma v Asda Stores Ltd* [1998] FSR 697, 701 (Ch) and *Taittinger SA v Allbev Ltd* [1993] FSR 641, 646 (CA) as to the number of cases brought to protect the 'Parma Ham' and 'Champagne' GIs.

[68] Indeed, it can be plausibly argued that 'free and fair imitation of the product often enhances the intrinsic value (and premium) of the genuine GI': WTO Doc IP/C/W/289 (29 June 2001), Attachment, para 8.

[69] For example, Australia introduced a certification trade mark regime in the mid-1950s (*Trade Marks Act 1955* (Cth), ss 83–92) and, before that, had a system for the registration of 'standardization' marks (*Trade Marks Act 1905* (Cth), s 22). Yet it was only in the mid-1990s that organisations such as the Stilton Cheese Makers Association and the Consorzio del Formaggio Parmigiano Reggiano took steps to avail themselves of this facility by registering their GIs as trade marks.

likely to become generic in countries relying on misrepresentation-based standards where GI owners have taken timely action to police their rights in those countries.[70]

If Art 23 protection is to be justified as a means of preventing future genericide, this can only be only be based on pragmatic grounds. That is, it would need to be accepted that for collectives that are only starting to develop a local reputation in goods marketed under a GI, the costs involved in seeking to take pre-emptive protective action abroad are so high that they justify the imposition of the Art 23 standard. Such a standard would prevent a third party from using that sign outright in order to ensure that if that collective one day wishes to trade internationally the sign has not become generic. This is not only a highly speculative concern, but raises a further question as to why GI owners should be privileged in this way. Trade mark laws do not provide this degree of pre-emptive protection for unknown marks in foreign markets. More generally, it is difficult to see why special allowances should be made to GI owners to restrain the potential genericide of their GIs when this is not done for trade mark owners, who are expected to be vigilant in enforcing their rights, throughout the world, at the risk of losing such rights. Enforcement costs and expediency are hardly sufficient reasons to increase minimum standards of GI protection at a global level.

As a final point, even if the progressive harm/genericide argument is taken at face value, there is a further problem in seeking to draw on trade mark anti-dilution laws to seek to justify GI extension in the form sought by the EU and its supporters. Anti-dilution protection under national or regional laws is always limited to 'famous' marks,[71] or marks with a reputation,[72] in the jurisdiction in question. The orthodox view is that only well-known marks are in need of protection against dilutive uses because of the investments that need to be made to develop and maintain the reputation and value in such signs. Yet the TRIPS Art 23 standard is not limited to GIs that have a particular reputation in the country where protection is sought. Again, the only reason for affording Art 23 level protection to signs without such a reputation would be for pre-emptive

---

[70] Cf. *Vine Products* (n 43), where the owner of the GI 'Sherry' was unable to restrain the use of 'British Sherry' because the GI owner had delayed for a number of years in enforcing its rights while the defendant established goodwill in its 'British Sherry'.

[71] See, under US law, 15 USC §1125(c).

[72] See, under European law, Directive 2008/95/EC of the European Parliament and of the Council of 22 October 2008 to Approximate the Laws of the Member States Relating to Trade Marks [2008] OJ L299/25, Art 5(2).

and/or pragmatic reasons, which are unlikely to be convincing to countries that would have to incur the costs of raising their domestic standards of protection accordingly.

In short, there are a number of problems in seeking to draw on anti-dilution law to justify expanded GI protection. The harms of dilution are, at best, difficult to perceive, and in some cases may well be attributable to the inaction of rights holders. Even if these harms are accepted, the significant costs that anti-dilution laws impose on other traders in restricting the language available to them in the (non-confusing) marketing of their own goods and services must always be remembered. It is for these reasons that trade mark anti-dilution laws remain so controversial, even in countries that might appear to have embraced them.[73] There would be something extraordinary about implementing in the TRIPS Agreement a minimum standard of protection for all GIs that draws on anti-dilution laws for normative support, when anti-dilution protection for famous trade marks remains so theoretically problematic and is not itself even mandated in the Agreement.[74]

The difficulties involved in drawing analogies with anti-dilution laws to justify the need for Art 23 protection might explain why much of the EU and its supporters' case for stronger protection has ultimately rested on bare claims about preventing 'misappropriation' of or 'free-riding' on the reputation of GIs (even if this reputation does not cross national boundaries).[75] Given the problems with relying on the economic justifications for extended GI protection outlined in the previous sections, it would appear that these sorts of appeals to commercial morality and fair play are the only remaining way of explaining why WTO Members should protect others' GIs in the absence of any misleading conduct.

---

[73] See, e.g., B Beebe, 'The Continuing Debacle of US Antidilution Law: Evidence from One Year of Trademark Dilution Revision Act Case Law' (2008) 24 Santa Clara Computer & High Technology Law Journal 449 (noting US courts' continuing scepticism towards anti-dilution law, even following significant legislative revision).

[74] See Michael Handler, 'Trade Mark Dilution in Australia?' [2007] EIPR 307, 308–310. For recent confirmation that TRIPS, Art 16.3 does not mandate anti-dilution protection for well-known marks, see the Singapore Court of Appeal's decision in *Novelty Pte Ltd v Amanresorts Ltd* [2009] 3 SLR 216.

[75] See, e.g., IP/C/W/247/Rev.1 (17 May 2001), para 11; IP/C/W/308/Rev.1 (2 October 2001), paras 17, 21; IP/C/W/353 (24 June 2002) at para 4. See also Lang (n 34) 490, 493 (referring to 'parasitic free-riding'); M Agdomar, 'Removing the Greek from Feta and Adding Korbel to Champagne: The Paradox of Geographical Indications in International Law' (2008) 18 Fordham Intellectual Property, Media and Entertainment Law Journal 541, 581 (referring to the need to prevent 'agropiracy').

Domestic laws that are aimed at preventing 'free-riding' without the need to show harm in the form of consumer confusion or even dilution are rare, but not unknown. For example, the ECJ has recently held that the prohibition in the European Trade Marks Directive on the taking of 'unfair advantage ... of the distinctive character or the repute' of a mark with a reputation would prevent a third party from attempting 'to ride on the coat-tails of that mark in order to benefit from its power of attraction, its reputation and its prestige, and to exploit, without paying any financial compensation and without being required to make efforts of his own in that regard, the marketing effort expended by the proprietor of that mark in order to create and maintain the image of that mark',[76] conduct that the ECJ also called 'parasitism' and 'free-riding'.[77] However, such laws are even more problematic than anti-dilution laws,[78] and do not provide a compelling argument in favour of extended GI protection.

The rhetoric of 'misappropriation', 'free-riding' and 'piracy' is commonly employed in intellectual property debates – such terms are designed to appeal to an intuitive sense of behaviour that is 'wrong' and thus ought to be prevented.[79] But it has also long been recognised that without further elaboration or justification such claims carry little force. If the problem with 'free-riding' on the reputation of GIs is viewed in economic terms, the issue is not whether one can point to the mere existence of some positive externality, such as the marketing advantage that flows to a third party in being able to use a foreign GI to convey the comparable qualities of its own goods. Rather, it is whether the law ought to intervene to ensure that that positive externality can be fully internalised by the GI owner. This can only be the case if to allow the non-confusing use of GIs by third parties that takes advantage of the GI's reputation or attractiveness would lead to an underinvestment in the creation or development of new GIs,[80] something there is no evidence to suggest would occur. In other words, even if it is accepted that some third party use of foreign GIs involves 'free-riding' on reputation or investment, advocates of extension cannot point to a

---

[76] Case C-487/07, *L'Oréal SA v Bellure NV* [2009] ECR I-5185, para 49.
[77] Ibid., para 41.
[78] For sustained criticism of the ECJ's decision, see Gangjee and Burrell (n 64) (on whose arguments I draw in the following paragraph) and Jacob LJ's judgment in *L'Oréal SA v Bellure NV* [2010] RPC 23.
[79] See D Franklyn, 'Debunking Dilution Doctrine: Toward a Coherent Theory of the Anti-Free-Rider Principle in American Trademark Law' (2004) 56 Hastings Law Journal 117.
[80] See generally D Barnes, 'Trademark Externalities' (2007) 10 Yale Journal of Law and Technology 1.

free-rider *problem* that justifies legal intervention, on economic grounds, in the form of the Art 23 standard. If the issue is to be looked at in moral terms, we need to move beyond the fact that there has been an act of 'copying' or 'borrowing' and ask what it is that makes such an act unfair.[81] Too often claims of unfairness or injustice or piracy in relation to the use of foreign GIs rest solely on the assumption that the user is appropriating the GI owner's 'intellectual property',[82] but since 'property' is the label given to what the law protects, such claims fail to address the question of why the law should grant protection to GI owners in the form of the Art 23 standard in the first place.[83]

Bald complaints about the theft or misappropriation of intellectual property should always be viewed sceptically, but there is a further reason why this is particularly true of complaints about the supposed misappropriation of GIs. In so far as terms such as 'champagne', 'port', 'feta' or 'haloumi' enjoy a positive reputation amongst consumers in certain countries, this cannot be attributed solely to the efforts of European wine or cheese makers. On the contrary, it was seen in Section 2 above that in countries such as Australia it was often locals or immigrants who popularised these beverages and foodstuffs amongst the broader community through goods produced *locally*. On this view, producers of Australian 'feta', 'haloumi' and the like are merely reaping the benefits of their labour and the labour of their predecessors. Any argument that such terms should become the exclusive property of European producers would involve transferring the benefit of this labour and investment to European producers without compensation – it is European producers who would be 'reaping without sowing'.

In summary, there are serious problems with attempting to rely on the idea that extended GI protection is needed to prevent dilution or misappropriation.[84] And the costs of such higher standards must always be kept

---

[81] See generally M Spence, 'Passing Off and the Misappropriation of Valuable Intangibles' (1996) 112 Law Quarterly Review 472.

[82] See, e.g., S Goldberg, 'Who Will Raise the White Flag? The Battle Between the United States and the European Union Over the Protection of Geographical Indications' (2001) 22 University of Pennsylvania Journal of International Economic Law 107, 140 (TRIPS-plus standards are needed 'to sufficiently protect . . . intellectual property rights'). See also WTO Docs IP/C/W/247/Rev.1 (17 May 2001), para 11; IP/C/W/308/Rev.1 (2 October 2001), paras 10, 14, 17, 21.

[83] Cf. E Weinrib, 'The Fiduciary Obligation' (1975) 25 University of Toronto Law Journal 1, 10–11.

[84] Dev Gangjee makes a similar point about the near-identical standard contained in Art 3 of the Lisbon Agreement 1958, noting that this standard cannot be justified on traditional 'unfair competition' grounds and instead owes more to a model based on bilateral treaties in which a party would agree to protect the GIs of the

in mind: as Barton Beebe has argued, '[f]rom the perspective of the public domain, freedom of commercial speech, and the goals of progressive intellectual property law, the terms of Article 23 are dismaying'.[85] That the EU and other advocates of extension have not been able to develop a cogent response to the critiques set out in this section merely reinforces the perception that the case for extension is little more than a grubby attempt to secure an unwarranted worldwide monopoly, applied consistently throughout the world, for established GI producers.[86]

3.5 Policy Arguments: Fostering the Production and Consumption of Quality Goods?

If none of the arguments used to justify trade mark protection can be used to justify the Art 23 standard of protection, and the 'misappropriation' argument is little more than an emotive appeal to commercial morality devoid of any independent normative foundation, the EU and its supporters need to turn more explicitly to policy arguments to support their extension agenda. Again, there are precedents in other areas of the law for imposing outright restrictions on the use of certain insignia on policy grounds alone. One example might be 'ambush marketing' legislation, often adopted by countries in the anticipation of hosting major sporting events, that proscribes the use of certain listed expressions or prevents parties associating themselves with the event in question. Such legislation involves a conscious (if not necessarily carefully weighed) decision to restrict commercial speech in exchange for an unrelated set of social benefits, such as the development of new infrastructure, increased revenue from advertising and tourism, and support for local sporting organisations.[87] Advocates of GI extension might be able to make broadly similar arguments in favour of higher standards of GI protection.

---

other in exchange for mutual benefits: Gangjee, *Relocating the Law of Geographical Indications* (n 3) ch 4. I return to this theme of bilateralism in the Conclusion.

[85] B Beebe, 'Intellectual Property Law and the Sumptuary Code' (2010) 123 Harvard Law Review 809, 873.

[86] To put it another way, it is not enough to complain that extended GI protection is needed to ensure predictability of outcome in relation to the same GI throughout the world if there is no independent justification for the Art 23 standard. For further criticism of the 'predictability of outcome' complaint, see M Handler, 'The EU's Geographical Indications Agenda and Its Potential Impact on Australia' (2004) 15 Australian Intellectual Property Journal 173, 186–188.

[87] For criticism, see L Longdin, 'Public Law Solutions to Private Law Problems: Major Events Legislation Subverts IP's Internal Balance' (2009) 4 JIPLP 726.

The most interesting policy reason that the EU and its supporters have raised in the cause of GI extension relates to the desirability of diversity in agricultural production, or the need to foster the production of 'quality' goods. By appealing to the uniqueness of place, GIs are said to offer the promise of something produced in accordance with traditional, perhaps artisanal, practices that are superior in quality to (or at least substantially different from) competing goods. In this respect GIs are put forward as affording an antidote to a globalised agri-food industry that tends to be characterised by an over-production of broadly homogenous goods, sourced as cheaply as possible from largely anonymous sources, which are then sought to be differentiated primarily by price.[88] At a time when many consumers are turning away from these sorts of mass-produced commodities and are expressing preferences for 'locally' or 'sustainably' produced goods as part of the 'new food movements',[89] GIs operate as a convenient shorthand for values such as heritage, cultural diversity and authenticity.[90]

Many would be sympathetic to the idea that there needs to be a change in the mindset of consumers in the developed world about the production and consumption of food. Notwithstanding this, the claim for a link between the generation of such a cultural change and higher standards of GI protection is weak at best. For a start, consumers who are concerned to locate goods produced at a particular place and/or in a particular manner can already do so readily. Parmigiano Reggiano, Feta produced in Greece, Prosciutto di Parma and Halloumi from Cyprus are readily available throughout the world, and any consumer keen to avoid mass-produced cheaper alternatives will have no trouble doing so.[91] Consequently, increased GI protection is in no way a precondition for the emergence of a changed attitude towards food production. The claim must therefore be

---

[88] For recent mainstream critiques of the American agri-food industry, see *Food, Inc.* (Magnolia Pictures, 2009) and J Safran Foer, *Eating Animals* (London: Hamish Hamilton, 2009).

[89] M Pollan, 'The Food Movement, Rising' The New York Review of Books (10 June 2010).

[90] See generally, D Giovannucci, E Barham and R Pirog, 'Defining and Marketing "Local" Foods: Geographical Indications for US Products' (2010) 13 JWIP 94.

[91] Even if there are cases where foreign GI owners might face difficulty in securing the registration of their names as certification trade marks because similar signs have already been registered as trade marks (as to which see D Gangjee, 'Quibbling Siblings: Conflicts Between Trademarks and Geographical Indications' (2007) 82 Chicago-Kent Law Review 1253, 1270–1276), consumers who are sufficiently motivated to do so should always be able to source the original product.

that higher international standards of GI protection will help generate a broader cultural shift in attitudes. In the case of terms that already enjoy a high level of international market recognition, the argument would be that GI extension would serve to reserve well-recognised signs for high-end products, thus encouraging consumers to try (better) quality goods. In the case of new GIs the argument would be that a high level of protection would encourage producers to market their goods by reference to geographical insignia, thereby encouraging greater respect for localism, tradition and the importance of place. But even if one is untroubled by the importance that this account assigns to law in changing cultural attitudes, or the idea that GI-branded goods are inherently superior to goods that are not so branded, attempts to link calls for extended GI protection to the rise of the new food movements are problematic.

The most obvious problem is that once they enter the international arena, GIs fit very uncomfortably alongside calls for the consumption of sustainable, locally-produced foods – judged in environmental terms, a regime that encourages people to buy foodstuffs grown or produced many thousands of kilometres away (as the EU clearly intends) has to be judged with extreme suspicion. There is also something disingenuous about any attempt to tie GI extension to 'artisanal' standards and practices. Whilst it is true that, within parts of Europe, many GIs are owned by small collectives with local markets,[92] on the world stage the GIs that are being disputed are predominantly employed by large corporate entities. However good Champagne may be, the French Champagne houses are very much part of, and not an answer to, global agri-business. Even the idea that GIs invariably reflect 'traditional' practices has come under strong criticism, particularly in the case of relatively new GIs.[93] A further problem with the 'cultural change' argument is that it does not necessarily follow that setting legal standards to encourage the greater adoption of GIs will necessarily assist consumers in making choices about their food purchases. Even if it is accepted that some of the 'information clutter' problems associated with having numerous competing certifying and standard-setting agencies in relation to foodstuffs can be overcome through a state-controlled

---

[92] These groups are themselves unlikely to benefit significantly from sales beyond their immediate region through the use of GIs: see Josling (n 14) 360.

[93] See T Broude, 'Taking "Trade and Culture" Seriously: Geographical Indications and Cultural Protection in WTO Law' (2005) 26 University of Pennsylvania Journal of International Economic Law 623, 676 ('In many cases, new GIs are indeed attempts to establish instant reputations through invented traditions that build a novel culture through self-reference to the distant or at least irrelevant past.').

GI-registration system, the problem remains that even these registered GIs convey relatively little meaningful information about the particular qualities of the goods in question, the precise standards used to assess these qualities, or what distinguishes such goods from non-GI comparators or even from other GI-branded goods.[94]

If the EU faces problems selling its agenda in terms of appeals to diversity in agricultural production even to supporters of the new food movements, for governments in countries like the US, Australia, Canada or New Zealand any agenda sold in these terms is obviously problematic, since it rests on hostility to the dominant model of agricultural production and distribution in those countries. There is, moreover, an implicit claim to the mystique and superiority of old world products and values contained in the EU's appeals that is unlikely to find favour with new world policy-makers. Insofar as such policy-makers are sceptical about the uniqueness or superiority of old world locales, they would have the support of geographers who are critical of the idea that 'place' should be considered as something immutable that can lend fixed, irreproducible characteristics to agricultural produce.[95]

3.6 Policy Arguments: Supporting Developing Country Farmers and Producers?

Can other policy factors then justify GI extension? While European producers would clearly be the primary beneficiaries of GI extension, it is also important to note that a significant number of developing countries support the extension of GI protection. Initially, such countries were those that could point to one or more locally-made products sold under GIs, such as tea, coffee, rice or chocolate, that had already secured significant international market access and had developed a strong global reputation.[96] More recently a broader range of developing and least-developed countries has given its support at least to further negotiations

---

[94] See M Chon, 'Marks of Rectitude' (2009) 77 Fordham Law Review 2311, 2339–2341. See also, in relation to French wine-labelling, Broude, ibid., 672 ('It has simply become too difficult for the casual, nonexpert consumer to maintain a working knowledge of French appellations and their association with the kinds of wine he or she wants most.').

[95] See, e.g., B Parry, 'Geographical Indications: Not All "Champagne and Roses"' in Bently, Davis and Ginsburg (n 32) 361.

[96] See (n 16) and the signatories to WTO Docs IP/C/W/204/Rev.1 (2 October 2000); IP/C/W/247/Rev.1 (17 May 2001) and IP/C/W/308/Rev.1 (2 October 2001).

on extension.⁹⁷ This raises the question of whether extension is justified on the basis of the benefits that might flow to farmers and producers from such countries, notwithstanding the other problems with increased GI protection outlined in the previous sections.

The argument that EU-style GI protection can benefit developing countries is rarely explained in careful detail. Proponents often go little further than pointing to the existence of goods from these countries, or the traditional knowledge of producers of such goods, that could be marketed under GIs.⁹⁸ The starting point for the more nuanced argument in favour of extension would seem to be that building up a reputation in agricultural produce is a time-consuming and expensive business. Much agriculture in developing countries is conducted by individual farmers or small collectives that will never have the resources to develop recognisable brands in foreign export markets. GI protection can be said to offer a means of allowing a large group of such farmers or collectives to secure strong rights in a geographical name quickly. Once the term is recognised as a GI, it is said that a high level of protection will present the GI owner with niche marketing opportunities in foreign markets (where the geographical name will already be protected) and will allow it to charge a price premium, with consequent rents flowing back to the farmers or small collectives.⁹⁹

While seeking to improve the livelihood of developing country farmers and producers is clearly an important and pressing development issue, there are a number of difficulties with the above arguments in favour of GI extension. As Justin Hughes has argued in a detailed study of developing country coffee and cocoa production, the essential problem with such arguments is that they tend to mistake 'the piling up of laws for the

---

⁹⁷ See especially WTO Doc TN/C/W/52 (9 July 2008) (negotiations on extension supported by the ACP and African Groups (see 'Groups in the WTO' [www.wto.org/english/tratop_e/dda_e/meet08_brief08_e.doc] for membership of these groups)). It is, however, arguable that the real interest of many of the ACP and African countries in adopting this position is to secure greater support for an amendment to the TRIPS Agreement to require the disclosure of the source of genetic resources and/or associated traditional knowledge in patent applications. For discussion of the EU's role in seeking to align 'disclosure of origin' and GIs as negotiation issues, see D Robinson and C Gibson, 'Governing Knowledge: Discourses and Tactics of the European Union in Trade-Related Intellectual Property Negotiations' (2011) 43 Antipode 1883, 1899.

⁹⁸ For a recent example, see Vittori (n 37) 306–307.

⁹⁹ See C Correa, 'Protection of Geographical Indications in Caricom Countries' (September 2002) 38–39, at: http://www.crnm.org/index.php?option=com_docman&Itemid=82; G Evans and M Blakeney, 'The Protection of Geographical Indications after Doha: Quo Vadis?' (2006) 9 Journal of International Economic Law 575, 607–608.

accumulation of reputational capital', with only the latter being 'the real way to help developing world agricultural products'.[100] Increasing global standards of protection would do almost nothing to improve market access for GIs from developing countries, given that almost none of these GIs have an already established global reputation.[101] Instead, producers and governments from these countries would first need to make significant investments in establishing and maintaining the quality of locally-made products to be sold abroad, as well as certifying compliance with quality and safety standards.[102] They would also need to make concerted efforts to market their products to retailers in export markets, so that their GIs become signs of commercial value: the mere fact that such products are marketed under a GI will not necessarily guarantee that they can be sold at a premium or that the product will be a commercial success.[103] Crucially, the precise level of legal protection that is afforded to GIs is largely irrelevant to the process of developing and maintaining a marketable international reputation. Indeed, developing country producers have already had substantial success in marketing goods such as 'Darjeeling' and 'Café de Colombia' under certification trade mark schemes around the world. The more pressing issues here are ensuring that rights in foreign markets are adequately enforced, and that any premiums that are obtained from the sale of goods under GIs are ultimately returned to the farmers and producers, rather than absorbed by government authorities or other private actors.[104]

---

[100] J Hughes, 'Coffee and Chocolate – Can We Help Developing Country Farmers Through Geographical Indications?' (International Intellectual Property Institute, Washington, DC, 2009) 7, at: http://papers.ssrn.com/sol3/papers.cfm?abstract_id=1684370. On the global cocoa trade generally, see Ó Ryan, *Chocolate Nations: Living and Dying for Cocoa in West Africa* (London: Zed Books, 2011).

[101] A Kur, 'Quibbling Siblings – Comments to Dev Gangjee's Presentation' (2007) 82 Chicago-Kent Law Review 1317, 1323–1324.

[102] See Gangjee, *Relocating the Law of Geographical Indications* (n 3) 285–286; Vivas-Eugui (n 49) 718; Correa (n 99) 39.

[103] See M Yeung and W Kerr, 'Are Geographical Indications a Wise Strategy for Developing Country Farmers? Greenfields, Clawbacks and Monopoly Rents' (2011) 14 JWIP 353. The accepted wisdom that consumers are more willing to pay a premium for GI-branded goods is not always borne out by the (admittedly limited) empirical evidence: see, e.g., A Tregear, S Kuznesof and A Moxey, 'Policy Initiatives for Regional Foods: Some Insights from Consumer Research' (1998) 23 Food Policy 383; M Loureiro and J McCluskey, 'Assessing Consumer Response to Protected Geographical Identification Labeling' (2000) 16 Agribusiness 309.

[104] See generally Hughes (n 100), 46–51, 115–122, 131–134. See also W McBride, 'GI Joe? Coffee, Location, and Regulatory Accountability' (2010) 85

Further, it should also be noted that adopting the Art 23 level of protection would entail significant costs for developing countries, both in terms of setting up new domestic laws, registration schemes and certification mechanisms, and in relation to protecting (an inevitably larger number of) European GIs under their domestic laws. A developing country would need to think carefully about whether the requisite resources would not be better spent in other ways.[105] In short, the process of securing higher incomes for farmers in developing countries is clearly far more complicated than merely encouraging the adoption of high GI protection standards, and an inordinate focus on levels of GI protection has the potential to deflect attention away from more important development issues that might genuinely improve the livelihood of developing country farmers.[106]

4. CONCLUSION: WHY MORE ATTENTION NEEDS TO BE PAID TO GIs IN PREFERENTIAL TRADE AGREEMENTS

It is ultimately unsurprising that the EU and its supporters have failed to develop a robust narrative to explain why the GI standards in the TRIPS Agreement ought to be increased. Whether looked at in economic terms,

---

New York University Law Review 2138 (arguing that while it might be thought that the adoption of domestic GI protection would empower developing country coffee producers by fostering participatory and transparent regulatory environments so as to facilitate the collective management of their reputation, developing countries without existing and well-developed institutional infrastructures will have difficulty corralling the many actors that are likely to seek to exploit GIs for private benefit).

[105] See, e.g., M O'Kicki, 'Lessons Learned from Ethiopia's Trademarking and Licensing Initiative: Is the European Union's Position on Geographical Indications Really Beneficial for Developing Nations?' (2009) 6 Loyola University of Chicago International Law Review 311 (on the Ethiopian government's rejection of *sui generis* GI protection in favour of reliance on standard trade mark laws to market geographically branded coffee). See also Hughes (n 100) 112–115 (arguing that the Ethiopian case is perhaps better understood as an illustration of the difficulties some developing countries might face in policing even minimal certification standards).

[106] Cf. B Mercurio, 'Resolving the Public Health Crisis in the Developing World: Problems and Barriers of Access to Essential Medicines' (2007) 5 Northwestern Journal of International Human Rights 1 (on the limitations of reforming the TRIPS Agreement in attempting to ensure access to medicine in the developing world).

or on policy grounds, the case for requiring all WTO Members to adopt the Art 23 level of protection is weak. Further, the EU and its supporters have done little to package their calls for extension in a way that would be attractive to countries that do not share the EU's agricultural policies or its attitudes towards GIs. While expanded GI protection is clearly likely to benefit established European producers in export markets, the same cannot be said for the countries that oppose the EU's expansionist agenda, yet such countries are not being offered any countervailing trade benefits in return for supporting GI extension. The EU's failure to address the relationship between GIs as intellectual property and as instruments of broader trade policy means that a multilateral solution to the GI extension debate that focuses exclusively on the provisions of the TRIPS Agreement, but which does not involve a genuine agreement to open up agricultural markets, is highly unlikely.[107]

Indeed, the TRIPS GI extension debate should best be considered as something of a sideshow. This is because real advances in securing higher levels of GI protection are happening through trade negotiations at the *bilateral* level – something that has not perhaps received the attention it has deserved. Over the past 20 years or so, the EU has managed to persuade a number of wine exporting countries (some of whom have been the most vocal critics of the EU's efforts to increase the TRIPS GI standards) to afford TRIPS-plus protection for European wine and spirit GIs, in return for granting producers from those countries increased access to European markets. This can be seen in the various agreements on trade in wine and/or spirits into which the EU has entered with Australia,[108]

---

[107] In part due to frustration with the stalemate over the TRIPS GI negotiations, efforts have been made over recent years to revitalise the Lisbon Agreement for the Protection of Appellations of Origin and their International Registration 1958. These led to the adoption of the Geneva Act of the Lisbon Agreement on Appellations of Origin and Geographical Indications 2015. This new agreement applies to GIs as well as appellations of origin and sets up an International Register for GIs. However, any suggestion that this new model might somehow accommodate some of the concerns of those opposed to the EU's extension agenda ignores the underlying causes of the global disagreement of GIs relating to market access and divergent agricultural policies.

[108] Agreement between the European Community and Australia on Trade in Wine [1994] OJ L 86/94, superseded by the Agreement between the European Community and Australia on Trade in Wine [2009] OJ L28/3. For consideration of the latter agreement, see V Waye, 'Wine Market Reform: A Tale of Two Markets and Their Legal Interaction' (2010) 29 University of Queensland Law Journal 211, and see also the Federal Court of Australia's decision in *Comité Interprofessionnel du Vin de Champagne v Powell* [2015] FCA 1110 (20 October 2015).

Mexico,[109] South Africa,[110] Switzerland,[111] Chile,[112] Canada[113] and the US,[114] each containing detailed provisions on the recognition and protection of specific, listed wine and/or spirit GIs.[115] A major feature of these agreements is that countries are obliged to protect or phase out the use of specific European GIs on wines or spirits from those countries, even if these are generic descriptors in such countries. This has allowed European producers to 'repropertise' valuable terms such as 'champagne', 'port', 'sherry' and 'tokay' in key export markets, putting local traders in those countries to the expense of finding new ways to describe their locally-made products.[116] In return, such countries have been provided with more stable access to European markets for locally-produced wines

---

[109] Agreement between the European Community and the United Mexican States on the Mutual Recognition and Protection of Designations for Spirit Drinks [1997] OJ L152/16.

[110] Agreement between the European Community and the Republic of South Africa on Trade in Wines [2002] OJ L28/4; Agreement between the European Community and the Republic of South Africa on Trade in Spirits [2002] OJ L28/113.

[111] Agreement between the European Community and the Swiss Confederation on Trade in Agricultural Products [2002] OJ L114/132, Annex 7 (wine) and Annex 8 (sprits).

[112] Agreement Establishing an Association between the European Community and its Member States, of the One Part, and the Republic of Chile, of the Other Part [2002] OJ L352/1, Annex V (wine) and Annex VI (spirits).

[113] Agreement between the European Community and Canada on Trade in Wines and Spirit Drinks [2004] OJ L35/3.

[114] Agreement between the European Community and the United States of America on Trade in Wine [2006] OJ L87/2. Although the Agreement refers only to 'names of origin', and Article 12.4 states that the names to be protected 'are not necessarily considered, nor excluded from being considered, geographical indications', under either US or European law, it is hard to see how the Agreement could be interpreted other than to impose conditions on the US's treatment of European GIs. For consideration, see B Rose, 'No More Whining about Geographical Indications: Assessing the 2005 Agreement between the United States and the European Community on the Trade in Wine (2007) 29 Houston Journal of International Law 731.

[115] For detailed consideration of these Agreements, see M Handler and B Mercurio, 'Intellectual Property' in S Lester, B Mercurio and L Bartels (eds), *Bilateral and Regional Trade Agreements: Commentary and Analysis* (2nd edn, CUP, 2015) 334–340. See also D Vivas-Eugui and C Spennemann, 'The Treatment of Geographical Indications in Recent Regional and Bilateral Free Trade Agreements' in M Perez Pugatch (ed), *The Intellectual Property Debate: Perspectives from Law, Economics and Political Economy* (Cheltenham: Edward Elgar Publishing, 2006) 305.

[116] See, e.g., Winemakers' Federation of Australia, 'Australian Fortified Wines: The Dawning of a New Era' (2009), http://www.wfa.org.au/assets/strategies-plans/

that are made in accordance with techniques and practices not otherwise recognised in the EU.[117] While the EU's efforts in the bilateral arena have been focused on wine and spirit GIs, its more recent energies have been directed at securing higher level protection for *all* types of GI. By way of a 2008 Economic Partnership Agreement with the EU, a bloc of Caribbean countries has agreed to afford the TRIPS Art 23 standard of protection to GIs for all goods under their domestic laws.[118] More recently, in the EU-Korea Free Trade Agreement, the Republic of Korea has agreed to provide the TRIPS Art 23 level of protection to a large number of listed European GIs for agricultural foodstuffs, a major effect of which will be that producers of generic goods such as 'feta' and 'parmesan' in third party countries will be forced to rebrand their goods for export to the South Korean market.[119]

It would be simplistic to say that the EU is likely to achieve through preferential trade agreements what it has failed to achieve at the WTO. One reason for this is that the US has also been addressing GIs in its preferential trade agreements over the past ten or so years. Rather than

---

pdfs/Fortified_Wines_Strategy.pdf (on plans to market 'apera' as a replacement for 'sherry' and 'topaque' as a replacement for 'tokay').

[117] For a summary of such benefits flowing to US wine producers, see http://www.ttb.gov/wine/itd_qas.shtml.

[118] Economic Partnership Agreement between the CARIFORUM States and the European Community [2008] OJ L289/I/3, Art 145B (but note Art 145C, which allows such countries to apply a 'genericism' exception).

[119] Free Trade Agreement between the European Union and the Republic of Korea [2011] OJ L127/6, Arts 10.18.4, 10.21.1 and Annex 10-A. But cf. Trade Agreement between the European Union and its Member States, of the one part, and Colombia and Peru, of the other part [2012] OJ L354/3, Art 210.1(b) of which requires that Peru and Colombia prevent the 'non-authorised use [which may cover any misuse, imitation or evocation] of geographical indications other than those identifying wines, aromatized wines or spirits drinks *that creates confusion*, including even in cases where the name is accompanied by indications such as style, type, imitation and other similar *that creates confusion to the consumer*' (emphasis added), which falls short of the TRIPS, Art 23 standard. For comment on the draft agreement, see P Covarrubia, 'The EU and Colombia/Peru Free Trade Agreement on GIs: Adjusting Colombian and Peruvian National Laws?' (2011) 6 JIPLP 330, 336–337. The issue of GIs is likely to remain controversial in other preferential trade agreements the EU is negotiating: see, e.g., C Viju, M Yeung and W Kerr, 'Geographical Indications, Conflicted Preferential Agreements, and Market Access' (2013) 16 Journal of International Economic Law 409 (on the planned Comprehensive Economic and Trade Agreement between the EU and Canada (CETA)); see further Consolidated CETA Text (26 September 2014), Art 7, http://trade.ec.europa.eu/doclib/docs/2014/september/tradoc_152806.pdf.

requiring its trading partners to protect a long list of US terms,[120] the US's approach has been to encourage such countries to adopt a trade mark model of protection for GIs and to manage potential conflicts between traditional trade marks and GIs by giving priority, wherever possible, to the former.[121] It is also likely that in future agreements the US will seek to impose requirements on its trading partners limiting their ability to enter into agreements with third parties that involve the automatic recognition of each other's GIs and the provision of the TRIPS Art 23 standard of protection for such GIs.[122] These approaches seem clearly designed to foster resistance towards the EU's attempts to make its model of GI protection a de facto global standard.[123]

What is more interesting about the treatment of GIs in the EU's bilateral agreements is they involve a far more explicit recognition of GIs as instruments of trade policy. That is, countries that might have little interest in increasing GI standards in the abstract are being offered some other trade benefit in return, with such countries being given the opportunity to consider whether such a trade-off would be in their overall national interests. This is something that is missing from the TRIPS GI extension debate. But it would be a mistake to suggest that bilateralism offers a simple way of resolving global disagreements over GIs. Trading levels of GI protection for market access is itself highly controversial:[124] it gives trade negotiators an extraordinary degree of power in being able to fix the meaning of certain terms, taking away from domestic courts and other tribunals the ability to assess whether particular signs qualify for

---

[120] The US has, however, sought specific recognition for the names 'Bourbon Whiskey' and 'Tennessee Whiskey' in the Market Access chapters of some of its preferential trade agreements.

[121] For consideration of these Agreements see Handler and Mercurio (n 115).

[122] See Section E of Chapter 18 (especially Art 18.36) of the Trans-Pacific Partnership Agreement 2015, https://ustr.gov/sites/default/files/TPP-Final-Text-Intellectual-Property.pdf. The negotiating parties are Australia, Brunei Darussalam, Canada, Chile, Japan, Malaysia, Mexico, New Zealand, Singapore, Peru, the US and Vietnam.

[123] See also M Handler, 'The WTO Geographical Indications Dispute' (2006) 69 Modern Law Review 70 (on the tactical reasons for the US and Australia bringing WTO dispute settlement proceedings against the EU in relation to its GI registration scheme).

[124] Indeed, negotiations between the EU and South Africa in the late-1990s over the much larger Trade, Development and Cooperation Agreement almost broke down over the EU's insistence that South Africa cease using certain generic wine and spirit denominations: see E Craven and C Mather, 'Geographical Indications and the South Africa-European Union Free Trade Agreement' (2001) 33 Area 312, 313–315.

GI protection at all, and depriving traders of language customarily used to market their goods.[125] As Antony Taubman has argued in relation to the work of such negotiators, 'any such extreme incursion on the public domain as a constraint on the common tongue should serve a public good of high priority'.[126] For those concerned about the consequences for particular countries of increased GI protection and the relinquishing of generic product descriptors for wine, foodstuffs and other goods through preferential trade agreements, what is needed is greater scrutiny of the work of such bilateral trade negotiators, to ensure that their decisions are made transparently and with careful explanations of how and why they serve this greater 'public good'.

---

[125] See Taubman (n 3) 262–263. See also Beebe (n 85) 873–875. Cf. T Battaglene, 'The Australian Wine Industry Position on Geographical Indications', WIPO Worldwide Symposium on Geographical Indications, Parma, 27–29 June 2005, at 9 (on how allowing the EU to monopolise certain 'generics' might ultimately advantage Australian wine producers as consumers look more to brand names and varietals).

[126] Taubman (n 3) 267.

8. International protection of Geographical Indications: the WTO multilateral register negotiations
José Manuel Cortés Martín

1. PROTECTION OF GEOGRAPHICAL INDICATIONS UNDER THE TRIPS AGREEMENT

One of the features of the TRIPS Agreement at the time of its adoption was that not all categories of intellectual property rights regulated therein had the same degree of legal or doctrinal development at the national level; neither had they the same degree of acceptance among countries. This was the case with Geographical Indications (GIs), a peculiar type of intellectual property asset because they do not confer individual rights (such as in the case of patents and trade marks) but rather 'collective rights'.[1] In such a case, the right over a geographical name does not belong to a single company, but to all producers in a given geographic area that respect a specific code of conduct. Furthermore, the exclusive rights granted by GIs do not extend to exclusivity over a certain category of products, as in the case of patents. The producers of Sherry wine are not entitled, nor do they wish, to prevent others from producing similar wine. The right conferred by the GI is limited to banning competitors outside the defined geographic area (or inside the geographic area, for those not respecting the product specification) from using the name 'Sherry' in connection with their products.

It is due to these peculiarities that the inclusion of GIs in the TRIPS Agreement caused heated debates during the Uruguay Round and continues to generate discussion between the new and old world.[2] Like many

---

[1] For a general overview of the TRIPS GI provisions, see Chapter 6 by Gervais in this volume. See also M Blakeney, *Trade-Related Aspects of Intellectual Property Rights: A Concise Guide to the TRIPS Agreement* (Sweet & Maxwell, 1996) 351–432; TP Stewart, *The GATT Uruguay Round: A Negotiating History (1986–1992)* (Kluwer, 1999) 110.

[2] On the one hand, some European countries, representing the Old World, have a long tradition of protecting this type of intellectual property. On the other hand, the United States, Canada, Australia and New Zealand, among others,

183

184 *Research handbook on intellectual property and GIs*

aspects of the Uruguay Round negotiations, the disagreement among Members impeded the creation of a comprehensive system for the international protection of GIs. As a result, there remain a series of outstanding issues with respect to GIs. The two most prominent issues, on which negotiations continue, are GI-extension[3] and the creation of a multilateral register. This chapter retraces the debates and proposals surrounding the creation of an international registration system for wine and spirit GIs. It explores the implications of a new draft text, which emerged in 2011, for the establishment of an international register, and which has the potential to make progress in this area.

In the case of GIs, the appropriate legal treatment and level of protection continued to be fiercely debated between WTO Members.[4] The debate over GI protection did not follow the usual North–South divide; instead, the dispute creates a different dichotomy of states, with 'emigrant' nations on one side and 'immigrant' nations on the other.[5] The 'emigrant nations' – the European Union, Switzerland and former Eastern bloc countries and a selection of developing nations – support extensive GI protection, while countries like Australia, New Zealand and the United States ally with Latin American nations and other 'immigrant' nations to oppose GI protection.[6] This explains why during negotiations, GI

---

representing the New World, have not historically had separate laws to protect GIs, apart from their respective systems of trademarks. See JM Cortés Martín, 'WTO-TRIPS Agreement: The Battle between the Old and the New World over the Protection of Geographical Indications' (2004) 7 JWIP 287.

[3] For a detailed study of the extension of protection see Chapter 7 by Handler in this volume.

[4] See L Lindquist, 'Champagne or Champagne? An Examination of U.S. Failure to comply with the Geographical Provisions of the TRIPs Agreement' (1999) 27 Georgia Journal of International and Comparative Law 309, 312 ('The inclusion of these [GI provisions] caused heated debates during the Uruguay GATT Rounds and continues to generate discussion. The article that causes most debate is Article 23 which deals with the protection of [GIs] for wines and spirits.').

[5] 'Emigrant' countries include those in Europe, Africa and parts of Asia, whereas 'immigrant' countries include the United States, Australia and Latin American countries. Especially for European countries, GIs have a long and proud tradition. Since antiquity, their existence has served to distinguish products and to indicate source, serving a similar function to that of present-day trade marks.

[6] See JM Cortés Martín, 'TRIPS Agreement: Towards a Better Protection for Geographical Indications?' (2004) 30 Brooklyn Journal of International Law 117; F Addor and A Grazioli, 'Geographical Indications Beyond Wines and Spirits – a Roadmap for a Better Protection for Geographical Indications in the WTO TRIPS Agreement' (2005) 5 JWIP 865, 883; affirming that the suitable protection of GIs has never been a conflict of interests between developed and developing countries, but between the countries of the Old World and the New World.

protection was a very sensitive issue. Only at the very end of the Uruguay Round was an agreement concerning GIs reached, and this was largely due to the parties' ability to link GIs with the agricultural negotiations taking place at the time.[7] According to the TRIPS Agreement, GIs are subject to the same general principles applicable to all categories of IPRs included in the Agreement, primarily the 'minimum standards',[8] the 'national treatment'[9] and the 'most-favoured-nation clause'.[10] As a result, one big difference from the pre-WTO situation is that GIs are now embedded in the WTO system, as they comprise one of the categories of intellectual property that are the subject of the TRIPS Agreement, which itself is an integral part of the WTO Agreement. Consequently, non-compliance with TRIPS obligations concerning GIs can be challenged under the WTO dispute settlement mechanism, and if a country fails to implement a panel ruling requiring compliance, it could eventually be faced with sanctions in areas of international trade governed by other parts of the WTO Agreement and lose benefits that accrue to it under that agreement for as long as it does not remedy the situation.[11]

2. NEGOTIATIONS FOR A MULTILATERAL SYSTEM OF GI REGISTRATION

At the Uruguay Round, proposals by the EU, the United States, and Switzerland were indispensable to framing eventual obligations concerning

---

[7] The proposals were presented by the European Union, the United States, Switzerland, Japan, and a group of developing countries consisting of Argentina, Brazil, Chile, China, Colombia, Cuba, Egypt, India, Nigeria, Peru, Tanzania and Uruguay. See, e.g., WTO Negotiating Group on TRIPS, *Draft Agreement on Trade-Related Aspects of Intellectual Property Rights*, presented by the European Union, GATT doc. MTN.GNG/NG11/W/68, 29 March 1990.

[8] TRIPS, Art 1: 'Members shall give effect to the provisions of this Agreement. Members may, but shall not be obligated to implement in their domestic law more extensive protection than is required by this Agreement.'

[9] TRIPS, Art 3: 'Each Member shall accord to the nationals of other Members treatment no less favourable than that it accords to its own nationals with regard to the protection.'

[10] TRIPS, Art 4: 'With regard to the protection of intellectual property, any advantage, favour, privilege or immunity granted by a Member to the nationals of any other country shall be accorded immediately and unconditionally to the nationals of all other Members . . . .'

[11] See M Geuze, 'The Provisions on Geographical Indications in the TRIPS Agreement' (2009) 10 The Estey Centre Journal of International Law and Trade Policy 1, 51.

GIs. For example, key elements like additional protection for wines and spirits (presently found in Art 23) and for a multilateral register for indications of wines and spirits were present in the EU Proposal. The eventual framework reflects 'a very sensitive compromise in an area that was one of the most difficult to negotiate'.[12] However, WTO negotiators did not resolve all issues on the table. Instead WTO Members agreed a 'Built-In Agenda' for future negotiations designed to facilitate international protection of this legal category. This means that under the TRIPS system, WTO Members must negotiate, *inter alia*, the establishment of a multilateral notification and registration system for GIs.[13] The precise terms of this obligation are contained in Art 23.4, which states:

> In order to facilitate the protection of geographical indications for wines, negotiations shall be undertaken in the Council for TRIPS concerning the establishment of a multilateral system of notification and registration of geographical indications for wines[14] eligible for protection in those Members participating in the system.[15]

---

[12] See M Geuze, 'Protection of Geographical Indications under the TRIPS Agreement and Related Work of the World Trade Organization' in WIPO, *Symposium on the International Protection of Geographical Indications in Eger, Hungary* (WIPO Publication No. 760, 1997) 199.

[13] TRIPS, Art 68.

[14] TRIPS, Art 23.4 exclusively addresses GIs for wine. However, the WTO Singapore Ministerial Conference in 1996 also included spirits. See WTO, 'Report (1996) of Council for TRIPS' 6 November 1996 (IP/C/8) at para. 34: ('In regard to GIs ... the Council will initiate ... preliminary work on issues relevant to the negotiations specified in Article 23.4 of the TRIPS Agreement [and] [i]ssues relevant to the notification and registration system for spirits will be part of this preliminary work. All of the above work would be conducted without prejudice to the rights and obligations of Members under the TRIPS Agreement.').

[15] This article was introduced in TRIPS at the request of the European Union, although its original proposal contemplated coverage applicable to all GIs. See Draft Article 21(3) in GATT Negotiating Group on TRIPS, 'Draft Agreement on Trade-Related Aspects of Intellectual Property Rights' 29 March 1990 (MTN. GNG/NG11/W/68). The Proposal was included in the text of the Chairman's Report to the GNG, 'Status of Work in the Negotiating Group' 23 July 1990 (MTN.GNG/NG11/W/76). During the autumn of that year, some countries were in favour of the creation of this register in the Uruguay Round. They even informally presented fairly detailed proposals which contemplated the creation of this register and these were debated by the Negotiations Group. However, other countries were committed solely to examining this question in the future. This disagreement was reflected in the project presented by the President of the Negotiations Group to the Ministerial Meeting of Brussels in December, 1990. See GATT Secretariat, 'Draft Final Act Embodying the Results of the Uruguay Round of Multilateral Trade Negotiations' 3 December 1990 (MTN.TNC/W/35/Rev.1).

In 2001, part of this work became part of the work programme of the Doha Development Agenda, as adopted by the WTO's Ministerial Conference.[16] In the Hong Kong Ministerial Declaration, adopted on 18 December 2005, Ministers took note of a progress report on the negotiations in the Special Session and agreed to intensify them in order to complete them within the overall timeframe for the conclusion of the negotiations that was foreseen in the Doha Ministerial Declaration.[17] Lastly, the Seoul Summit Document, agreed at the G20 Seoul Summit, 11–12 November 2010, asked for a broader and more substantive engagement in order to bring the Doha Development Round to a successful, ambitious, comprehensive and balanced conclusion consistent with the mandate and built on the progress achieved.[18]

The submissions presented at the TRIPS Council for the establishment of this register can be divided into two camps. On the one hand, there is the minimalist approach proposed by the United States together with a coalition of 19 countries like Canada, Australia, New Zealand, along with many Latin American nations (Joint Proposal Group).[19] The second proposal, presented by Hong Kong, China, also advocated a modest approach, in which the registration of a GI would lack a process of substantive examination or opposition at the multilateral level.[20] By contrast, a more comprehensive approach is favoured by the European

---

[16] The mandate of the Special Session is set out in the first sentence of paragraph 18 of the WTO Doha Ministerial Declaration, 20 November 2001 (WT/MIN(01)/DEC/1) (hereafter, Doha Declaration).

[17] WTO, 'Doha Work Programme – Ministerial Declaration', 22 Dec 2005 (WT/MIN(05)/DEC) at para. 29.

[18] Once such an outcome is reached, G20 members commit to seek ratification, where necessary, in their respective systems. See WTO Trade Negotiations Committee, 'Seoul Summit Document' 26 November 2010 (TN/C/W/57) at para. 43.

[19] Argentina, Australia, Canada, Chile, Costa Rica, Dominican Republic, Ecuador, El Salvador, Guatemala, Honduras, Japan, Rep. Korea, Mexico, New Zealand, Nicaragua, Paraguay, Chinese Taipei, South Africa, the US ('Joint Proposal Group'), 'Proposed Draft TRIPs Council Decision on the Establishment of a Multilateral System of Notification and Registration of Geographical Indications for Wines and Spirits' 1 April 2005 (TN/IP/W/10) and Addenda 1–3. In March, 2011, the Joint-Proposal Group circulated a revision of its proposal, adding a section on special treatment for developing countries, aligning the formatting with the structure of the draft and using its preferred wording from the composite draft. See TRIPS Council Submission, 31 March 2011 (TN/IP/W/10/Rev.4).

[20] The proposal is contained in Annex A of the Communication from Hong Kong, China, 23 April 2003 (TN/IP/W/8).

188 *Research handbook on intellectual property and GIs*

Union as part of a coalition of 108 WTO Members[21] (the so-called W52 sponsors), which includes Switzerland, former Eastern bloc countries, and a huge number of developing countries.[22] This approach embraces a modified and stripped-down version of the EU's original proposal for the multilateral register. Previously the EU had proposed that if a term is registered the assumption – the legal phrase was 'irrebuttable presumption' – would be that it should be protected in all WTO Members except those that have successfully challenged the term.[23] As matters presently stand, the negotiated compromise among the sponsors envisages a system applying to all WTO Members where they have to take a term's registration 'into account' and treat it as *'prima facie'* evidence (first sight, or preliminary, before further investigation) that the term meets the TRIPS definition of a GI in Art 22.1. As a result, this new proposal reflects an attempt at compromise, with the EU conceding ground in an attempt to bring an end to the deadlock that has continued for so many years.

In the rest of this chapter, an attempt will be made to elaborate upon the different negotiating proposals concerning the notification and registration system of GIs. The following issues are of particular concern: (i) the legal form of the register; (ii) the nature of participation; (iii) mandatory or optional elements to be notified; (iv) the registration process, including examination; and (v) the legal consequences of registration.

2.1 The Joint Proposal

The register proposed in this approach, which can be described as the least common denominator for the negotiation, is characterized solely by its informative nature. As a result, a GI registration would not create legal rights and consequently it would not require protection by other WTO

---

[21] The EU-led proposals are contained in the following WTO documents: (i) 28 July 1998 (IP/C/W/107); (ii) 14 June 2005 (TN/IP/W/11); and (iii) 19 July 2008 (TN/C/W/52).

[22] This proposal is sponsored by Albania, Brazil, China, Colombia, Croatia, Ecuador, the European Union and its 27 Member States, Georgia, Iceland, India, Indonesia, the Kyrgyz Republic, Liechtenstein, Moldova, the Former Yugoslav Republic of Macedonia, Pakistan, Peru, Sri Lanka, Switzerland, Thailand, Turkey, the Africa, Caribbean and Pacific Group (The ACP Group) and the African Group. Both these groups together consist of a further 61 WTO Members. See Albania et al., 'Draft Modalities for TRIPS Related Issues' 19 July 2008 (TN/C/W/52) and Addendas 1–3.

[23] See Bulgaria et al., 'Negotiations Relating to the Establishment of a Multilateral System of Notification and Registration' 24 June 2002 (TN/IP/W/3).

Members. Additionally, this system is strictly voluntary, which means that no WTO Member shall be required to participate.

The joint proposal is succinct and has not been significantly modified since it was tabled in 2002. Subsequent revisions have been made to add co-sponsors but no substantive modifications have taken place. It is based on the assumption that GIs are territorial rights and, therefore, the conditions for granting and exercising them must be established in national fora. This means that under this system, any GI established in accordance with national legislation would be entitled to protection, regardless of whether it is registered in the WTO database. Moreover, the proposal states that, in accordance with Art 23.3, identical or similar GIs may be submitted by more than one WTO Member, provided such designations have been recognized by each of them in accordance with their national systems.

With regard to legal effects under national legislation, participants would be legally bound to consult the database, along with other sources of information, while non-participants would be encouraged to refer to the database.[24] Registration itself would not generate obligations for protection, but would merely create a source of information. As a result, multilateral registration would not give rise to any presumption regarding protection outside of the 'home' country. Thus, Members' participation would be limited to receiving these lists of recorded GIs, among other sources of information, to be referred to when they must make decisions on the protection in their territories of GIs of other Members. With regard to appeals or objections, the proposal sets out that decisions concerning protection for GIs, regardless of whether the WTO is notified, should occur at the national level at the request of interested parties.

2.2 The Hong Kong, China Proposal

The second proposal was presented by Hong Kong[25] and it attempted to reconcile the minimalist approach of the US-led Group with the initial EU maximalist approach. Having evolved over time, it is no longer truly a middle-ground proposal. Instead, it is much closer to the European-led

---

[24] Thus Draft Art 6 states that non-participants are 'encouraged, but are not obliged, to consult the Database in making decisions under their domestic law involving registration or protection of trademarks and geographical indications for wines and spirits'. See TN/IP/W/10/Rev.2.

[25] Hong Kong, China has no interest in GIs, but it has a systemic interest in fulfilling the negotiating mandate under Art 23.4 of the TRIPS Agreement. With this objective in mind Hong Kong, China tabled the alternative proposal in 2003, in TN/IP/W/8.

Group, except for participation in the system. Concerning legal effects, registration would take effect only after a cursory, formal examination of the GI subject to questions relating to its conformity with the general definition of Article 22.1. If a term is registered, this would be preliminary evidence ('*prima facie*') – which could be rebutted – about who owns the term and that it is protected in the country of origin. But the TRIPS exceptions would remain applicable in the local jurisdiction in accordance with domestic law.

As regards participation, protection only embraces those WTO countries choosing to participate in the system. However, those participating in the WTO system would be compelled to give legal effects to registrations. The proposed multilateral system would involve only a formal examination of the notified GIs. This would involve checking the documents submitted by a WTO Member to ensure they met the minimum formal requirements. The register would contain the information on the registered GIs and would be made available to participating Members, for example through a database available on the Internet. Concerning legal effects, the Hong Kong proposal states that the registration would be accepted by any domestic court, tribunal or administrative body of the participating WTO Members as *prima facie* evidence for proving three issues: (i) ownership; (ii) conformity with the TRIPS definition in Article 22.1; and (iii) protection in the country of origin, as specified in Article 24.9. These three issues would be deemed to have been proved unless evidence to the contrary was produced by WTO Members challenging the validity of the registration. In short, registration would give rise to a rebuttable presumption on these three issues. As a result, legal effects would be restricted to altering the burden of proof, in favour of GI owners, during domestic litigation proceedings.[26]

2.3 The EU-led Proposal

Initially, the European Union proposed a full registration system, combining elements from the Lisbon Agreement[27] and the EU Agricultural Products Regulation.[28] This ambitious proposal tried to favour legal

---

[26] TRIPS Council, 'Special Session – Minutes of Meeting' 19 October 2009 (TN/IP/M/22) at para. 45 ('Consequently, according to the Hong Kong Proposal, this would facilitate GI protection through Members' domestic legal systems.').

[27] Lisbon Agreement for the Protection of Appellations of Origin and their International Registration, 31 October 1958, 923 UNTS 205 (1974). See also Chapter 5 by Geuze in this volume.

[28] Under Council Regulation (EC) No. 510/2006, which replaced Regulation 2081/92, the protected designation of origin (PDO) allows agricultural producers

certainty by advocating the creation of a system which would be opened to GIs on all products that would benefit from unconditional protection in all WTO markets upon entry in the register. However, in order to put an end to the deadlock, the EU-led Group has subsequently made a considerable effort to reduce the legal consequences of the register. Accordingly, in 2008 the EU relaxed its position concerning the legal effects of registration, while maintaining that registering should imply the GI will be protected in all the WTO Member's markets.

As part of the concessions made, the proposed register would not be an automatic protection system. Instead, GI right holders would still have to go through the procedures provided by each country, meaning that all exceptions would remain available and that it would be the national authorities who would decide whether to grant protection. Consequently, the first concession is related to the consultation of the register by domestic authorities when making decisions on registration and protection of trade marks and GIs. The entry of a name on the register would trigger two legal effects.

Firstly, it would be considered *prima facie* evidence that, in the 'home' country, the notified GI met the definition laid down in TRIPS Art 22.1. This means that the notifying WTO Member would have checked that there was a product associated with the designation, a defined place of origin, and that the link between product and place was satisfied – for example, a characteristic of that product was linked to its geographical origin. The EU's present position is very reasonable, the more so as it would apply 'in the absence of proof to the contrary'. In other words, domestic authorities retain the freedom to decide that they would consider that a notified GI did not meet a GI definition in their country since there was evidence to the contrary.

The second consequence of a registration is also reasonable: as generic status is mentioned as an exception to protection in Article 24.6, it would be expected that those who claim a registered term is generic would have

---

in EU Member States an exclusive right to the name of a particular foodstuff that is determined to be unique because the production, processing or preparation takes place in a specific area using local expertise and resources. The protected geographical indication (PGI) also gives an exclusive right to a name for foodstuffs, but unlike the PDO, it does not require such a strong link to the region and also grants protection on the basis of a product's reputation. See JM Cortés Martín, *La Protección de las Indicaciones Geográficas en el Comercio Internacional e Intracomunitario* (Ministerio de Agricultura, Pesca y Alimentación, 2003) 325–451. Regulation 510/2006 was in turn recently replaced by Regulation (EU) No 1151/2012 of the European Parliament and of the Council of 21 November 2012 on Quality Schemes for Agricultural Products and Foodstuffs [2012] OJ L343/1.

to substantiate this.[29] Thus, domestic authorities should consider assertions of generic status only if these were substantiated by appropriate and adequate evidence.[30] Beyond that, domestic authorities would have full latitude to decide for or against protection of a term on the basis of contrary evidence provided by themselves or brought by any third party.

This means that this proposal gives careful consideration to domestic legal systems. It would be up to national authorities to act within the framework of their domestic system in accordance with their own law. Furthermore, the checks and balances of the exceptions provided in Article 24 TRIPS would remain unchanged.

2.4 A Critical Appraisal of the Proposals

As an overall assessment, although fundamental positions have not changed, in the present author's view it seems there exists a genuine and sincere desire on the part of delegations to move forward and resolve the remaining differences in these negotiations, so as to be ready to contribute to any movements in the wider context of the Doha Round negotiations. This section reflects on the state-of-play and two remaining challenges. These relate to the legal effects of registration and the extent of participation.

With respect to the consequences that should flow from an entry on the international register, two general approaches are on the table, namely that: (a) an entry should merely result in better information being available, to be used by decision-makers and national systems; and (b) that an entry should result in a legal presumption in national systems.

According to the minimalist approach group favouring a merely informative register, a legal presumption is not acceptable for a number of reasons: firstly, a legal presumption would increase the *scope of legal protection* for GIs, and this would be outside the scope of this negotiation, which is only about *facilitating* protection; secondly, a legal presumption would violate the principle of territoriality; and, thirdly, a legal presumption would alter the balance of rights and obligations in the TRIPS Agreement. By contrast, these countries prefer a limited information system in which national GIs would be notified and incorporated automatically. However, WTO Members sponsoring this system have not yet

---

[29] See Albania et al., 'Draft Modalities for TRIPS Related Issues' 19 July 2008 (TN/C/W/52) Annex at [2].
[30] For possible categories of such evidence, see Chapter 18 by Gangjee in this volume.

explained how they would implement the obligation to 'consult and take into account' the information on the register. As legal uncertainty for market operators, including registered GI users, their competitors and trade mark owners, ought to be avoided, there is a need to elaborate on these issues.

The Joint Proposal calls for the establishment of a simple database as a source of information that Members' authorities may or may not consult; however, even if they consult it, it is unclear what the legal import of this consultation would be. In this author's view, it should not be up to WTO Members to decide whether or not to take into account information on the register, essentially for the following reasons: firstly, leaving countries to decide would create legal uncertainty and discrepancies, and would of course not be in the interest of the right holders or of business in general; and, secondly, this approach does not fulfil the mandate which called for a registration system, not a database system, which would amount in practice to a duplication of the information already provided by the applicants and therefore would not add any value.

As regard legal effects, it would seem that a multilateral IPR register clearly must imply multilateral protection and this should be the key requirement when establishing such a register. However, the US-led Joint Proposal is limited to creating a record rather than a true registration regime. The system does not provide for a mechanism to filter out names that should not be protected and therefore risks creating more confusion than clarity. The proposal is silent on the need for elements of proof, for the assessment of eligibility, or for an opposition procedure – elements which seem indispensable in the context of an IPR register. Under this approach, it is impossible to ensure that terms which do not meet the provisions of Art 22.1 (the definition) or Art 24.9 (requiring protection in the country of origin as a precondition), or which fall under one of the exceptions provided for in Art 24, are denied eligibility. The US-led Proposal also does not establish procedures to resolve possible legal disputes, an indispensable procedure for any future multilateral register. The great uncertainty regarding legal effects may thus increase litigation and, consequently, administrative costs. It also does not provide for any monitoring mechanism which requires national authorities to 'refer' to the lists of GIs on the database. As a result, these national authorities will not know whether to rely on the information included in the system when making a determination on the protection of a GI. For all these reasons, it is difficult to understand how the mandate to facilitate the protection of GIs established in Art 23.4 would be fulfilled through this system.

As to the participation in the system, the minimalist approach group also does not provide for a system with a truly multilateral character. In reality, it is unclear whether non-participating Members would be bound to give protection according to Art 23. If non-participating Members were not bound, the mandate to facilitate GIs protection – as mandated in Art 23.4 – would be undermined. The literal meaning of the US-led Proposal seems based on a political commitment without legal force: authorities would be 'encouraged' to refer to the register, yet the register gives rise to no national legal commitment. In the author's opinion, Art 23.4 calls for more ambitious action than this proposal offers. The proposal concentrates on the first part of the job, namely the establishment of a notification system, where the register would simply compile participating Members' information. At the time of writing, it is unclear whether this would satisfy the requirement of producing legal effects that registration inherently should entail in the context of IPRs.[31] If transparency alone is the only advantage offered by the proposed US-led system, it might not be sufficient to justify its costs. To meaningfully 'facilitate' legal protection for GIs under Art 23.4, a multilateral system should help administering bodies implement, and producers and consumers avail themselves of, legal protection. To respond to this mandate, it seems essential after so many years of negotiations to foresee at least that, beyond the simple obligation to consult a source of information, WTO Members should provide some clear assurances that the national authorities responsible for GIs – judges, trade mark examiners or other authorities – would have the obligation not only to consult the information in the register but also to take due account of this information when making decisions by giving it all the necessary weight.

By contrast, the modified and less ambitious proposal sponsored by the EU could, in this author's view, help to facilitate GI protection as prescribed by Art 23.4. This proposal strikes a balance between different interests and would be the appropriate tool towards a register which would truly facilitate GI protection, and not duplicate what would already be available essentially via the Internet. The proposal would not entail any automatic protection, would not oblige WTO Members to change their protection system and would not generate excessive administrative costs and burdens. Arguably, these are the reasons why this proposal presently enjoys the support of nearly two-thirds of the WTO Membership.

---

[31] This can be deduced from other sections of TRIPS which employ the word 'registration', most notably in Part II, Section 2, on trade marks. See TRIPS Arts 15, 18 and 19.

3. THE ROAD AHEAD

Early in 2011, a drafting group of negotiators produced a comprehensive draft composite text on the proposed multilateral register for GIs for wines and spirits, which was presented to a formal TRIPS Council negotiation meeting on 3 March 2011. This text comes entirely from the WTO members and covers six main areas:

1. Notification – e.g., how a GI term would be notified and which WTO Member would do it (also related to 'participation');
2. Registration – e.g., how the system would be run and the WTO Secretariat's role;
3. Legal effects/consequences of registration, in particular any commitments or obligations on WTO Members arising from a term's registration (also related to 'participation');
4. Fees and costs – including who would bear these burdens;
5. Special treatment for developing countries (officially, 'special and differential treatment'); and
6. Participation – whether the system will be entirely voluntary and effective only for those who opt in, or whether a term's registration would have some implications for all WTO members.

This text is to be welcomed as it can form the basis for moving ahead after over 13 years of talks that have seen many views exchanged but little movement in positions.[32]

According to the core of the Doha Development Agenda, which seeks to place developing countries' needs and interests at the heart of the Work Programmes, this composite text addresses for the first time special and differential treatment. Indeed, there are details in this new text of proposed special treatment for developing and least-developed countries. This would include delays in implementing the system and technical assistance from developed countries. A key difference is over the delays; namely, the 'transition periods'. Some developing countries who are part of the EU-led proposal favour giving developing countries ten years after the system comes into being before they have to consult terms in the register, and 20 years for least-developed countries. By contrast, the Joint-Proposal

---

[32] See TRIPS Council, 'Multilateral System of Notification and Registration of Geographical Indications for Wines and Spirits – Draft Composite Text', doc. WTO JOB/IP/3/Rev.1, 20 April 2011. The text is available as an Attachment to the Chairman's Report to the Trade Negotiations Committee, 21 April 2011 (TN/IP/21).

group bases its suggested transition periods on the totally voluntary form it is proposing for the system – the delays, at this stage for an unspecified numbers of years, would start from when a developing or least-developed country volunteers to participate in the system. The composite text also contains a new section on costs and fees, with views differing on whether the cost should be borne by the WTO's budget – meaning all members would fund the system – or whether 'user fees' should be charged to countries registering terms.

However, all the present divergent positions are included in this new text. As a result, the document contains around 208 rival provisions, separated by square brackets. This means that the real challenge still lies ahead and WTO Members must continue negotiating to narrow down differences and work towards removing the square brackets. Thus, the ongoing drafting seems fragile and delicate and the process is still a work in progress.[33]

This fragility is especially evident for certain fundamental questions, where positions remain more or less unchanged. This occurs with respect to the consequences/legal effects of registration. All WTO Members seem to accept an obligation to consult the information on the Register. They also seem to be willing to take the information on the Register into account 'when making decisions regarding registration and protection of trade marks and geographical indications' under their national procedures. However, views differ significantly as to how such information should be taken into account, what weight and significance should be given to it, and whether there should be a specific legal obligation to take the information into account. While some WTO Members are of the view that the mere obligation to consult the Register is not enough to ensure meaningful facilitation of protection of wine and spirit GIs, others are concerned about the extra-territorial effects of GI protection.

As a result, the Joint Proposal Group continue to reject the proposition that if a WTO Member registers a term, this would be *prima facie* evidence that the term meets the definition of a GI under the TRIPS Agreement in other countries, in the absence of proof to the contrary. The main critique in recent debates relates to the weight given to this legal presumption. This term *prima facie* evidence raises doubts about its real meaning, which could be: 'on first sight', 'preliminary', 'before further investigation', and so on. The Joint Proposal Group objects on the basis that the EU system would create obligations in other countries' legal systems, in turn described as 'extraterritorial'. Opponents are concerned that the EU-led Proposal

---

[33] Ibid.

would shift the burden of proof away from the GI rights holder on the issue of whether a term qualifies as a GI. By contrast, the EU-led W52 Group denies that its proposal would be extraterritorial because countries could still use their own national GI regimes to decide whether to protect the term. As a result, the rights enshrined in the TRIPS Agreement would be respected, as the final decision on whether or not to protect a GI will be left to national authorities.

Given these criticisms, we might consider the following question. Is a system which creates *prima facie* legal effects at the international level really necessary to facilitate GIs protection? There are several reasons why I believe it is necessary. The most important is that international legal effects would make GI protection easier by providing registered GIs with a presumption of eligibility for protection.[34] However, it is doubtful if the present EU proposal would enable producers to reduce costs as they would have to continue to enforce protection in each national market. It is true, however, that at least occasional free-riding would be discouraged because potential free-riders would have to bear the burden of proof and incur litigation costs. This means that in the case of litigation, the EU's current proposal for a register would be a tool for these producers to 'facilitate' GIs' protection by a reversal of the burden of proof. But it is doubtful if this could be particularly valuable for producers in developing countries who might not have the means to assert their rights in all markets. The notification, examination and opposition phases should therefore be considered an investment in the system's viability; the costs involved would be offset by the benefits that would be derived from effective protection. Without a presumption of eligibility, it would be difficult in most cases, if not impossible, for the average GI right holder to enforce their rights, because they would have to build a case from scratch before local courts. In certain cases, litigants would be far from home and operating within the context of unfamiliar legal systems. This inconvenience would threaten the clear intention expressed in Art 23.4 to facilitate GIs' protection. Additionally, it is doubtful whether under the present EU-led Proposal, producers with a policy of international expansion would be able to save costs when defending their geographical designations around the world. On the other hand, under the EU-led Proposal

---

[34] But see B Goebel, 'Geographical Indications and Trade Marks – The Road from Doha' (2003) 93 TMR 986 (arguing that most of the existing multilateral systems of notification and registration, such as under Art 6*ter* of the Paris Convention, the Hague Agreement in the field of industrial designs and the Madrid Protocol in the field of trade marks, all rely ultimately on determinations under domestic law to determine eligibility and protection).

public administrations would have timely information that would allow them, for instance, not to register trade marks containing such GIs, as prescribed by Art 23.2 TRIPS. As a result, usurpation should diminish and, in turn, litigation and administration costs would decrease. In any case, the current EU-led Proposal is the only one that would make GI protection easier to implement because registered GIs would benefit from a presumption of eligibility for protection; moreover, piracy would be discouraged. These two features seem to benefit all parties: producers, consumers and administrations.

Similar debates arise around the issue of whether participation in the system should be voluntary or mandatory; and whether allowing countries to opt out of the system entirely conforms to the use of 'multilateral' in the mandate. Some WTO Members interpret the reference to 'a multilateral system' in Art 23.4 to mean that the system should apply to all. By contrast, other WTO Members interpret the words 'those Members participating in the system' to mean that not all Members are expected to participate. This raises the issue of whether a system based on voluntary participation could bind WTO Members once a GI has been registered. The logical answer seems to be that a voluntary system could only bind participating WTO Members.

In my opinion, the use of the words 'those Members participating in the system' does not necessarily mean that participation must be voluntary. Due mainly to efficiency reasons, an ideal system would require all WTO Members to participate, even if a literal reading of Art 23.4 only establishes a voluntary participatory system. Otherwise, a system whereby GIs are recognized and protected in some WTO Member markets but not in others, would surely create legal and economic uncertainty, thus undermining the objectives of TRIPS protection. We must not forget that Art 23.4 TRIPS calls for the establishment of a multilateral system of notification and registration of GIs. And a multilateral system can only be understood as requiring all parties to be bound to protect registered GIs. Unlike plurilateral trade agreements, which bind only signatories, a multilateral system should be understood as including all WTO Members.[35] Therefore, in the lexicon of the WTO, 'plurilateral' must be understood as referring to a system in which participation is

---

[35] The Agreement establishing the WTO expressly affirms that although the four Plurilateral Trade Agreements (Agreement on Trade in Civil Aircraft; Agreement on Government Procurement; International Dairy Agreement and the International Bovine Meat Agreement) are part of the WTO Agreement, they create neither obligations nor rights for Members that have not accepted them.

entirely voluntary, whereas 'multilateral' should be understood to bind all Members.[36] Certainly, the phrase 'Members participating in the system' at the end of Article 23.4 TRIPS Agreement seems to refer to voluntary participation. However, this reference can be interpreted as meaning no more than those WTO Members who chose to participate by registering their GIs in the system. Under this interpretation, a participating WTO Member would still be obligated to afford protection to GIs registered by other countries, even though it chooses not to register its own GIs. Nonetheless, the protection of registered GIs must be obligatory for all by virtue not only of the mandate of Art 23.4, but also via the other GI provisions, particularly Art 24.1.[37] According to this interpretation, this provisional intersection results in a system of obligatory GI protection for all because Art 24.1 TRIPS Agreement prohibits any country from refusing to negotiate to improve the protection of individual GIs. Thus, it would seem that the notification of a GI to the future register could be equivalent to a request to open negotiations.

Last but not least, there remains the question relating to the product coverage of the register. Could it include GIs for products other than wines and spirits? The Joint Proposal group and some others say there is no mandate to extend the system to other products and in doing so, to link these registration talks to the debate on whether to extend to other products the higher level of protection now given to wines and spirits ('GI extension'). However, other WTO Members, among them, the EU-led Group and China, are insisting that the multilateral register must not discriminate in favour of wines and spirits. Nowadays, it seems that interventions on

---

[36] WTO Council for TRIPS, Special Session, *Minutes of Meeting*, WTO doc. TN/IP/M/4, 6 February 2003, para. 21.

[37] Adopting a broader trade law perspective, it must be recalled that the WTO Agreement has eliminated the imbalances caused by the collateral agreements associated with GATT, also referred to as 'Codes', concluded after the Tokyo Round (1973–1979). In most cases these Codes differentiated between the norms and procedures for decision-making and dispute resolution and their acceptance among the Contracting Parties was limited. However, certain countries which had assumed only the minimum level of obligations had previously tried to benefit from the Most Favoured Nation (MFN) clause of Article I of GATT, demanding the advantages resulting from these Codes, which they themselves had ignored. To avoid these imbalances, Article II.3 of the WTO Agreement states specifically that the MFN Clause is not applicable to the four Plurilateral Agreements. It must be observed, however, that nothing equal has been established in the section of the TRIPS Agreement in relation to GIs' protection. Therefore, if we admitted that the multilateral register of GIs only must bind participant countries, nothing in it would prevent the other WTO Members from demanding the application of the MFN Clause established specifically in TRIPS, Article 4.

200 *Research handbook on intellectual property and GIs*

this aspect of opening the register to all products are relatively low-key on both sides due mainly to the mandate in this forum, which is limited to negotiations on the establishment of a multilateral system of notification and registration of GIs for *wines and spirits*. However, the drafting group spent considerable time in discussing whether or not the composite text should be neutral with respect to possible GI extension or not, without any definitive result.

4. CONCLUSIONS

To what degree should GIs be protected by international law? Do current international rules provide sufficient safeguards, or should governments implement another system of more effective protection? These are the questions that need to be solved by WTO Members in the framework of the TRIPS Agreement's Built-In Agenda for GIs. The peculiarity of the GI debate is that it exhibits not so much a North–South division, but rather a more unusual and interesting split: that between the New World and the Old World. And the difficulty is perhaps that GIs stand at the intersection of three sets of national interests, related to international trade policy, intellectual property protection and agricultural policy.[38] To enhance GI protection seems to emphasize the importance of local culture and tradition in the face of globalization.[39]

Even though the TRIPS provisions in Arts 22–24 offer a legally binding framework for the protection of future GIs, the work on TRIPS is far from finished and the debate between the New and Old Worlds continues to be divisive in the furtherance of the TRIPS goal of protecting IPRs and global economic interests. By virtue of the programme incorporated into the text of the Agreement, GI protection needs to evolve and improve. However, in spite of the major concessions and considerable efforts made

---

[38] In fact, the EU submitted in 2004 a proposal in the context of WTO agriculture negotiations that is relevant to the GI debate. It concerns a list of names that, in the EU constitute GIs but in other countries are used generically to indicate a type or kind of product. The proposal aimed to 'claw back' such names by reserving their use for EU producers in the geographical locations. Some other WTO Members argued that the Doha text on Agriculture did not provide a mandate for such a proposal. The proposal intended to be complementary to the work on GIs in the context of TRIPS on the Doha Work Program. Now, it seems that the link between agriculture negotiations and GIs' protection does not exist anymore according to revised draft modalities for agriculture, TN/AG/W/4/Rev.2, 19 May 2008.

[39] See Chapter 3 by Barham and Chapter 4 by Bérard in this volume.

by some WTO Members to review their previous positions, for now the deadlock continues. However, there has been some movement in the past months towards a single composite text as the basis for further discussion, identifying elements of convergence. As of April 2011, the TRIPS Council Chairperson's report contains the Draft Composite Text in an Annex.[40] This is the first single text to contain the full range of Members' views since talks began in 1997, with competing text options marked by square brackets. While there still is a long way to go, this Draft Composite Text seems to provide a good basis on which to continue negotiations towards a multilateral system of notification and registration for GIs. Despite this, the real challenge still lies ahead and WTO Members must continue negotiating to narrow down differences and remove the square brackets. Thus, any achievements are fragile and delicate; the process is still a work in progress.

Be that as it may, in the present author's view, the TRIPS Register should be an accurate, reliable and authentic source of information.[41] However, WTO Members sponsoring the minimalist approach do not consider the legal presumption acceptable for a number of reasons: firstly, it would increase GIs legal protection, and this would be outside the scope of this negotiation, which is only about facilitating protection; secondly, it would violate the principle of territoriality; and, thirdly, it would alter the balance of rights and obligations in the TRIPS Agreement.[42] In spite of this criticism, I think that only the EU-led Proposal could contribute to facilitating multilateral GI protection as the Art 23.4 TRIPS Agreement prescribes because it strikes a balance between different interests and would be the appropriate tool to create a register which would truly facilitate GI protection, and not duplicate what would already be available on the Internet system.

---

[40] See WTO doc. TN/IP/21, 21 April 2011.
[41] For an alternative in the framework of an evolved Lisbon Agreement, see D Gervais, 'Reinventing Lisbon: The Case for a Protocol to the Lisbon Agreement (Geographical Indications)' (2010) 11 Chicago Journal of International Law 1, 67.
[42] See, WTO Council for TRIPS, Special Session, *Minutes of Meeting*, TN/IP/M/25, 4 March 2010.

9. Thinking locally, acting globally: how trade negotiations over Geographical Indications improvise 'fair trade' rules

Antony Taubman

> The work of the legislator is to give names, and the dialectician must be his director if the names are to be rightly given? (390d) . . . And therefore a wise dictator . . . should observe the laws of moderation and probability (414e) . . . But if this is a battle of names, some of them asserting that they are like the truth, others contending that they are, how or by what criterion are we to decide between them? For there are no other names to which appeal can be made, but obviously recourse must be had to another standard which, without employing names, will make clear which of the two are right, and this must be a standard which shows the truth of things. (438e)
>
> (Plato, *Cratylus*)

> Strictly speaking there are no signs but differences between signs.
> (Ferdinand de Saussure, *Third Course of Lectures on General Linguistics* [1910–11])

> But if you say 'How am I to know what he means, when I see nothing but the signs he gives' then I say 'How is *he* to know what he means, when he has nothing but the signs either?'
> (Ludwig Wittgenstein, *Philosophical Investigations*)

1. OVERVIEW[1]

Geographical Indications (GIs)[2] are at the centre of contemporary multilateral trade negotiations that seek to recalibrate and reframe the

---

[1] This chapter is a revised version of the author's article 'Thinking Locally, Acting Globally: How Trade Negotiations over Geographical Indications Improvise "Fair Trade" Rules' [2008] IPQ 231, which is reproduced with the kind permission of the publishers of the *Intellectual Property Quarterly*. The background to this chapter includes the author's own earlier paper: 'The Way Ahead: Developing International Protection for Geographical Indications: Thinking Locally, Acting Globally' 15 November 2001 (WIPO/GEO/MVD/01/9), which is cited where appropriate. The revision does not take account of all multilateral, bilateral and domestic developments since 2008. No view or analysis in this chapter can or should be attributed to the World Trade Organization, its Secretariat or its Members.

[2] The term 'Geographical Indications' as such was first defined in multi-

terms of trade in the global economy. Recognizing past cumulative innovation embedded in traditional products, GIs link the mainstream trade interests associated with agricultural commodities and new conceptions of a 'knowledge economy' which privileges the intangible component of trade.[3] GIs, and their conceptual precursors,[4] have been controversial for as long as there have been international negotiations on intellectual property (IP) standards, for essentially the same reasons that drive the debate today: proponents seek to forge a rigid relationship between a linguistic symbol, a product and the geographical origin of the product that is legally enforceable in foreign markets, but communities worldwide differ in language use and in the signification they attach to signs. And when newer producers use terms to evoke or connote the traditional products of commercial rivals, countries differ over whether this is legitimate competition or unfair appropriation of reputation.

Trade negotiators working to advance the cause of traditional producers aim to protect their GIs by reining in the natural synchronic diffusion and diachronic evolution of language, and by converting inherently contingent linguistic connections into legally binding constraints on the denotation and connotation of certain evocative words. But the logic of trade negotiations means that 'new world' producers are reluctant to give up this linguistic flexibility without buying access to 'old world' markets in other forms. These negotiating dynamics favour pragmatic trade-offs between disparate trade interests over a policy-driven approach of objective rulemaking.

GI protection creates incursions into the public domain of the common language, purportedly to promote the creation of high-level public goods such as consumer protection, protection of traditional producers and cultural diversity. But any intrusion on the public domain raises questions about which interests and which public goods will be prioritized. GIs and trade marks are typically constructed to represent polar opposites in the

---

lateral treaty law in the Agreement on Trade-Related Aspects of Intellectual Property Rights ('TRIPS'), Annex 1C to the Agreement Establishing the World Trade Organization (WTO), done at Marrakesh, 15 April 1994, entry into force 1 January 1995. The term is now found also in the Geneva Act of the Lisbon Agreement on Appellations of Origin and Geographical Indications, adopted on 20 May 2015 ('Geneva Act of Lisbon').

[3] European Commission, Why do Geographical Indications Matter to Us? (Brussels, 30 July 2003).

[4] For the sake of clarity, depending on context, 'GI' is not necessarily used in this paper in the precise legal sense but as a general descriptive term embracing appellations of origin, designations of origin and indications of source, although the latter term may not fit within the formal TRIPS definition of a GI.

policy debate, even betokening a cultural divide.[5] WTO dispute settlement has explored the intricate linkage between them.[6] But, paradoxically, GIs may be protected by trade marks,[7] and the rationale for and mode of protection of GIs are following the trajectory earlier charted by trade marks and the associated unfair competition doctrines, such as the law of passing off, from strict consumer protection against deception to the formulation of property that is considered of value in itself and therefore entitled to protection against various forms of encroachment, usurpation or dilution by rival traders.

International trade negotiations over GI protection equivocate between two methodologies.[8] The first, a 'fix-rules' approach, settles on the objective rules expected to apply in the domestic regulatory context, so that national authorities interpret the internationally-agreed rules and apply them to the facts at the level of municipal law.[9] The second, a variant of the classic 'managed trade' or 'fix-outcomes' approach in conventional trade negotiations, would pre-empt domestic regulation by settling in advance what the outcomes should be: for instance, bilateral trade negotiations,[10] and potentially multilateral ones,[11] have actually settled on what words

---

[5] See Chapter 3 by Barham in this volume.

[6] WTO disputes DS290, European Communities – Protection of Trademarks and Geographical Indications for Agricultural Products and Foodstuffs (Complainant: Australia) and DS174 European Communities – Protection of Trademarks and Geographical Indications for Agricultural Products and Foodstuffs (Complainant: United States).

[7] See Australia – US Free Trade Agreement, signed 18 May 2004, entry into force 1 January 2005.

[8] For the articulation of this distinction in conventional trade negotiations see in particular Jagdish Bhagwati, *The World Trading System at Risk* (Princeton: Princeton University Press, 1991) 3–5.

[9] E.g. TRIPS Arts 22, 23 and 24.

[10] An early example: the bilateral Exchange of Notes on Commercial Relations between the United States of America and Portugal (28 June 1910):

> attention having been called . . . to the final protocol annexed to the treaty of commerce recently concluded between Portugal and Germany whereby the names of 'Porto' and 'Madeira' are recognized as being strictly designations of origin and whereby it is agreed to prevent the sale in the German Empire under these names of wines not originally from the Portuguese districts of Douro and of the island of Madeira . . . the United States of America will fully exercise the powers vested in it by law in order to protect in the United States of America the names 'Porto' and 'Madeira'; and that, with this end in view, it will apply strictly laws and ruling forbidding labeling or branding of wine so as to deceive or mislead the purchaser concerning the nature or the origin of the product.

[11] European Commission Press Release, 'WTO talks: EU steps up bid for better protection of regional quality products' (28 August 2003, Brussels, IP/03/1178)

fit the definition of 'geographical indication', and when they would cease to have generic significance in the common language, pre-empting the normal function of registrars and courts in weighing the linguistic significance of terms so as to determine the bounds of legitimate exclusions from the common domain of language.[12] In no other field of IP would trade negotiators agree on the eligibility for protection – patentability, registrability, distinctiveness, originality – of individual titles. Such improvised outcomes are more difficult to negotiate multilaterally than bilaterally, but mechanisms already exist that create a presumption of eligibility for protection following international registration,[13] and have been proposed within the framework of TRIPS.[14]

GIs lend themselves to such fix-result negotiations because they resemble the kind of 'national champions' that symbolize the type of precise, distilled sectoral interests that typically establish managed-trade desiderata. Indeed, for agricultural trade, the interests may be one and the same. The 'new economy' or 'trade and' aspect of IP protection can approach – even touch upon or overlap with – the traditional preoccupations of trade negotiations. Further, GIs, uniquely in IP law, unite global protection with an intrinsically localized basis of protection, linking cultural diversity with global markets: thinking locally, acting globally.

This – literal – common ground between international IP rules and classical trade law suggests a possible reconciliation between these presumed

---

(advising of a 'short list of 41 EU regional quality products whose names the EU wants to recuperate . . . in the agriculture negotiations within the [WTO] Doha Development Agenda').

[12] See Agreement between Canada and the European Community on Trade in Wine and Spirits, 16 September 2003, Art 12 (Customary Terms and Transitional Arrangements).

[13] As exists under the Lisbon Agreement for the Protection of Appellations of Origin and their International Registration of 31 October 1958, as amended on 28 September 1979 (hereafter 'Lisbon'), and the Geneva Act of Lisbon when the latter enters into force. For further details, see Chapter 5 by Geuze in this volume.

[14] The negotiations mandated under TRIPS Art 23.4 have led, for instance, to a Communication from the European Communities, 'Geographical Indications' 14 June 2005 (TN/IP/W/11) proposing a mechanism whereby a WTO Member 'which has not lodged a reservation in respect of a notified geographical indication within [a certain] 18-month period . . . or which has withdrawn such a reservation, shall . . . not refuse protection of the registered geographical indication'. Other WTO Members have opposed this approach. At the time of publication, negotiations had made no substantive progress since April 2011 (see Report by the Chairman, Ambassador Darlington Mwape (Zambia) to the Trade Negotiations Committee, document TN/IP/21, 21 April 2011). See also Chapter 8 by Martín in this volume.

opposites, despite concerns that GIs may symbolize not so much a new economy as a new agricultural protectionism.[15] This chapter looks for a common doctrinal root between mainstream trade law and international IP laws, in the form of a conception of fair competitive relations between traders, potentially uniting the Paris Convention[16] and GATT.[17] GI protection hinges on the semantic effect of linguistic signs. Negotiations that seek to fix the linguistic functions of those signs need also to construct a practicable conception of fairness in trading relations. Settling terms for GI usage may be an empirical exercise guided by an objective, even Rawlsian,[18] conception of equity, or an improvised, negotiated outcome, a realist compromise between diverse pragmatic interests – a 'politically expedient instance of semantic reloading'[19] – or both. Can negotiated outcomes on GI protection produce a fairer trading system, that delivers on the promise of enhanced public welfare that the IP system offers in principle?

2. THERE GOES THE NEIGHBOURHOOD? GIs, LOCALITY AND GLOBAL TRADE

'Location, location, location'[20] is the informal benchmark for assessing the value of land – real property. But the value of location also guides the assessment of trade interests in intangible property as well. GIs can enhance the value of products whose distinctive characteristics are associated with their geographical origin, and can protect the linkage between a product and its physical source, in dispersed international markets. Distinctive local characteristics provide the elusive 'value added' that agricultural exporters products strive for, faced with a long historic trend of

---

[15] S Laing, 'EU on GIs: Free Trade or Protectionism?' *Trade Briefs, Trade Law Centre for Southern Africa*, 9 September 2003, at http://www.tralac.org/

[16] Paris Convention for the Protection of Industrial Property (of 20 March 1883, as amended on 28 September 1979) ('Paris').

[17] General Agreement on Tariffs and Trade (entry into force on 1 January 1948) ('GATT').

[18] J Rawls, *A Theory of Justice* (Harvard University Press, 1971). (Central to Rawls's argument developing 'justice as fairness' is the selection of principles that free and rational persons seeking to further their own interests would accept in an initial position of equality as defining the fundamental terms of their association.)

[19] D Gangjee, 'Say Cheese! A Sharper Image of Generic Use through the Lens of Feta' [2007] EIPR 172, 173.

[20] Attributed to the American businessman William Dillard (http://www.famous-quotes.com/author.php?aid=2035).

declining commodity prices.[21] 'Thinking locally' in terms of the distinctive indigenous qualities of their produce, traditional producers 'act globally' in seeking protection worldwide for the GIs that signify those qualities. The wine trade used geographically localized and immovable *terroir*[22] to leverage privileged access to globalized markets: intangible and real property interests overlap. Intense disputes over the demarcation of locations denoted by valuable GIs in the past,[23] and in new jurisdictions currently creating GI systems,[24] is a measure not only of the commercial value of these GIs, but of the value of ensuring that land falls within the authoritatively determined boundaries.

The protection of GIs therefore links global trade in physical produce with the 'new economy' of borderless trade in intangibles[25] and with local property markets as well. But there is another way of 'thinking locally' about GIs: GIs are signs used in commercial discourse, and acting globally to protect GIs in the interests of original producers creates tension with how linguistic communities think locally about their own use of language. Unmoored from their origin, geographical terms may develop secondary meanings among foreign linguistic communities,[26] just as non-geographical terms imported from foreign languages are argued to have developed exclusive geographical denotations.[27] This leads to conflicting

---

[21] LP Mahé, F Ortalo-Magné and K Löfgren, 'Five Proposals for a European Model of the Countryside' (1999) 14 Economic Policy 87, 118.

[22] The claim often put forward is that while some processes of wine production may be duplicated, the *terroir* is unique and impossible to replicate. For detailed analysis, see Chapter 3 by Barham and Chapter 4 by Bérard in this volume.

[23] Such as the Champagne Riots in 1910 and 1911 precipitated by early decrees establishing the *Appellation d'Origine Contrôlée*, which first excluded and then included the district of Aube and the town of Troyes. See A Knoll, 'Champagne' (1970) 19 International and Comparative Law Quarterly 309.

[24] *Coonawarra Penola Wine Industry Association Inc and Others v Geographical Indications Committee* [2001] AATA 844 (5 October 2001) ('Coonawarra'); D Gangjee, 'Melton Mowbray and the GI Pie in the Sky: Exploring Cartographies of Protection' (2006) 3 IPQ 291.

[25] Hansard, Australian House of Representatives, Agriculture, Fisheries and Forestry Legislation Amendment Bill (No. 1) 1998 Second Reading (Mr Connor), 688: 'Mr Brian Croser, past President of the Winemakers Federation ... challenged the audience not to view his industry as a rural industry but as part of the entertainment industry.'

[26] See e.g. *Institut National des Appellations D'Origine v Vintners Int'l Co., Inc.*, 954 F2d 1574, 22 USPQ2d 1190 (Fed. Cir. 1992) ('Chablis with a Twist').

[27] See e.g. Joined Cases C-465/02 and C-466/02 *Germany and Denmark v Commission of the European Communities* [2006] ETMR 16 at [46] ('It is common ground in the present proceedings that the term "feta" is derived from the Italian word "feta", meaning "slice", which entered the Greek language in the seventeenth

local discourses about the 'correct' signification of signs that form part of the 'general cultural and gastronomic stock'.[28]

3. TAKING LINGUISTIC TURNS

The regulation of GIs and other commercial signs is ultimately a linguistic matter, and may be assisted by the analytical tools developed in linguistics. The law of GIs determines whether and how certain prescribed signs can be used in connection with a product, its origin, its producer and/or its characteristics. According to the TRIPS definition,[29] a GI[30] seems both to name a product ('identifies'), and to convey information about it ('as originating in' a certain location). This is a complex linguistic function. 'Identify' suggests a strict naming or denoting function – either pointing at the good itself, or expressing an intrinsic link between the good and its geographical origin, rather than a contingent or mutable linkage. Ladas, in contrasting ('grammatically, as well as juristically') the more precise terms 'indication of source' and 'appellation of origin', notes that appellation 'means a name given to a person or a thing'.[31] From this act of naming he infers a legal consequence: 'it evokes the idea of susceptibility of appropriation or the idea of a property right'.[32] By contrast, indication 'is what serves to indicate or point at something, or informs'.[33] Can an 'indication' also be a name, and thus be a fit subject of property? A GI directly identifies the product while also pointing to its origin and hinting about its characteristics. If 'burgundy' were a true GI in English, 'this Burgundy wine is from the region of Bourgogne' would be a tautology. The use of the term to denote the wine also communicates its origin.[34]

---

century. It is also common ground that "feta" is not the name of a region, place or country.').

[28] Case C-317/95 *Canadene Cheese Trading v Hellenic Republic* [1997] ECR I-4681 at [28] (per Advocate-General Colomer).

[29] TRIPS, Art 22.1.

[30] The Oxford English Dictionary Online defines 'indication' as 'a sign, token or symptom'. Notably, TRIPS is the first multilateral treaty on IP for which English is an authoritative language; cf. Paris Convention, Art 29(1)(c) which specifies that the French text prevails.

[31] S Ladas, *Patents, Trademarks and Related Rights, National and International Protection* (Vol. 3, Harvard University Press, 1975) 1574.

[32] Ibid.

[33] Ibid.

[34] This distinction is clear from two alternative definitions given for 'Burgundy' in Webster's Third New International Dictionary (1966): 'any of the red or white

The word 'identify' suggests this kind of necessary relationship, not the contingent linkage connoting or alluding to a geographical location (for example 'suède gloves from Sweden, the true home of suède'). Still, actual linguistic usage blurs distinctions between naming and describing, and these nuances evolve diachronically and vary across linguistic communities. Even if strictly communicating a product's geographical origin, a GI may connote further properties that are typical of the product, but not exclusive to it ('this wine has the flinty finish of a true Champagne'). Since the TRIPS definition requires that certain qualities must also be 'attributable' to the product, the GI may also connote these qualities even if close analysis is difficult to sustain.

In mapping from this definition to specific legal measures, the essential distinction in linguistics between the denotation and connotation[35] of a sign becomes a crucial tool: a profound difference between the proponents and opponents of stronger international rules on GI protection is whether the law should intervene to govern merely the denotation[36] of the sign – the product that it actually points to in the eyes of the consumer – or its connotation[37] as well – the penumbra of associations and qualities that GI proponents claim are 'usurped', 'appropriated,' 'diluted' or 'imitated'.[38] Between connotation and denotation lies a dichotomy that has riven international debate on IP since at least 1880. What is unfair or illegitimate use of a geographical name for competing products? Pointing a sign to the original product with deceptive effect or fraudulent intent? Or is it enough merely to appropriate some of its connotative effect, even if the denotation remains accurate (for example Parma-style ham, made in Canada)?

Viewing GI negotiations as a process of legislating on the referential effect of signs helps explain the peculiar intractability of the debate. After

---

table wines from vineyards in the departments of Côte d'Or, Yonne, and Saône-et-Loire, France' (a strict *denotation*, using the term as a GI) and 'a table wine that resembles the red Burgundy of France but is produced elsewhere and that is usu. darker red and heavier-bodied than claret' (a *connotation* of the qualities of the wine denoted Burgundy in the first definition).

[35] Based on the original distinction introduced by JS Mill and Frege's distinction of *Bedeutung* and *Sinn* (*On Sense and Reference*, 1892).

[36] The Oxford English Dictionary Online defines 'denote' as 'to mark; to mark out (from among others); to distinguish by a mark or sign'.

[37] The Oxford English Dictionary Online defines 'connote' as 'to signify secondarily or in addition; to include or imply along with the primary or essential meaning'.

[38] Cf. Lisbon Art 3, referring to 'usurpation or imitation' (and Geneva Act of Lisbon, Art 11, with references to imitation as well as use that would impair or dilute reputation in an unfair manner).

all, it touches on long-established theoretical paradoxes: language is a set of consensual rules that in turn distil the community's actual usage. The signification of linguistic signs is a kind of '*opinio juris*'[39] within a linguistic community. Geneva[40] is the epicentre of multilateral negotiations seeking to forge inflexible links between the GI as signifier and the original product as signified. In a nice irony, it was Saussure's path-breaking lectures at Geneva[41] that characterized the contingency and fluidity of language use, and the arbitrary relationship between signifier and signified.[42] Yet he observed that language users are rule followers, so linguists confront a paradox of the 'mutability and immutability of the sign . . . the fact that language changes in spite of the inability of [individual] speakers to change it'.[43] The difficulty in overlaying a distinct normative element on language was explored by Plato[44] and crystallized by Wittgenstein: '[T]his was our paradox: no course of action could be determined by a rule, because every course of action can be made out to accord with the rule.'[45]

4. LANGUAGE COMMONS, PUBLIC GOODS AND NORMATIVE SEMANTICS

The regulation of GIs (like that of trade marks) skewers Wittgenstein's paradox: the legislator decides that a defensible public interest, the

---

[39] By analogy with the concept of *opinio juris* in international law, which transforms custom into binding law: '[n]ot only must the acts concerned amount to a settled practice, they must also be such, or be carried out in such a way, as to be evidence of a belief that this practice is rendered obligatory by the existence of a rule of law requiring it'. See *Federal Republic of Germany v Denmark and Federal Republic of Germany v The Netherlands (North Sea Continental Shelf Cases)*, (1969) ICJ 3 (20 February 1969).

[40] Seat of the WTO, including the meeting place of the Council for TRIPS, and of WIPO, responsible for administration of key treaties including Paris, the Madrid Agreement and the Lisbon register of Appellations of Origin, as well as earlier attempts to craft a treaty on GI protection.

[41] C Bally and A Sechehaye (eds), W Basking (trans), F de Saussure, *Course in General Linguistics* (McGraw-Hill, 1956).

[42] A Taubman, 'The Way Ahead: Developing International Protection for Geographical Indications: Thinking Locally, Acting Globally' 15 November 2001 (WIPO/GEO/MVD/01/9).

[43] De Saussure (n 41) 74.

[44] Plato, *Cratylus*, available at: http://classics.mit.edu/Plato/cratylus.html

[45] Ludwig Wittgenstein, *Philosophical Investigations* (Blackwell, 1951) at 201a; see also Saul Kripke, *Wittgenstein on Rules and Private Language* (Blackwell, 1982).

generation of a certain public good, justifies the arbitrary imposition of a linguistic rule that is influenced by, but not wholly determined by, actual linguistic usage, and seizes the authority to make that imposition: the 'wise dictator' of Plato's *Cratylus*.[46] Where a term is generic in a particular jurisdictional context, the core task of an IP policymaker is to determine what exclusions from the public domain or constraints on *publici iuris*[47] (in this instance, the language commons) are required in order to generate higher-level public goods (in this instance, the nature and ordering of public goods is contested, but range from prevention of fraud,[48] to suppression of deception or confusion of consumers,[49] to protection of the producer's reputation within a particular trade,[50] to absolute protection of a form of collective property against any diminution of its value through unauthorized commercial use[51]).

Hence, GI trade negotiations proceed on the assumption that some form of constraint on individual sets of interests is inevitable, as in any legal or policy compromise, recalling the very compromise that defines a linguistic community: 'Language furnishes the best proof that a law accepted by a community is a thing that is tolerated and not a rule to which all freely consent.'[52]

Judicial analysis of the function of trade marks and GIs typically moves from the descriptive to the normative, in charting the bounds of the public domain of the terms available in the common language to name or describe quotidian products. *Chablis with a Twist*[53] hinged on the determination of the public domain status in US commerce of the contested

---

[46] See the introductory quotation to this chapter.
[47] E.g. *Syncom Formulations v SAS Pharmaceuticals* 2004 (28) Patent and Trade Mark Cases 632 (Del) (holding that certain linguistic elements are *publici iuris*).
[48] See Paris Convention 1883 (original text), Art 10 restricting protection of designations of origin used with fictitious or fraudulent trade names.
[49] See TRIPS, Art 22.2(a)
[50] See Joined Cases T-53/04 to T-56/04, T-58/04, T-59/04 *Budejovicky Budvar narodni podnik v OHIM (intervener: Anheuser-Busch, Inc)*, CFI (12 June 2007) (Unreported) at [190], [203] (proof of likelihood to misappropriate or weaken reputation of the GI, or at least evidence of nature of reputation, required for conflict with a trade mark in distinct commercial fields).
[51] *Société Empresa del Tabaco Cubatabaco v Aramis Inc & Ors* Paris Court of Appeal, 4th Chamber Reg No 1998/10814 (17 May 2000) 13 (Unreported). (Restraining trade mark use of appellation *Havana* on dissimilar goods); *Institut National des Appellations d'Origine v Yves Saint Laurent*, Cour d'Appel de Paris (28 October 1993) [1994] EIPR D74 (Champagne used as trade mark for perfume).
[52] De Saussure (n 41) 71.
[53] *INAO v Vintners* (n 26).

element of the trade mark. The decision highlighted the empirical quality of the court's duty under the relevant law, pivoting on 'whether the relevant portion of the American public, that portion being consumers of wine and wine products, would perceive the mark *Chablis with a Twist* or even the term "Chablis" to indicate that the product came from the Chablis region of France' and the requirement 'to present evidence to establish the additional requirement . . . that the alleged geographic misrepresentation would be a material factor in the decision of consumers to purchase Vintners' product'. The court concluded that 'based on the record no reasonable factfinder could have found the term "Chablis" to be used in the United States as anything other than a generic name for a type of wine with certain general characteristics'.

Yet this 'reasonable factfinder' is also a norm-setter, their observations ultimately determining what is the right usage of the term under scrutiny. This is the role of the law envisaged in *Cratylus*: 'the work of the legislator is to give names, and the dialectician must be his director if the names are to be rightly given'.[54] Yet the 'wise dictator . . . should observe the laws of moderation and probability'.[55] This captures the role of administrative or judicial intervention in linguistic usage – to fix a certain usage as 'right', even to exclude further evolution of usage,[56] while observing 'the laws of moderation and probability'. In this light, then, the creation of the category of 'semi-generic'[57] may seem an attempt to render a judgment of Solomon, splitting the difference of a sign's significations – suggesting that

---

[54] Plato, *Cratylus*, 390d.
[55] Ibid (414e).
[56] E.g. Lisbon, Art 6 (no genericization permitted); see also Geneva Act of Lisbon, Art 12 and accompanying Agreed Statement.
[57] 27 CFR § 4.24 (1991) (BATF regulations) provides in relevant part:

(a)(1) A name of geographic significance which is also the designation of a class or type of wine, shall be deemed to have become generic only if so found by the Administrator.
(2) Examples of generic names, originally having geographic significance, which are designations for a class or type of wine are: Vermouth, Sake.
(b)(1) A name of geographic significance, which is also the designation of a class or type of wine, shall be deemed to have become semi-generic only if so found by the Administrator. Semi-generic designations may be used to designate wines of an origin other than that indicated by such name only if there appears in direct conjunction therewith an appropriate appellation of origin disclosing the true place of origin of the wine, and if the wine so designated conforms to the standard of identity, if any, for such wine contained in the regulations in this part or, if there be no such standard, to the trade understanding of such class or type. . .
(2) Examples of semi-generic names which are also type designations for grape wines are Angelica, Burgundy, Claret, Chablis, Champagne, Chianti, Malaga,

certain terms may indicate an origin but may also serve as a designation of type, depending on the context of use. But responsiveness to context is clearly the most fundamental influence on the relationship between signifier and signified, just as the mechanisms that govern gene expression assume greater meaning than a bare DNA sequence: 'the idea or phonic substance that a sign contains is of less importance than the other signs that surround it; thus the value of a term may be modified without either its meaning or its sound being affected, solely because a neighbouring term has been modified'.[58] In the actual practice of commerce, the signification of GIs will ultimately be determined by context: the community perceiving the sign, its presentation on a label, the linkage to a physical product, the penumbra of associations, and the surrounding language.[59]

The international debate over GI protection can therefore be distilled into the following essential issues.

Given the sign's inherent mutability, given the contingent relation between the signifier and the signified, and given the acceptance that certain public goods are created when this mutability is constrained by law:

- Which 'wise dictator' or 'legislator' can assume authority for which language communities?
- By which expert 'dialectician' and standards of moderation and probability are they to be guided?
- How should the context – linguistic, jurisdictional, commercial – of actual use influence that assessment?

5. GI NEGOTIATIONS, TRIPS AND UNFAIR TRADE

The negotiation of legal mechanisms that aim to fix certain language usage in commerce is an endeavour to determine that certain forms of commercial behaviour are, and are not, legitimate trading activities. Negotiations over GIs are therefore yoked together with negotiations over other concerns about unfair trade – such as subsidies, tariff protection, or regulatory discrimination.[60] International GI protection – as a distinct set of IP

Marsala, Madeira, Moselle, Port, Rhine Wine (syn. Hock), Sauterne, Haut Sauterne, Sherry, Tokay.

[58] De Saussure (n 41) 71.
[59] TRIPS Art 23.1
[60] Regarding Basmati rice, see dispute WTO WT/DS17 – European Communities – Duties on Imports of Rice, Complaint by Thailand (Thailand claims EC has violated the most-favoured-nation requirement under GATT Art I in their

214 *Research handbook on intellectual property and GIs*

standards,[61] in its overlap with trade marks and as a subject of TRIPS litigation,[62] in its uncertain linkage with domain names[63] – remains one of the most complex and contentious issues in international IP law,[64] a complexity enhanced by its linkage with agricultural trade and market access issues[65] and the broader 'fair trade' debate.[66] Despite the standards established by TRIPS and bilateral deals on GIs,[67] multilateral negotiations on GIs confront major, protracted differences.[68] Yet progress on these issues

---

preferential treatment of *basmati* rice from India and Pakistan.); see also WTO, Trade Policy Review, Hong Kong, China, 16 February 2007 (WT/TPR/M/173) (India raises restricted imports of rice including Basmati under the Rice Control Scheme) and Specific Trade Concern 328, India's concerns regarding U.S. default MRLs, limits of determination or limits of quantification on basmati rice under the Agreement on Sanitary and Phytosanitary Measures G/SPS/R/64, paras 47–48.

[61] GIs as such fall within the definition of intellectual property in TRIPS, and within the spectrum of intellectual property protection their scope can overlap with trademarks, particularly given the possibilities for the recognition of collective interests within the trademark system (see, for example, Article 7*bis*(3) of the Paris Convention). Consultations convened by the former Director General of the WTO have 'clarified that trademark systems [are] legitimate forms of protecting GIs, in line with the general principle that Members are entitled to choose their own means of implementing their TRIPS obligations,' document WT/GC/W/633, TN/C/W/61 of 21 April 2011. Some forms of GI protection do not confer distinct ownership of property rights. On whether TRIPS determines the issue of 'property' status, see D Gangjee, *Relocating the Law of Geographical Indications* (Cambridge: CUP, 2012) 202–206.

[62] WTO disputes DS290, European Communities – Protection of Trademarks and Geographical Indications for Agricultural Products and Foodstuffs (Complainant: Australia) and DS174, European Communities – Protection of Trademarks and Geographical Indications for Agricultural Products and Foodstuffs (Complainant: United States).

[63] WIPO Secretariat, 'Internet Domain Names and Geographical Indications' 3 April 2003 (SCT/10/6).

[64] TRIPS Arts 23.4 and 24.2 reflect this; in some readings, Art 24.1 is seen as a built-in agenda provision, and in other readings as a permissive provision for the kind of bilateral and plurilateral negotiations that allow for more restrictive protection of GIs.

[65] European Commission, 'WTO Talks: EU Steps up Bid for Better Protection of Regional Quality Products' (28 August 2003, Press Release IP/03/1178).

[66] European Commission, 'Fact Sheet: Geographical Indications' (Hong Kong: WTO Ministerial, 2005).

[67] See e.g. EU bilateral agreements relating to wines, spirits and/or agricultural products with Australia, Canada, the United Mexican States, South Africa, US, Albania, Switzerland and Chile, at: http://ec.europa.eu/trade/policy/countries-and-regions/agreements/index_en.htm

[68] The recent procedural debates in the course of the diplomatic conference on the Geneva Act of Lisbon in May 2015 gave evidence of the force of these

may be one essential ingredient for an overall accord on agricultural trade in the framework of the current WTO Doha Round of trade negotiations.[69] Multilateral debate on GIs was already difficult and tendentious prior to Doha. There is a sense that fundamental interests are at stake, but those interests are not clearly identified, running the risk that some of the remedies proposed may prove to be fruitless; or that goals are too boldly identified, prioritizing protection of national GI champions over formulating objective rules; or, intriguingly, both – changes in the rules that would not protect their intended target.

The new TRIPS standards on GI protection build on over a century's progressive international normative development.[70] Yet the TRIPS provisions appear to have more the character of an armistice or a provisional settlement, than a negotiated outcome: TRIPS has a certain instability thanks to the built-in agenda provisions that cluster around GI standards,[71] further enhanced by the debate over an 'implementation' mandate given at the Doha Ministerial,[72] suggesting that the standards are permanently open to review, rather than a definitive expression of multilateral concord on substantive standards, laying out an agenda for continuing controversy rather than settling it. TRIPS obliges WTO Members to undertake bilateral negotiations on GI protection[73] – an obligation since fulfilled[74] – in tension with the critique of bilateralism in IP norm-setting[75] and with the view in international trade law of multilateralism as an end in itself.[76] The debate over current and future GI standards has been a central part of the work of the WTO TRIPS Council and related bodies for a decade and

---

differences (http://www.wipo.int/meetings/en/details.jsp?meeting_id=35202). A further, ongoing illustration is the protracted debate in the WTO over whether to extend TRIPS Art 23 levels of protection to all products, the subject of consultations mandated to the former Director General of the WTO (document WT/GC/W/633, TN/C/W/61 of 21 April 2011); see also Chapter 7 by Handler in this volume.

[69] WTO Doha Ministerial declaration, 20 November 2001 (WT/MIN(01)/DEC/1), para 18.

[70] For a discussion of the Paris, Lisbon and Madrid treaties, see Chapter 5 by Geuze in this volume.

[71] TRIPS Arts 23(4), 24(1), 24(2), 71.

[72] WTO Doha Ministerial declaration, 20 November 2001 (WT/MIN(01)/DEC/1), paras 12 and 18.

[73] TRIP Arts 24.1, 24.2.

[74] See discussion *infra*.

[75] Peter Drahos, 'BITS and BIPS: Bilateralism in Intellectual Property' (2001) 4 JWIP 791.

[76] WTO Dispute Settlement Understanding, Art 23: Strengthening of the Multilateral System.

a half, but at the time of writing appears to be no closer to resolution – neither resolution in the sense of an agreed, consensual outcome, nor even resolution in the sense of greater optical clarity.

6. ROOTS

Despite the topicality of the issue and the novelty of the concept of 'geographical indication' in TRIPS,[77] GIs (in some form or another) are governed by long-established laws internationally and in many national jurisdictions. A geographically descriptive indication of origin is an intuitive, natural distinguishing sign in commerce, its heritage traced to antiquity.[78] *Paris*, the foundational international treaty on industrial property, has required some sanctions against false indication of source since the original act of 1883; in 1891, Madrid[79] bolstered this basic protection; in 1958 Lisbon established a plurilateral mechanism for more rigorous and structured protection among a relatively small group of like-minded countries.

And GI protection has deep roots in domestic IP law and policy. The ostensible policy objectives of GI protection – suppressing unfair competition, preventing deception and confusion – are uncontroversial in themselves, and indeed attract more public support than relatively abstruse policy rationales for other forms of IP. The political intensity of the GI debate partly stems from a visceral popular instinct that foreign traders are appropriating indigenous cultural identities. But the deep roots of GI protection do not lend it stability. It can seem curiously unsettled, even foundationally indeterminate, at the international level. The vernacular sense that an evocative term such as *Basmati* is misappropriated or misused can be difficult to capture in national law,[80] let alone in an operational set of rules with global effect. Tracing a

---

[77] But see earlier work on draft treaty language on GIs in WIPO, such as the draft Revision of the Paris Convention, 1975 (PR/DC/4), draft art 10*quarter*, and concurrent work on a treaty for international protection of GIs, WIPO Committee of Experts on the International Protection of Geographical Indications, GEO/CE/I/2, 1990.

[78] See e.g. the use of Geographical Indications applied to amphorae of fish sauce (*garum sociorum*) in ancient Roman trade 'G HISP^' (for 'Garum Hispanicum'), to distinguish this Spanish product from imitation sauce from North Africa and Southern France, Alex R Furger, 'Vom Essen und Trinken im römischen Augst' (1985) 8 Archäologie der Schweiz 168.

[79] The Madrid Arrangement for the repression of False Indications of Origin, 1891, as revised until the Additional Act of Stockholm of 14 July 1967.

[80] HV Chandola, 'Basmati Rice: Geographical Indication or Mis-Indication' (2006) 9 JWIP 166.

similar path for the law of trade marks,[81] the tort of passing off[82] and unfair competition,[83] a right to take action against misleading or deceptive behaviour by competitors, ostensibly in the interests of the consumer, evolves into a distinct property interest[84] that generates its own rationale for protection against dilution or usurpation in the producer's interests.

Several long-running disputes over the legitimate use of certain GIs have taken on the same structural quality and longevity of the international legal instruments ostensibly established to deal with them.[85] And the following points of controversy in the current international debate have recurred in diplomatic conferences for over a century:

- Should the country of origin or country of use determine the GI status of a term?[86]
- Should wine and spirits enjoy higher levels of protection?[87]
- Should GI protection go beyond suppression of actual or likely deception to create absolute or *per se* exclusivity?
- What weight attaches to actual linguistic usage by the relevant public as against producer interests in the country of origin?

---

[81] E.g. United States *Trademark Dilution Revision Act* of 2006, overturning *Moseley v Victor's Secret Catalogue, Inc.*, 537 US 418 (2003) and providing for dilution if defendant's mark is likely to dilute a famous mark; see also Ilanah Simon, 'The Actual Dilution Requirement in the United States, United Kingdom and European Union: A Comparative Analysis' (2006) 12 Boston University Journal of Science & Technology Law 271.

[82] J Davis 'Why the United Kingdom Should Have a Law Against Misappropriation' [2010] Cambridge Law Journal 561 (arguing that flexible judicial interpretations of what counts as misrepresentation as well as damage have moved passing off closer to misappropriation in recent years).

[83] *Inter partes* actions under the consumer protection provisions of the Australian Trade Practices Act 1974 (ss 52 and 53) are principally between competitors and not initiated by consumers.

[84] See B Sherman and L Bently, *The Making of Modern Intellectual Property Law* (CUP, 1999) 168–172 (on the 19th-century controversy over the transformation of trade mark law into intellectual property).

[85] For a detailed consideration of the long-running Budweiser disputes, see Chapter 14 by Heath in this volume.

[86] See the debate on this point at the London Conference of the Paris Union in *Actes de la Conférence Réunie a Londres* (Bureau International de l'Union, Berne 1934) 411–417.

[87] The 1891 Madrid Arrangement in Art 4 provided for higher protection for 'products of the vine', and concern about maintaining this higher level of protection led France to oppose strengthening of Art 10 of Paris at the 1911 Washington Conference. *Actes de la Conférence de Washington* (Bureau International de l'Union, Berne 1911) 104, 303).

Further, the tension between a *laissez faire* approach to allowing existing trade practices in agricultural products to continue, and reining in contested uses of GIs – the pivotal question of 'legitimacy' and 'fairness' of trade that this paper addresses – dates back to the first Paris Conference.[88] Settling this controversy was a prominent negotiating objective for the European Union in the Uruguay Round trade negotiations,[89] but TRIPS has defined the terms of engagement for ongoing wrangling over GI protection, rather than performing its intended core function – the resolution of bilateral trade-related disputes on IP through a multilateral dispute settlement process based on transparent and predictable rules,[90] consistent with an equitable and welfare-promoting conception of IP protection.[91] Hence bilateral and regional negotiation on GI protection and on individual GIs has intensified since this unstable multilateral platform was created.

7. IDENTITY PRESERVATION

Yet the difficult international debate on GIs is infused with stirrings of hope that developing countries – home to many geographical references of powerful resonance and association, and custodians of highly-prized traditions, heritage and customs deeply imbued with distinctive local characteristics – may reap the tangible benefits from the introduction of IP rules into the domain of trade law through the vector of TRIPS,[92] benefits that text books and IP advocates have offered in principle but can appear elusive in practice, especially for developing economies.[93] Pressure to

---

[88] See *Actes de la Conférence internationale pour la protection de la propriété industrielle réunie a Paris du 4 au 20 Novembre 1880* (Ministère des Affaires Etrangères, Impr. Nationale, Paris 1880) 84–88.

[89] See e.g. GATT Doc., *Guidelines Proposed by the European Communities for the Negotiation of Trade-related Aspects of Intellectual Property Rights*, 20 November 1987 (MTN.GNG/NG11/W/16).

[90] See WTO Dispute Settlement Understanding and TRIPS preamble: 'Emphasizing the importance of reducing tensions by reaching strengthened commitments to resolve disputes on trade-related intellectual property issues through multilateral procedures.'

[91] TRIPS, Art 7.

[92] E.g. Swiss statement at the 'Special Session of the Committee on Agriculture' 13 May 2005 (TN/AG/R/19) at [37] ('All WTO Members have much to gain from "GI-extension", least-developed, developing and developed alike. GI-extension is one of the elements which helps ensure a fair overall balance in the results of the Doha Round and it is fully in line with the Doha Development Agenda').

[93] Commission on Intellectual Property Rights, *Integrating Intellectual Property Rights and Development Policy* (London: DFID, 2002) 87–91.

dismantle conventional forms of agricultural protection – for fiscal, policy or trade diplomacy reasons – fuels interest in compensating producers in international trade through stronger protection of their GIs: in the past, an express European negotiating objective.[94]

GI protection helps transform traditional agricultural produce from a low-value raw commodity into an added-value 'new economy' bundle of know-how, reputation and branding. The traded physical product (GATT 'trade-in-goods' paradigm) becomes a carrier for commercializing intellectual capital and cultural memes (TRIPS 'IP-is-trade' paradigm). As a globalized way of recognizing inherently local characteristics, GI protection offers indigenous communities a conception of IP more in tune with their values and interests than other forms of IP. GIs offer a practical option for protecting products that embody traditional knowledge and traditional cultural expressions,[95] and even as a vector for the *de facto* recognition in foreign jurisdictions of customary law of indigenous communities.[96]

Attempts through trade negotiation to exert exclusivity over terms such as 'basmati'[97] may therefore be seen as addressing the erosion or loss of distinctive cultural identities of traditional producer communities. GI protection is linked to 'identity preservation' in its technical sense, ensuring the genetic integrity of specific basmati rice traits[98] (guarding against genericization or dilution of its distinctive qualities through cross-breeding).[99] The EU has applied non-tariff regulatory

---

[94] See e.g. WTO Committee on Agriculture 'EC Comprehensive Negotiating Proposal' 14 December 2000 (G/AG/NG/W/90) at [3].

[95] See e.g. Statement by HE Mrs Rini MS Soewandi, Minister of Industry and Trade, Indonesia, WTO Cancun Ministerial Conference 11 September 2003 (WT/MIN(03)/ST/24); numerous examples cited in the *Consolidated Survey of Intellectual Property Protection of Traditional Knowledge*, 4 April 2003 (WIPO/GRTKF/IC/5/7); Chapter 17 by Sherman and Wiseman in this volume.

[96] A Taubman, 'Saving the Village: Conserving Jurisprudential Diversity in the International Protection of Traditional Knowledge' in K Maskus and J Reichman (eds) *International Public Goods and Transfer of Technology under a Global Intellectual Property Regime* (CUP, 2005) 521.

[97] See e.g. TRIPS Council, 'Minutes of the Meeting on 25–27 June 2002' 10 September 2002 (IP/C/M/36/Add.1), *passim*.

[98] For attempts to identify genetic traits, see D Marie-Vivien, 'From Plant Variety Definition to Geographical Indication Protection: A Search for the Link between Basmati Rice and India/Pakistan' (2008) 11 JWIP 321.

[99] Directorate of Rice Development, *Problems and Prospects of Rice Export from India* (Patna, 2003) at para 11 ('In [the] absence of genetically pure seed of basmati varieties, in [the] majority of basmati rice fields, a variation in plant height, grain size and maturity of the crop is found. This is one of the major reasons for poor quality of basmati rice.').

trade barriers to ensure the genetic integrity of basmati rice.[100] But 'identity preservation' also has a figurative sense – the concern that basmati is integral to the cultural integrity and self-conception of traditional producers, potentially undermined by broader exploitation of the reputation of the rice. This leads to a potential bridge between the protection of cultural diversity and the use of GIs to protect distinctive products with local characteristics,[101] a beachhead into the debate about 'trade and culture'.[102]

The 'development agenda' for GIs highlights how changing expectations of legitimacy of commerce and 'fair trade' between trading partners now reach well behind the border. At issue is the fairness of domestic transactions conducted in the common language of one country, where this conflicts with the cultural value and historic connotations of a term in its country of origin. For instance, political pressure continues for the US and others to recognize developing countries' terms such as *Basmati* and *Jasmine Rice*, even where they have been found by domestic decision-makers to be unambiguously generic.[103]

---

[100] See EC Minutes of Meeting (Addendum), Trade Policy Review: European Communities, 24 January 2005 (WT/TPR/M/136/Add.2) 17, concerning derogation of tariffs on basmati rice and the need for shipments to include one basmati variety only. The EU, India and Pakistan continue to work towards developing a genetically based control system. See *Final Report on the 'Plant Authentication Workshop'*, IRMM, Geel, on 21 and 22 September 2009 (26 Feb 2010) at: http://ec.europa.eu/agriculture/analysis/external/basmati/fulltext_en.pdf

[101] F Addor and A Grazioli, 'Geographical Indications beyond Wines and Spirits: A Roadmap for a Better Protection for Geographical Indications in the WTO/TRIPS Agreement' (2002) 5 JWIP 865.

[102] T Broude, 'Taking "Trade and Culture" Seriously: Geographical Indications and Cultural Protection in WTO Law' (2005) 26 University of Pennsylvania Journal of International Economic Law 623; D Gangjee, 'Geographical Indications and Cultural Heritage' (2012) 4 WIPO Journal 92; T Voon, *Cultural Products and the World Trade Organization* (CUP, 2007).

[103] US Federal Trade Commission, declining a request for a trade regulation to prevent US rice growers from using the terms 'basmati' and 'jasmine', citing existing standards for rough rice (Code of Federal Regulations 868.212 (e), established by the Agricultural Marketing Act of 1946, and amended in 1993 to include a special grade introduced for aromatic rice, specifically including these two varieties; at http://www.ftc.gov/opa/2001/05/fyi0131.htm and http://www.gipsa.usda.gov/fgis/standards/ricestandards.pdf

8. COLLECTIVE MANAGEMENT OF TRIPS

This broadening range of interests engaged – and the prospect of yoking GI protection to domestic identity politics[104] – intensifies the focus on GIs as a trade issue, both tactically and strategically, but also accentuates the contentiousness of the debate. Negotiations on GI protection, centred on the WTO but present in regional and bilateral fora,[105] disclose conceptual indistinctness and uncertainty over the legal and policy basis for GI protection: the categories of (mis)use against which a GI is to be protected, in whose interests, to what end, and to the exclusion of what other interests? If exclusions from the common domain of language are to be mandated to promote public goods, what public goods are to be prioritized: does the information function of GIs for consumers support or stand in tension with the producers' quasi-property interests? Is there a higher order public good, that of sustaining product integrity and 'identity preservation'? Isn't the fluidity of language itself a public good? Yet, as trade negotiations, the process takes on a more pragmatic hue: individual GIs come to be viewed as distinct national assets, to be championed in international negotiations with the object of securing specific forms of protection, in a manner that would be exceptional, say, for a single patent or an individual copyright work, encouraging a fix-outcome viewpoint.

Ideally, progress in negotiations would be founded on a stronger understanding of the nature and purpose of *international* protection of GIs and a clearer sense of the full interests at stake and how they should be optimally reconciled. There is a fatal presumption that new international rules on GI protection would be, in effect, self-executing. There has been comparatively little attention to the fundamentals of domestic protection in developing countries. This is critical, since, unlike any other form of IP under TRIPS, a GI must be protected in its country of origin to trigger any obligation to protect it elsewhere.[106] A more settled international framework would reduce the diversion of policy-makers from the crucial task of identifying and effectively protecting valuable national GIs assets at the national level,[107] as the basis for recognition in domestic and foreign

---

[104] M Sunder, 'IP3' (2006) 59 Stanford Law Review 257.
[105] See also WIPO analyses, e.g. 'The Definition of Geographical Indications' 1 October 2002 (SCT/9/4), 'Geographical Indications and the Territoriality Principle' 1 October 2002 (SCT/9/5).
[106] TRIPS Art 24(9).
[107] E.g. K Das, 'International Protection of India's Geographical Indications with Special Reference to Darjeeling Tea' (2006) 9 JWIP 459.

markets. BIRPI[108] and WIPO respectively sponsored the development by developing country experts of model laws for developing countries on:

- marks, trade names and acts of unfair competition (1965);[109] and
- appellations of origin and indications of source (1975).[110]

The commentary for both model laws stressed the particular value for developing countries of appellations and indications, with stress on traditional products such as handicrafts,[111] anticipating by over four decades the current interest in GIs as development-friendly, TK-oriented IP. Yet in practice, the GIs which are actually protected overwhelmingly originate in developed countries, in spite of the wealth of potential candidates in the developing world,[112] and the drift towards genericization. Together with the implementation of TRIPS obligations to protect GIs, the current international controversy – whatever its outcome – is apparently stimulating some developing countries to pay greater attention to domestic protection of their own GIs as the foundation for protection in foreign markets.[113]

Nonetheless, many developing countries have joined the European Union in placing high priority on higher international standards for GI protection in the review, implementation, and further negotiation of TRIPS.[114] Others counter these ambitions as a key defensive concern

---

[108] United International Bureaux for the Protection of Intellectual Property (the predecessor of WIPO).

[109] BIRPI, *Model Law for Developing Countries on Marks, Trade Names, and Acts of Unfair Competition* (BIRPI, Publication 805(e), Geneva, 1967).

[110] WIPO, *Model Law for Developing Countries on Appellations of Origin and Indications of Source* (WIPO Publication 809, Geneva, 1975). During proposed amendments to the Paris Convention, the insertion of a new provision – Art 10*quarter* – favouring developing countries, by 'reserving' certain terms for them, was also considered. See Gangjee (n 61) 196.

[111] E.g. Commentary on the BIRPI *Model Law*, s 51 ('Misuse of Indications of Source and Appellations of Origin') 83.

[112] Within the Lisbon system, by 2005, European countries had registered 784 GIs out of a total of 834 registered GIs, and France had registered 77 per cent of all GIs protected under the Agreement. See WTO Secretariat, 'Issues Related to the Extension of the Protection of Geographical Indications Provided for in Article 23' 18 May 2005 (TN/C/W/25, WT/GC/W/546) 5. This pattern is slowly diversifying: by January 2015, 101 of 896 registrations were held by non-European Parties (developing countries and transitional economies), and the French share had fallen to 56 per cent, see WIPO, *Appellations of Origin*, No 43 (January 2015), at 211.

[113] As required, for instance, under Article 24.9 of TRIPS.

[114] Most concretely in the co-sponsorship of a number of developing countries, along with the European Union and Switzerland, of a proposal tabled in the WTO Trade Negotiations Committee for negotiating modalities on TRIPS that would

in trade negotiations.[115] Both 'sides' have expressed their negotiating positions in terms of principle, equity and fairness. But the starkness of positions struck – an egregious blend of stereotypical North–South and transatlantic/old–new world dichotomies[116] – belies the diversity of actual economic interests and contemporary practices.[117]

A key negotiating issue is the push to extend to products other than wines and spirits: the so-called 'absolute'[118] protection afforded under Art 23.1 of TRIPS.[119] This issue reflects the century-long debate over whether international rules should restrain only the use of GIs to denote origin in a way that actually deceives or defrauds the consumer purchasing the labelled item, or whether they should prevent any commercial gain derived from the second-hand connotative qualities of a GI. But this negotiating ambition begs the question of what tangible benefits would flow from the prevention of non-deceptive 'evocation' or 'usurpation' of GIs that would otherwise be protected at the general level established by Art 22.2.[120] Typically this debate is constructed in terms of equity and

---

include extension of higher GI protection to products other than wine and spirits, see document TN/C/W/52 of 19 July 2008.

[115] D Spencer, 'A Way Forward for Geographical Indications' WIPO Symposium on Geographical Indications, San Francisco, 11 July 2003.

[116] JM Cortés Martín, 'The WTO TRIPS Agreement: The Battle between the Old and the New World over the Protection of Geographical Indications' (2004) 7 JWIP 287.

[117] Note, for instance, the division within the EU over genericization of certain GIs: *Germany and Denmark v Commission of the European Communities* (n 27) relating to 'feta' and pitting Germany, Denmark, the UK and France against the European Commission and Greece. Others are also caught in the middle: TRIPS Council, 'Special Session, Minutes Of Meeting on 16–17 March 2006', 19 May 2006 (TN/IP/M/16), para 100 ('South Africa was a producer and exporter of "feta", a generic term in its territory. Currently the European Communities, which had been its main export market until recently, were now blocking such exports and sending South African producers threatening letters.').

[118] Communication from Bulgaria et al., 'Work on Issues Relevant to the Protection of Geographical Indications: Implementation of Article 24.1 – Extension of Additional Protection for Geographical Indications to Products Other Than Wines and Spirits' (IP/C/W/204/Rev.1).

[119] For a detailed review, see Chapter 7 by Handler in this volume. This provision prevents the 'qualified' use of GIs, i.e. even where the true origin of the goods is indicated, or the GI is used in translation or accompanied by expressions such as 'kind', 'type', 'style', 'imitation' or the like.

[120] This prevents uses that mislead the public as to geographical origin of the product, or otherwise constitute an act of unfair competition within the meaning of Article 10*bis* of Paris.

general principle[121] rather than specific trade gains that would be yielded by filling this normative gap.[122]

Conventionally, the difference is construed as lying between protection against misleading use of GIs in the consumer's interest, and a form of absolute protection as a collective property right defended against all usurpation and evocation.[123] Negotiating ambitions are driven by a perception of the economic interest derived from expanding the scope of that area of trade that is considered illegitimate by virtue of the value gained from not misleading or confusing. It is accepted wisdom that the difference between the two provisions is not merely legally significant, but has sizeable – positive or negative – economic implications. Crucially, however, the economic benefit should not be assessed in terms of the value of trade in protected products, but the enhanced rent accruing from the legal distinction between the two provisions.

But general GI protection under TRIPS goes beyond protection against consumer deception or misleading the public,[124] and covers 'any use which constitutes an act of unfair competition within the meaning of Art 10bis' of Paris.[125] The stress laid on TRIPS Art 23.1 as a higher or 'absolute' form of protection implies that this extra limb[126] adds little to the conventional consumer deception test. Yet Paris Art 10*bis* has been given generous readings to provide a doctrinal basis for protection of undisclosed information[127] and regulatory test data,[128] and potentially also to restrain misappropriation of TK.[129] The rules of treaty interpretation would suggest that TRIPS Art 22.2(b) is neither otiose vis-à-vis TRIPS Art 22.1(a), nor fully co-extensive with TRIPS Art 23.1, so that it would go beyond simple consumer protection but not as far as absolute protection with no

---

[121] The positions are reviewed in: WTO Secretariat, 'Issues Related to the Extension of the Protection of Geographical Indications Provided for in Article 23 . . .' 18 May 2005 (TN/C/W/25, WT/GC/W/546).

[122] For research addressing this point, see: J Hughes, 'Coffee and Chocolate – Can We Help Developing Country Farmers through Geographical Indications?' International Intellectual Property Institute, Washington, DC (2009).

[123] Contrast TRIPS Art 22.2(a) with TRIPS Art 23.1.

[124] TRIPS Art 22.2(a).

[125] TRIPS Art 22.2(b); see history and analysis of Paris Art 10*bis* below.

[126] TRIPS Art 22.2(b)

[127] TRIPS Art 39(2).

[128] TRIPS Art 39(3).

[129] Norway: 'Memorandum on Documents WIPO/GRTKF/IC/9/4 and WIPO/GRTKF/IC/9/5', 20 April 2006 (WIPO/GRTKF/IC/9/12) 6-10; WIPO 'Glossary of Key Terms Related to Intellectual Property and Genetic Resources, Traditional Knowledge and Traditional Cultural Expressions' 27 April 2012 (WIPO/GRTKF/IC/22/INF/8) Annex, 47.

reference to context of use – so the much-debated 'extension' need not be so great a leap. The implications of TRIPS Art 22.2(b) have been curiously neglected in an otherwise exhaustive debate, but in invoking Paris Art 10*bis* it clearly provides a general safeguard against dishonest use of GIs, and possibly also acts of misappropriation and usurpation – it certainly extends the range of protection beyond consumer interests to include producer interests. Conceivably, the only distinction between the two levels of protection may be that between 'rule of reason' and *per se* protection. In principle, Paris Art 10*bis* adds a safety-net, reinforcing international expectations of fair competitive relations concerning the representation and identification of traded goods. TRIPS Art 22.2(b) links this general standard explicitly to the protection of GIs as such, in turn repairing the gap in Paris Art 10bis (concerning designations of origin) left open in previous negotiations.

9. TOWARDS FAIR TRADE?

This legacy of Paris[130] recalls that TRIPS as a trade agreement hinges in general on a conception of 'legitimate trade'.[131] As a trade law instrument, TRIPS should be read as a means of defining and reconciling legitimate expectations[132] of a fair competitive relationship in international trade,[133] sustaining a conception of 'legitimate trade'[134] and reducing 'distortions'[135] to trade. In common with general WTO law, it regulates competitive relationships on an ostensibly fair and non-discriminatory manner, rather than promoting 'free trade' or the protection of IP as an end in itself. The nominal cause of action in TRIPS dispute settlement is not compliance with its substantive standards, but rather the nullification or impairment

---

[130] Signifying deeper and broader roots: see also the review of the early diplomatic history of Paris, *infra*.

[131] Cf. US Trade Act 1930 § 1337 ('Unfair practices in import trade'), defining as unlawful both 'unfair methods of competition and unfair acts in the importation of articles' ((a)(1)) and 'importation into the United States, the sale for importation, or the sale within the United States after importation by the owner, importer, or consignee, of articles' that infringe a 'valid and enforceable' patent, copyright, trade mark, semiconductor chip product, or registered design.

[132] See C Carmody, 'WTO Obligations as Collective' (2006) 17 European Journal of International Law 419, 421 (WTO obligations 'are about expectations concerning the trade-related behaviour of governments').

[133] See the panel report in *Japan – Film* (WT/DS44/R).

[134] TRIPS, first preambular paragraph.

[135] Ibid.

of an expected benefit.[136] Hence the regime should be seen as structuring mutual expectations of fair access to trading partners' economies. Strictly, non-compliance with TRIPS is only indirectly a cause of action, as a *prima facie* instance of nullification and impairment – in principle one would still need to establish some form of damage to legitimate trading interests to sustain a complaint.[137] By this logic, TRIPS functions within the WTO not to compel compliance as such, but rather to preserve legally defined mutual expectations regarding the fair or legitimate competitive relationship between competitors, and in turn to reduce bilateral trade frictions through objectifying ideas of fairness in access to markets. This analysis would create a more coherent doctrinal framework for the incorporation of IP standards into international trade law, a development that had been viewed by many trade experts as a kind of legal and policy miscegenation. In fact, however, GI protection under TRIPS can instead force a reassessment of the essential nature of the WTO as a regime.

The essential logic of multilateral trade law is to remove unfair or illegitimate barriers to trade, such as discriminatory regulation, rather than to eliminate all barriers. IP protection is traditionally construed as a potentially legitimate exception to trade law obligations, rather than as a positive obligation in itself, reflecting a basic philosophy that any regulatory intrusion into trade needed a valid justification. Thus the core multilateral trade law text, the GATT, identifies IP as an exception to its provisions: GATT should not be 'construed to prevent the adoption or enforcement by any contracting party of measures . . . (d) necessary to secure compliance with laws or regulations which are not inconsistent with [GATT provisions], including . . . the protection of patents, trade marks and copyrights, and the prevention of deceptive practices'.[138] This did not, however, provide a blanket exemption for IP, as a 'necessity test' applied, requiring IP to be regulated in a fair and non-discriminatory manner: thus the exception is subject to the proviso that 'such measures are not applied in a manner which would constitute a means of arbitrary or unjustifiable discrimination . . . or a disguised restriction on international trade'.[139] Even if IP protection has a legitimate policy objective, these core trade law principles require the competitive relationship

---

[136] TRIPS Art 64.1, invoking GATT Arts XXII and XXIII as the basis of the cause of action in dispute settlement.
[137] WTO Dispute Settlement Understanding, Art 3.8: infringement of obligations under, *inter alia*, TRIPS, is 'considered prima facie to constitute a case of nullification or impairment'.
[138] GATT 1948, Art XX (General Exceptions).
[139] Ibid.

between enterprises to be maintained free of unfair bias towards any group of producers: GATT jurisprudence[140] confirmed, for instance, that the protection of IP did not overrule the principle of national treatment.[141] This provision is the seed of reconciliation between mainstream trade law and IP law.

This reconciliation comes into still clearer focus concerning GIs. GATT Art IX ('Marks of Origin') requires that national regulations on origin marking, in effect, be 'least trade restrictive', in the classic mode of international trade law as a ceiling on regulatory intrusion in trade. But it also sets a regulatory minimum, a positive obligation to act, relating to marks of origin – an embryonic form of the TRIPS obligations on GIs:

> The contracting parties shall co-operate with each other with a view to preventing the use of trade names in such manner as to misrepresent the true origin of a product, to the detriment of such distinctive regional or geographical names of products of the territory of a contracting party as are protected by its legislation. Each contracting party shall accord full and sympathetic consideration to such requests or representations ... regarding the application of [this] undertaking ... to names of products which have been communicated to it by [an] other contracting party.[142]

This provision was cast as an obligation to cooperate in resolving claims of unfair behaviour, rather than as a substantive standard. In a defence of existing, potentially controversial, trade practices, the Havana Conference[143] agreed that this provision should not affect 'the present situation as regards certain distinctive names of products, provided always that the names affixed to the products cannot misrepresent their true origin',[144] and made provision for 'the governments concerned to proceed to a joint examination of particular cases which might arise if disputes occur as a result of the use of distinctive names of products which may

---

[140] See GATT Panel Report on Section 337 of the US Tariffs Act 1930, (BISD 345) (s. 337 enabled the International Trade Commission to take action against imports that infringed IP rights and threatened damage to US industry interests, under the general rubric of unfair competition. Despite the Art XX exception, and though it had a legitimate objective, this provision was found to have breached the national treatment requirement of GATT as it was a means of arbitrary or unjustifiable discrimination between countries where the same conditions prevail, or a disguised restriction on international trade).
[141] There is no in-principle tension with international IP law, as the principle of national treatment is foundational in the core multilateral IP treaties.
[142] GATT 1948, Art IX, para 6.
[143] UN Conference on Trade and Employment, Havana, 1948.
[144] Reports of Committees and Principal Sub-Committees, *UN Conference on Trade and Employment* (Havana 1948) 79 at [48]:

have lost their original significance through constant use permitted by law in the country where they are used'[145] (that is pragmatic resolution of disputes over genericization of GIs). This provision was therefore on a par with multilateral IP law, which was similarly permissive of past potential misuse of GIs. It did not require the protection of GIs where the origin was not misrepresented[146] and did not require the significance in the country of origin to determine protection elsewhere.

A GATT dispute panel[147] considered this provision prior to the entry into force of TRIPS. The case pivoted on mainstream GATT national treatment issues, but also included a typical difference over legitimacy in the use of allusive geographical terms: in this case, in Japan bottles of wines, whiskies and brandies bore labels using English, French or German terms, such as *Chateau* or *Reserve* or *Village*, which the EC claimed breached GATT Art IX.6, which 'did cover not only the usurpation of a specific regional or geographical name but also the way in which a trade name could mislead as to geographical or regional origin'.[148] The panel found that labelling made clear that domestic products were of Japanese origin, and that there was no evidence that use of European terms 'had actually been to the detriment of 'distinctive regional or geographical names of products'. This case – determined under the mainstream trade law of GATT – dealt with two core issues still before the WTO today:

- Is it illegitimate trade to allude to or 'usurp' GIs when there is no actual evidence of consumer deception; and
- Is it necessary to suppress the evocative use of GIs when the true origin is indicated?

---

the text of [the provision of the Havana Charter corresponding to GATT Art IX:6] should not have the effect of prejudicing the present situation as regards certain distinctive names of products, provided always that the names affixed to the products cannot misrepresent their true origin. This is particularly the case when the name of the producing country is clearly indicated. It will rest with the governments concerned to proceed to a joint examination of particular cases which might arise if disputes occur as a result of the use of distinctive names of products which may have lost their original significance through constant use permitted by law in the country where they are used.

[145] Ibid.
[146] Paris, Art 10*bis*.
[147] GATT panel decision, Japan – Customs duties, taxes and labelling practices on imported wines and alcoholic beverages (BISD 34S/83).
[148] Ibid. 10.

European law deals tellingly with the relationship between mainstream trade law, and its liberalizing trend,[149] and the protection of GIs, as a potential constraint on trade.[150] This was at issue in *Parma Ham*,[151] when the ECJ found that GI protection was a quantitative restriction on trade, but was still justifiable. The ECJ identified the rationale of GI law in a way that united consumer protection and producer interests:

> [D]esignations of origin are protected 'against improper use ... by third parties seeking to profit from the reputation which they have acquired. They are intended to guarantee that the product bearing them comes from a specified geographical area and displays certain particular characteristics. They may enjoy a high reputation amongst consumers and constitute for producers who fulfil the conditions for using them an essential means of attracting custom. The reputation of designations of origin depends on their image in the minds of consumers. That image in turn depends essentially on particular characteristics and more generally on the quality of the product. It is on the latter, ultimately, that the product's reputation is based. For consumers, the link between the reputation of the producers and the quality of the products also depends on his being assured that products sold under the designation of origin are authentic.[152]

The court found that the EC Treaty

> prohibits all measures which have as their specific object or effect the restriction of patterns of exports and thereby the establishment of a difference in treatment between the domestic trade of a Member State and its export trade, in such a way as to provide a particular advantage for national production or for the domestic market of the State in question.[153]

The dispute in question related to whether otherwise genuine Parma ham could be sliced outside the region of origin in Italy, and GI protection that prevented the slicing and repackaging of ham outside the designated region was just such a restriction. Yet it was justified as an exception under Art 30 since it was 'necessary and proportionate and capable of upholding

---

[149] The former Art 28 of the Treaty of Rome (now Art 34 of the Treaty on the Functioning of the European Union (TFEU)) prevented quantitative restrictions, or their equivalents, on imports.

[150] Under Art 30 of the Treaty of Rome (now Art 36 of the TFEU), the prohibition on quantitative restrictions or their equivalents did not apply to measures for the protection of industrial and commercial property.

[151] Case C-108/01, *Consorzio del Prosciutto di Parma v Asda Stores Ltd* [2004] ETMR 23.

[152] Ibid., para 64 (internal citations omitted).

[153] Ibid., para 54.

230 *Research handbook on intellectual property and GIs*

the reputation'[154] of the GI, and 'there are no alternative less restrictive measures capable of attaining'[155] the objective pursued.

This line of analysis of GI protection under trade law recalls the gradual development of rules against the misleading use of geographical references under the aegis of the Paris Convention and subsequent multilateral treaties. This development similarly shows a progressive construction of a conception of 'legitimate' or 'fair' trade that shifts the conception of what level of intervention is required, from a bare safeguard against fraud to the recognition of distinct producer rights that are of inherent value and are protected accordingly.

Thus the first draft Paris Convention submitted to the 1880 diplomatic conference provided for an absolute protection against false indications of origin, but this did not survive the conference, meeting resistance from those delegations who did not accept obligations to prevent existing practices in commerce, even if they may appear fraudulent.[156] The agreed text only required suppression of a false indication of origin when this was joined with a name that was fictitious or used with fraudulent intent;[157] the narrowing of this provision was seen as a safeguard for continuing the use of genericized terms in third parties. This outcome was an effective agreement that other uses of designations of origin were not sufficiently grave to warrant an international rule against them, even if they were reprehensible and suppressed by law in some countries. This outcome established the tension that runs through international negotiations to the present day, between the defenders of the original nomenclature and defenders of a more *laissez faire, caveat emptor* approach to labelling that is evocative or otherwise linked to a designation of origin, where trade interests impelled a defence of genericized use. The trend has moved glacially towards more stringent protection of GIs, but international consensus on constraining the evocative use of claimed GIs is no easier to secure today than it was in 1880.

Disappointment at this limited conception of illegitimate trading activity[158] led to a sustained move to strengthen international standards, in

---

[154] Ibid., para 64.
[155] Ibid., para 79.
[156] *Actes de la Conférence internationale pour la protection de la propriété industrielle réunie a Paris du 4 au 20 Novembre 1880* (Ministère des Affaires Etrangères, Impr. Nationale, Paris 1880) 84, 85, 88; see also Gangjee (n 61) 41–52.
[157] Paris, Art 10 (1883): requiring seizure at the border by ex officio act or civil action of 'every production bearing falsely as indication of origin, the name of a stated locality when this indication shall be joined to a fictitious commercial name or a name borrowed with fraudulent intention'.
[158] In 1926, the Director of the United International Bureaux of Industrial, Literary and Artistic Property, Berne (BIRPI) observed:

two broad directions – Britain sought to strengthen general rules against unfair competition[159] and France sought rigour on appellations of origin in particular, including through a separate special agreement under *Paris*.[160] These interests converged in the amendment of Art 10*bis* at the 1911 Washington Act of the Paris Convention, to alter it from a simple requirement for national treatment in the law of unfair competition, to a positive obligation to provide effective protection.[161] Equally, the two countries pursued this protection at the Versailles peace conference[162] and in the League of Nations[163] (as an element of 'equitable treatment for the commerce of all Members'[164] in its covenant), before securing further revisions to *Paris* itself, in the form of more elaborate descriptions of acts of unfair competition. The United States, in particular, continued to oppose any suggestion that there should be specific rules on false designations of origin, up to the 1958 review conference in Lisbon when it alone voted against.[165] It sponsored the development of Pan-American industrial

---

[o]f course such a very limited protection could not suffice in the fight against those frauds which constitute a plague to honest trade. All countries where celebrated goods come from have a vital interest in protection against such frauds, not alone for goods manufactured in a well-known place, but especially for the products of the soil, such as, for example, French wines, Swiss cheeses, and so on. (Cited in M Ostertag, 'International Unions for the Protection of Industrial, Literary and Artistic Property' (1926) 25 Michigan Law Review 107, 115)

[159] Described in C Wadlow, 'The International Law of Unfair Competition: The British Origins of Article 10*bis* of the Paris Convention for the Protection of Industrial Property' (Oxford IP Research Centre, 19 November 2002).
[160] In the form of the Madrid Agreement in 1891.
[161] 'All contracting states undertake to assure to *ressortissants* of the Union effective protection against unfair competition' Actes de Washington (1911), at 254, 305 and 310. This followed unsuccessful attempts to introduce a general rule against unfair competition at the conferences of 1883 and 1900.
[162] Treaty of Versailles (28 June 1919), Chapter III of Part X ('Unfair Competition'). Under Art 274, Germany undertook the obligation to prevent acts of unfair competition, including providing border measures, while under Art 275 it agreed to reciprocal obligations to recognize administrative or judicial decisions protecting GIs in their country of origin. For commentary, see: '*Les abus de la législation allemande en ce qui touche les marques et les appellations d'origine sont supprime par les deux articles relatifs aux méthodes déloyales de concurrence*' 27 *Actes de Versailles* vol.1, 237.
[163] League of Nations, *Report to the Assembly from the Economic and Financial Organization* (1922), A.59.
[164] The Covenant of the League of Nations, Art 23(e).
[165] Art 10*bis*(3), providing for the prohibition of 'indications or allegations, the use of which in the course of trade is liable to mislead the public as to the nature, the manufacturing process, the characteristics, the suitability for their

property conventions that pointedly omitted references to designations of origin.[166] However, Paris Art 10, on fraudulent indications of source, was ultimately broadened with the removal of the restrictive requirements for outright fraud and the inclusion of *indirect* use of false indications.[167]

This gradual normative development within Paris, while falling short of the kind of rigorous protection of GIs that proponents sought, represented a progressive expansion of the conception of 'unfair competition' such that the requirement to suppress '[a]ny act of competition contrary to honest practices in industrial or commercial matters'[168] is not considered to be 'limited to honest practices existing in the country where protection against unfair competition is sought' but rather is extended to 'honest practices established in international trade'.[169] This provision therefore posits a broad common expectation of 'legitimacy' in international trade, in particular trade that unfairly usurps or damages the reputation of other traders as a means of damaging their competitive position. Yet the explicit opposition, sustained over successive diplomatic conferences, to any inclusion of false designations of origin as a specific example of unfair competition in Paris Art 10*bis* would suggest that multilateral norms remained relatively permissive. Thus *Paris* barely intruded into regulating the relationship between signifier and signified of a GI as sign, so that new world producers were in principle free to evoke or allude to the geographical penumbra surrounding their traditional competitors' goods provided that there was no outright deception of the consumer: a notion of 'legitimacy' in trade centred on consumer rather than producer interests, and an implicit ordering of the public goods served by GI protection that did not satisfy the interests of certain old world producers, exemplified by France.

Such advocates of a more rigorous protection of GIs therefore took a parallel path, preferring tighter norms among a smaller number of countries to diluted multilateral rules, establishing first the special Madrid agreement on false indications of origin and later forming a 'Special Union' under the Lisbon Agreement, forging an international system that

---

purpose or the quantity of the goods'. The US voted against the inclusion of the words 'the origin': see *Actes de Lisbonne* (BIRPI, 1963) 777–779; see also Knoll (n 23) 314.

[166] SP Ladas, 'Pan American Conventions on Industrial Property' (1928) 22 American Journal of International Law 803, 821.

[167] Paris, Art 10 now refers to the 'direct or indirect use of a false indication of the source of the goods or the identity of the producer, manufacturer, or merchant'.

[168] Paris, Art 10*bis* (as amended).

[169] GHC Bodenhausen, *Guide to the Application of the Paris Convention* (Bureau de l'Union, 1968) 144.

expressed a strikingly different view of legitimate trading activities, and of the role of the 'wise dictator' in predetermining legitimate language usage:

- unique deference to the linguistic usages of the wine trade in 'home' or source countries, national courts being otherwise deemed competent to determine the generic quality of appellations for all products other than 'products of the vine';[170]
- a prohibition on the diachronic evolution of the linkage between signifier and signified, with an outright ban on genericization,[171] regardless of the empirical state of affairs;[172]
- a preclusion of the linguistic context of use and what is actually denoted or connoted to the actual consumer of a product, in favour of an absolute protection of a protected term against 'any usurpation or imitation',[173] even if the consumer is accurately informed about the true origin;
- automatic presumption of eligibility for protection in foreign jurisdictions unless steps are taken within a specific deadline to refuse it.[174]

TRIPS was the next major step in the formulation of multilateral rules governing GIs. The two divergent perspectives on legitimacy of trading activities evoking geographically-related terms remained largely unresolved throughout the negotiations.[175] The TRIPS provisions have been

---

[170] Madrid Agreement, Art 4: 'The courts of each country shall decide what appellations, on account of their generic character, do not fall within the provisions of this Agreement, regional appellations concerning the source of products of the vine being, however, excluded from the reservation specified by this Article.'

[171] Lisbon, Art 6: 'An appellation which has been granted protection in one of the countries of the Special Union pursuant to the procedure under Article 5 cannot, in that country, be deemed to have become generic, as long as it is protected as an appellation of origin in the country of origin.' (See also Geneva Act of Lisbon, Art 12.)

[172] Cf. the court as factfinder in *Chablis with a Twist* (n 26).

[173] Lisbon, Art 3: 'Protection shall be ensured against any usurpation or imitation even if the true origin of the product is indicated or if the appellation is used in translated form or accompanied by terms such as "kind," "type," "make," "imitation," or the like.'

[174] Lisbon, Art 5.

[175] J Keon, 'Intellectual Property Rules for Trademarks and Geographical Indications: Important Parts of the New World Trade Order', in C Correa and A Yusuf (eds), *Intellectual Property and International Trade: The TRIPS Agreement* (Kluwer Law International, 1998), 165, 174; see also Chapter 6 by Gervais in this volume.

exhaustively analysed and debated:[176] no attempt is made here to review them in detail. Notably, however, TRIPS contained some imprints of the trends identified above – the unique deference to 'products of the vine',[177] but subject to safeguards of generic use;[178] and preclusion of the consumer context in protecting some GIs.[179] But other 'old world' desiderata were cast in terms of forward commitments to negotiate – on a stronger international framework for recognition of GIs *as* GIs,[180] as well as negotiating away exceptions based on such factors as good faith generic use[181] through an uncharacteristic (for the WTO) legitimization of bilateral deal-making.

10. BILATERAL BARTERING

Taking a cue from this prompt in TRIPS, bilateral bartering about GI protection – long part of bilateral trade negotiations[182] – has proceeded apace.[183] Bilateral negotiations on GIs are typically cast as reaching across two distinct cultures (notably new and old world), mirroring the two philosophies of GI protection in international debate:[184] one viewing a GI as a national or communal asset (typified by an appellation system), the other favouring private IP (typically trade marks). But some bilateral agreements have aimed to entrench one form of protection or another between like-minded countries.[185] Unlike other fields of IP, differences are arrayed not merely on the familiar reductionist linear scale of 'strong' as against 'weak'

---

[176] See e.g. Addor and Grazioli (n 101) 101.
[177] TRIPS, Art 23.1; but for the subsequent WTO debate on 'extension' of this provision, see Chapter 7 by Handler in this volume.
[178] TRIPS Art 24.6.
[179] TRIPS Art 23.1 prevents the use of a GI for wines or spirits even where use is qualified, such as Champagne-style wine. In this, it closely mirrors Lisbon, Art. 3.
[180] TRIPS Art 23.4.
[181] TRIPS Art 24.1.
[182] See e.g. the 1910 US-Portugal deal on 'porto' and 'madeira', in Exchange of Notes on Commercial Relations between the United States of America and Portugal (28 June 1910).
[183] E.g. Free Trade Agreement between Chile and Mexico, 27 August 2001 WTO document (WT/REG125/1), Article 15−24 and Annex 15−24).
[184] D Vivas-Eugui, 'Negotiations on Geographical Indications in the TRIPS Council and their Effect on the WTO Agricultural Negotiations: Implications for Developing Countries and the Case of Venezuela' (2001) 4 JWIP 703.
[185] See e.g. Australia-United States Free Trade Agreement, Article 17.2 ('Trademarks, *Including* Geographical Indications'): 'Each Party shall . . . provide that geographical indications are eligible for protection as marks.'

IP protection, but also over the relative value of (and even cultural assumptions behind) these two forms of protection. There is often assumed to be a cultural divide and a zero-sum trade-off between the camps,[186] or at least that trade mark law and GI protection are inherently at odds.[187] Yet in practice, and at the fundamental level of principle, GIs are protected by both means. Roquefort, a celebrated GI, was recognized by royal letters patent, registered in France as a distinctive oval trade mark in 1863 and recognized as an appellation of origin in 1924.[188] It has been successfully protected overseas, nationally, through specific GI laws or the law of trade marks[189] while, regionally, it is protected by trade mark[190] and as a protected designation of origin;[191] it is registered in the international registers both for appellations[192] and for trade marks.[193] Curiously, this functional commonality only serves to highlight the subjective, value-laden flavour of the multilateral debate.

By contrast, in bilateral negotiations, pragmatic sectoral interests are

---

[186] E.g. SD Goldberg, 'Who Will Raise the White Flag? The Battle between the United States and the European Union over the Protection of Geographical Indications' 22 University of Pennsylvania Journal of International Economic Law 107.

[187] S Stern, 'Geographical Indications and Trade Marks: Conflicts and Possible Resolutions', 13 June 2003 (WIPO/GEO/SFO/03/13).

[188] *Décret du 22 janvier 2001 relatif à l'appellation d'origine contrôlée 'Roquefort'*, JO no. 21 du 25 Janvier 2001, amended by *Décret du 17 mai 2005 modificatif relatif à l'appellation d'origine contrôlée 'Roquefort'*, JO no. 120 (25 May 2005) page 9023 texte n° 32, referring to 'la loi du 26 juillet 1925'.

[189] US Certification Mark 0571798 of 10 March 1953 (Community of Roquefort): 'the certification mark is used upon the goods to indicate that the same has been manufactured from sheep's milk only and has been cured in the natural caves of the Community of Roquefort, Department of Aveyron, France'; see *Community of Roquefort v William Faehndrich, Inc*, 303 F. 2d 494 (CA 2 1962) (a geographical name does not require a secondary meaning in order to qualify for registration as a certification mark); see also USPTO, Trademark Manual of Examining Procedure (TMEP) (April 2014) § 1306.02 'Certification Marks That Are Indications of Regional Origin'.

[190] OHIM, Community Trade mark 001514124, Filing date: 17/02/2000 ('Roquefort Société').

[191] Council Regulation (EC) No 510/2006 of 20 March 2006 on the protection of geographical indications and designations of origin for agricultural products and foodstuffs; now replaced by EC Regulation 1151/2012.

[192] Appellation ROQUEFORT; Lisbon Registration Number 459; Registered on 20.12.1967.

[193] Madrid Register: 139304 ROQUEFORT SOCIÉTÉ; 139305 ROQUEFORT SURCHOIX; 170077 ROQUEFORT; 262762 ROQUEFORT; 391725 ROQUEFORT; 442869 ROQUEFORT; 477479 ROQUEFORT; 594993 GALICHONS AU ROQUEFORT; 595424 ROQUEFORT RIGAL; 672250 ROQUEFORT SOCIÉTÉ (various label marks, dating from 1909).

closer to the surface than in multilateral discourse. For instance, bilateral negotiators over trade in wine barter directly between more rigorous GI protection for old world producers and more permissive market access for new world producers using non-traditional production methods (confronting such conventional trade negotiation issues as regulatory barriers to market entry). They improvise an *ad hoc* conception of legitimate trade or fair competition, working within the boundaries of a multilateral policy space which has TRIPS minimum standards as a floor, and GATT restraints on illegitimate discrimination as a ceiling.[194] New world wine producers seek to legitimize new oenological practices[195] and secure simplified certification; they directly trade on GI protection to secure such access: once Europe agreed on mutual acceptance of oenological practices 'the US industry ... concluded that ... they'd be willing to relinquish the use of European names';[196] '[i]f the [Australian] industry wants to sign this [bilateral] agreement, then it has to be prepared to trade away some of these terms'.[197] The 2005 US-EU bilateral agreement made this trade-off explicit when the US agreed to prospectively change the legal status of 17 semi-generic names in favour of the EU.[198] As part of the deal, the agreement provides for mutual recognition of winemaking practices, removing 'the continuous uncertainty of temporary derogations [from EU Regulations], providing more stable market conditions for US wine exporters'.[199]

Such terms are 'relinquished' in a very direct manner, precluding normal domestic decision-making: the negotiators are – to use Plato's terminology – in effect the 'wise dictators' and their own 'dialecticians'. Bilateral negotiators haggle over and settle on a list of terms to be deemed as GIs, pre-empting independent review of the actual denotation or connotation of the terms.[200] Agreements ostensibly set rules, providing for substantive definitions of 'GIs', but their real effect is in the listed GIs and the

---

[194] See the above discussion also with reference to European trade law.
[195] Such as cation-exchange resins for wine stabilization, and oak chips in steel casks.
[196] T Carter, 'US Vintners Hold Names Hostage in Trade Dispute', Washington Times, 29 November 2002.
[197] Leon Gettler, 'Port, Sherry Down the Drain', *The Age* (19 July 2004); for details, see Chapter 10 by Stern in this volume.
[198] Agreement between the European Community and the United States of America on Trade in Wine [2006] OJ L87/2.
[199] Foreign Agricultural Service, U.S. Mission to the European Union, US-EU Wine Trade Issues: EU-US Wine Agreement, 9 August 2007.
[200] Several such agreements are considered in Chapter 7 by Handler in this volume.

conditions for their protection, applied as direct bilateral obligations. In effect, negotiators agree that a term will cease to have generic effect in the common language as of a certain date, from which it will resume having an exclusive reference to its historic geographical origin. Under its bilateral agreement with the EU, 'Canada shall no longer deem that [certain contested terms] are customary in the common language of Canada as a common name for wines as foreseen in [TRIPS, Art 24.6]'.[201] 'Port', for instance, would be deemed to change its semantic character by December 2013, regardless any denotative or connotative link between the signifier 'port' and the signified bottle of fortified wine perceived by the relevant linguistic community in Canada. The agreement stipulates directly that listed names 'are eligible for registration as protected geographical indications'[202] rather than allowing a distinct domestic decision-making process to apply general rules in practice. This direct settlement of what counts as 'legitimate trade' may be defensible as an agreed functional resolution, an improvised sense of equity, which realistically responds to specific interests. But it risks losing consistency and balance in the application of underlying principles – the 'moderation and probability' and empirical grounding of the wise dictator.[203]

11. NEGOTIATING LEGITIMACY

Trade negotiators aim to reconcile divergent views as to what forms of trade are legitimate: in the above examples, what standards should apply to imported wine, and what terms may be permitted to designate a bottle and to connote the properties of its contents. This chapter has discussed two general ways of establishing a sense of legitimate trade, a 'fix-rule' trading regime and a 'managed-trade' (or 'results-oriented') trading regime, a tension long present in more traditional areas of trade negotiations.[204] In trade negotiations over GIs, the choice lies between:

- the articulation of international standards that guide domestic processes, which in turn determine whether and how a claimed GI is to be protected, with a clear distinction between public

---

[201] Agreement between Canada and the European Community on Trade in Wine and Spirits (16 September 2003); Art 12 (Customary Terms and Transitional Arrangements).
[202] Ibid., Art 10.
[203] Plato's *Cratylus*, 414e.
[204] For this distinction, see Bhagwati (n 8).

international law and the operation of municipal law within a framework of international standards (this approach would limit itself to such roles as clarifying the definition of GI[205] or the nature of the 'unfair competition' against which WTO Members must protect GIs),[206] or negotiating to extend the scope of higher protection beyond wines and spirits or to eliminate exceptions and limitations;[207] and

- the 'managed trade' or 'results-oriented' option of determining *a priori* and at the international level which terms are to be protected as GIs and how, pre-empting domestic legal and administrative processes and fact-based determinations.

This distinction also finds an analogue in the choice between a rule of reason (a claim for GI status and the scope of protection assessed on a case-by-case basis and on merits) and a *per se* approach to determining geographical significance and finding damage, for instance even in the absence of evidence of deception or confusion;[208] it is echoed also in administrative arrangements for enforcement of GIs, including pre-emptive *ex officio* enforcement. A choice also lies between *ex ante* or prior determination of the protected status of a GI (typically associated with GI registration systems), and a case-by-case determination of GI status (associated with general use of unfair competition or passing off remedies).[209] The factors guiding the choice lie essentially between:

- administrative efficiency, predictability and transparency, and a lighter practical burden on interested parties, through *per se* rules; and
- a closer, more closely analyzed correlation between actual outcomes and the ostensible policy goals of GI protection, through rule-of-reason findings.

---

[205] By contrast, see WIPO, *Joint Recommendation Concerning Provisions on the Protection of Well-known Marks* (1999), in effect a gloss on the TRIPS and Paris obligations.

[206] TRIPS, Art 22.2(b).

[207] TRIPS, Art 24.1.

[208] D Barjolle and B Sylvander, 'Protected Designations of Origin and Protected Geographical Indications in Europe: Regulation or Policy?' (European Commission, June 2000) 13.

[209] See 'Need for prior recognition' in WTO, 'Review under Article 24.2 of the Application of the Provisions of the Section of the TRIPS Agreement on Geographical Indications' 24 November 2003 (IP/C/W/253/Rev.1) 12.

The trend in anti-trust law[210] has been towards rule of reason, or optimally differentiated rules,[211] and away from *per se* determinations, on the basis of greater welfare gains. Yet arguments have been put forward that pre-emptive determinations should be favoured for GI protection, both pragmatically, as being less burdensome on complainants and therefore more equitable in practice,[212] and on broader policy and cultural grounds.[213] The core question is whether trade negotiations should make deals which reach directly into domestic law and determine specific outcomes, or should they be restricted to clarifying and strengthening the policy basis for the international standards that govern a distinct layer of domestic regulation. This tension also keys into the broader debate about whether the WTO system is (or should be) at core a concentrated diplomatic process, closer to the early GATT system, aimed at the pragmatic settlement of trade differences through consultation and negotiation, or a legalistic regime, aimed at the articulation, interpretation and application of objective rules to disputes on their individual merits, regardless of broader trade interests and political relations.[214]

Strikingly, unlike other substantive IP standards,[215] TRIPS provides for several scenarios in which rules give way to outcome-oriented, diplomatically-negotiated solutions:

---

[210] See e.g. U.S. Department of Justice and the Federal Trade Commission, Antitrust Guidelines for the Licensing of Intellectual Property, 6 April 1995.

[211] A Christiansen and W Kerber, 'Competition Policy With Optimally Differentiated Rules Instead of "Per Se Rules vs Rule Of Reason"' (2006) 2 Journal of Competition Law & Econ. 215.

[212] E.g. Proposal from Bulgaria et al., 'Work on Issues Relevant to the Protection of Geographical Indications: Extension of the Protection' 29 March 2001 (IP/C/W/247) ('[I]n order to defend a geographical indication for a product under TRIPS:22, a plaintiff must prove to judicial or administrative authorities that the public has been misled, or that there has been an act of unfair competition. This is complicated and expensive.').

[213] For example, Barjolle and Sylvander (n 208) 14 argue that (i) AOs and GIs are applied to goods that are 'tied to and inseparable from their geographic provenance', so that it is a 'question of requiring respect for accepted and established local customs', and (ii) they are 'collective property' and 'public goods', so the intellectual property in the name and the intrinsic properties of the goods, 'include[] a patrimonial aspect which justifies public intervention against misuse'.

[214] Key contributions to the debate include John Jackson, *The Jurisprudence of GATT and the WTO: Insights on Treaty Law and Economic Relations* (CUP, 2000) and Deborah Cass, *The Constitutionalization of the World Trade Organization* (OUP, 2005).

[215] But see TRIPS, Art 40 on anti-competitive practices.

- negotiations 'aimed at increasing the protection of individual geographical indications under Art 23',[216] bilaterally or multilaterally, that could entail waiving some or all of the balancing exceptions or limitations[217] – for instance, banning the use of terms 'customary in common language as the common name',[218] akin to past bilateral agreements;[219]
- 'practical conditions' to differentiate homonymous indications, as a pragmatic bilateral settlement instead of zero-sum exclusivity;[220]
- bilateral and plurilateral consultations dealing with concerns about compliance with TRIPS GI obligations.[221]

Despite these openings and the negotiating goals of some WTO Members, TRIPS is likely to retain a rule of reason approach to protection of individual GIs,[222] although Art 23.1 is effectively a *per se* determination as to the existence of damage to the original producers. TRIPS still operates at the level of general principles to be applied and implemented domestically according to local factors within the agreed scope of flexibility, and does not currently prescribe specific outcomes or pre-empt domestic determinations. Domestic tribunals can still determine whether a certain term conforms with the definition of GI at all, and whether it has generic signification or a secondary meaning as a trade mark. In making these determinations, a tribunal is still entitled to function as an objective factfinder guided by actual usage of the term within the relevant linguistic community. The harmonizing impact of TRIPS has yielded a massive critical literature.[223] Yet it remains far from an international *per se* or *ex ante* approach to determining the IP status of individual cases. Despite the strong imprint of 'diplomatic' or 'managed trade' methodologies in the TRIPS text, and despite the bilateral recourse to this approach, the TRIPS Council remains focused on a 'fix-rule' approach. This choice underscores the inherent corrective mechanism of multilateralism, by contrast with the more arbitrary bilateral deals which move readily within *per se* territory in terms both of the absolute status of GIs and the terms and timeframe

---

[216] TRIPS, Art 24.1.
[217] TRIPS, Art 24.4–24.8.
[218] TRIPS Art 24.6.
[219] See the bilateral examples discussed above.
[220] TRIPS, Art 23.3.
[221] TRIPS, Art 24.2.
[222] As pointed out by Barjolle and Sylvander (n 208).
[223] An early contribution: M Hamilton, 'The TRIPS Agreement: Imperialistic, Outdated, and Overprotective' (1996) 29 Vanderbilt J. of Transnational Law 613.

for their protection, regardless of the empirical state of affairs among the relevant language communities.[224] And this pre-emptive mode may also be the consequence – *de facto* or *de jure* – of the implementation of the notification/registration system foreseen in Art 23.4, which is the chief reason why the European model[225] is resisted strongly by others;[226] while Hong Kong attempted an elegant compromise,[227] under which a registered GI would benefit from only a limited level of presumption of eligibility for protection – a frank political compromise between two antithetical ways of viewing the international level of protection.

12. CONCLUDING OBSERVATIONS

A 'Utopian trade negotiator' is either an oxymoron or unemployed; the term is unknown even on Google.[228] Trade deals are not principled exercises in Rawlsian equity or macroeconomic theory, but pragmatic improvisations of politically workable settlements that provide sufficient confidence to disparate interest groups for the outcome to be politically sustainable.

Two broad approaches to securing such outcomes have been identified – fix-rule and fix-outcome. The 'fix-rule' is plainly the most appealing in principle. But the long experience of disproportionate outcomes in GI protection for developing countries suggests that establishing bare rules favours countries with more resources, entrenching existing inequities. Ironically, though, fix-outcome deals typically also favour those with the resources and the clout to pursue their national champions. One concern that has arisen in TRIPS negotiations over a GI register has been that it would effectively swamp developing countries' capacities to exercise whatever discretion is left to them.[229] The argument for practical equity

---

[224] E.g. 'port' in Canada, *supra*.
[225] EU proposal, WTO document (TN/IP/W/11).
[226] Argentina et al., Joint Proposal, WTO document (TN/IP/W/10).
[227] Hong Kong, China, WTO document (TN/IP/W/8). For a more detailed discussion, see Chapter 8 by Martín in this volume.
[228] 'No standard web pages containing all your search terms were found. Your search – utopian trade negotiator – did not match any documents', at www.google.com; the search engine does, however, presumably without irony, propose in place of 'Utopian trade negotiator' the following: 'Did you mean: "*european* trade negotiator"?' (emphasis in the original) (visited 23 August 2007).
[229] See WTO Secretariat, 'Side-By-Side Presentation of Proposals, Addendum' 4 May 2007 (TN/IP/W/12/Add.1) 6, citing views put forward by Argentina and Guatemala.

therefore cuts both ways, as despite the romance of their claimed GIs, developing countries are also exporters of 'generic' products and have limited capacity to establish rigorous protection of their own GIs domestically. This suggests the optimal approach multilaterally would still be to elaborate firmer rules, providing clearer guidance to all parties, rather than bartering over lists and conditions in the manner of early GATT rounds of negotiations trading bilateral tariff concessions on vast lists of products – 'managed trade' in the guise of multilateralism.

Rules that regulate denotation of GIs are inherently more defensible and consumer-oriented than more rigorous protection that would suppress free-riding on connotation as well. Trade negotiators should ensure they are paid handsomely in terms of market access on other fronts if they trade away, not merely their producers' capacity to evoke the qualities of their rivals' goods, but even the terms customarily used in the market to designate their own goods: any such extreme incursion on the public domain as a constraint on the common tongue should serve a public good of high priority.

To negotiate on GIs – rules or outcomes – is to construct a working conception of fairness in trade, a self-justifying determination as to what behaviour is legitimate. TRIPS was a massive bet by negotiators that international rules requiring exclusive IP rights would advance the objective of the multilateral trade regime to preserve fair competitive opportunities for traders and to promote equitable economic development. The development dimension of GIs gives their proper protection special significance, and illustrates the unexpected doctrinal roots shared by trade law and IP. Yet negotiators have struggled for over a century, in peace conferences, trade negotiations, diplomatic conferences and numerous bilateral processes, to settle on a common conception of what ways of using GIs count as fair trade and what do not. A decade of TRIPS Council debate has only entrenched this intractability. The drift to bilateral deal-making, trading away GIs for market access, is therefore inevitable, creating new facts on the ground which will retrospectively be deemed equitable, just as negotiators – wise dictators – 'deem' that words will cease to have their actual denotation and connotations among the users of the common language on an appointed date.

PART III

DOMESTIC PROTECTION MODELS

10. A history of Australia's wine Geographical Indications legislation
Stephen Stern

1. INTRODUCTION

The story of Australia's first forays into the complex area of the recognition, registration and protection of Geographical Indications (GIs) has been limited to the area of GIs for wines. As is described below, the legislative scheme that first introduced the term 'geographical indication' into Australian jurisprudence dates from only 1993.[1] Before that time, no Australian legal textbook dealt with the concepts of geographical indications, appellations of origin or the like. It was only when Australia became involved in the WTO negotiations that led to the TRIPS Agreement that this term became known within Australian legal circles, outside of a handful of specialist legal practitioners. Apart from a provision of Australia's trade mark legislation that imposes obstacles to registration of trade marks 'containing or consisting of a false geographical indication', even as at 2014 there remains no Australian legislative system dealing specifically with geographical indications outside of the wine sector.[2]

That is not to say that this legal concept was unknown in Australia. In 1918, Percy Wilkinson, who was then the Director of the Commonwealth Laboratory presented a paper to the Royal Geographical Society of

---

[1] Before the Australian Parliament introduced legislation in December 1993 designed specifically to protect wine GIs, there was no legislation of that nature in Australia. The then-existing wine legislation, namely the *Australian Wine & Brandy Corporation Act 1980* (now known as the *Australian Grape and Wine Authority Act*), did incorporate provisions relating to what is known as the Label Integrity Program (described below). Whilst falling far short of specific protection for GIs, this program was the first wine-specific step towards ensuring that, amongst other things, Australian winemakers did not mis-describe their wines.

[2] On alternative protection options, see MR Ayu, 'How does Australia Regulate the Use of Geographical Indication for Products other than Wines and Spirits' (2006) 3 Macquarie Journal of Business Law 1.

Australia[3] in which he encouraged the Australian wine industry to cease using foreign wine names. His paper stated:

> The description of Australian wines under European geographical wine names, in defiance of the express terms of international conventions, may prove a stumbling block to the expansion abroad of the Australian wine trade: and this objectionable practice constitutes a potential weapon capable of being used against us by foreign competitors. Fortunately for the Australian wine industry in the matter of the misuse of European geographical wine names, it is never too late to rectify a wrong...

It only took another 92 years for European GIs to obtain full protection in Australia! Interestingly, Mr Wilkinson advocated using – and even registered as trade marks, available for use by Australian vignerons – a series of wines' names 'capable of replacing as trade descriptions for Australian wines the European geographical wine names, including "Burgalia" to be used instead of "Burgundy", "Chabalia" instead of "Chablis", "Champalia" instead of "Champagne"' and so forth.[4] In any event, the materials set out below deal only with the wine sector, although this can, of course, be seen as a case study for other sectors or industries.

The Australian wine GI history can be described as having six phases:

(i) Pre-1987: There was no legislative GI system in place and no significant attempts from any European authority to protect European wine designations.
(ii) 1987–1993: In this period, there was both major GI litigation, principally commenced by the INAO[5] of France, as well as negotiations between the Australian Government and the EC on a wine treaty.
(iii) 1993–2005: The first Australian GI system was enacted into Australian law. Australian GIs were registered and the first Australian disputes were fought. There were ongoing negotiations of the Joint Committee established by the 1993 EC-Australia Wine Agreement, which *inter alia* agreed to the final phase out date of European GIs.

---

[3] W Percy Wilkinson, *The Nomenclature of Australian Wines in Relation to Historical Commercial Usage of European Wine Names, International Conventions for the Protection of Industrial Property* (Royal Geographical Society of Australasia, Victoria Branch, 1919), reproducing a paper initially published in the (1918) 34 Victorian Geographical Journal.

[4] Whilst I applaud Mr Wilkinson's foresight, I am somewhat relieved that the Australian wine sector did not take up Mr Wilkinson's suggested alternative wine names!

[5] INAO: Institut national de l'origine et de la qualité, previously known as l'Institut national des appellations d'origine.

(iv) 2005–2010: The Australia-United States Free Trade Agreement (AUSFTA),[6] which impacted on the wine trade, came into effect, bringing with it a series of amendments dealing with conflicts between GIs and trade marks.
(v) 2011–2014: In December 2008 the EU and Australia signed a new treaty replacing the 1993 Wine Agreement. The new Treaty came into effect in 2010,[7] and Australia thereafter amended its wine legislation to implement the new Treaty.
(vi) 2014 onwards: The *Wine Australia Corporation Act 2010* was amended and became the *Australian Grape & Wine Authority Act 2013*, which came into force on 1 July 2014.

2. THE LEGAL LANDSCAPE PRIOR TO THE EU-AUSTRALIA 1994 WINE TREATY

The earliest recorded attempts to protect what we would call a wine GI occurred in the early 1920s, although that was merely a skirmish between two wine producers over a single regional name.[8]

2.1 The First Australian Attempts to Protect a Wine Region's Name – 1925

The earliest case of the name of a wine region being protected was the High Court decision in *Thomson v B Seppelt and Sons Ltd*.[9] In the 1920s, Australia's High Court had original jurisdiction in trade mark matters. The dispute involved an opposition by a leading vigneron – Mr Thomson – in the grape growing and wine making area known as 'Great Western' against an application by one of Australia's then largest wine companies, Seppelt, which had applied to register the trade mark GREAT WESTERN for still and sparkling wines on 8 December 1923. The evidence before the court was that the name Great Western had been the name of a township since 1860 'in the neighbourhood of which a number

---

[6] http://www.dfat.gov.au/fta/ausfta/final-text/index.html
[7] http://www.daff.gov.au/agriculture-food/wine-policy/trade-in-wine
[8] Interestingly, as described below, it was the same two wine sector players who fought out another skirmish over the same regional name 80 years later, in what was Australia's first court decision dealing with a conflict between a trade mark and a GI.
[9] *Thomson v B Seppelt & Sons Ltd* [1925] HCA 40; (1925) 37 CLR 305 (29 October 1925).

of vineyards had been established, and the industry of wine making [had] been carried on [by 1925] for... sixty years or more'. Therefore, Mr Thomson as well as a number of other vignerons from the district of Great Western in Victoria and J. Richardson, a wine and spirits merchant in Melbourne opposed Seppelt's application. In January 1924, the Registrar of Trade Marks gave his decision, dismissing the opposition, essentially on the basis of acquired distinctiveness. He held that the words 'Great Western' as applied to wine had 'ceased merely to denote a place, and ha[d] come to mean to the public of Australia generally wine of [Seppelt] produced from grapes grown in the district of Great Western'. The decision was justified on the basis that Seppelt had established that its use of the words 'Great Western' was so extensive and so widely recognized in the trade and by the public more generally in the Commonwealth that when they were used with wine products, they were understood automatically to refer to Seppelt and none other.

Mr Thomson appealed to the High Court, where he argued that the name 'Great Western' was a geographical term describing the place where many persons grew grapes and made wine. The court held that in order to be registrable, the name 'Great Western' must be a distinctive mark. Justice Rich stated:

> I am always reluctant to grant anyone a monopoly in a geographical name. I doubt whether 'Great Western' is a proper registrable name to denote wines, any more than in England 'Leicester' would be registrable in connection with boots, or 'Burton' in connection with ale . . . However this may be, the evidence shows that the name Great Western is, according to its ordinary signification a 'geographical name', and has not acquired a secondary meaning. It is not distinctive of the respondent's wines, that is, is not adapted to distinguish those wines from those of other vignerons in Great Western.[10]

Justice Higgins noted:

> If we grant to the applicants the exclusive right to the use of the words 'Great Western' in connection with wines . . . [this] grant will, on any construction, seriously interfere with [the] legitimate use of the words. It would be as absurd in the same way, though not to the same degree, as if a winegrower in Champagne in France were forbidden to say that his wine is a Champagne wine.[11]

Chief Justice Knox had similar concerns:

> On the evidence I think it is clear that the words 'Great Western' are, according to their ordinary signification, a geographical name, and that as applied to

---

[10] Ibid 315–316.
[11] Ibid 315.

wine they mean that the wine was produced from grapes grown in the district of Great Western. It follows that, in my opinion, the words are not adapted to distinguish the wine of the respondent from that made by other persons from grapes grown in that district.[12]

The appeal was allowed and the trade mark application was rejected.[13]

The High Court decision simply resolved the dispute before it and did not set in train any legislative or other drive to regulate protection for wine (or any other) GIs. In the specific context of trade mark law, both the former *Trade Marks Act 1955* and the *Trade Marks Act 1995*[14] contain provisions that were specifically designed to prevent registration of names whose meaning was primarily geographical. Further, *Thomson* doesn't go as far as to suggest that geographical names, as a class (which might refer to obscure places unfamiliar to the Australian public), are inherently barred from registration. However, it was not until 1993 and the negotiations leading to the signing of the EC-Australia wine agreement,[15] that Australia put in place a legislative scheme for wine GIs. Over the 20 years of its existence this system has already seen some major changes. The value of regional names for wines has clearly been recognized within Australia, even if only for the last 15 or so years, as the following case study illustrates.

2.2 How Much is an Australian Wine GI Worth?

It was only a year or two after the implementation of Australia's GI system in 1993 that the seeds of the first serious litigation were planted. A group of vignerons and winemakers in the Coonawarra area of South Australia started drawing lines on maps, planning out the region that they wanted to be entitled to use the name 'Coonawarra' for wines from the eponymous area. At the time of the incident described below, the Coonawarra litigation was about six months old, having started in May 2000. By the end of that calendar year, there were over 70 parties locked in dispute in front of the Federal Administrative Appeals Tribunal. The parties

---

[12] Ibid 311.
[13] However, 80 years later, Mr Thomson's grandson and the successor corporation to Seppelt would lock horns again, fighting once more over the name 'Great Western'.
[14] Much of the legislation described in this chapter has been significantly amended over the years, with several Acts even changing their names. For consistency, each Act is described in this chapter using the name current at the time referred to in the text.
[15] See Section 2.5 below.

included 45 vignerons endeavouring to be included in what was to become the Coonawarra wine region as originally mapped out by the Australian Geographical Indications Committee (GIC), and 25 vignerons within the region as already mapped by the GIC, who were fighting to keep the other 45 out. The novelty and complexity of the litigation as well as the enormous amount of ongoing coordination between the two dozen lawyers representing the 45 'outsiders' had resulted in significant legal expenditure.

The lawyer[16] representing Mildara Blass Wine Estates (as Treasury Wines was then known), the company's in-house IP Director instructing him and the company's Financial Controller met to discuss whether or not the company would continue with what was becoming expensive litigation. The Financial Controller was concerned that the case was costing a fortune and wanted to know why the company shouldn't pull out of the case. He listened for a few moments to the explanation as to the likelihood of success and then cut in, saying: 'Look, just give me some figures.' He literally pulled in front of him an envelope, turned it to the blank side and asked: 'How many acres do we have outside the region that we're trying to get in?' '200 acres' he was told. 'And how many tonnes of fruit are produced in an average year by this vineyard?' he then asked. 'At an average of 8 tonnes per acre, about 1600 tonnes.' 'At what price can we sell the grapes if they can be sold as Coonawarra grapes?' '$1800 per tonne.' 'And if we can't get the vineyard into the Coonawarra region, what price per tonne then?' 'About $1300 per tonne – $500 a tonne less.' His pen flashed across the back of the envelope, '1600 x $500 = $800K' he wrote. 'Per year' he mused. 'Okay, get on with the case. It'll clearly pay for itself in a couple of years at the most, leaving aside entirely the increase in property value!'

And so Mildara Blass fought on, and ultimately its 'Robertson's Well' property became the northern-most vineyard in the Coonawarra wine region, because Australian wine GIs had been recognized by grape purchasers, by winemakers, by corporate executives and ultimately, by consumers as having real value. Each was prepared to pay for the name Coonawarra. So, turning back the clock, what is Australia's history of protecting wine regional names? I use that terminology deliberately, because it has been said that it was not until the legislative amendments to Australia's wine laws in July 2010, that Australia really could claim that it had a system of true wine GIs.

---

[16] In the interests of full disclosure, it should be noted that the lawyer was the author of this chapter.

2.3 The *Australian Wine and Brandy Corporation Act 1980*

2.3.1 The history of Australian wine production

Before going further, it is important to provide the historical background for the Australian wine sector, as this puts into context the sector's slow development.[17] Australia can date its wine industry back to the First Fleet of convicts to arrive in Australia from England in 1788. Vine cuttings were brought out and planted in the first white settlement in what is now Sydney. The industry waxed and waned, and in the early 20th century, like Europe, it was devastated by phylloxera.[18] The Australian wine sector limped along for the next four or five decades, not making many inroads into what was principally a beer-drinking nation. You needed to be a brave man in the 1960s or even 1970s to walk into an Australian public bar and order a glass of wine.[19] Nevertheless, by the 1970s, the wine sector started to grow significantly,[20] and by the end of that decade, was important enough to warrant Government assistance and intervention. As a result, the Government enacted the *Australian Wine and Brandy Corporation Act 1980* (AWBC Act).

2.3.2 The objectives of the AWBC Act

The AWBC Act was enacted for the principal purpose of establishing the Australian Wine & Brandy Corporation. This organization was subsequently renamed as the Wine Australia Corporation and then after a merger with the Grape and Wine Research and Development Corporation was renamed again in 2014 as the Australian Grape & Wine Authority (AGWA).[21] This is a Government statutory corporation, whose goals are to *inter alia*:[22]

---

[17] For further background, see R Osmond and K Anderson, *Trends and Cycles in the Australian Wine Industry, 1850–2000* (Centre for International Economic Studies, University of Adelaide 1998); J Beeston, *A Concise History of Australian Wine* (Allen and Unwin 1994).

[18] L. Evans, *Len Evan's Complete Book of Australian Wine* (Wedson Publishing 1990) 23.

[19] And ordering a glass of wine was not the same as actually receiving one!

[20] See e.g. K Anderson (ed), *The World's Wines Markets: Globalization at Work* (Edward Elgar Publishing 2004); R Jordan, P Zidda and L Lockshin, 'Behind the Australian Wine Industry's Success: Does Environment Matter?' (2007) 19 International Journal of Wine Business Research 14.

[21] Throughout this chapter, this organization will be referred to by its name at the time in question. The same principle applies to the name of the wine legislation.

[22] See Commonwealth of Australia, Parliamentary Debates, 11 September 1980, 1195 [Peter James Nixon].

(i) promote and control the export of grape products from Australia;
(ii) promote and control the sale and distribution, after export, of Australian grape products; and
(iii) promote trade and commerce in grape products among the Australian States and Territories.

2.3.3 Label Integrity Program

As part of amendments made to the AWBC Act in 1989, the legislation incorporated provisions implementing a wine Label Integrity Program. This program requires wine producers to keep records relating to the vintage, grape varieties used and source of origin of their wine and to make those records available for audit. The purpose of auditing is to ensure truth in labelling, thereby increasing consumer confidence in the reliability of label claims. The legislation gave the power to the AWBC to carry out audits of wine producers' records and stocks.[23] However, this legislation stopped short of making it an offence to make false statements on wine labels. Rather, information obtained by the AWBC showing that false label claims have been made can be the subject of prosecutions under other State, Territory and Commonwealth consumer protection laws, such as the *Trade Practices Act 1974* (TPA 1974), now known as the *Australian Competition and Consumer Law 2010*, prohibiting the making of false or misleading statements. On the other hand, it should be stressed that the Label Integrity Program, as enacted in 1989, was merely the first step in the process of identifying wines by regions, which step has been surpassed by further statutory amendments.

2.4 European Attempts to Protect GIs in Australia

2.4.1 Spanish Champagne and the generic problem: passing off and the TPA

The next relevant Australian court decision to protect wine GIs was initiated in 1981. It concerned an action by the Comité Interprofessionel du Vin de Champagne (CIVC) to prevent the importation into and sale in Australia by NL Burton of Spanish sparkling wine, produced by Freixenet, and which was labelled and sold as Spanish Champagne.[24] By 1981, there had still not been any legislative move to protect GIs in Australia. Thus

---

[23] Part VI A of the Wine Australia Corporation Act (as amended). Available at: http://www.comlaw.gov.au/Details/C2012C00401/Html/Text#_Toc322685321

[24] *Comité Interprofessionel du Vin de Champagne and Anor v NL Burton Pty Ltd and Anor* (1981) 38 ALR 664.

when the CIVC tried to prevent the use of the name Champagne on Spanish sparkling wine imports, it had to rely on the common law tort of passing off as well as on the then relatively new consumer protection and competition statute, the TPA 1974, as described below.[25] As we will see, each has its limitations.

Passing off is a common law tort which seeks to prevent misrepresentations in trade that damage goodwill or reputation. As a result of the famous 'Spanish Champagne' litigation in Britain in the 1960s,[26] 'extended' passing off remained in the 1980s as the principal method of non-registration-based GI protection in the common law world. Unlike the conventional tort, in the 'extended' version the misrepresentation relates not to the individual trade or commercial source of the product but to membership in a class of products which suggests that it has certain qualities.[27] To succeed in a passing-off action, a plaintiff must prove, amongst other things, that it possesses a reputation (whether alone or jointly with other persons) in a name, mark get-up or style which has been used by the defendant in such a way as to mislead members of the public into the false belief that products from the defendant emanate from the plaintiff. It is also necessary to prove that the plaintiff has suffered damage as a result of the defendant's misconduct. A significant limitation to a passing-off action is the need to establish that the GI has a reputation in the jurisdiction where the action is being taken. In Australia this might be relatively straightforward for several dozen prominent EU wine GIs, but it would be very difficult to establish such a reputation for the overwhelming majority of the thousands of lesser known wine GIs recognized in the EU. Thus it is almost impossible under the laws of passing off to prevent the misuse of most European GIs. An additional problem is that no Australian court will protect a GI (or mark or trade name) which is found, by evidence, to have fallen into the public

---

[25] It is important to note that Australia has never had any general law of unfair competition, and the prohibitions contained in the TPA 1974 are the closest equivalent. Equally, Australia has never had any laws dealing with trade mark dilution.

[26] *J. Bollinger v Costa Brava Wine Co Ltd* [1960] Ch 262 (for the preliminary points of law to decide whether an action could be maintained in principle); *J. Bollinger v Costa Brava Wine Co Ltd* [1961] 1 WLR 277; [1961] RPC 116 (At trial).

[27] The authorities are conveniently summarized by Arnold, J in *Diageo v Intercontinental Brands* [2010] EWHC 17 (Ch); [2010] ETMR 17 at [1] (which describes 'a line of cases stretching back nearly 50 years in which suppliers of products of a particular description have sought to restrain rival traders from using that description, or a confusingly similar term, in relation to goods which do not correspond to that description on the ground of passing-off').

domain, that is become generic.[28] On that issue, it has been asserted by the Australian wine industry, over many years, that a handful of European GIs have become generic, by reason of lengthy and widespread (mis)use in respect of Australian wines. Accordingly, the protection of GIs in Australia has been very difficult under the law of passing off, which provides only a limited degree of protection.

Turning to the TPA 1974,[29] it was considered to be Australia's first (effective) competition and consumer protection legislation. That legislation contains provisions designed to protect consumers against false or misleading or deceptive conduct of traders, but which are commonly used by traders against their competitors. Section 52 provided: 'A corporation shall not, in trade or commerce, engage in conduct that is misleading or deceptive or is likely to mislead or deceive.' This section is used as a modern alternative to passing off, as it gives broader protection against any type of misleading or deceptive conduct in trade. Although there are several differences between the TPA 1974 action for misleading or deceptive conduct and the tort of passing-off, their shared elements include:

i. the need for the plaintiff to establish an Australian reputation for the relevant GI; and
ii. the fact that the TPA 1974 will not protect GIs held to have fallen into the public domain.

In other words, the statutory action under the TPA 1974 offers only limited substantive assistance to those wishing to protect GIs in Australia above that offered by the tort of passing off.[30]

In November 1981, the CIVC commenced an action in the Federal Court of Australia seeking an interlocutory injunction to prevent the use of the name 'Champagne' on Spanish sparkling wines imported by NL Burton. This was met, in part, by an application made by B Seppelt & Sons, S Wynn, Penfolds Wines and the Australian Wine & Brandy Producers Association Inc. to be joined as respondents to the court action in order to protect what they asserted was the right of Australian producers to continue to use the name 'Champagne' for Australian sparkling wines. Although the application to be joined was refused, the CIVC's Australian lawyers amended the pleadings, limiting the CIVC's claim to

---

[28] See Chapter 18 by Gangjee in this volume.
[29] The TPA 1974 has now been amended and is now called the Australian Competition and Consumer Act 2010.
[30] One of the main advantages in an action for misleading conduct is that there is no requirement for the plaintiff to prove damages.

the particular Freixenet wine that was the subject of the action. So as to try to head off any further application by interveners from the Australian wine sector, the CIVC's lawyers 'made it perfectly clear that they [did] not seek to restrict the use of the word "champagne" by Australian manufacturers manufacturing wine by the "méthode champenoise", but [only] where the wine is imported'.[31] In order to succeed, the CIVC had to establish that the use of 'champagne' in connection with Freixenet wine would be misleading or deceptive for the relevant class of persons likely to be purchasers, either at the wholesale or retail level. That in turn required the CIVC to establish that the name 'Champagne' was distinctive of the sparkling wines of the Champagne district of France. The key question in the hearing thus became what the meaning of the name 'Champagne' was to Australian purchasers.

There was considerable evidence presented to the Federal Court concerning extensive use of the name 'Champagne', both in respect of non-French imported sparkling wines and in respect of Australian sparkling wines. Indeed, Seppelt provided evidence that in 1981 it had produced approximately 30 million bottles of Great Western Champagne, that is about 45 per cent of all wines sold in Australia that year using the name Champagne. The Federal Court found, on an interlocutory basis, that the CIVC had not established a prima facie case that the name 'Champagne' meant to Australian purchasers Champagnes from France, and thus the CIVC's application for an interlocutory injunction was refused. In addition, the Federal Court found that even if a prima facie case had been established by the CIVC, the balance of convenience mitigated strongly against making any interlocutory injunction, because the respondent had been marketing its Spanish Champagne in Australia for 15 years, 'and its business should not be interrupted unless the benefit to the public would be significant'.[32] The Federal Court did note that some measure of consumer confusion was possible as a result of the use of the name 'Champagne' on the Freixenet wines, but that there would not be a significant number of members of the public likely to suffer harm if the interlocutory injunction were refused. The court case was subsequently withdrawn by the CIVC, and there was no final judgment given by the court. Thus the name Champagne continued (until 1 September 2011) to be used in Australia, both in respect of imported wines not from France

---

[31] See *Comité Interprofessionnel des Vins de Champagne v NL Burton Pty Ltd* (1981) 38 ALR 664 at 666. However the CIVC has long maintained that it never authorized its lawyers to make any such a concession.

[32] *Comité Interprofessionnel des Vins de Champagne v NL Burton Pty Ltd* (1981) 38 ALR 664 at 671.

and in respect of Australian wines.[33] It was only through diplomatic negotiations that France managed to (re)gain exclusive rights to use the name 'Champagne' (see Section 2.5 below).

2.4.2 The Beaujolais litigation – 1987

The next serious step came in 1987, when the Institut National des Appellations d'Origine (INAO),[34] the Union Interprofessionnel du Vin de Beaujolais (UIVB)[35] and 19 Beaujolais growers, producers and negoçiants took court action in Australia to prevent the sale of Australian wines labelled as Beaujolais. The wines were being labelled, amongst other names, as Beaujolais, Beaujolais style, Australian Beaujolais and Australian Beaujolais style. Beaujolais is a controlled appellation of origin in France.[36] As a result of long use, the Beaujolais region's name came to be synonymous with a particular light wine made there according to traditional constant local usages. The Beaujolais litigation in Australia was based on the argument that Beaujolais wine had acquired a reputation in Australia over many decades and the use of that name by Australian wineries for their own product (which commenced in about 1984), would mislead or deceive purchasers into the belief that Australian wines of that name were 'true' Beaujolais wines, having the same characteristics, as the French ones.

Before commencing the litigation, the INAO and the UIVB had attempted for more than six months to negotiate with the Australian wineries and producers misusing the Beaujolais name. Many of them agreed to cease their misuse, but a core group, consisting of the seven largest Australian producers, eventually dug in their heels and refused to negotiate. At the time when the INAO first took steps to protect the Beaujolais AOC in Australia, that name had been adopted by over thirty Australian winemakers in the previous two-year period. Another ten or so were to adopt the name in the midst of the Federal Court litigation commenced by the INAO and the UIVB. Of the 40 or so Australian producers who had adopted it, all but seven agreed to cease using the term without having to be sued. However, the remaining seven produced around 80 per cent of Australia's wine.

---

[33] Provided that the wines complied with an Australian food law standard that required that sparkling wines using the name 'Champagne' must comply with certain production methods.

[34] The National Institute of Appellations of Origin now called l'Institut National de l'Origine et de la Qualité but still known by statute as the INAO.

[35] The Interprofessional Union of Beaujolais Wines. See: http://www.beaujolais.com

[36] On the history of the French appellation system, see Chapter 2 by Stanziani in this volume.

The Federal Court action was not based on any suggestion that consumers would believe that, for example, a Rosemount wine labelled Australian Beaujolais or a Woodley's Queen Adelaide Australian Beaujolais Style, were French wines. Rather the action had the same basis as the English Spanish Champagne litigation of 1961. It was pleaded that by the use of the name Beaujolais, the Australian producers were misrepresenting that their wines were Beaujolais, even if produced in Australia. The Beaujolais producers' case was that no matter how good the Australian wines were, or how similar to the French Beaujolais, the simple fact remains that they were not Beaujolais and using the name was misleading as to the characteristics and qualities of the wines.

Whilst in 1985, the sales of French Beaujolais were 700,000 bottles, within two years, those figures declined to about 15,000 bottles, whereas the annual sales of Australian wines misusing Beaujolais climbed above 2 million![37] Had the INAO and the UIVB not acted so promptly and efficaciously, there was a considerable risk that within a short period of time, the Beaujolais AOC could have become regarded as generic and unable to be protected. Unfortunately for the lawyers involved, the Federal Court case settled in 1992 before it reached trial. Each of the Defendants settled and agreed to cease using the Beaujolais name after a short phase-out period, and no court decision was ever received.

2.4.3 The bigger picture – actions against other misuses

Indeed, there was a much bigger picture, at that time, than the single Beaujolais case. The INAO started from about 1989 taking steps in Australia to obtain the cessation of misuse by Australian producers of a range of other French controlled appellations of origin. These included:

- St Julien
- Bordeaux
- Sancerre
- Pineau de Charentes
- Saumur
- Cognac
- Calvados
- Muscat de Beaumes de Venise
- Graves
- St Emilion

---

[37] Sales statistics on file with the author.

No litigation was commenced in respect of some of the best-known French appellations of origin, such as Champagne, Chablis, Burgundy and Sauternes. Rather, protection of these names was left to the political arena, given the fact that there would be strong arguments advanced by Australian producers that these names had fallen into the public domain. The author is personally aware[38] of over 80 separate matters which were commenced by the INAO in Australia between 1987 and 1999, seeking to protect French wine AOCs, whether in court or otherwise.

2.5 The EC/Australia Wine Treaty – the 1994 Treaty

However, there was still an even bigger picture in respect of which the Australian protection campaign by the INAO was merely a small skirmish. Mid-way through the Beaujolais litigation, Australia started negotiating with the European Community (EC), as the European Union was then called, a treaty on wine. The discussions that preceded the 1994 Treaty were protracted and lasted for almost four years. The Treaty was initialled in January 1993 and signed in January 1994. As we will see, the 1994 Treaty removed or reduced many of the non-tariff barriers to entry into the EC of Australian wines and dealt with issues such as multi-regional blending and labelling, and so forth.[39] The most pertinent question is what each party gained from the 1994 Treaty that was seen as a sufficient benefit to warrant signing it.

2.5.1 What was the EC gaining and Australia conceding?
The Australian wine industry agreed to a Treaty that resulted in legislation prohibiting the use in Australia of many thousands of EC wine geographical indications that did not emanate from the region in question and comply with local laws. Although the obligations undertaken in the 1994 Treaty were entirely mutual, what the EC achieved was an agreement that Australia would protect all EC wine geographical indications, and the 1994 Treaty listed several thousand of them in the Annexes. Most importantly, the Annexes included the EC GIs that the Australians claimed were generic and thus unable to be protected by court action. It was through the 1994 Treaty that the EC achieved what it could not achieve through litigation.

---

[38] The author should declare that he is aware of these matters as he acted for the INAO in these matters.

[39] Agreement between Australia and the European Community on Trade in Wine (replaced by the 2010 treaty); at: http://www.wineaustralia.com/australia/LinkClick.aspx?fileticket=3ZzM%2fIYMdbo%3d&tabid=279

The 1994 Treaty protected 'geographical indications' as described in Article 2(b):

> 'Geographical indication' shall mean an indication as specified in Annex II, including an 'Appellation of origin', which is recognized in the laws and regulation of a Contracting Party for the purpose of the description and presentation of a wine originating in the territory of a Contracting Party, or in a region or locality in that territory, where a given quality, reputation or other characteristic of the wine is essentially attributable to its geographical origin.

This term thus includes but is by no means limited to French controlled appellations of origin (AOCs) and Italian Controlled Denominations of Origin (DOCs). The language of the 1994 Treaty when defining GIs thus covered all forms of wine GIs from all of the member States of the EC and of Australia.

Examples of GIs are:

- for Australia, Barossa Valley, Coonawarra, Mornington Peninsula and Hunter Valley;
- for Portugal, Porto and Madeira;
- for Italy, Barolo and Chianti;
- for France, Beaujolais, Burgundy, Champagne and Chablis (all of which are, of course, AOCs); and
- for Spain, Sherry/Jerez and Malaga.

As a matter of commercial reality, probably 95 per cent of protected EC geographical indications are unknown in Australia, so to agree to protect them in Australia was not a major concession. However, there was one very significant concession made by the Australian wine industry. Amongst these thousands of now protected EC geographical indications were 23 wine geographical indications from France, Portugal, Italy and Spain whose use has been so extensive and protracted in Australia that the Australian wine industry regarded these names as generic. These included:

- Champagne
- Burgundy
- Chablis
- Sauternes
- Hermitage
- Claret
- Port
- Sherry
- Madeira.

The importance of this handful of geographical indications should not be underestimated. Indeed, it was asserted by the Winemakers Federation of Australia (WFA) that use of these GIs by the Australian industry constituted a significant percentage of all wines sold in Australia at 1994.

2.5.2 What was Australia gaining and the EU conceding?
Stated briefly, as a direct result of the 1994 Treaty, Australia gained much easier access to the European wine market. Australia's home market for wine was, in the early 1990s, considered saturated with the per capita wine consumption remaining static in the preceding years. The Australian wine industry was thus looking to its exports in order to increase sales. Indeed, Australian exports of wines had grown from $21 million in 1985 to $344 million in 1993 and then, thanks largely to the new ability to gain access to the EC markets, to around $2.1 billion in 2003 and to almost $3 billion in 2008. The concessions made by the EC related to what were, in essence, non-tariff barriers to the importation of Australian wines into the EC. The actual concessions are described below.

2.5.3 Description of the 1994 Treaty
In summary, the 1994 Treaty:

i. provided for the mutual recognition and protection of the EC's and of Australia's GIs;
ii. reduced the number of analyses the EC requires of Australian wines for export into the EC from eight to three;
iii. allowed Australian winemakers to market wines in the EC labelled with multi-varietal and multi-origin blends;
iv. allowed for the export to the EC of Australian dessert wines with an alcoholic strength by volume previously prohibited;
v. prevented either the EC or Australia from introducing additional certification requirements on imports for the other's wines.

As regards the protection of each party's GIs, that term was defined in Article 2 of the 1994 Treaty to refer to GIs which were 'recognised in the laws and regulations of a Contracting Party'. From Australia's perspective, until amendments were made to the AWBC Act in December 1993, no laws or regulations specifically recognized Australia's wine GIs. Thus the 1994 Treaty played a pivotal role in causing the Australian Parliament to enact with all haste legislation providing for the recognition and protection of Australia's wine GIs.

More importantly, Article 8 of the 1994 Treaty dealt specifically with the 23 EC wine GIs (the majority of which are French AOCs) which were considered generic by the Australian wine industry. These 23 GIs were of crucial importance in the negotiations (for the reasons described above) and were dealt with by setting phase-out periods for their use by Australian vignerons. In respect of some of them the phase-out period was only brought to an end by the enactment of the 2010 amendments to the AWBC Act which provided that the phase-out period finished on 1 September 2011, over seventeen years after the 1994 Treaty was signed. Since most of the French GIs ceased being misused before 2000,[40] this final phase-out date was, in fact, most important for the GIs Port and Sherry.

Whilst it was said by Australian winemakers that the three phase-out tranches (described below) dealt with geographical indications that were allegedly generic in Australia, in reality not all of the names accorded a phase-out were indeed even widely misused, let alone generic. By way of example, in respect of the AOCs Sancerre and White Bordeaux, whilst in each case there was one Australian producer misusing that GI in respect of Australian wines, as a result of agreements with the INAO, those producers had ceased misusing the names before the 1994 Treaty was signed. Similarly, in respect of the name Graves, there were only two Australian producers misusing that name and both had also reached an agreement with the INAO to phase-out their misuse. Further, there were no Australian producers at all misusing the AOC Saint-Emilion in 1993.[41]

The phase-out periods agreed between Australia and the EC and dealt with by the 1994 Treaty were divided into three tranches. The 1994 Treaty provided, in two of the three cases, a final date such that use of the listed geographical indications up to that final date would not constitute a breach of the Australian laws protecting the GIs of the EC.

i. The first group had a transitional (or phase-out) period which ended on 31 December 1993, and comprised: Beaujolais; Cava; Sancerre; Frascati; Saint-Emilion/St Emilion; Vinho Verdi/Vino Verde; White Bordeaux.
ii. The second group had a transitional (or phase-out) period which ended on 31 December 1997, and comprised: Chianti; Madeira; Frontignan; Malaga; Hock.
iii. The third group had a transitional period whose final date was not determined by the 1994 Treaty. Rather, the 1994 Treaty provided

---

[40] By way of example, the last known Australian sparkling wine labelled as 'Champagne' ceased being sold after 2000.
[41] Personal records, on file with the author.

that the EC and Australia would make 'every endeavour ... to agree by 31 December 1997 at the latest on transitional periods for these names'. These were: Burgundy; Moselle; Chablis; Port; Champagne; Sauternes; Claret; Sherry; Graves; White Burgundy; Marsala.

As discussed, those negotiations over a phase-out date for the third group did not reach fruition until 14 years later, when Australia and the EU reached agreement on a new wine treaty. In regard to the transitional period for Beaujolais, as that AOC was the subject of the significant litigation described above, and as there were agreements between the INAO, the UIVB and more than 45 Australian winemakers requiring them to cease using the Beaujolais name, the 1994 Treaty specifically stated that: 'The transitional period for "Beaujolais" ... shall be subject to the terms of any agreement between the Australian producers and the competent French authorities representing the producers of "Beaujolais" and to any court order relating thereto.'

The 1994 Treaty also dealt with the protection of the traditional expressions[42] used in the EC and Australian wine industries. A large number of the traditional expressions in respect of which the 1994 Treaty was expected to extend are set forth in the Annexure to the 1994 Treaty. They include names such as:

(a) Qualitatswein & Liebfraumilch (Germany)
(b) Cru, Clos & Premier (France)
(c) Gran Reserva, Amarillo & Amontillado (Spain)
(d) Classico & Amarone (Italy).

Pursuant to the provisions of the 1994 Treaty, the traditional expressions would not be protected until the date when the Joint Committee (established pursuant to provisions of the 1994 Treaty between representatives of the EC and representatives of Australia) agreed to protect the traditional expressions. The new wine treaty signed in 2008 put in place the final steps necessary to protect each party's traditional expressions.[43] The negotiations over the protection of the EC traditional expressions took many years to complete, largely because:

---

[42] The term 'traditional expression' means a traditionally used name referring, in particular, to the method of production or to the quality, colour or type of the wine.

[43] In fact, Australia does not purport to have any traditional expressions for wines, as a result of which the new treaty does not protect any such names.

A history of Australia's wine GIs legislation 263

- the original list included terms such as 'vintage' and 'réserve' which the Australian wine industry said were standard English words long used for Australian wines; and
- the list included terms such as 'tawny' and 'ruby' that the Australian producers of fortified wines were insistent should not be given up.

3. THE NEW AUSTRALIAN SYSTEM FOR PROTECTING GIs

As stated above, before the necessity to comply with its obligations under the 1994 Treaty, Australia had no laws specifically protecting GIs. Thus the introduction of such laws was a major step for Australia and one that required a significant amount of work to formulate an entirely new system of laws dealing with legal concepts never before the subject of Australian judicial or statutory treatment. Given, however, that these new laws were dealing with wine, the logical repository for them was in the AWBC Act.

Obviously acting with the intention of protecting consumers and of encouraging the wine industry to act responsibly, the 1993 amendments to the AWBC Act enshrined in Australian legislation not only the language of the 1994 Treaty (which requires the protection of the listed EC GIs) but also its spirit, by extending the protection to 'words or expressions that so resemble a registered geographical indication as to be likely to be mistaken for it'. The protection was thus granted to all EC wine GIs contained in an Annex to the 1994 Treaty and to all Australian wine GIs which are determined in accordance with the processes described below. This was achieved by creating a Register of Protected Names in which were listed all EC wine GIs contained in the Annex to the 1994 Treaty, and in which were also listed those Australian wine GIs which were subsequently recognized in law.

3.1 Protection of Registered Geographical Indications

The Australian legislative scheme under the Australian Wine and Brandy Corporation Act 1980, as amended, prohibited the sale, import or export of wine: (a) with a false description[44] or (b) with a misleading description.[45] A false description was defined as one using: (a) the name of a country[46] or

---

[44] WCA s 40C.
[45] WCA s 40E.
[46] WCA s 40D(1)(a).

(b) a registered GI[47] and from where the wine did not originate. It was no defence to using a false description if it was 'accompanied by another word or expression such as "kind", "type", "style", "imitation" or "method" or any similar word or expression'.[48] A misleading description was defined as one using: (a) a registered GI;[49] (b) a translation of a registered GI (such as Burgundy);[50] or (c) a word or expression that so resembles a registered GI as to be likely to be mistaken for it.[51]

3.2 The First Test of the New Legislation – Protecting a Foreign GI

One of the first tests of the new legislation related to an Australian wine called 'La Provence'.[52] In this litigation, the INAO joined with the Comité Interprofessionnel des Vins de Côtes de Provence (CIVCP) to act against Mr and Mrs Bryce, in order to stop them from selling their wine under this label. The Bryce's vineyard was established in Tasmania in 1956 by Jean Miguet, a winemaker originally from Provence. The vineyard had a number of subsequent owners until it was purchased by the Bryces in 1980. In 1985 the respondents registered the name 'La Provence Vineyards' as a business name. In fact when the Bryces had purchased their vineyard, they had been offered but had refused to pay for the goodwill in the name 'La Provence'. However, they adopted it unilaterally several years after the purchase had been finalized. From 1986 wine was made for the Bryces and marketed under this name.

Interestingly, Provence is not recognized by French law as a wine GI, and the only French wine AOCs that use the name Provence are Côtes de Provence, Côteaux d'Aix-en-Provence and Côteaux d'Aix-en-Provence – Les Baux de Provence. However, as part of their court claim, the INAO and the CIVCP referred to the EC/Australia Wine Treaty and to the Australian Register of Protected GIs which, when listing all of the protected EC wine GIs in the Annexe, grouped the GIs into their regions. The GIs from the regions of Provence and Corsica are listed under the heading 'Provence and Corsica'. The claim against the Bryces was that the name

---

[47] WCA s 40D(1)(b).
[48] WCA s 40D(4).
[49] WCA s 40F(1)(a).
[50] WCA s 40F(1)(b).
[51] See s 40F(4)(a).
[52] *Comité Interprofessionnel des Vins de Côtes de Provence & Anor v Stuart Alexander Bryce & Anor* [1996] FCA 742; 69 FCR 450 (23 August 1996). Again, in the interests of full disclosure, the author acted for the INAO and the CIVCP in this matter.

Provence appeared in the Register, even if it looked like a heading. The Bryces' response, that Provence was not a GI itself but only a heading in the Register for the specific GIs registered in 'the Provence Area' did not find favour with Judge Heerey. He reasoned as follows:

> Provence is well-known in Australia as a wine producing region. . . . 'Provence' is therefore a word used in the description and presentation of wine to indicate the region in which it originated and thus a 'geographical indication' within the meaning of the Act. Moreover, it is not in my opinion a valid argument to call the words 'Provence and Corsica regions' a heading and then say that as a heading they are merely directions to the reader and not an operative and substantive part of the Register. The word 'heading' is not used in the extract from the Register which was in evidence nor, as far as I could see, in Annex II or in any other part of the Agreement. There is no reason for inferring that the word 'Provence' was intended to be any less protected than the AOC names appearing immediately under that word . . . I find therefore that wine sold in bottles bearing the respondents' labels would be sold with a false description and presentation within the meaning of s 40C(1) by reason of the inclusion on the labels of the word 'Provence', which is a 'registered geographical indication'.

La Provence was thus held to be in breach of the AWBC Act, because of the appearance in the Register of Provence. However, the INAO and the CIVCP were unsuccessful in so far as they argued that the use of La Provence was also in breach of the AWBC Act by reason of the registration of the Côtes de Provence AOC. In regard to this part of their claim, Justice Heerey held:

> 'Resemble' means 'to be like or similar to' (Macquarie Dictionary). The word 'Provence' does not in my opinion resemble the words 'Côtes de Provence' or any of the other AOC names any more than the word 'Australia' resembles the words 'South Australia'. Considered as words, the words 'Australia' and 'South Australia' have a connection because the geographical areas they denote are connected, the latter being part of the former. But for that very reason the one word is not like or similar to the other. They both convey different meanings, that difference being that one refers to the whole, and the other a part only, of Australia . . . In any case, if 'Provence' does resemble 'Côtes de Provence' I do not think the former so resembles the latter as to be likely to be mistaken for it. 'Provence' denotes Provence, the region of France known by that name. 'Côtes de Provence' would convey a smaller geographical entity, albeit one that is in Provence. Does the name 'Provence' resemble the name 'Côtes de Provence' to such a degree that a wine purchaser or consumer in Australia would be likely to mistake the former for the latter, that is to say to think that when he or she saw the word 'Provence' (or 'La Provence') on the respondents' labels, what he or she saw were the words 'Côtes de Provence'? I think not.[53]

---

[53] Ibid 469–470.

The Court held (somewhat strangely) that the Bryces, whilst having misused a protected French GI, had not done so 'knowingly', as they had not been specifically referred in the letter of demand to the Register of Protected Names. The fact that they were told that use of the name 'Provence' would be in breach of the AWBC Act was held to be inadequate warning. Knowledge was one of the essential elements of the constitution of the offence in s 40C of the AWBC, which provided: 'A person must not, in trade or commerce, knowingly sell wine with a false description and presentation.' The Court finally decided that the respondents only acquired knowledge of the false description after the court proceeding had commenced. The offence was thus not made out. However, Justice Heerey concluded:

> My findings in relation to the issue of knowledge means that although the respondents have won this battle, they may lose the war. I say that because the applicants would be able hereafter to bring to the respondents' knowledge the critical fact that the name 'Provence' is on the Register. Therefore any future claim under the Act by the applicants would not fail for lack of the respondents' knowledge and would seem likely to satisfy all the other elements under s 40C.[54]

3.3 What is an Australian Wine GI?

When the AWBC Act was amended to incorporate provisions relating to the creation and protection of Australian wine GIs, the *Australian Wine & Brandy Corporation Regulations*[55] were also amended by the introduction of Regulations 24 and 25. These had the specific purpose of enabling the means of creating, registering and protecting Australian wine GIs, such as Margaret River, Coonawarra, Barossa Valley, Hunter Valley and Mornington Peninsula. The Australian Wine & Brandy Corporation Regulations define a wine region as:[56]

- a single tract of land that is discrete and homogeneous in its grape growing characteristics to a degree that is measurable;
- including not less than five vineyards of no less than five hectares each;
- producing not less than 500 tonnes of fruit each year.

---

[54] Ibid 471.
[55] http://www.comlaw.gov.au/Details/C2012C00401
[56] See Regulation 24.

The regulations further provide that in defining a region (or sub-region), the GIC may take into account, amongst other things:[57]

- the existence of natural features in the proposed region;
- the history of grape and wine production in the area;
- climatic uniformity;
- grape harvesting dates;
- the geology of the proposed region;
- the history of the proposed region.

However, whilst these Regulations superficially gave the appearance that Australia was moving along the well-trodden international path of true GIs, the manner in which the AWBC Act defined GIs was quite disingenuous.

There were in fact two different definitions of GIs for wine in s 4 of the AWBC Act. One definition mirrored the definition in the 1994 Treaty and was very similar to the eventual TRIPS GI definition in Art 22.1, whilst the alternative bore no similarity to any internationally recognized GI definition. The alternative definitions for a GI in relation to wine read as follows:

(a) a word or expression used in the description and presentation of the wine to indicate the country, region or locality in which the wine originated; or

(b) a word or expression used in the description and presentation of the wine to suggest that a particular quality, reputation or characteristic of the wine is attributable to the wine having originated in the country, region or locality indicated by the word or expression.

The definition in paragraph (a), when coupled with the definition of 'originate' meant that for a winemaker to comply with these rules, all that was required was to source 85 per cent of their grapes from within the region. It was irrelevant what variety of grapes, how or where they were planted or tended, how or where the wine was made and so on. This appeared to be nothing more than an indication of provenance of 85 per cent of the raw materials of the wine and *not* a true geographical indication, at least in so far as TRIPS defines it. It is this paragraph (a) GI definition that has been enforced by the AWBC since 1994. However, the 2010 amendments to the AWBC Act, which renamed it the *Wine Australia Corporation Act 2010*, have changed the definition of

---

[57] See Regulation 25.

a 'geographical indication' to remove this strange anomaly, adopting a definition that is in line with both the 1994 Treaty and with the definition in the TRIPS Agreement.

There are a number of problems raised by the criteria contained in the Regulations that are, after all, unique to Australia. These problems can best be illustrated by a series of hypothetical situations and issues that are not dealt with or resolved by the legislation.

- What happens if, after a five-vineyard region is registered, the owner of one vineyard buys one or more of the others and there cease to be five separate vineyards?
- Can one wine producer divide his vineyard into five areas, for example in order to pass them on to his children, and in this way (subject to all of the other criteria being met) make an application for his 'own' region and thereby obtain a monopoly?
- The provision that 'a single tract of land must be discrete' raises the question in what way must it be 'distinct'. Is it only a question of reputation or a question of grape growing attributes? What is the degree of distinction you have to provide?
- What if climate changes result in the total annual tonnage produced in a region dropping below 500 tonnes?
- What if there were ten vineyards each of 6 hectares within the region in question, but each vigneron wanted, for quality reasons, to obtain a maximum of 3 tonnes of grapes per acre (which is the equivalent of about 7 tonnes per hectare)? That would result in the area producing 10 x 6 x 7 = 420 tonnes only. Why should that area be deprived of being entitled to registration whereas if the same vignerons were prepared to produce, say 6 tonnes per acre, the area would be entitled to registration?

The concept of a 'single tract of land' means that there cannot be 'islands' or pockets of areas entitled to use the same name. Concepts used in France to classify individual fields or vineyards as being suitable for use under a particular GI are totally foreign in Australia. As is described below, this concept was responsible, in part, for the truly bizarre shape of the Coonawarra wine region.

There are several French controlled appellations of origin that are each in the hands of a single producer. These include the very famous 'Romanée-Conti' and 'La Tâche' AOCs under the control of Domaine de la Romanée-Conti, 'Clos de Tart' under the control of Mommessin SA, 'La Romanée' under the control of Bouchard Père et Fils and 'Château Grillet' controlled by the Neyret-Gachet family. Why should our GI system not permit the same situation of single ownership to exist in Australia?

3.4 The Australian Wine GI System – States, Zones, Regions and Sub-regions

Australia's wine GIs are divided by the legislation into a hierarchy consisting of States (federal units), zones (such as the Limestone Coast of South Australia), regions (such as Yarra Valley) and sub-regions. There is even a so-called super-zone called 'South Eastern Australia' which allows wines from the States of Queensland, Victoria and South Australia to be blended into a wine entitled to use the same geographical indication. As to whether wine geographical indications are considered valuable in Australia, one has only to look at the Coonawarra litigation (see Section 2.2 above and also below).

3.5 Penalties

There were significant penalties fixed by the AWBC Act for persons convicted of breaching the provisions protecting geographical indications. The legislation provided for the imposition of both jail sentences and of fines and quite significant ones at that. Individuals were made liable for up to two years imprisonment and of pecuniary penalties of up to AU$12,000 for each offence. Companies convicted of breaching these new provisions of the Wine Australia Corporation Act were made liable to fines of up to AU$60,000 per offence.

3.6 Injunctions

The AWBC Act contained provisions in s 44AB empowering the Federal Court of Australia to grant injunctions:

(i) preventing continuing conduct such as the misuse of geographical indications; and
(ii) requiring persons to do particular acts (such as to place retractive advertising in newspapers or journals informing the public that a particular winemaker had no right to use a specific geographical indication).

3.7 Rights of Interested Persons

As history has shown in Australia, the misuse of an EC GI by Australian winemakers was unlikely to be the subject of a prosecution or civil action by other Australian winemakers, even if the misuse gave a competitive advantage to the misuser. By way of example, the continuing use of the AOC Frontignan by various Australian winemakers after the phase-out period for use of that name had concluded constituted a clear breach of the AWBC Act. However, no steps were taken to prosecute the miscreants. The 1993 amendments to the AWBC Act gave the right to seek injunctions to:

(i) the Australian Wine & Brandy Corporation;
(ii) a declared winemakers' organization;
(iii) a declared wine grape growers' organization;
(iv) a wine producer or wine grape grower; and
(v) an organization responsible for wine promotion or protection of the interest of persons engaged in winemaking.[58] Clearly organizations such as the INAO, the CIVCP, the UIVB and the CIVC, which have all commenced litigation in Australia seeking to protect French AOCs would qualify.

Where an Australian vigneron misuses an Australian GI, one can clearly envisage that person bearing the brunt of civil proceedings, seeking damages and injunctions, instituted by a number of persons. Undoubtedly the AGWA would be interested in preventing the misuse of Australian GIs, as would any local vignerons' organization. Individual vignerons may also feel sufficiently incensed to commence their own action. However, where the misuse is of an EC GI protected pursuant to the 1994 Treaty, history has shown that in the nineteen years since the 1994 Treaty was signed, the AGWA has not taken any court proceedings against the misuse of EC GIs (of which there were many).[59] Accordingly, the provision giving the rights to organizations responsible for, amongst other things, the protection of the interests of persons engaged in winemaking has been essential, effectively giving the right to such organizations to seek injunctions against persons misusing 'their' geographical indications. The 1993 amendments to the AWBC Act also spell out that particular persons have the right to launch private criminal prosecutions for breaches of the provisions dealing with the protection of GIs, in addition to civil proceedings seeking injunctions and/or damages. The statute includes both domestic and foreign

---

[58] See s 40L(3).
[59] It should, however, be recorded that in the last few years, the current team at the AWBC (now known as the Australian Grape and Wine Authority) has interceded and convinced a small number of misusers of French AOCs to cease their usurpations.

representative associations amongst those who can commence such proceedings.[60]

3.8 Geographical Indications Committee

The 1993 amendments to the AWBC Act also established the Geographical Indications Committee (GIC), which has the role of determining the names and boundaries of Australia's wine regions.[61] Although there was, prior to the system being instituted, a fairly clear understanding within the wine industry of what the geographical boundaries were for the overwhelming majority of Australia's GIs, they were not formally recognized under Australian law until their boundaries had been fixed by the GIC. For one or two regions, the boundaries had been a constant source of minor disputation.

The GIC, comprised of three persons,[62] is given wide consultation powers and the right to call on third parties (such as relevant experts) to assist it in its deliberations. In making determinations of the boundaries of Australia's wine GIs, not only may the GIC make a determination on an application made by winemakers or wine grape growers, but it also has the power to make determinations acting on its own initiative, without the need for a formal application by members of the industry.[63] In making a determination, the legislation requires the GIC both to identify the boundaries of the GI in question as well as the word or expression used to describe the GI.[64] Unlike countries such as France and Italy, the Regulations made pursuant to the AWBC Act for determining Australia's GIs do not impose any limits on matters such as the grape varieties used in any particular GI. Similarly, the Regulations do not deal with maximum yields, minimum alcoholic levels and so forth. The principle question in determining whether a wine is entitled to use an Australian GI is whether the wine was made from grapes grown in the

---

[60] See s 40K.
[61] See s 40N.
[62] The three persons are a Presiding Member appointed by the Chairperson of the Australian Wine and Brandy Corporation, one representative appointed by the Chairperson on the nomination of a declared winemakers' organization (which is the Winemakers Federation of Australia) and one person appointed by the Chairperson on the nomination of a declared wine grape growers' organization (which is Wine Grape Growers Australia). It is not a requirement that any of these persons have any legal or even viticultural skills or knowledge. See the Schedule to the *Australian Grape and Wine Authority Act*.
[63] See s 40Q.
[64] See s 40T.

designated area. Regulation 24[65] provides that in making such determinations, the GIC must take into account, amongst other things, a series of matters described below. When the GIC has made a determination, such a determination is only considered to be an interim determination and will not be finalized until a process of public consultation has been carried out. The AWBC Act requires that once an interim determination has been made, that fact must be appropriately publicized in a manner chosen by the GIC. At that point, written submissions are called for from members of the public in response to the interim determination. Finally, on considering those written submissions, the GIC will make a final determination.

3.9 Appeal Process

Given the importance of fixing stable boundaries for GIs and the fact that such determinations can have a significant economic impact, especially in the better-known and more valuable regions, the legislation specifically deals with the means by which an appeal process may be undertaken. Final determinations may be appealed to a Federal tribunal known as the Administrative Appeals Tribunal (AAT).[66] The AAT can hear appeals against decisions of the GIC based on a wide range of claims, including those based on allegations that:

(i) the GIC took into account irrelevant considerations; and
(ii) the GIC failed to take into account relevant considerations.

The AAT also has the power to review decisions referred to it on their merits. In other words, this Tribunal has a very wide discretion in hearing appeals. The AWBC Act also provided for appeals from the AAT to be taken to the Federal Court of Australia.

3.10 Register of Geographical Indications

The 1993 amendments to the AWBC Act made provision for the establishment of a Register of Geographical Indications.[67] This Register is open to public inspection and is available for copying. It contains the names of all protected GIs, including those European wine GIs registered pursuant to

---

[65] Regulation 24 of the Australian Wine and Brandy Corporation Regulations (Amendment) – Statutory Rules No 338 of 1994.
[66] See s 40Y of the AWBC Act.
[67] Available at: http://www.wineaustralia.com

the 1994 Treaty and those Australian GIs registered as a result of determinations made by the GIC.

4. LITIGATION IN AUSTRALIA: BOUNDARY DETERMINATION AND CONFLICTS WITH TRADE MARKS

The first litigation regarding the recognition and delineation of an Australian wine region concerned the area in South Australia known as Coonawarra.[68] The litigation described below concerned the determination of the boundaries of the region; essentially who was inside and who wasn't. There were three levels of determinations or hearings – before the GIC, the AAT and finally the Full Federal Court. The six-week hearing at the AAT was complex and involved over 75 parties. It is estimated by some that the total legal and witness costs of the dispute were well over $5 million. We may be amused when we hear of a single row of vines in the prestigious Montrachet vineyard in Burgundy selling for millions of dollars. However, whilst our vines are not yet that valuable, many companies did indeed spend hundreds of thousands of dollars if not more, to retain or gain the right to use the particular name. So the naming game has certainly hit our shores in a big way.

4.1 Coonawarra – The Geographical Indications Committee

The original application to register the Coonawarra region as a GI and to determine its boundaries was made to the GIC by local grape growers' and winemakers' associations in the area in 1995. After much consultation, the GIC published its Interim Determination in April 1997 and its Final Determination in May 2000.[69] The basis of the GIC's final determination, as published in May 2000, was that the key factor in determining the borders of Coonawarra was a resolution passed by a local viticultural association in 1984 which purported to restrict the region to some local administrative boundaries (known as the 'Hundreds' of Comaum and

---

[68] Again the author must disclose that he acted before the full Federal Court for Beringer Blass Wine Estates Limited, which led to the successful appeal against the AAT decision.

[69] A detailed consideration of both these GIC decisions is found in G Edmond, 'Disorder with Law: Determining the Geographical Indication for the Coonawarra Wine Region' (2006) 27 Adelaide Law Review 59.

Penola).[70] The successor organizations had promoted this resolution to the GIC as the determinative factor for deciding as to which parties should be within the region. The GIC accepted that recommendation and defined the region's borders accordingly. In doing so, the GIC also noted that it prioritized the proximity to the area of the region known as the 'cigar' as being the historic area of the Coonawarrra region which should then form the boundaries of the Coonawarra GI.[71] This is a (roughly) cigar-shaped area of clay-based loam or *terra rossa* soil, raised about a metre or so above the rest of the surrounding area, particularly rich and suitable for grape growing, and on which, historically, most of the region's vineyards had been located.

Interestingly, some believe that had the original application to the GIC been limited to the so-called 'cigar' area, then it might have been somewhat more defensible. However, the two associations that made the applications to the GIC each included members whose vineyards were located off the 'cigar', and who would have been excluded from the region had the application been limited to it. Thus for internal political reasons, the two associations had no realistic alternative but to seek to register a larger area as the Coonawarra wine region. This in turn led to the submissions to the GIC having to deal with the somewhat nonsensical concept, from a viticultural perspective, of 'proximity to' what was otherwise argued to be a key regional characteristic.

4.2 The Federal Administrative Appeals Tribunal

Within a month of the GIC handing down its determination, 46 applications for review had been lodged at the AAT by persons outside of the Coonawarra GI as determined by the GIC, each party obviously wanting the region's boundaries altered so as to incorporate their vineyard. The principal opponents to these 46 applications were Southcorp[72] and a group of 25 vignerons whose properties were within the Coonawarra GI

---

[70] When South Australia was surveyed in the 19th century, the maps of the colony were drawn on a square grid system, with the lines between the grids being 10 miles (approximately 16 km). Thus each square in the grid occupied 100 square miles and the name 'Hundred' was given to these areas. Each was also given an administrative name, with the two relevant 'Hundreds' in the Coonawarra area being known as Penola and Comaum.

[71] Geographical Indications Committee, *Supplementary Statement of Reasons: Final Determination of the Coonawarra Region*, p. 10.

[72] Southcorp was acquired, several years after the conclusion of the litigation, by Foster's, which subsequently divided into two companies, one of which, Treasury Wine Estates, is the wine company.

as determined by the GIC. The appellants who were trying to have their vineyards included placed great reliance on two arguments:

- the area that they established by evidence from, amongst others, leading viticulturalists, could demonstrate grape-growing homogeneity, a key aspect of the definition of a wine region and which was referred to in Regulation 24 of the Australian Grape and Wine Authority Regulations 1981; and
- the history of the wider region that was historically known as 'Coonawarra'.

After a six-week hearing, the AAT published its decision on 5 October 2001.[73] The AAT determined a new Coonawarra GI, about twice the size of the Coonawarra GI as determined by the GIC, retaining only about one-third of the original boundary line. Of the 46 applicants to the AAT proceedings, 24 became part of the new Coonawarra GI, with the other 22 remaining outside it.

The parties to the AAT proceedings placed great weight on Regulations 24[74]

---

[73] *Coonawarra Penola Wine Industry Association Inc & Ors and Geographical Indications Committee* [2001] AATA 844. Available at: http://www.austlii.edu.au/au/cases/cth/aat/2001/844.html

[74] Regulation 24 of the Wine Australia Corporation Act provides that:

region means an area of land that:
(a) may comprise one or more subregions; and
(b) is a single tract of land that is discrete and homogeneous in its grape growing attributes to a degree that:
  (i) is measurable; and
  (ii) is less substantial than in a subregion; and
(c) usually produces at least 500 tonnes of wine grapes in a year; and
(d) comprises at least 5 wine grape vineyards of at least 5 hectares each that do not have any common ownership, whether or not it also comprises 1 or more vineyards of less than 5 hectares; and
(e) may reasonably be regarded as a region.

subregion means an area of land that:
(a) is part of a region; and
(b) is a single tract of land that is discrete and homogeneous in its grape growing attributes to a degree that is substantial; and
(c) usually produces at least 500 tonnes of wine grapes in a year; and
(d) comprises at least 5 wine grape vineyards of at least 5 hectares each that do not have any common ownership, whether or not it also comprises 1 or more vineyards of less than 5 hectares; and
(e) may reasonably be regarded as a subregion.

and 25[75] of the Australian Wine and Brandy Corporation Regulations (Amendment) – Statutory Rules No 338 of 1994, which describe the criteria for defining a wine region and a wine sub-region.

The AAT's decision nevertheless adopted much of the GIC's decision, namely that an important factor in defining the region was the 'cigar',[76] as was proximity to it:

> The adoption of the two hundreds as the wine region border was also, in our view, recognition by the local wine industry that the boundary of the cigar shaped strip of *terra rossa* soil at the heart of the region was no longer an adequate marker for the boundary of the Coonawarra in 1984. It remained at that time, however, a significant part of the expanded region. That, in our view, remains the position today. Whether one characterises it as a 'marketing tool' or even challenges the homogeneity of the cigar itself, it is historically and scientifically the signature of the Coonawarra Wine Region. Proximity of this strip of arable soil would be in our view, an important factor in the determination of the boundary. Because of this, we do not consider, at this time, unless an overwhelming countervailing reason was demonstrated that land outside the two hundreds and not proximate to the topography of the cigar could justify inclusion in a Coonawarra Wine Region.[77]

4.3 Full Federal Court of Australia

Five of those remaining 22 applicants whose properties were outside the Coonawarra region as determined by the AAT lodged an appeal to

---

wine grape vineyard means a single parcel of land that:
(a) is planted with wine grapes; and
(b) is operated as a single entity by:
  (i) the owner; or
  (ii) a manager on behalf of the owner or a lessee, irrespective of the number of lessees.

zone means an area of land that:
(a) may comprise one or more regions; or
(b) may reasonably be regarded as a zone.

[75] Regulation 25: ' a single tract of land that is discrete and homogeneous in its grape growing attributes to a degree that: (i) is measurable; and (ii) is less substantial that in a subregion . . .'.

[76] *Beringer Blass Wine Estates Limited v Geographical Indications Committee* [2002] FCAFC 295 (20 September 2002) at [22] Cigar Shape: 'The area had been defined historically by reference to a cigar shaped limestone ridge which runs north and south through the Coonawarra township, and on which can be found an abundance of terra rossa soil.'

[77] *Coonawarra Penola Wine Industry Association Inc & Ors and Geographical Indications Committee* [2001] AATA 844 at [137].

the Full Federal Court in November 2001. This appeal was opposed by Southcorp. The three-judge Full Federal Court panel handed down its decision on 20 September 2002,[78] stating that the basis on which the AAT had made its decision was not correct in law, and specifying the new criteria to be taken into account. The Full Federal Court said that the new boundary was either to be agreed by the five successful applicants with Southcorp and the GIC or the matter would be referred back to the AAT for another hearing, but this time based on the correct legal principles as determined by the Full Federal Court. The outcome was that the parties reached agreement on the new boundaries of the Coonawarra region, which then included the five successful applicants. Significantly, the Full Federal Court also provided some clarification on the factors by which the boundaries of a GI should be appropriately determined, stressing that the expert evidence of geologists, viticulturalists and the like is more important than reliance on historical events or local politics.

The Full Federal Court adopted the appellants' arguments and held that the administrative boundaries (namely the Hundreds that had formed a key aspect of the GIC's and the AAT's decisions) were not relevant, nor was any concept of 'proximity' to the 'cigar'. The Court stated:

> The characteristics of wine essentially attributable to the region where the grapes are grown will not be influenced by the location within that region of local government or land survey boundaries administratively fixed for reasons unrelated to soil, climate or other conditions which bear on grapevine horticulture. Whilst boundaries of this kind may have a role to play in the selection of an appropriate name, word or expression to describe a region, to use them to identify the region is likely to introduce a wholly irrelevant consideration.

Indeed, the Full Federal Court referred to materials submitted during the hearing which showed that the vineyards of each of the five appellants were closer and thus more proximate to the 'cigar' than much of the region determined by the GIC and the AAT to be within the Coonawarra region. A key aspect of the court's decision was that the region had to be homogeneous in its grape-growing attributes, in accordance with the Regulations, stating that: '... the central issue ... is to identify a single tract of land that is discrete and homogeneous in its grape growing attributes to the requisite degree'.[79]

---

[78] *Beringer Blass Wine Estates Limited v Geographical Indications Committee* [2002] FCAFC 295.
[79] Ibid at [69].

5. THE US-AUSTRALIA FREE TRADE AGREEMENT

5.1 Further Amendments to the AWBC Act: Trade Mark v GI Disputes

The next significant development in Australian GI law occurred in 2004, when Australia and the USA signed a Free Trade Agreement (the FTA). The FTA contains provisions designed to protect US brand owners from having their Australian trade mark rights adversely impacted by foreign GIs that become protected in Australia. The FTA provides that a GI cannot be registered using a particular name if there are, in Australia, pre-existing trade mark rights to the same name. While this sounds straightforward, the implementation of this provision into Australian law was bizarre, to say the least.

The AWBC Act was amended by the addition of provisions that operate in the following manner.[80] Whenever an application is made to the GIC for determination of a GI, or whenever the GIC is considering determining a GI, it must appropriately advertise this, notifying the public of the proposed GI and of the possibility for interested persons to object to the determination. An objection to the GI under consideration may be made based on the objector's trade mark rights. The objector may have trade mark rights in one of three ways.

- First, the objector may have an existing trade mark registration.
- Second, the objector may have a pending trade mark application, lodged in good faith.
- Third, the objector may have 'common law' trade mark rights as a result of having adopted and used in good faith the trade mark (i.e. rights based on actual use and without any application for or registration of a mark).

Such an objection will be considered by the Registrar of Trade Marks. It is at this point that the process becomes complex and potentially problematic.

The Registrar of Trade Marks is required to determine if such an objection is 'made out'. The Act provides no guidance as to what factors should be taken into account in determining this. Thus, if the Register of Trade

---

[80] See Division 4 – Australian Geographical Indications – Subdivision D – Objections to determination of geographical indications based on pre-existing trade mark rights, ss. 40RA to 40RG of the AWBC Act.

Marks is satisfied that one of the trade mark rights set out above exists, the objection is automatically 'made out'. On the face of the Act, it is not at all clear that to 'gazump' (or prevent the registration of) a proposed GI, the trade mark on the basis of which the objection has been made must have been used 'first in time', that is, prior to the use of the GI. Indeed, this issue is only addressed directly in a footnote to s 40RC(3) of the Act. That section provides that if the Registrar of Trade Marks decides that the ground of objection is made out, the Registrar may nevertheless still recommend to the GIC that the GI be registered if the Registrar 'is satisfied that it is reasonable in the circumstances to recommend to the Committee that the proposed GI be determined despite the objection having been made out'. The footnote to this provision states: 'For example, it may be reasonable for the Registrar of Trade Marks to make such a recommendation if the Registrar of Trade Marks is satisfied that the proposed GI was in use before the trade mark rights arose.' This is, of course, a 'first in time, first in right' approach of the type contemplated by the FTA, although it is put backwards.[81] In other words, rather than provide that the GI cannot be registered if the trade mark rights arose first in time, the legislation states the obverse! The effect is the same, but its implication is that if the proposed GI was *not* in use before the trade mark rights arose, then the Registrar should not recommend to the GIC that the proposed GI be determined.

Insofar as the reference in section 40RC to 'it [being] reasonable in the circumstances to recommend to the [GIC] that the proposed GI be determined despite the objection having been made out' is concerned, the AWBC Act gives no guidance as to what other circumstances may persuade the Registrar to still recommend that the GI be registered. The Act therefore grants to the Registrar of Trade Marks a very broad discretion to make a recommendation that a GI be registered irrespective of an objection having been 'made out'. This discretion would appear to fly in the face of the FTA, although it may be limited by future regulations or decisions. On its face, it is very positive that such decisions are referred for determination to the Registrar of Trade Marks. The GIC is not an appropriate organization to consider the balancing of trade mark rights against the registration of a GI. The question of the priority in time between competing interests (such as between two trade marks) is a fundamental issue that Trade Mark Examiners and Hearing Officers deal with daily.

---

[81] For a detailed consideration of the 'First in Time' principle in GI v trade mark conflicts, see Chapter 13 by Goebel and Groeschl in this volume.

280 *Research handbook on intellectual property and GIs*

5.2 Use of an EU Geographical Indication in a Non-infringing Manner – 'First'

Apart from boundary determination issues, conflict between GIs and trade marks is another controversial area of wine law. The *Trade Marks Act 1995* (Cth) defines GIs in s 6 and also sets out a ground of opposition in s 61, which prohibits the registration of a mark containing or consisting of a false GI where registration is sought for goods that are similar to those for which the GI is used or if the use would deceive or cause confusion. Additionally, s 42 of the Trade Marks Act prohibits the registration of marks which are scandalous or 'contrary to law'. The reference to 'contrary to law' includes the unlawful use of GIs recognized under the AWBC Act. In 2005, the Trade Marks Office handed down an interesting decision of a Hearing Officer to the effect that the use of common English words in the description and the presentation of a wine, which is also a protected wine designation, if used in good faith and not as a way to indicate that the wine originated in this country, is not a false or misleading description or presentation.[82]

Ross and Veronica Lawrence had applied to register the trade mark 'FEET FIRST' for wines, but their application was rejected by the examiner on the basis that the trade mark incorporated the German GI 'First'. The Examiner had requested that the applicants consent to an endorsement that 'the trade mark will be used only in relation to wines originating from the FIRST sub-region within the Einzellagen wine growing area within Germany'. The Lawrences refused to consent to such an endorsement and requested a hearing, where they convinced the Hearing Officer to accept their application. Their argument was that in using the word 'First', they were not using the German GI, but were using the word as part of the ordinary and common English expression 'feet first' as in 'to jump or dive in somewhere, feet first'. In accepting this argument, the Hearings Officer held as follows.

> [W]ords which include WICKER, FIRST, DOCTOR, SAND, WOLF, HORN and LUMP are words which have ordinary English significations and which are also included in the Register of Protected Names as being German localities or regions in which wines are grown. There is a tension where Australian traders wish to use such words for the sake of their ordinary English significations and the words are on the Register of Protected (Names) as being the names of localities in other parts of the world where wines are grown . . . Thus I consider that where the context of a geographical indication makes it obvious that it is

---

[82] *Application by Ross & Veronica Lawrence* [2005] ATMO 69 (21 November 2005).

being used within a trade mark for the sake of its ordinary English signification as a word contained in an English dictionary, with no potential reference to the geographical location, it is appropriate that the application should be accepted for possible registration...

Such instances of foreign geographical indications where acceptance should ensue (if there are no other valid grounds for rejection) might include such examples as <u>WICKER</u> BASKET, TIMBER <u>WOLF</u> (but not WHITE <u>WOLF</u> or <u>WOLF</u> WHITE), RAM'S <u>HORN</u> (but not RED <u>HORN</u> or <u>HORN</u> RED), <u>SAND</u> CASTLE (but not <u>SAND</u> CHATEAU or CHATEAU <u>SAND</u>) or WITCH <u>DOCTOR</u> (but not <u>DOCTOR</u> GRAPE). Instances of Australian registered GIs where the contextual use of a word is within a commonplace expression or name and is such that it is apparently not the intention of the applicant to use it within the statutory definition of a geographical indictor include, for example <u>ORANGE</u> TREE (but not <u>ORANGE</u> VALLEY), APPLE <u>PEEL</u> (but not <u>PEEL</u> CREEK) or HAMLET, PRINCE OF <u>DENMARK</u> (but not DENMARK SWEET).

This principle should apply where the geographical indication forms a part of a known and commonplace English expression or name which has no reference to the geographical location. If the protected expression is not part of a known and commonplace English expression or name, such as, for example, <u>WICKER</u> WOMBAT, or STONE <u>DOCTOR</u>, consideration will need to be given as to whether the term is being used for the sake of its ordinary English signification.[83]

Interestingly, as explained below, when the AWBC Act was amended in 2010, a new defence to the misuse of a registered GI was included in the legislation,[84] to the same effect as this decision, namely where the person was using the designation as a common English word or term.

5.3 Great Western and the Possibility of Co-existence

The name 'Great Western' is likely to have two entirely different meanings to wine consumers. On the one hand, many will know it as the famous wine region north west of Melbourne where wines have been made continuously since the early 1850s. For others, the name 'Great Western' will call to mind the (unregistered) trade mark of Seppelt, the subsidiary of Treasury Wine Estates (formerly Foster's Wine Group). Seppelt has for many decades used the name 'Great Western' for its sparkling wine (previously called 'Champagne'). The facts are that the region had been named as Great Western in about 1855, and already had grapes planted in the area at that time. About five years later, the same name 'Great Western' was adopted by the predecessors in business to Seppelt, as a result of

---

[83] Ibid [9], [32]–[34].
[84] See s 40FA(2).

which the two names had co-existed for over 140 years at the time of the dispute. The 1920s dispute between Seppelt and the vignerons of the Great Western region described above arose when Seppelt wished to register the name as a trade mark. Corporations must have long memories, because in 1998, when the vignerons of the Great Western region wanted to register that name as a GI, Seppelt (then owned by Southcorp) objected.

The issue between the parties was this. Whilst the team at Seppelt was basically in support of a wine region named 'Great Western' being registered, the result of such a registration would have been that, in order to continue to use the name 'Great Western' in respect of wines, 85 per cent or more of the grapes used in such wines would have had to be sourced from that region. That may have been possible for some of the Seppelt wines that used the name 'Great Western', but certainly would not have been possible for its famous sparkling wine. The overwhelming majority of the grapes used for that wine were shipped into the region. Accordingly it was not in Seppelt's commercial interests to allow for a region to be determined using the Great Western name. The company thus objected to registration of the name as a wine region. Southcorp then persuaded the GIC to not make any decision on the original application, which lay quietly gathering dust on someone's desk for several years. Subsequently the AWBC Act was amended to incorporate the provisions described above arising from the US-Australian FTA, allowing for a mechanism to resolve disputes between trade mark owners and those persons wishing to register identical or similar GIs. Further, Southcorp, which had been the principal party that had fought Foster's Wine Group in the Coonawarra litigation, was acquired by Foster's.

At that point, partly because of a coincidence of personalities who were involved in Foster's, Southcorp, their respective lawyers and the lawyers for the vignerons in the Great Western region who wanted to register the GI 'Great Western', a dialogue commenced between the parties with a view to resolving the dispute without litigation.[85] After some intense negotiations, an agreement was reached between Foster's and the vignerons of the Great Western region which would allow the region to be registered under the Great Western name, but subject to certain conditions (a mechanism allowed by s 40G of the AWBC Act). The parties agreed that the matter would be referred to the Registrar of Trade Marks under section 40RC of the AWBC Act (which relates to

---

[85] Boring though this further disclosure may be, the author should disclose that he acted in this matter for the vignerons of the Great Western region, negotiating against his usual client Foster's.

objections to trade marks containing GIs) on the basis that they would ask the Registrar to accept that:

(i) Seppelt had long used the name Great Western in good faith;
(ii) the name was (obviously) identical to the name which had been proposed as the name of a wine region;
(iii) accordingly the objection to registration of the GI had been made out;
(iv) as the name Great Western had been used as the name of the region since 1855, that meant that it had been used prior to use of that name as a trade mark, but, more relevantly, it had co-existed with the trade mark usage for 140 years;
(v) as a result, there were circumstances that warranted the registration of the region under the name Great Western, but to avoid any risk of confusion between use of the name as a trade mark and use of the name as a wine region, all use of the name as a wine region had to comply with the following conditions:

The name of the wine region must only be used:

(a) as a GI and not as a trade mark;
(b) except for bulk wine or cleanskin wine, in conjunction with a brand name;
(c) in a manner that is less conspicuous (such as in colour or typestyle) than the brand name;
(d) in a font size no greater than 50 per cent of the average font size of the letters in the biggest word in the brand name;
(e) for bulk wine or cleanskin wine, only as part of the phrase 'from the Great Western region'.

The Registrar of Trade Marks made an interim decision to this effect on 20 December 2006, and the decision matured into a final decision early in 2007.

The result was that the name Great Western could continue to be used as a trade mark by Seppelt, allowing its usage that had started 140 years ago, and equally the region's vignerons were allowed to use the name as the name of a registered wine region (albeit under defined conditions). The status quo was generally maintained, the result having been obtained without any real litigation.

5.4 Rothbury: A Questionable GI Application Threatening Prior Trade Mark Rights

The next dispute between persons wishing to register a name as a GI and a party claiming that it was a trade mark concerned the name 'Rothbury' and played out over three years from 2005. A small group of vignerons in the Lower Hunter Valley Region wanted to have two sub-regions determined in the area. Of the two proposed sub-regions, the smaller one (to the south of Broke Road) would be called Pokolbin, and that was the area where each of the applicants had their own respective vineyards. The sub-region to the north of Broke Road would be called Rothbury.

This plan struck two major problems. The first was that the name Pokolbin was the name that had been used by vignerons within the entire region, and not just those to the south of Broke Road. This pitted the applicants against several dozen vignerons who had used the name Pokolbin for their wines for many years. The second problem was that, although there was indeed an ancient parish of that name, and which name actually appeared on maps of the region, the name Rothbury:

(i) was not known by local vignerons to be the name of their parish; and
(ii) had not been used by anyone as the name of a GI.

Most importantly, Rothbury was the name of a subsidiary and winery of (yes, once again we encounter . . .) Foster's, whose predecessors in business had used the name Rothbury since about 1969, had several trade mark registrations consisting of or containing the name Rothbury and which was the owner of trade mark applications for that name.

Accordingly, after the application for the creation of the two sub-regions was submitted to the GIC and advertised, the Foster's subsidiary, Rothbury Wines Pty Ltd, lodged an objection under s 40RC.[86] The matter was argued before the Registrar of Trade Marks pursuant to s 40 RC of the AWBC Act after significant evidence had been filed by both sides (although it was the opponent, Rothbury Wines Pty Ltd that actually filed the overwhelming majority of the evidence).[87]

The summary of the decision by Deputy Registrar Michael Arblaster set out succinctly why he upheld that the objection to registration had

---

[86] If the author noted here that he did not act for either party in this matter, he would be lying, as he acted for Rothbury Wines Pty Ltd.

[87] See *Objection by Rothbury Wines Pty Ltd to determination of Geographical Indication filed in the names of Murray Tyrell, Tyrell's Vineyards Pty Ltd and Trevor Drayton* (2008) ATMO 13 June 2008 (Dy. Registrar Arblaster).

been made out and why there were no circumstances that would justify recommending the determination of the GI.

1. The evidence shows quite limited use of the word 'Rothbury' as a place name to designate an ill-defined area in the Lower Hunter Valley.
2. Most people and businesses within the area covered by the proposed geographical indication (GI) appear to identify their geographical location by use of the word 'Pokolbin'.
3. The administrative Parish of Rothbury, and its boundaries, are unknown to many of its residents and the term appears to be used only in relation to identifying real property in the transfer of land title.
4. The word 'Rothbury' thus has very little force as a geographical word.
5. The evidence also shows that the word 'Rothbury' has developed a secondary meaning as a trade mark denoting the wines of the Objector, distinct from whatever geographical origins it might have had.
6. Because a GI can be used in a trade-mark-like manner to promote the term as denoting the source of origin of the wine, the use of 'Rothbury' as a GI would be confusing with the registered trade mark 'The Rothbury Estate'.
7. The word 'Rothbury' has acquired a secondary meaning to denote the Objector and it wines. The Objector therefore has trade marks rights in the word 'Rothbury' (in relation to wine). The use of 'Rothbury' as a GI would be confusing with the term 'Rothbury' which is contained in the registered mark 'The Rothbury Estate' and which consists of the Objector's pending mark Rothbury.
8. Because a GI is territorially based, and its use is open to all that live within it, but the Objector's trade marks are not, the use of the word Rothbury as a GI would, because of the reputation of the trade marks, be confusing.
9. The likelihood of confusion is not ameliorated by any force that the word 'Rothbury' has otherwise developed ...

Interestingly, because s 40RC does not limit the circumstances pursuant to which the Registrar of Trade Marks can form the view that it is reasonable to recommend that a name be registered as a trade mark, that unlimited discretion had a significant impact on the evidence collected and submitted by the objector, Rothbury Wines Pty Ltd. The argument was also made that Rothbury was ineligible as a GI, under GI legislation requirements. Evidence was collected by it from both a leading viticulturalist and from a geologist and was put forward to establish that, even if its trade mark objection was not made out, the application for the sub-regions should fail. The evidence was submitted in order to demonstrate that there was no factual basis that would allow either or both proposed sub-regions to be distinguished from the surrounding areas, or either proposed sub-region from the other, on the basis of grape-growing attributes. In the event, Deputy Registrar Arblaster held that it was not his role to consider the relevance of geology or landforms for the purposes of determining a boundary, and this evidence played no substantive role.

6. NEW EU-AUSTRALIA WINE TREATY

In 2008, Australia and the EU finalized their latest round of negotiations on the wine accord between them. The changes since the 1994 Treaty were so significant that the parties signed a new Treaty which superseded its predecessor (the new Treaty). One of the most crucial aspects of the new Treaty was that it (effectively) set a final date for the phase-out by Australia of the third tranche of EU GIs containing key names such as Champagne, Chablis, Burgundy, Port and Sherry. The final date was 12 months after Australia implemented the provisions of the new Treaty in domestic legislation, which turned out to be 1 September 2011, that is 18 years after the phase-out had been agreed to in principle. Whilst that phase-out date was irrelevant for almost all EU GIs, which had been voluntarily phased-out by the Australian industry in previous years (such as Champagne and Sauternes), other names such as Port and Sherry had continued to be used by Australian producers. Thus, for these names, the phase-out date had real meaning.

6.1 Amendments Made to the AWBC Act

When implementing the new Treaty, the Australian Government made some sweeping changes to the AWBC Act, later renamed in 2010 as the *Wine Australia Corporation Act* and, more recently in 2014, as the *Australian Grape and Wine Authority Act*.[88] During this transition, the Australian Wine and Brandy Corporation was itself renamed as the Wine Australia Corporation and subsequently the Australian Grape and Wine Authority.

Whilst the other amendments were important, new rules for the protection of GIs were implemented. The new legislation:

(i) added the use of registered translations of a registered GI to the definition of 'false description and presentation';[89]
(ii) clarified that certain uses of GIs were not a 'false description and presentation' or a 'misleading description and presentation' in certain defined circumstances;[90]
(iii) reduced dramatically the number of German GIs that are protected (as a result of the EU limiting the number of such GIs in contrast to the 1994 Treaty);

---

[88] See Financial Framework Legislation Amendment Act 2010 (Act No 148 of 2010) – Schedule 4.
[89] See s 40D(2)(c).
[90] See ss 40DA, 40DB, 40FA and 40FB.

(iv) inserted a new definition[91] of a 'geographical indication' which is now consistent with the definition as set out in the new Treaty and TRIPS;
(v) included a number of new defences against alleged breaches of the provisions protecting registered GIs (such as where the use is of a common English word or term like 'Feet First');[92]
(vi) removed the need to establish that a person had breached the legislation 'knowingly' in order to establish a breach of the provisions that protect registered GIs and traditional expressions;[93]
(vii) renamed the Register of Protected Names in order to cover protected traditional expressions and quality wine terms, so that it became the 'Register of Protected Geographical Indications and Other Terms'. It was also reorganized into four parts to include: (a) GIs for both Australian and foreign wines, as well as the translation of foreign GIs; (b) traditional expressions for wines originating in foreign countries; (c) quality wine terms for wines originating in Australia; and (d) other terms in relation to wines (e.g Méthode Champenoise).[94]

6.2 Breadth of Protection for Geographical Indications Narrowed

One interesting aspect of the 2010 amendments relates to the breadth of protection for registered GIs. Arguably, the amendments narrowed the protection that is available. The 1993 amendments to the AWBC Act prohibited not only the use of names or words which were the same as registered GIs, but also the use of names or words which were similar. The language which the legislation used (in s 40F) was that it was an offence to use any name or word 'which would be likely to be mistaken for the

---

[91] See s 4.
[92] See s 40FA(2).
[93] In the Act before the 2010 amendments, the penalty provision for selling a wine with a false or misleading description and presentation was subject to the fault element of intention. The amended Act no longer includes any reference to the mental element of intention. Proving that the misleading or false description was used without any knowledge of the law and in good faith will no longer allow a person to avoid liability.
[94] The term 'Methode Champenoise' should have been registered as a traditional expression but was for some unknown reason, omitted from the list of expressions to be registered. When that fact was realized, a flurry of negotiations ensued that resulted in an exchange of letters pursuant to which Australia agreed that the term would be protected. The method chosen was to create a new Part of the Register of Protected Terms, to be used for 'Other Terms', and to enter 'Methode Champenoise' in that Part.

registered geographical indication'. The expression 'likely to be mistaken for' was language that has not appeared in any other Australian intellectual property or consumer protection legislation or jurisprudence. The law of passing off refers to 'deceptive or confusing' use; trade mark law uses the terms 'substantially identical or deceptively similar' and 'deceive or confuse'; and the consumer protection law uses language such as 'mislead or deceive'. Thus there was no guidance for Australian courts as to what the expression 'likely to be mistaken for' means. When Counsel for the Applicants in the *Bryce* dispute used the trade marks and consumer protection tests as an analogy to the language used in the AWBC Act, Justice Heerey noted:

> [A]lthough the regime established by [the relevant part] of the [AWBC] Act is based on registration of names, in my respectful opinion it does not follow that doctrines of trade mark law necessarily apply. Although comparisons may be made between some provisions . . . and trade mark legislation, and between such provisions and some in Part V of the Trade Practices Act, the fact remains that in providing the 'legal means' contemplated by Article 6.1 of the [Australia-EC Wine] Agreement Parliament has chosen a means of enforcement which is sui generis. [. . .This] is notably different from trade mark legislation and Part V of the Trade Practices Act in that it prescribes a sanction of imprisonment.

In 2009 submissions were made to the committee looking at making changes to the AWBC Act when implementing the provisions of the new Treaty, to the effect that the statutory language should be amended to more closely follow consumer protection legislation. Thankfully the 2010 amendments followed those suggestions, and removed the words 'likely to be mistaken for' and replaced them with the expression 'as to be likely to mislead'. That has given far greater certainty to the test for breach.

On the other hand, in making such changes to s 40F, the government substantially watered down the protection given by that section. This is because whilst under the original language of s 40 the legislation referred solely to a comparison between the word to which objection was taken on the one hand and the registered GI on the other, the new provision requires a complainant to additionally establish that persons be misled as to whether the wine (bearing the misleading term) originated in another territory. This provision now states: 'the description and presentation of wine is misleading if it includes an indication or term that so resembles a registered GI as to be likely to mislead that the wine originated in a country, region or locality in relation to which the indication is registered'. That would, arguably, allow a producer to sell a wine bearing the label 'Australian Bordaux', and to argue that as the label makes it clear that the wine emanates from Australia, no one could possibly be misled into believing that the wine originated in a foreign region. If a wine is adequately labelled as 'Australian', it is hard to see how such a new requirement as is set out in s 40F could be met.

6.3 Traditional Expressions

Whilst the AWBC Act, as amended in 1993, provided for the protection of traditional expressions, those provisions were not enacted nor made operative as there were no traditional expressions entered onto the Register of Protected Names. The EU and Australia had agreed to make the protection of traditional expressions a part of their ongoing negotiations, which covered a wide range of topics, including the phase-out dates for the third tranche of the so-called generic names. These negotiations were successfully concluded, as a result of which an agreed list of EU traditional expressions that would be protected in Australia was registered. Words such as 'vintage', 'tawny' and 'ruby' disappeared from the list, and the balance were entered as registered traditional expressions in Part 2 of the Register of Protected Geographical Indications and Other Terms.

7. NO MORE GIs PLEASE

Since mid-2010, the Australia Grape and Wine Authority has been effectively discouraging the lodging of any further applications for Australian GIs by the imposition of a very significant application fee.[95] Whilst the fee is said to be a 'cost recovery' exercise (and indeed may well be since significant work goes into considering a new application, including regional visits and work with independent experts), the AU$27,500 fee[96] will certainly slow the enthusiasm of regional associations wishing to register new regions or sub-regions. AGWA does retain a discretion to waive or significantly reduce the fee.[97]

---

[95] The fee is implemented by Regulation 22A of the Australian Grape and Wine Authority Regulations 1981.
[96] Current as at 1 August 2014.
[97] Regulation 22A(2).

8. CONCLUSION

When Australia signed the 1994 Treaty and then set about establishing its own system of GIs for wines, the wine industry was adamant that it did not want an 'appellation system'. It saw no place in Australia for a system that is as administratively burdensome and, as some believe, as over-regulated as the European systems.[98] Rather, what was desired was a simple and easy-to-use system that imposed few compliance burdens. Critics might say that the result was that Australia did not really institute a system of GIs at all. Rather, Australia instituted a system of indications of source for 85 per cent of the grapes within a wine.

Whether the criticism is valid is open to debate. What is clear, however, is that Australia does not yet have a wine sector that knows enough about our country, our regions, our climate, our *terroir*, the suitability of grape varieties for the various regions, the best viticultural practices for each area and so forth. Although founded in the late 18th century, our wine sector was so decimated by phylloxera that it really only started again about 40 years ago. We have not had the experience or history of the almost two thousand years of viticulture (like countries such as France, Italy and Spain), from which we have learned. Australia needs time to experiment, as much with its viticulture and winemaking as it does with its legislative processes. Remember that even a country such as France took decades to perfect its GI legislation.[99] The *Loi de Consommation* of 1 August 1905 was not sufficient to establish an effective system for the protection of wine regional names, and the laws of 1919 and 1927 were also unsatisfactory. The Law of 30 July 1935 that created the INAO was the first law that truly worked in France to protect wine GIs, but less than 70 years later, the various Decrees that established the 400+ French wine controlled appellations of origin have been revisited, updated and amended. Thus, it should not be surprising that Australia's wine laws, only some 30 years old, are considered by many to be good first steps on the right path. We do not have a mature wine law system but one that is growing and evolving, especially under the influence of pressures from important markets like the EU and the USA. Thus we can expect that our legal system dealing with the recognition, delimitation and protection of wine GIs will continue to evolve and to present plenty of interesting

---

[98] Evidence given to the AAT in the Coonawarra case by the then-current chairman of the AWBC.
[99] For the history of the French appellation system, see Chapter 2 by Stanziani in this volume.

challenges in the years ahead. Whether those following the current practitioners will have as much fun as we have had with the Beaujolais litigation, the Coonawarra dispute, the La Provence litigation, the Rothbury fight and the Great Western negotiated settlement (amongst others), is hard to say. Only time will tell.

11. A comparative analysis of GIs for handicrafts: the link to origin in culture as well as nature?
*Delphine Marie-Vivien**

1. INTRODUCTION

A Geographical Indication (GI) identifies a good as originating in a region, where a given quality, reputation or other characteristics of the good is essentially attributable to its geographical origin.[1] This definition found in the WTO's TRIPS Agreement recognises a very old concept: the existence of a link between a product and its place of origin. As embedded quality constructs, regional names have for centuries provided a tool for competitive positioning and the signalling of reputation. Besides concerns related to consumer and producer protection, GI protection supports territorial and rural development, biodiversity and traditional knowledge.[2] However far we go back in history, this concept of products with regional reputations (that is origin products) applied to all types of products; for example to minerals (marble), art objects (bronze, ceramics, pottery or terracotta), textiles (silk), perfumes (incense) and processed agricultural products (honey).[3] Such examples from ancient times suggest that the range of products that have acquired notoriety linked to their place of origin is in fact limitless. Nowadays, however, in some countries and more particularly in Europe, GIs are still limited to agricultural products and

---

* The author sincerely wishes to thank Estelle Biénabe for her very relevant critical review of the chapter and Laurence Bérard and Erik Thevenod-Mottet for their insightful comments.
[1] TRIPS, Art 22.1.
[2] See E Barham and B Sylvander (eds) *Labels of Origin for Food: Local Development, Global Recognition* (CABI, 2011).
[3] INAO. *Une réussite française: lappellation d'origine contrôlée* (INAO, 1985) 11.

foodstuffs,[4] as part of the Common agricultural policy, along with wines and spirits.[5]

The European vision can be explained by the interpretation of the link to a given territory primarily through the concept of *terroir*, which emphasises the land and the soil. This interpretation is a result of the monitoring system for wine production in France, the birthplace of the modern appellation of origin, in the early 20th century. Such a vision influenced the TRIPS Agreement, whereby wines and spirits benefit from a higher level of protection compared to other goods.[6] However TRIPS protects GIs for all kinds of goods, beyond agricultural products, thus offering many WTO members the opportunity to protect non-agricultural goods such as handicrafts.[7] The progression towards GIs for handicraft products also results from the absence of a suitable intellectual property tool to protect the traditional knowledge held by a specific group of artisans. GIs are therefore used to document such knowledge, even if the object of GI protection is the name alone.[8]

In India,[9] for example, GIs are predominantly registered for non-agricultural and non-foodstuff products. These made up two-thirds of the

---

[4] Products are listed in Annex I of the European Community Treaty and in Annexes I and II of the European Regulation No. 510/2006, replaced since 2012 by Regulation No. 1151/2012 of the European Parliament and of the Council of 21 November 2012 on Quality Schemes for Agricultural Products and Foodstuffs [2012] OJ L323/1. These Annexes may be amended in order to extend or reduce the list of products eligible for the registration of PDOs and PGIs as long as they constitute agricultural products or foodstuffs.

[5] In 2008 wines and spirits were integrated into the PDO and PGI categories; see Council Regulation (EC) No 479/2008 of 29 April 2008 on the Common Organisation of the Market in Wine [2008] OJ L148/1. Previously, the production of wines and spirits was regulated by a system of specific rules, due to the then extant structure of the European market in wine.

[6] F Addor and A Grazioli, 'Geographical Indications beyond Wines and Spirits: A Roadmap for a Better Protection for Geographical Indications in the WTO/TRIPS Agreement' (2002) 5(6) JWIP 865.

[7] J Audier, 'Passé, présent et avenir des appellations d'origine dans le monde: vers la globalisation' (2008) *Bulletin de l'O.I.V.*, 405; B O'Connor, *The Law of Geographical Indications* (Cameron May, 2004).

[8] M Blakeney, 'Protection of Traditional Knowledge by Geographical Indications' (2009) 3 Int. J. Intellectual Property Management 357; A Kamperman Sanders, 'Incentives for Protection of Cultural Expression: Art, Trade and Geographical Indications' (2010) 13 JWIP 81.

[9] See e.g. on the Indian Act, S Balganesh 'Systems of Protection for Geographical Indications of Origin: A Review of the Indian Regulatory Framework' (2003) 6 JWIP 191.

215 registered GIs as of December 2014.[10] In Colombia, half of the recognised Denominations of Origin are for handicraft goods (11 out of 22 in December 2014).[11] In Switzerland a register for GIs for non-agricultural goods exists since 2013. In Thailand, nine GIs for handicraft goods are registered, out of which four are for silk, an emblematic product of the country.[12]

The international recognition of GIs was settled through the TRIPS framework, which not only led to new domestic legislations protecting GIs but also to wider international recognition of GIs beyond the country of origin. Whether it is achieved via the registration of a non-Member's individual GI in the EU or through free trade agreements including a list of GIs to be protected by the signatories, the practices of countries protecting handicraft GIs is challenging and influencing the EU's vision. Countries such as India, Brazil, Vietnam and Malaysia are requesting GIs for their handicrafts to be protected in the EU, be it for the famous Kashmir Pashmina[13] from India, or the two non-agricultural GI products, Guacamayas (spiral basketwork) and Chulucanas (pottery) to be protected in Europe following the bilateral agreement with Peru and Colombia in 2012.[14] In general, the preference is for protection within the harmonised *sui generis* EU GI regime,[15] and not according to a trade mark regime.[16] Due to the absence of a uniform framework for all types of goods, negotiations in bilateral agreement are less smooth and international trade is hampered.

European producers themselves are now demanding a unitary protection framework at the EU level, according to a recent study on Geographical Indications protection for non-agricultural products in the

---

[10] See http://www.ipindia.nic.in/girindia/. For a comprehensive analysis of the Indian GI system, see D Marie-Vivien, *Le droit des indications géographiques en Inde: un pays de l'Ancien monde face aux droits français, européen et international* (Ecole des Hautes Etudes en Sciences Sociales, 2010).

[11] See http://www.sic.gov.co/drupal/denominacion-de-origen

[12] See http://www.ipthailand.go.th/

[13] Indian GI application No.46, filed on 9 December 2005, Geographical Indication Journal No. 13.

[14] See Trade Agreement between the European Union and Colombia and Peru at: http://trade.ec.europa.eu/doclib/docs/2011/march/tradoc_147704.pdf

[15] No foreign GIs for non-agricultural goods have been protected in the Member States of the EU. See Insight Consulting, REDD & OriGIn, *Study on Geographical Indications for Non-Agricultural Products in the Internal Market* (Final Report, 18 Feb 2013) (Hereafter, *EU Study*).

[16] For a comparison of *sui generis* GI protection versus trade marks, see D Gangjee, 'Quibbling Siblings: Conflicts Between Trademarks and Geographical Indications' (2007) 82 Chicago-Kent Law Review 1253.

internal market. The study, commissioned by the European Commission, identified several candidates within the EU. For example in France alone around 100 possible candidates, such as Stone from Bourgogne or Marseilles Soap were identified.[17] Subsequently in 2014 France passed a law recognising Geographical Indications for non-agricultural goods, which has been enforceable since 4 June 2015.[18] These discussions have raised questions regarding the basis for GI protection in general, as well as whether GIs should be treated differently according to the classification of subject matter in question.

The divide between agricultural and non-agricultural products resides in the absence of physical elements linking non-agricultural goods to the soil. While natural factors besides the soil such as the climate, the origin of raw materials, or environmental elements such as water can indeed influence product quality, the territorial link for non-agricultural products or handicraft goods is based above all on the producers' know-how, skills and practices – that is, on human factors. The issue is whether geographical names designating goods linked to their place of origin essentially via human practices can be registered as GIs. Can GIs be considered to apply to cultural as much as to natural products?[19] The underlying question is how to assess the link between the quality, reputation or characteristics of a product and its place of origin.

This chapter sets out to analyse the validity of such a link to origin, since this is the central legal criterion setting GIs apart as a distinct intellectual property right. This is especially true of *sui generis* GI systems, where this criterion is evaluated *a priori* in order to decide whether to grant exclusive rights to the use of the name. The first section explores the concepts underpinning GIs, with a historical discussion of the evolving legal definitions of GIs over time. The chapter then proposes an original framework for better scrutinising the different types of links with the origin, drawing from different jurisdictions on empirically informed case studies for both agricultural and non-agricultural GIs. It extensively reviews GI specifications resulting from the practices both of producers/applicant and of examining authorities (that is registrars) from two comparators, namely

---

[17] Savon de Marseille is one of the 834 non-agricultural GI products over the 31 countries identified in the *EU Study* (n 15) 42.

[18] Décret 2015-595 du 2 juin 2015 relatif aux indications géographiques protégeant les produits industriels et artisanaux et portant diverses dispositions relatives aux marques.

[19] See D Gangjee, 'Geographical Indications and Cultural Rights: The Intangible Cultural Heritage Connection?' in C Geiger (ed), *Research Handbook on Human Rights and Intellectual Property* (Edward Elgar Publishing 2015) 544.

India and France, with additional references to other jurisdictions.[20] The analysis draws on the concepts of natural and human factors introduced by the Lisbon Agreement for the Protection of Appellations of Origin and their International Registration,[21] where they were used cumulatively to qualify the appellation of origin. However, in order to embrace the looser definition of GI in TRIPs, the analysis uses these two concepts both as cumulative and alternative criteria. The cases selected highlight the various possible combinations of natural and/or human factors in linking products to their origin, beginning with products linked to the territory exclusively through human factors in Section 2 and then with products linked to the *terroir* through human factors combined with natural factors in Section 3.

Section 4 discusses the key findings of this theoretical and empirical analysis. First, one result is that the existing categorisation of goods in the EU does not allow for clear discrimination between products that deserve GI protection and those that do not, or between different types of GIs, as the link to the territory can widely differ within the same category of products, while there may be similarities in the nature of the link across different categories of products. Second it is proposed considering in the international and national GI legal frameworks the concepts of natural and/or human factors in determining the strength of the link to the origin, regardless of whether the product is classified as agricultural, a foodstuff, handicraft or industrial. Such insights cast new light on the existing legal regimes for GIs and suggest potential avenues for reform. This paper argues that a link to the origin based solely on human *or* natural factors might be qualified as weaker than a link based on the combination of natural *and* human factors. It consequently develops the rationale for establishing two levels of GI protection based on the strength of that link. The argument contributes to the evaluation of the EU Commission on the feasibility of recognising GIs for non-agricultural goods as well as to the implementation of French

---

[20] The examples were selected from the analysis of specifications for the entire list of Indian GIs registered by 2010 and a broad sample of French PDOs and PDIs chosen outside the wine and spirit sector. For more details, see D Marie-Vivien, Le droit des indications géographiques en Inde: un pays de l'Ancien monde face aux droits français, européen et international (Thèse de Doctorat, Ecole des Hautes Etudes en Sciences Sociales, Paris, 2010), at https://tel.archives-ouvertes.fr/tel-00587307/document

[21] Lisbon Agreement for the Protection of Appellations of Origin and their International Registration of 31 October 1958. For details, especially regarding its revision in 2015, see Chapter 5 by Geuze in this volume.

law, while also providing guidance for GI examination throughout the world.

2. CONCEPTS UNDERLYING GIs

A look back at the history of GIs reveals that, in France,[22] the first modern law on the protection of appellation of origin of 1905, designed to combat fraud in the sale of goods and adulteration,[23] was created to take account of and protect place names for qualifying natural products whose specificity was dependent on natural elements. Prior to this, place names were protected only for 'manufactured' goods.[24] However, the subsequent law of 1919,[25] still in force, which gave the Court the task of defining appellation of origin in case of conflicts between users, did not discriminate between various kinds of products and allowed for handicraft appellations to be protected. Most important was the consideration of the concept of '*terroir*'[26] which ultimately included, in addition to the natural environment, human skills, know-how, practices and knowledge of producers. Appellations designating non-agricultural goods were recognised by the judiciary, in accordance with local, fair and constant use of the appellation of origin. Finally, in 1935, a law was passed which provided for the *appellation d'origine controlée* (AOC), initially only for wines and spirits, but subsequently extended to cheeses in 1990, and then to agricultural, forestry and food products. The French AOC has therefore never been applied to non-agricultural goods.

In recent years, following the misappropriation of the name of the city Laguiole, famous for its knives, and in a context of debate about

---

[22] N Olszak, Droit des Appellations d'Origine et Indications de Provenance (TEC & DOC 2001) 1–187.

[23] *Loi du 1er août 1905 sur la repression des fraudes dans la vente des marchandises et des falsifications des denrées alimentaires et des produits agricoles* (Law of 1 August 1905 on the repression of fraud in the sale of goods and adulteration of foodstuff and agricultural products) (5 August 1905) Journal Officiel No. 210.

[24] *Loi du 28 juillet 1824 Relative aux Altérations ou Suppositions de Noms dans les Produits Fabriques* (1825) 7 *Bulletin des Lois* no 19, 65 (Law of 28 July 1824 on the misuse of names for manufactured products). See M Plaisant & Fernand-Jacq, *Traité des noms et appellations d'origine* (Librairie Arthur Rousseau 1921) 18.

[25] *Loi du 6 Mai 1919 Relative à la Protection des Appellations d'Origine* 8 May 1919 Journal Officiel 4726 (Law of 6 May 1919 on the Protection of Appellations of Origin).

[26] J Audier, 'Réflexions juridiques sur la notion de terroir' (1993) *Bulletin de l'OIV* 423; see also Chapter 3 by Barham in this volume.

re-localising manufacture and industry in France, the French authorities wished to extend GIs to cover processed products originating in a specific territory. In that sense, the French Law on consumption passed on 17 March 2014 provides for the creation of GIs for non-agricultural products,[27] whose protection regime is introduced in the Intellectual Property Code: it applies to

> the name of a region or a specific place used to describe a product, other than agriculture, forestry, food or marine products, originating from that place and which has a specific quality, a reputation or other characteristics that can be essentially attributed to its geographical origin. The conditions of production or processing of this product, such as the cutting, extraction, or manufacture must comply with a specification.[28]

The specification shall describe 'the process of preparation, production and of processing, elaboration, including the steps of production of processing which shall take place in the geographical area or the specific place as well as the steps which ensure the specific characteristics'.[29]

At the international level, the Lisbon Agreement, signed in 1958 within the framework of the Paris Convention[30] provides a very precise definition of appellations of origin which has to be applied by all members (presently 28): 'Appellation of Origin means the geographical name of a country, region or locality, which serves to designate a product originating therein, the quality and characteristics of which are due exclusively or essentially to the geographical environment, including natural and human factors.'[31]

The Lisbon Agreement clearly points to the link between the quality and characteristics of the product and its geographical environment. The *travaux préparatoires* of the Lisbon Agreement reveal that the first proposal did not mention the geographical environment, referring rather to 'the place and method of production, manufacturing, extracting or assembling of products'.[32] This development in the text makes it possible

---

[27] Loi No. 2014-344 *du 17 mars 2014 relative à la consommation*, Journal Officiel de la République Française No 65, 18 March 2014, p. 5400.
[28] Art L. 721-2 of the Intellectual Property Code.
[29] Ibid.
[30] See Chapter 4 of D Gangjee, *Relocating the Law of Geographical Indications* (Cambridge University Press, 2012).
[31] Art 2 of the Lisbon Agreement. The same definition was introduced in France in 1966, via Art 1 of the Law of 6 July 1966 amending and completing the Law of 6 May 1919, JO 7 July 1966, p. 5781, which is now Art L.115-1 of the Consumer Code.
[32] As proposed unanimously by the 4th Committee of the work and reports of the Lisbon Conference in 1956.

to interpret the concept of human factors: the practices, skills and know-how of the producers incorporated in obtaining the product. The term 'the place' has been replaced by 'natural factors', which are more restrictive, focusing on nature, which comprises the climate and the elements of the natural environment (soil, water, clay, caves, and so on) that can influence the quality of the raw material or processing for processed goods. Based on a mandatory combination of human and natural factors, the Lisbon Agreement is very restrictive regarding the strength of the link, but does not, however, discriminate according to the kind of products, which can be agricultural or not.[33] The Lisbon Agreement provides a very high level of protection against any usurpation or imitation, even if the true origin of the product is indicated or if the appellation is used in translated form or accompanied by terms such as 'kind', 'type', 'make', 'imitation', or the like.[34] The system provides for a notification and opposition procedure within its membership, and has led to the international registration of more than 800 appellations of origin.

This definition of the appellation of origin was introduced in those European countries which were signatories to Lisbon and later 'Europeanised' with the introduction of the harmonised European regime for agricultural and foodstuffs GIs in 1992, subsequently replaced by a new regulation in 2006[35] and more recently by Regulation (EU) No 1151/2012 in 2012.[36] While not the principal focus of this chapter, wines are regulated by Council Regulation (EC) No 1308/2013 while spirits are regulated by Regulation (EC) No. 110/2008 of 15 January 2008, and aromatised wines by Regulation (EU) No. 251/2014.

The EU's Protected Designation of Origin (PDO),[37] is very similar to the AO of the Lisbon Agreement, albeit with greater precision on the

---

[33] Art 2.1 of the Lisbon Agreement refers to product, with no further details. Consequently in practice, non-agricultural products have been registered under the Agreement.

[34] Art 3 of the Lisbon Agreement.

[35] Council Regulation 510/2006 of 20 March 2006 on the Protection of Geographical Indications and Designations of Origin for Agricultural Products and Foodstuffs [2006] OJ L93/12, which replaces Council Regulation 2081/92 of 14 July 1992 on the Protection of Geographical Indications and Designations of Origin for Agricultural Products and Foodstuffs [1992] OJ L208/1.

[36] Regulation No 1151/2012 of the European Parliament and of the Council of 21 November 2012 on Quality Schemes for Agricultural Products and Foodstuffs [2012] OJ L323/1.

[37] See Art 2(2)(a) of Regulation 510/2006; similar wording is evident in Art 5(1) of Regulation 1151/2012. For wines, the designation of origin is defined in Art 93.1(a) of Regulation 1308/2013 as:

origin of raw materials that must originate in the geographical area where the production, processing and preparation occur.[38] In order to take into account the tradition of Northern Europe countries, the Court of Justice of the European Union (CJEU) recognised protection for geographical names for products 'which cannot be shown to derive a particular flavour from the land, but which may nevertheless enjoy a high reputation amongst consumers and constitute for producers established in the places to which they refer an essential means of attracting custom'.[39] This distinction has led to the creation of the Protected Geographical Indication (PGI), for products that possess 'a specific quality, reputation or other characteristics attributable to that geographical origin, and the production and/or processing and/or preparation of which take place in the defined geographical area'.[40] Regarding the link to the territory, the PGI definition does not expressly require that natural factors such as the raw materials be sourced in the defined geographical area. From legal practice and jurisprudence, however, it appears that the demarcation of the geographical area of origin of raw material is authorised only if the source of the raw material influences the quality of the final product. Indeed, the CJEU has decided that for PGIs, a foodstuff may be treated as originating

---

the name of a region, a specific place or, in exceptional cases, a country used to describe a product referred to in Art 92.(1) that complies with the following requirements: (i) its quality and characteristics are essentially or exclusively due to a particular geographical environment with its inherent natural and human factors; (ii) the grapes from which it is produced come exclusively from this geographical area; (iii) its production takes place in this geographical area; (iv) it is obtained from vine varieties belonging to *Vitis vinifera*.

[38] See Art 3 of Regulation 510/206; Arts 5(1) and 5(3) of Regulation 1151/2012.
[39] *Exportur SA v LOR SA and Confiserie du Tech SA* (C-3/91) [1992] ECR I-5529 at [28].
[40] Art 2(2)(b) of Regulation 510/2006; similar wording is evident in Art 5(1) of Regulation 1151/2012. Regarding wines, according to Art 93.1(b) of Regulation 1308/2013, a GI means 'an indication referring to a region, a specific place or, in exceptional cases, a country, used to describe a product. . .' which complies with the following requirements: '(i) it possesses a specific quality, reputation or other characteristics attributable to that geographical origin; (ii) at least 85% of the grapes used for its production comes exclusively from this geographical area; (iii) its production takes place in this geographical area; (iv) it is obtained from vine varieties belonging to *Vitis vinifera or* a cross between the *Vitis vinifera* species and other species of the *Vitis* genus'. Regarding spirits, according to Art 15(1) of Regulation 110/2008, a geographical indication shall be 'an indication which identifies a spirit drink as originating in the territory of a country, or a region or locality in that territory, where a given quality, reputation or other characteristic of that spirit drink is essentially attributable to its geographical origin'.

from the geographical area concerned if it is processed or produced in that area, even if the raw materials are produced in another region.[41]

Thus for PGI, compared to the PDO, the strength of the link to the origin is weak, less stringent and less exclusive.[42] This weaker conception of the link to origin is a result of the lower importance attached to natural factors. This flexibility of the European regulation regarding the link to the origin contrasts with the strict conception regarding the kind of products that may benefit from GI protection, which is restricted to a list of agricultural products and foodstuffs, as part of the Common Agricultural Policy of the EU. Such a list can be amended, but additions must fit into the general definition of agricultural products or foodstuffs.[43] The level of protection is high in Europe, with the prohibition of any misuse, imitation or evocation, even if the true origin of the products or services is indicated or if the protected name is translated or accompanied by an expression such as 'style', 'type', 'method', 'as produced in', 'imitation' or similar, including when those products are used as an ingredient.[44]

Finally in 1994, TRIPS defined GIs very broadly, to include the concept of the appellation of origin. The general protection provided by the TRIPS Agreement for GIs differs from the Lisbon Agreement by enlarging the scope of the link between the product and its place of origin. In addition to product qualities being causally attributable to origin, characteristic and reputation are also recognised criteria. As a result, there are three alternative criteria for qualifying for recognition and protection as a GI under TRIPS. The criterion of reputation introduced both in TRIPS and in the EU PGI, provides greater opportunities while opening the door to innovative experiences.[45]

---

[41] *Carl Kühne & Others v Jütro Konservenfabrik GmbH* (C-269/99) [2001] ECR-I 9517. For analysis, see GE Evans, 'The Strategic Exploitation of Geographical Indications and Community Trade Marks for the Marketing of Agricultural Products in the European Union' (2012) WIPOJ 159.

[42] L Bérard and P Marchenay, 'From Localized Products to Geographical Indications: Awareness and Action' (CNRS, 2008) 14–19.

[43] An example is the addition of salt, whereas mineral waters have been withdrawn. Council Regulation (EC) No 692/2003 of 8 April 2003 Amending Regulation (EEC) No.2081/92 [2003] OJ L99/1.

[44] Art 13 of Regulation 1151/12.

[45] The difference between the Lisbon Appellation of Origin and the TRIPS GI is mainly focused on the additional criterion of reputation found in the latter, according to D Gervais, 'Traditional Knowledge: Are We Closer to the Answers? The Potential Role of Geographical Indications' (2009) 15 ILSA Journal of International and Comparative Law 551. However, the mandatory existence of both natural and human factors in Lisbon is underestimated.

Two different levels of protection are provided for GIs, depending on the category of goods.[46]

The first level (Art 22) is a minimum standard of protection for all products. It prohibits any use that constitutes an act of unfair competition in the sense of Art 10*bis* of the Paris Convention, which misleads the public as to the geographical origin of the goods. It also prohibits the registration of a trade mark that would contain or consist of a GI for goods not originating in the territory indicated but only where this is also misleading as to the true place of origin. Art 22 also applies where the sign in question is literally true as to the territory in which the goods originate, but would falsely indicate to the public that the goods originate in another territory. The minimum standards in this provision focus on preventing consumers from being misled, which requires to be proved, and unfair competition, which has to be judged by a court. Such a case-by-case application of protection through judicial decisions often generates problems.

The second level of protection (Art 23) is only available for wines and spirits. It strictly prohibits the use of an untrue GI, even if it is used in translation or accompanied by a qualifying expression such as 'kind', 'type', 'imitation', and so on. And the registration of a trade mark containing a GI for wines or spirits not having this origin is prohibited, even if the public is not misled as to the true origin of the product. Moreover, Art 24.1 mentions that members will enter into negotiations aimed at increasing the protection of individual GIs for wines and spirits.

The TRIPS Agreement does not, however, provide any guidelines for evaluating the existence of a link between the product and its geographical origin. The TRIPS GI definition uses the terminology of 'geographical origin' *per se*, whereas the original European proposal of July 1988 contained the requirement of 'including natural and human factors'.[47] This restriction did not survive subsequent rounds of negotiations and in today's TRIPS definition – the fruit of a consensus between Old and New World countries – 'geographical origin' is open to many interpretations since it does require the mandatory combination of human *and* natural factors. When implementing TRIPS many countries introduced its broad definition into their domestic legal framework. For example in India, the

---

[46] E Thévenod-Mottet & D Marie-Vivien, 'Legal Debates Surrounding Geographical Indications' in Barham and Sylvander (n 2) 13.

[47] GATT, 'Guidelines and Objectives Proposed by the EC', 7 July 1988 (MTN.GNG/NG11/W/26).

Geographical Indications of Goods (Registration and Protection) Act, 1999 defines GI in terms identical to the TRIPS definition. Furthermore, the Indian GI Act defines 'goods' to mean any agricultural, natural or manufactured goods or any handicraft or of industry goods and includes foodstuffs.[48] It thus formalises the validity of GIs for handicraft goods in India.

The successful dissemination of the concept of GIs of TRIPS has consequences for the Lisbon Agreement and its restrictive definition of appellation of origin, considered by many as the reason for the limited number of signatories. In 2009, a working group on the development of the Lisbon system was established and it has since worked extensively on a draft for a new Act of the Lisbon Agreement, finally adopted on 20 May 2015 in Geneva, with an eye to improving the system, *inter alia*, by incorporating the TRIPS definition of the GI, while maintaining the principle of an international register and a high level of protection.[49] Nevertheless, in the absence of any detailed provisions – in TRIPS or national adaptations of it – on the substantive examination of the link to the origin, practice and case law provide valuable guidance when attempting to define the basis for qualifying as a GI, both for non-agricultural goods as well as for agricultural goods. Lessons from existing national experiences will form the basis for the rest of this chapter.

3. LINK TO ORIGIN BASED ESSENTIALLY ON HUMAN FACTORS

Non-agricultural goods and particularly handicrafts are usually linked to a given place via human factors such as the practices and know-how of the producers, with relatively few natural factors. Unexpectedly, however, an emphasis on human factors in the absence of natural factors also occurs with foodstuffs in Europe more generally and in France in particular.

3.1 Handicraft Goods

Handicraft goods are usually characterised by sophisticated know-how, skills and practices, with different kinds of know-how contributing to their uniqueness and anchorage in a given place.

---

[48] S 2 of the Indian Geographical Indications of Goods Act, 1999.
[49] For a detailed overview of this redrafting process, see Chapter 5 by Geuze in this volume.

3.1.1 Sophisticated know-how

For many GIs in the handicraft sector, the method of production is highly intricate and based on the consideration that only manual methods lead to desired results, whilst machine-made copies are of inferior quality. This is illustrated by the Indian GI Kancheepuram Silk,[50] for silk woven in the ancient, royal town of Kancheepuram, famous for its temples. The method of production is characterised by the use of thick silk yarn which gives it its heavy weight and bright colours as well as by the use of two extra shuttles on each side of the loom, besides the shuttle used for the main body of the sari, to weave the borders in contrasting colours. The reputation of Kancheepuram saris also lies in the use of silver, gold and red silk threads known as 'zari'. It is generally accepted that fake Kancheepuram saris are made of thinner silk yarn, have only one border and do not contain gold.[51] The method of production includes dressing the yarn, sizing, the degumming of the yarn, including the number of times it is rinsed and the time required to do so, as well as a meticulously detailed dyeing process. The intricate details of the description demonstrate the sophistication of the skills involved. In France, meanwhile, the Dentelle du Puy (lace) appellation of origin has been recognised by the Court according to the law of 1919. The numerous certificates offering guarantees of the independence, impartiality and sincerity that proved the existence of very traditional, local, constant and fair usage were the motivations for the Court to reserve the appellation exclusively for handmade lace by artisans whilst excluding machine-made lace.[52] In India, woodcraft products are also protected by GIs, with the product being entirely hand-chiselled then painted. Take, for example, the GI Kondapalli Bommalu for painted wooden figurines.[53] Their production requires know-how about selection of wood, cutting and seasoning, woodcarving and painting.

Besides the know-how involved in making an object, assembling it and treating the raw material, GIs are also registered on the grounds of traditional designs and drawings. While the method of production may

---

[50] GI application No. 14, filed on 7 October 2004, Geographical Indication Journal No. 4.

[51] Interview with a manager of the weaving service centre of Kancheepuram, December 2006. On file with the author.

[52] *Chambre Syndicale des Fabricants de Dentelles et Passementeries de la Haute-Loire v Gouteyron et Jérôme* [1931] Propriété Industrielle 188 (Le Puy-en-Velay Civil Court, 19 February 1931).

[53] GI application No. 44, filed on 10 November 2005, Geographical Indication Journal No. 13.

be widespread, motifs and drawings are often specific to a region. Such GIs raise issues about their 'uniqueness' – embedded essentially in the designs – which, according to intellectual property laws, are normally protected as 'models and designs'.[54] Nevertheless, exclusive rights on designs and models are only granted for a limited period of time and only for *new* models and designs, which does not apply to designs in traditional products.

The sophistication of the design may ensue from the technique used and vary according to the degree of its mastery, as illustrated by the Indian GIs Pochampally Ikat[55] and Orissa Ikat.[56] Ikat – a Malaysian-Indonesian word that means tie-dye – involves the sequence of tying or wrapping and dyeing sections of bundled yarn to a predetermined colour scheme prior to weaving. Two elements distinguish imitations from the originals: they are machine-woven and above all the motifs are printed after weaving and not created during the weaving process with pre-dyed yarn.[57] The GI Pochampally Ikat is entirely based on the Ikat technique and is described in a very detailed manner. The diamond shaped motifs, or 'chowka', which, because of the recent history of Ikat production in Pochampally, are simpler than Ikat made in other regions of India, are only briefly mentioned in the specification. However, such motifs distinguish Pochampally Ikat from Ikat produced in other places,[58] for example, the GI Orissa Ikat[59] is characterised by floral motifs with sophisticated shaded effects. Pochampally Ikat and Orissa Ikat GIs highlight the situation of creative know-how intertwined with the Ikat technique. The creation of designs cannot therefore be dissociated from the mastery of the technique. Their 'uniqueness' lies in the particular way the technique is applied in a specific place and not only in the type of design.

---

[54] See e.g. Section IV of TRIPS and the Hague Agreement Concerning the International Registration of Industrial Designs of 1925.

[55] GI application No. 4, filed on 15 December 2003, Geographical Indication Journal No. 13.

[56] GI application No. 22, filed on 1 February 2005, Geographical Indication Journal No. 12.

[57] A case between the owner of the Pochampally Ikat IG and an infringer who manufactures saris with machine printed pattern. Complaint to the High Court of Delhi, 887/2005.

[58] These simplified modern designs have proved more popular with new generations of Indian consumers.

[59] Applicant's reply, dated 3 November 2005 to the letter from the GI Registry dated 21 October 2005, accessed in the file available at the GI Registry, Chennai.

3.1.2 Absence of a specific origin requirement for raw material

In many cases the raw materials used for non-agricultural GI products are not sourced locally, but must nevertheless be of high quality, known as 'generic quality'.[60] In Europe, for example, out of 127 products considered in the EU Study, 46 non-agricultural GI products do not exclusively source their raw materials from within the defined geographical area.[61] In India quality raw material implies traditional raw material, such as natural dyes;[62] or, for woodcrafts, it is timber quality.[63] The use of 'fake' raw material, such as synthetic thread, helps identify counterfeits. However, the quality of the raw material is not attributed to local natural factors. For Indian handicraft GIs, either the geographical origin of raw material is not described, or this origin is far removed from the product's manufacturing zone and indicated on a purely documentary basis without being mandatory. The silk yarn used for Kancheepuram Silk according to GI specification is bought from Gujarat, situated in North India. The 'uniqueness' of the Indian Konark Stone Carving[64] GI registered for sculptures of traditional dancers from Orissa is due to the raw material, snake stone. The mineral composition of the stone is described but its geographical origin is not documented, demonstrating the extent to which this aspect is considered superfluous.

The source of raw material explains the localisation of production but local sourcing is not mandatory as it is not seen as a condition for obtaining a quality product. For example, the Indian GI Kondapalli Bommalu, designating wooden figurines, indicates that the wood comes from the surrounding region of Kondappalli, but the area identified corresponds to the area where the figurines are sculpted – a village of 1.5 km^2 – and does not include the nearby forests. Similarly, the reputation of GI Mysore Silk[65] is mainly due to the silk yarn produced in the ancient kingdom of

---

[60] G Allaire, 'Quality in Economics: A Cognitive Perspective' in M Harvey, A McMeekin and A Warde (eds), *Qualities of Food* (Manchester University Press, 2004) 61.

[61] *EU Study* (n 15).

[62] Presentation of the Assistant Registrar of the GI Registry, Mr Natarajan, 17–18 September 2008, Delhi. The six criteria are: 'Quality of raw fibre, Natural dye, Quality of water, Colour fastness, Durability, Professional skill.'

[63] E.g. the 'Sankheda Furniture' GI specification insists on the use of 100 per cent teak wood.

[64] GI application No. 87, filed on 9 April 2007, Geographical Indication Journal No. 15.

[65] GI application No. 11, filed on 22 July 2004, Geographical Indication Journal No. 3.

Mysore where there is a tradition of silk worm farming.[66] The specification emphasises its 'uniqueness', attributed to the superior quality of the silk yarn used, a zari consisting of 65 per cent silver and 0.65 per cent gold and a special process of twisting the yarn which gives the fabric its wavy effect. However, there are no provisions specifying the source of the silk.

While such handicrafts also exist in Europe, experience shows that they may not be granted protection according to EU GI law. In France, in the 1990s the 'faïence de Moustiers' producers' association drafted an appellation of origin to protect the revival of earthenware production in Moustiers, which attracted many avid prospectors. But since the raw material (fuller's earth, enamel, and so on) was no longer directly sourced from Moustiers, as it had been in the 18th century, the application was not accepted. Indeed, French law has since 1966 incorporated the definition of the appellation of origin of the Lisbon Agreement, which provides for a combination of natural and human factors.[67] Furthermore, as earthenware is neither an agricultural product nor a foodstuff, a PGI application was impossible, revealing the limitations of French and European regulations. For the same reasons, Porcelaine de Limoges and Porcelaine de Nevers applications were rejected.

In Colombia, where the legal framework provides for the protection of GIs through denomination of origin, natural factors are mandatory, with the raw material originating from the geographical area (for the *Mochila Arhuaca*, a handwoven bag, some dyed wool comes from Peru and therefore a collective trademark was registered),[68] even if the link with the origin can be considered as very strong thanks to human factors. For the same reason, Harris Tweed from Scotland is defined in s 7(a) of the Harris Tweed Act of 1993 as 'a tweed which has been handwoven by the islanders at their homes in the Outer Hebrides, finished in the Outer Hebrides, and made from pure virgin wool dyed and spun in the Outer Hebrides'. It is only protected as a certification trademark, in the absence of a mandatory local wool requirement.[69]

---

[66] Personal interview with Mr Vijayan, the General Director of KSIC (Karnataka Silk Industries Corporation), the GI applicant. On file with the author.

[67] See 'Faïences de Moustiers' PIBD (1992) No. 509.I.85, Ministerial Answer No. 15479 of 6 June 1991 and official journal of the Senate debates, 26 September 1991, p. 2088; F Pollaud-Dulian, *Droit de la Propriété Industrielle* (Montchrestien, 1999) 734.

[68] 'Protecting Traditional Knowledge and Culture: Colombia Gives an Example' (9 Jan 2012), at http://tktotem.blogspot.co.uk/2012/01/protecting-traditional-knowledge-and.html

[69] Dan Anthony, 'From Croft to Catwalk: The Harris Tweed Collective Mark' (2013) 1 WIPO Magazine 24.

3.2 Foodstuffs

Worldwide, most GIs for agricultural goods and foodstuffs are widely perceived to satisfy the link requirement on the basis that natural factors determine the specificity of the product. According to WIPO: 'Agricultural products typically have qualities that derive from their place of production and are influenced by specific local, geographical factors such as climate and soil. It is therefore not surprising that a majority of GIs throughout the world are applied to agricultural products, foodstuffs, wine and spirit drinks'.[70] However, a careful analysis of European registered PGIs and Indian GIs do not support such an argument. Many agricultural products and foodstuffs are in fact linked to their origin mainly through human factors. Yet in France, this acknowledgement has been relatively ignored.

In India, the first GI registered for a liqueur, Feni,[71] is an example of an application based on traditional know-how. The cashew tree was introduced to Goa by the Portuguese in the 16th century and the use of cashew apples for the production of the liquor only exists in this Indian state. The apples are collected and crushed to extract the juice, which is then fermented and distilled without the addition of any foreign ingredient, using traditional tools.[72] The GI specification does not highlight natural factors, mentioning only that the quality of apples varies, depending on soil characteristics and the place of cultivation, which is not demarcated. In contrast, the distillation zone is restricted to Goa. Even though Feni is undoubtedly an agricultural product, its link to origin is based only on the human factor of the distillation techniques taking place within Goa.

The UK's 'Melton Mowbray Pork Pie' PGI provides another illustration where the emphasis is on the quality of the raw materials rather than on their origin within the defined region. The pie was made in Melton Mowbray and its surrounding areas, being historically linked to the established practice of hunting in this area. In the specification, the emphasis is on traditional methods of production. Melton Mowbray Pork Pies have a traditional bow walled pastry giving them their characteristic shape. The meat content of the whole product must be at least 30 per cent.

---

[70] WIPO, *Geographical Indications: An Introduction* (WIPO Publication No. 952(E)) 10.
[71] GI application No. 20, filed on 19 December 2007, Geographical Indication Journal No. 27.
[72] See D Rangnekar, 'Geographical Indications and Localisation: A Case Study of Feni' (ESRC Report 24 September 2009); see also Chapter 16 by Rangnekar & Mukhopadhyay in this volume.

A comparative analysis of GIs for handicrafts 309

The pies must be free of artificial colours, flavours and preservatives.[73] A preliminary question referred to the ECJ by the High Court in London regarding the validity of this PGI application was filed but eventually withdrawn. There will therefore be no further clarification by the ECJ of the requirements for registration of PGIs without natural factors, which has been considered valid by the European Commission.[74]

In France, where GIs are used for rural development, the presence of natural factors and the requirement for locally sourced raw material in particular, is a hotly debated issue. Very few GIs are linked to origin solely through human factors, even though the situation is changing. One case is the PGI Bergamote de Nancy, for sweets made traditionally with sugar, glucose syrup and natural bergamot essence, a recipe used in Lorraine cuisine since the 18th century, which are renowned in the region.[75] Yet it is difficult to assess the conformity to European criteria for this PGI, since it was registered via the fast-track procedure for incorporating lists of pre-existing national GIs within the newly harmonised EU-wide GI regime.[76] The PGI Pâtes d'Alsace represents another controversial example due to the mechanisation of the processing and the consequently weak human factor, while no local natural factors are involved as the raw materials are not sourced locally. The GI specification explains that since ancient times, Pâtes d'Alsace[77] have been produced from flour and eggs. This distinctively Alsatian tradition is based on a production method that, until the 19th century, was essentially domestic and rural, the special domain of the housewife, with recipes and know-how handed down from mother to daughter, using eggs which were easily available from farms. Such pasta is eaten with traditional dishes such as jugged hare, fish matelote or Rhine salmon. According to Professor Olszak, the application was made largely because a small Italian producer of pasta had begun to produce an imitation that did not contain the same ingredients and involved only

---

[73] See Dossier No (UK/PGI/0005/0335) at http://ec.europa.eu/agriculture/quality/door/list.html. See also A Tregear and G Giraud, 'Geographical Indications, Consumers and Citizens' in Barham & Sylvander (n 2) 63.

[74] *Northern Foods Plc v The Department for Environment, Food and Rural Affairs & Anor* [2005] EWHC 2971 (Admin); see also D Gangjee, 'Melton Mowbray and the GI Pie in the Sky' (2006) 3 IPQ 291.

[75] Dossier No (FR/PGI/0017/0195).

[76] Established in Commission Regulation (EC) No 1107/96 of 12 June 1996 on the Registration of Geographical Indications and Designations of Origin under the Procedure Laid Down in Art 17 of Council Regulation (EEC) No. 2081/92 [1996] OJ L148/1. The PGI application was based on the specifications in the French Dossier LA/19/90 (Red Label).

[77] Dossier No (FR/PGI/0005/0324).

two producers. Yet he suggests that 'Pâtes d'Alsace' does not deserve GI registration.[78] The PGI specification essentially indicates that pasta produced on an industrial scale in the region respects the traditional balance of ingredients: 'This organoleptically optimal mix [of eggs and durum-wheat semolina] is today unanimously accepted by Alsatian producers, enabling them to carry on the tradition of eggpasta'. Indeed, in this case the link to origin does seem too weak to be eligible for the rights conferred to GIs.

Apart from these rare examples of dubious validity, French tradition is reluctant to protect GIs for products where raw materials are not sourced locally. One case is the PGI Calisson d'Aix (lozenge shaped sweets made out of ground almonds),[79] where 70–80 per cent of these sweets are made by eight calisson makers. The local public authorities wanted to boost almond production in the Aix region by introducing into the PGI a large but localised almond supplying zone around Aix, whereas calisson makers at the time were using almonds coming from California. The European Commission objected to the localisation of almonds in the Aix area as there was no justification for a link between the quality of the almonds and their place of cultivation.[80] The final specification attributed the link to origin entirely to the know-how of the calisson makers.[81] This case is still under examination at the French level. Another controversial case concerns the Alsace geranium. Horticulturists wanted to protect the traditional geranium from Alsace, which flourished in the 1950s and is a resistant type which flowers quickly and abundantly. The planting of cuttings in pots and the monitoring of their growth until they are ready to be sold takes place in Alsace. The know-how of local horticulturalists is essential in adapting the cuttings to the cold climate. However, the application was rejected by the French authorities because cuttings that previously originated from Alsace now come from Kenya and are chosen according to generic quality criteria such as good health.

In conclusion, it appears that the French tradition of GI protection does not promote products linked to their origin primarily via human factors, whereas European legislation (in terms of recognising the validity of PGIs) allows for this, meaning that there are different approaches to assessing the link to origin for protecting GIs even within the European Union.

---

[78] Interview with Professor Olszak, on file with the author.
[79] Draft specifications for the PGI 'Calissons Aix', Union of Manufacturers of Calissons Aix en Provence, version 4 of 4 January 2006, p. 3, under application for registration but not yet accepted at the Community level at the time: http://www.inao.gouv.fr/repository/editeur/pdf/CDC-IGP/calissons-d-aix.pdf
[80] Interview with E Monticelli, on file with the author.
[81] Ibid.

4. THE LINK TO ORIGIN VIA A COMBINATION OF NATURAL AND HUMAN FACTORS

Most agricultural and foodstuff products are linked to their origin through both natural and human factors. The importance of natural factors mainly depends on whether the product is a raw or a processed good. Practice shows that handicraft goods can be linked to a given territory in the same way.

4.1 Agricultural Goods and Foodstuffs

4.1.1 Agricultural raw products obtained from a local variety

When the final product is subject to relatively little processing, as in the case of horticultural goods and cereals, the natural factors influencing the quality of the product are the soil and the climate, while the human factors consist of cultivation methods. Products resulting from the cultivation of an ancient local variety are strongly anchored in their environment, as those varieties are particularly well adapted to their surroundings. GIs can also help maintain a diversity of varieties as illustrated by the Indian Navara Rice GI,[82] which designates a rice resulting from the crossbreeding of two indigenous varieties from Kerala in South India. Navara, like Basmati[83] or Rooibos,[84] is not a geographical name but the name of a rice variety, named after its short cultivation cycle, endemic to Kerala according to the GI application. Historical sources from 2500 BC testify to the Ayurvedic medicinal properties of this rice, which is used in Ayurvedic formulations.[85] The cultivation process is exclusively organic, since it is used for medicinal purposes, thus involving certain know-how and a specific *modus operandi*. This rice is slowly becoming extinct because of the difficulty of cultivation and low yield.[86] 'Fake' Navara rice is grown in the same geographical zone but from other varieties.

Similarly, the French appellation of origin Châtaigne d'Ardèche is

---

[82] GI application No. 17, filed on 25 November 2005, Geographical Indication Journal No. 17.
[83] D Marie-Vivien, 'From Plant Variety Definition to Geographical Indication Protection: A Search for the Link Between Basmati Rice and India/Pakistan' (2008) 11 JWIP 321.
[84] E Biénabe, (2009) 'Le Rooibos d'Afrique du Sud: Comment la Biodiversité s'Invite dans la Construction d'une Indication Géographique' (2009) 2 Autrepart 117.
[85] 'Susruta Samhita', Susrutacharya, 2500 BC.
[86] See http://njavara.org; 'Njavara Facing Extinction', The Hindu, 21 December 2007.

characterised by the cultivation of local varieties. For centuries local society was organised around chestnut groves, which began to deteriorate. New varieties resulting from the hybridisation of plants from different regions, more suited to certain technical and marketing criteria, were proposed. Since such an innovation would have radically changed the cultivation of chestnut groves, shifting it from the domain of agroforestry to intensive orchard farming, producers rejected it and applied for an appellation of origin to protect local varieties, a mode of cultivation and the landscape. From a census of 66 local varieties, the Châtaigne d'Ardèche appellation selected 19 main varieties, specifically ruling out hybrid varieties.[87] To counter the European commission argument that a common characteristic between the different varieties of chestnuts had not been established, the producers decided to use the 'local ecotype concept' found in all 66 varieties.[88] In such cases, the GI specificity is strongly related to its natural features based on local varieties and shaped by human practices.

4.1.2 Processed products obtained from local raw material

The Darjeeling Tea GI is a significant Indian example of a processed product linked to origin through human and natural factors,[89] now also registered in Europe, despite some EU members opposing the prohibition of the use of the indication for blended tea.[90] Its qualities are attributed to an exceptional natural environment, including the soil, but above all from the altitude of between 600 and 2000 metres at which it is cultivated, the steep slopes and very specific weather conditions: wet and cool in summer and dry and cold in winter. Yields are much lower than in non-Darjeeling districts. The species *Camellia sinensis*, is used for cultivating all Indian tea. Darjeeling tea leaves are processed in the production area, in the traditional 'orthodox' way initially invented by the British, who adapted the processes from China. In Europe, and especially in France, processed GI products obtained from local raw material are common features. Such is the case of the olive oil from Nyons processed from

---

[87] See Art 4 of the Decree of 28 June 2006 concerning the 'Chestnut Ardèche' appellation of origin: 'Chestnuts from local varieties of the species *Castanea sativa* Miller listed in the technical regulations are provided for in Art 1 of this Decree. Hybrids are prohibited.'

[88] Letter from the European Commission, General Directorate for Agriculture and Rural Development, 9 October 2009 and interview April 2010. The PDO was registered in 1994. See Dossier No (FR/PDO/0005/00874).

[89] GI application No.1, filed on 27 October 2003, Geographical Indication Journal No. 1.

[90] See Dossier No (IN/PGI/0005/0659). The objections are recorded in Commission Implementing Regulation (EU) No. 1050/2011 of 20 October 2011.

olives of the Tanche variety grown in the departments of Drôme and Vaucluse. The 'tanche' is a variety typical of this region, particularly well adapted to its mixed climate. The smallest olives are crushed and mixed using traditional methods; the paste is either pressed or centrifuged to extract the oil.[91]

4.1.3 Processing under special environmental factors

A very famous French case is the Roquefort appellation of origin, protected since the 15th century through a royal charter. Protection was granted to the inhabitants of Roquefort along with a monopoly over the maturing process, carried out in well-guarded cellars.[92] The first appellation of origin for cheese in France, it contains provisions on production and refining conditions, as well as 'natural factors' referring to both the source of raw materials and the processing environment.[93] The particular character of Roquefort lies, on the one hand, in the characteristics of the milk from traditional breeds of sheep fed according to traditional farming, and on the other, in the originality of the natural caves of Roquefort sur Soulzon, entirely dug from fallen rocks at the foot of Combalou's limestone cliffs 'where a miracle of nature takes place that gives Roquefort its unique flavour'.[94] The raw material was initially sourced from a very large area but subsequent to the EU definition of the appellation of origin that provides for mandatory local sourcing of raw material, the milk supplying region was limited to the mid-mountain ranges south of the Massif Central, where there has been a long and particular tradition of sheep farming.[95]

In the absence of local raw material, quality might be related to environmental factors during processing. The Indian GI Monsooned Malabar Coffee[96] is processed following a technique directly linked to climatic conditions. The coffee was once stored for a long time before being shipped, but then underwent a transformation that was initially unexpected because of the weather, which is hot and wet on the Malabar Coast

---

[91] For Huile d'olive de Nyons, see Dossier No (FR/PDO/0117/0142).

[92] M-H Bienaymé, 'L'Appellation d'Origine Contrôlée' (1995) Revue de Droit Rural 419, 420.

[93] *Loi du 26 juillet 1925 Ayant pour but de Garantir l'Appellation d'Origine du Fromage de Roquefort* (30 Juillet 1925) JO 7190 (Law on the Appellation of Origin Roquefort, 26 July 1925).

[94] Ibid.

[95] Dossier No (FR/PDO/0217/0131).

[96] GI application No. 85, filed on 5 April 2007, Geographical Indication Journal No. 21.

during the monsoon season. This process was called 'monsooning' and the resultant coffee became famous. The geographical origin of the coffee processed is non-specific and includes coffee cultivated anywhere in India. It can be compared to the French PGI 'Saucisse de Morteau', linked to origin by the slow smoking practice and know-how, with wood obtained from softwood forests found throughout Franche-Comté, in particular from the mountainous areas. This practice gives the Morteau sausage its amber colour and smoky taste and is inextricably linked to its manufacturing region.[97] This example is quite unique in France, where the strategy generally pursued for specifying the product is to localise the source of the raw material. For the Saucisse de Morteau, the producers and the French government initially sought to do so, but such a limitation was rejected by the European Commission on the grounds that the zones supplying pork are different from the sausage production zones and it is not possible to establish a link between the source of the pork and the quality or reputation of the Saucisse de Morteau.[98] Yet Saucisse de Morteau is linked to origin via natural factors such as the local wood used for the smoking of pork.

4.2 Non-Agricultural Goods – Handicraft Goods

4.2.1 Natural factors are the source of raw material

Under the Lisbon Agreement 28 appellations of origin for handicrafts based on human and natural factors such as raw materials have been registered at the time of writing. Examples include Emaux de Limoges from France, Marbre Lepenica for natural products, or Trojanska Keramika for processed products from Bulgaria.[99] However, they are very few in number compared to wines and spirits. In France, the Poterie de Valauris appellation was recognised through a judgment of the Court of Cassation on 18 November 1930 for pottery made with local clay.[100]

---

[97] Dossier No (FR/PGI/0005/0556).

[98] INAO, Comité National des IGP, Labels rouges et STG. PGI application 'Saucisse de Morteau ou Jésus de Morteau' request for additional information from the European Commission. Answer from the ODG, File No. 2008-414, 27 November 2008. The first letter of the Commission is dated 24 August 2007, the response of the applicant group is of 13 August 2008 and the response of the Commission is dated 3 October 2008.

[99] See the Lisbon Express Registry database at http://www.wipo.int/ipdl/en/search/lisbon/search-struct.jsp

[100] In France, following the implementation of the 1919 law which allowed for the recognition by the Court, some appellations of origin for handicraft were recognised by the Courts, e.g., the 'Emaux de Limoges' appellation through

In Colombia, the preparation of the spiral woven handicrafts designated by the denomination of origin 'Guacayamas', is from a fibre extracted from the fique, a plant cultivated in the geographical area by the specific indigenous community of Laches.[101] In Vietnam, one GI designates a conical hat from Hué, prepared from local variety of leaves.[102]

In India, a rare GI example specifying local raw materials is the Kashmir Pashmina[103] for shawls made in Kashmir from the undergrowth of fleece from the mountain goat Capra Hiracus with a fineness of 12–16 microns. The specification outlines the following steps: procurement of the raw material, combing, cleaning; soaking in clean cold water, draining of the water and mixing with rice flour; the storage of the delicate pads of wool in deep stone pots, spinning of the yarn on the traditional 'chakra'; warping, dressing and reeling the yarn; weaving; clipping the loose threads; dyeing; washing and packing. The GI specification retraces the geographical route of the different phases of production and identifies three sub-regions: Ladakh, the grazing ground of the goats where the wool is collected, the outskirts of Srinagar, the capital of Jammu and Kashmir, where the wool is spun and the entire region of Jammu and Kashmir, where the wool is woven. The link to territory is characterised by different types of know-how, including weaving, and by local raw materials. However, the Ladakh zone is not demarcated in the same level of detail, especially in terms of latitude and longitude, as the overall GI zone. It suggests that collecting the wool is accorded less importance, combined with a certain disregard for this activity compared to spinning and weaving.

Here India may have a different perspective from that of Europe, where a PDO was registered for Native Shetland Wool[104] which designates only the wool, collected and spun in a specific area according to traditional methods. The weaving phase is not included in the PDO, so the fabric can be produced anywhere, according to any method.

---

a judgment of the Court of Appeals of Limoges on 18 February 1946 and the 'Cholet' appellation through a judgment of ruling of the Angers Court of Appeals on 17 November 1936.

[101] See Cestería Guacamaya, File no. 06-85475, registered on 19 June 2009.
[102] GI Certificate No. 00020, National Office of Intellectual Property, Vietnam (19 July 2010).
[103] GI application No. 46, filed on 9 December 2005, Geographical Indication Journal No. 13.
[104] Dossier No (UK/PDO/0005/0737).

4.2.2 Natural factors influence the processing

As with agricultural goods, non-agricultural goods can derive their reputation from natural factors which impact upon the processing. In France, the appellation of origin Toiles de Cholet has been protected since 1936 by judicial recognition,[105] which noted that the reputation of Cholet sheets and textiles was due to the bleaching techniques, consisting of stretching the textiles out on green, wet and clayey meadows. Textiles are bleached with water pumped directly from the ground, the quality of which is seen as exceptional for bleaching and not available anywhere else in the region. Their special weight and strength also depend on natural elements from the soil and specific local climate conditions.[106]

Local clay used to mould objects can also be seen as critical in linking the product to its territory. It is not a final element constituting the product, and therefore not a raw material, but is rather similar to an environmental factor. The Indian GI Aranmula Metal Mirror[107] uses the specific local clay gathered from the river of Aranmula to cast the metal mirror.

5. DISCUSSION

5.1 Consistency When Evaluating the Link Requirement

5.1.1 No justification for discriminating on the basis of subject matter categories

The implementation of TRIPS in countries which have recently adopted GI protection and the practice at the EU level as well as in France considered above shows that the link with the place of origin can be analysed on the basis of human factors, individually or combined with natural factors, not only for agricultural products and foodstuffs but also for

---

[105] Commercial Court of Cholet, 8 January 1936, Etablissements Béra c./ Syndicat patronal des industries textiles de la région de Cholet, regional archives, Conseil général, Département Maine et Loire, côte 143a63.

[106] *La Blanchisserie de la rivière Sauvagean et le blanchiment des toiles à Cholet* (Cahors: Association des Amis du Musée du Textile Cholerais, REMPART 1992). The protection of this appellation is still effective because the same Angers Court of Appeal decided in 1992 that the trademark 'Création Maret Cholet France' used to designate woven textile was misleading since it could be confused with the 'Cholet' appellation of origin, if the methods of manufacturing did not meet the appellation criteria. See Court of Appeal of Angers, Chamber 1B, 17 February 1992.

[107] GI application No. 3, filed on 8 December 2003, Geographical Indication Journal No. 3.

Table 11.1 Subject matter categories and types of links

	Natural factors + human factors	Human factors
Agri-food products	Roquefort Nyons Olive Oil Darjeeling Tea Monsoned Malabar Coffee	Melton Mowbray Pork Pie Feni alcohol Calisson d'Aix
Handicraft products	Pashmina shawls Toile Cholet Conical Hat Guacayamas	Kancheepuram silk Pochampally Ikat Dentelle du Puy Faience Moustiers

non-agricultural goods. The review of existing practice demonstrates that, whatever the nature of the product is, GI links have been recognised in the following situations (see Table 11.1).

For any kind of goods, the analysis of the existence of human and/or natural GI factors provides a useful way to approach and assess the link to origin. As a result, there is no reason to treat categories of products differently.

5.1.2   The validity of GIs based on human factors

In practice GIs based on human factors are widely acknowledged and registered. However, this requires further assessment. Is a link to geographical origin via know-how alone legally valid with respect to the definition of a GI? Can know-how be rooted in an area – 'essentially attributable' to origin in the language of TRIPS – and can it confer on a product a quality or reputation linked to its geographical origin? Indeed, this link to territory based on human factors is often questioned, with know-how and methods of production seen as easily transmissible, including beyond the region of origin. Then there is the well-documented migration of artisan communities, whose history suggests that their roots in a given area can prove to be transient.[108] For example, the Kancheepuram GI application states that the weavers were originally from another state, Andhra Pradesh, and migrated four hundred years ago when their village was swept away by the sea. In the event of artisans migrating outside their area of origin, would a reservation of the name only to goods produced in the

---

[108]   Amongst the Indian GIs, see in particular Orissa pattachitra, Nirmal paintings and Kota doria.

demarcated area be an unfair restriction? Can an artisan, or a group of artisans who migrate, produce an identical product in their new environment? Will the know-how evolve in the course of this migration? How can this be assessed?

First, several criteria are possible. Beginning with the environment, it might be suggested that producers are influenced by their surroundings, including both natural factors and human interaction. As an illustration, climate (a natural factor) influences the type of produce, for example light cotton saris in South India, shawls in the mountainous regions of the Himalayas, specific types of geranium flowers in Alsace. For our purposes it can be accepted that, though exceptions might still exist, that know-how is not blindly 'transferred' as is, but adapted when used in new surroundings.

Second, the definition of geographical origin requires some elaboration: origin refers to a place having a certain meaning thanks to history and shared know-how.[109] The collective nature of know-how that results from its being shared within a community located for a long time in a specific area ensures its continued existence in this place. Individuals isolated from the community, or even small groups cut off from the main community, will not be able to execute this know-how with equal proficiency. This underlies the prominent collective dimension of GIs, resulting from a group of producers sharing their know-how and cross-controlling the quality of the product. In India, as a consequence of the caste system that associates a given community with a specific activity, know-how linked to handicrafts is held by specific groups and passed down from generation to generation.[110] For example, the GI Pipli Appliqué Work[111] is made by artisans from the caste of Darjis, who hand down their knowledge and titles in this way. The idea of a 'basin of skills' is introduced when skills are developed over a large area.[112] The know-how might well then spread among the locals. In the Kancheepuram Silk case, the silk was originally woven only by the Salia community, but now all the communities within eight kilometres of Kancheepuram, representing 75 per cent of the population of Kancheepuram, weave silk.[113]

---

[109] Bérard and Marchenay (n 42) 10.
[110] M-C Mahias, 'Les Sciences et les Techniques Traditionnelles en Inde' (1997) 37 L'Homme 105.
[111] GI application No. 86, filed on 9 April 2007, Geographical Indication Journal No. 15.
[112] Bérard and Marchenay (n 42).
[113] NS Gopalakrishnan, PS Nair and AK Babu, *Exploring the Relationship between Geographical Indications and Traditional Knowledge: An Analysis of the*

The historical depth of the localisation of the community of producers is the criterion of validity for such GIs. Indeed, the Indian GI Rules state that the link to origin must be demonstrated through the history, in a section specially designed for this purpose: 'Proof of Origin (Historical Records)'. This principle has been expressly implemented in India for the GI Chanderi sari: producers who want to use the GI must have resided in Chanderi for at least 15 years.[114]

In France and at the EU level too, proof of historically verifiable human factors or pre-existing know-how is required to objectivise the link to territory.[115] Such criteria are provided in the new French GI regime for non-agricultural goods 'the specification . . . gives details about the quality, the reputation, *the traditional know-how* or other characteristics'.[116] As a result, GI goods linked to origin through mainly human factors such as handicraft goods or foodstuffs cannot be disqualified *a priori*, when know-how is shared by a community and has existed for a certain period of time, as part of the local culture and tradition. It may be worth considering a level of sophistication or refinement as an additional requirement, which would then impart real 'uniqueness' to the product and compensate for the absence of natural factors.

For GIs using raw materials which are not exclusively sourced from the area demarcated by the Geographical Indication, any risk of consumer confusion on the origin of the raw materials should be avoided. The mandatory mention of this 'external' source in the labelling of the processed good is recommended.[117]

Besides, the link to origin of products characterised by specific designs and resulting from creative know-how seems rather fragile, and subject to the risk of changing patterns, following fast-changing fashion in the textiles domain.[118] However, for traditional designs, GIs are a more desirable tool of protection than the mere protection of design and models, since

---

Legal Tools for the Protection of Geographical Indications in Asia (Geneva: ICTSD, 2007) 35–38.

[114] Rules of the Chanderi Foundation; interview with the GI applicant, on file with the author.

[115] L Bérard and P Marchenay, *Les produits de terroir, entre cultures et règlements* (Paris: CNRS Edition, 2004).

[116] Art. L. 721-7 of the Code of Intellectual Property.

[117] Whether to require such a clarification has recently been considered in the EU even for agricultural PGIs. See D Gangjee, 'Proving Provenance? Geographical Indications Certification and Its Ambiguities' (2015) World Development [forthcoming].

[118] Statement of A Mohamed Jamuluddin regarding Kancheepuram Silk GI, on file with the author.

Table 11.2 Scope of protection depending upon the nature of the link

	Human factors	Natural factors
STRONG LINK	X	X
WEAK LINK	X	

GIs are protected without time limitation.[119] GIs are also much more relevant when it is the protection of a product's name and reputation that is sought. For example, the Pochampally designation is famous for its Ikat, and it seems very unlikely at present that the weavers of Pochampally will decide to weave other motifs which are typical of Orissa: they will not abandon their own motifs and above all their name, because the Pochampally appellation is as well known as the Orissa appellation.

5.1.3 A bifurcation according to the strength of the link

Second, the question remains of whether a link to origin solely via human factors should be governed by the same legal regime as the rules applying to the more robust link to *terroir*, which involves a combination of natural and human factors. It is argued that the concept of *terroir* shall not be applied to GIs where soil and nature are not cumulatively influencing the quality of the product. In such cases, the more restricted concept of origin is arguably more appropriate. Here, given the absence of natural factors, the link to origin is undoubtedly weaker. Maintaining the principle of two legal options is proposed, as is the case in Europe with the two categories of PDO and PGI, based on the criteria of human and natural factors – whether individually or combined – which remain relevant irrespective of the nature of the product (see Table 11.2).

5.2 The Implications in France, for the EU and in International Treaties

5.2.1 France

This chapter shows the need for a consistent and uniform legal regime for GIs regardless of the type of goods, with the criteria of validity of GIs being assessed by the existence of natural and/or human factors. Nevertheless, recent developments in France cast doubts upon such a recommendation. The French Law of 2014 recognising GIs for non-agricultural goods maintains a product-based approach at the institutional level. GIs for non-agricultural goods will be scrutinised and approved by the National

---

[119] Subodh Kumar, CII Personal interview, April 2010.

Institute of Intellectual Property (INPI), alongside other intellectual property rights, whereas PDOs and PGIs for agricultural goods and foodstuff continue to be scrutinised and regulated by a dedicated body, the National Institute of Quality and Origin (INAO) under the aegis of the Ministry of Agriculture.

When making the decision to approve or reject a GI specification application, the INPI conducts an open, public inquiry, and also consults the national organizations that represent the enterprises, craftsmen and professional organizations which make up the conformity assessment of industrial products and handicrafts bodies.[120] But if there has been no response to the application within a period of two months, opinions submitted are deemed favourable.[121] Consultation of the INAO in the process of registering GIs for non-agricultural goods is provided for by the law, but only in cases of a risk of confusion with a PDO/PGI for agricultural goods.[122] There is no requirement for a general collaboration between the two institutions.

It is thus questionable whether the expertise of assessing the link with the origin will be sufficient at the INPI. Indeed, it is not clear if any substantive examination will be conducted by the INPI besides establishing a synthesis of public inquiry and consultation of interested stakeholders, which contrasts with PDO/PGIs for agricultural goods. Indeed, for assessing the validity of agricultural goods and foodstuff PDO/PGIs, INAO has 18 local branches working with producers and decision-making bodies, as well as five national committees in charge of the approval of labels of quality and PDO/PGIs.[123] All the national committees bring together representatives of producers/processors engaged in the relevant products and benefiting from PDO/PGI (or other quality labels), who account for at least 50 per cent of the members of the committee, qualified experts and members of Ministries and governmental agencies. INAO is also responsible for checking the representative nature of the applicant and the plan of inspection/control prepared by the certification body together with the producers.

Such a dichotomy in the recognition of GIs according to the kind of good is all the more regrettable if we consider that in 2006 INAO expanded its mandate to include all voluntary quality labels in agriculture

---

[120] Décret 2015-595 du 2 juin 2015 relatif aux indications géographiques protégeant les produits industriels et artisanaux et portant diverses dispositions relatives aux marques; JORF, 3 June 2015.
[121] Art. L 721-9(3) of the Intellectual Property Code.
[122] Art. L 721-9(3) of the Intellectual Property Code.
[123] See Article R642-6 of the Rural Code; see also http://www.inao.gouv.fr/.

and foodstuffs such as organic farming, guaranteed traditional specialties, Label Rouge, with the risk of diluting INAO's specific skills for appellation of origin and Geographical Indication that have been acquired progressively throughout its long history. INAO could have profited from its unique skills concerning the link to the origin by extending its activities to GIs for non-agricultural products, transcending the Ministries of Agriculture and Industry, and thus promoting the consistency and visibility of GIs.

Not only did the French government choose a sectorial policy instead of a global GI policy, but in so doing, it only provided for the registration of a GI and not of an appellation of origin. By way of a clarification, recognition and protection as an appellation of origin is already available for non-agricultural goods, but only according to the judicial procedure provided by the Law of 1919, which is far more complicated in terms of implementation. Being non-registration based and dependent upon judicial recognition, it does not extend to the drafting of a specification, organising producers into a collective body or to enforcing controls. With one of the first GIs ready to be protected, the 'Pierre de Bourgogne', a stone extracted from Bourgogne whose soil is distinctive for winemaking, qualifying for an appellation of origin, it is unfortunate not to have created the possibility of registering appellations of origin alongside Geographical Indications. If France does not defend the concept of the appellation of origin it invented, it might jeopardise the ongoing relevance of the appellation of origin at the EU level, where it may be overtaken by subsequent development.

5.2.2 Europe and other national frameworks

The possibility of extending GI protection to non-agricultural goods was identified by the European Commission in its 2011 Communication on a single market for intellectual property rights,[124] which proposed a review of the existing legal framework for GI protection of non-agricultural products across the EU, and its implications for the internal market. Member States offer different legal systems for the protection of non-agricultural products: for instance, through unfair competition or consumer protection law, or via collective or certification marks. The Commission considered that this fragmentation of the legal framework to grant protection for GIs for non-agricultural products may negatively affect the functioning of the internal market and trade negotiations with third countries.

---

[124] European Commission 'A Single Market for Intellectual Property Rights' COM(2011) 287 final (24.05.2011).

The ensuing Study on Geographical Indications protection for non-agricultural products in 2013[125] recommended a fourth legal instrument for non-agricultural goods, besides those for agricultural products/foodstuffs, wines and spirits. The Commission eventually decided to pursue its analytical work through a Green Paper published in July 2014.[126] It aims to consult with all stakeholders in the broadest possible manner on whether there is a need in the EU to increase GI protection for non-agricultural products, and if so what approach should be taken. Recently 136 responses were collected originating from 18 Member States.[127] In October 2015, the European Parliament adopted a resolution on the possible extension of GI protection of the European Union to non-agricultural products, in response to the Commission Green Paper.[128]

The Commission was open to exploring a possible supervisory, operational and administrative role by the Office for Harmonization in the Internal Market (OHIM). The Study is of the view that it would probably be difficult for a department of the European Commission, such as DG Internal Market to do it. Indeed, it would be required to have new human and financial resources, which can be difficult to secure. The Study considers that an office could be better positioned to manage such a system of GIs and, as the main EU-wide intellectual property office, the OHIM could be considered as a suitable candidate in that respect, having legal, administrative and financial independence. An important point to bear in mind is that the establishment of a new EU-wide system will not be free of costs. Moreover, as conflicts or potential conflicts between GIs and other IP rights often involved trademarks, there could be an interest in having the trade mark register and the GI register managed by the same office. The EU parliament supports such a proposal. Yet the Study reports the disadvantages of the OHIM, which lacks specific expertise when it comes to Geographical Indications.[129] Indeed, the examination of the link to the region of origin is completely different from the examination of the grounds of validity of trademarks, such as distinctiveness.

---

[125] *EU Study* (n 15).

[126] European Commission, 'Making the Most Out of Europe's Traditional Know-How: A Possible Extension of Geographical Indication Protection of the European Union to Non-Agricultural Products' COM(2014) 469 final (15.07.2014).

[127] M Koenig, S Czerska and V Marie d'Avigneau, *Public Consultations on Geographical Indication Protection for Non-agricultural Products in the EU: Summary of Results* (Industrial Property Unit J1, DG Growth, 19 Jan 2015).

[128] 2015/2053(INI), 'Possible extension of geographical indication protection of the European Union to non-agricultural products'.

[129] *EU Study* (n 15) 318.

Regarding the need to maintain two levels of geographical origin, the Study recommends a single definition for a geographical designation, and, therefore, when considering which GI definition(s) should be used, the options on offer did not include both PDOs and PGIs, which we think biased the answers.[130] The Study also considers that a double definition approach could pose problems in its implementation. For instance, the regulating body would have to verify the strength of the link between the product and its geographical area to define whether it can be registered as a PDO or a PGI. Addressing precisely this issue, the recommendations provided in the present chapter would assist in examining the strength of the link with the origin by using the criteria of natural and/or human factors, even if 'each protected product is to be seen as an original with its own history, composition and specific quality'.[131] Indeed, at present, it remains unclear on what grounds the distinction between PDO and PGI is based in the present EU regime under Regulation 1151/2012.[132]

The notion of two distinct types of geographical designations is supported by the majority of producers in Europe,[133] who have reaffirmed their commitment to maintaining this distinction, whereas the EU Commission had proposed merging the two categories of GIs when drafting Regulation 1151/2012,[134] following a report from the European Court of Auditors, 2011,[135] which considered that two different links might confuse consumers. The preference for two types of geographical designation continues to be evident in responses by stakeholders to the Green Paper of 2014.[136] The author also wishes to emphasise that with PDOs and PGIs enjoying

---

[130] Ibid., 298.

[131] A Profeta, R Balling, R Schoene and A Wirsig, 'The Protection of Origins for Agricultural Products and Foods in Europe: Status Quo, Problems and Policy Recommendations for the Green Book' (2009) 12 JWIP 622, 632.

[132] C Geiger, D Gervais, N Olszak and V Ruzek, 'Towards a Flexible International Framework for the Protection of Geographical Indications' (2010) 1 WIPO Journal 147.

[133] Proposal for a Regulation of the European Parliament and of the Council on Agricultural Product Quality Schemes COM(2010) 733 final (10.12.2010). This also reiterates the position of maintaining the concept of the appellation of origin.

[134] See European Commission, Green Paper on Agricultural Product Quality: Product Standards, Farming Requirements and Quality Schemes COM(2008) 0641.

[135] European Court of Auditors, *Do the Design and Management of the Geographical Indications Scheme Allow it to be Effective?* (Special Report No 11, 2011).

[136] The figures being: 33.3 per cent are in favour of two links with the origin; 26.5 per cent are against and 42.2 per cent do not have an opinion.

the same level of protection at the EU level, the attractiveness of PDOs is diminishing. One way of avoiding this would be to create a meaningful difference between the two types of GIs, by providing a different level of protection in accordance with the strength of the link.

Finally, having two distinct link requirements which vary according to the strength of the connection to origin is not entirely new in the broader intellectual property context. If we look at patent law, in many countries alongside the regular inventive patent there is the utility model patent (also known as the petty patent or innovation patent), which benefits from a shorter period of protection.[137] With GIs claiming to support cultural and biological diversity, it would be regrettable not to provide for a diversity of legal references and to abandon the originality of the South European countries' vision, where the appellation of origin was born.

Looking beyond Europe, in other countries which use the concept of Denomination of Origin such as Colombia, providing two levels of geographical references would allow the protection of GIs for products which have raw material coming from outside the area and that otherwise can only be protected as trademarks under current regimes.

5.2.3 International recognition of this distinction

At the WTO, the unprincipled distinction provided by the TRIPS Agreement between wines and spirit and other goods[138] should clearly be abandoned and replaced by a distinction based on the strength of the link. Such a distinction for wines and spirits reflects a history that has been largely modified with the implementation of GI laws in non-wine producing countries and is no longer justified. The proposal to base the protection on the strength of the link to territory and not on the kind of goods sheds new light on the current WTO negotiations around the extension of additional protection, currently granted only to wines and spirits, to all products.[139]

---

[137] WIPO, 'Protecting Innovations by Utility Models' at: http://www.wipo.int/sme/en/ip_business/utility_models/utility_models.htm.

[138] F Addor and A Grazioli, 'Geographical Indications beyond Wines and Spirits: A Roadmap for a Better Protection for Geographical Indications in the WTO/TRIPs Agreement' (2002) 5 JWIP 865.

[139] GE Evans & M Blakeney, 'The Protection of Geographical Indications After Doha: Quo Vadis?' (2006) 9 Journal of International Economic Law 575; T Kongolo, 'Any New Developments with Regard to GIs Issues Debated under WTO? ' (2011) 33 EIPR 83; E Thevenod-Mottet, 'Avenir des indications géographiques dans le contexte international' (2009) 41 Revue Suisse Agric. 331. For a comprehensive review of this debate, see Chapter 7 by Handler in this volume.

Finally, the potential revitalisation of the Lisbon Agreement,[140] which for the first time introduced these criteria of human and natural factors to define appellations of origin, suggests an optimal path towards a global system for the protection of GIs. First, let's remember that the Lisbon Agreement is open to any kind of goods. Second, the Geneva Act of the Lisbon Agreement on appellations of origin and geographical indication, adopted at a diplomatic conference on 20 May 2015,[141] introduces the definition of GI as provided for in the TRIPS Agreement alongside the definition of the appellation of origin, thus confirming that a two-level approach is relevant.[142] Regarding the level of protection conferred, the new Act provides uniform protection for AOs and GIs. However, the new Act of the Lisbon Agreement still does not address the issue of how to assess the link with the place of origin for GIs and AOs, which is a shame since it is the main international reference point for using the concepts of natural and human factors. The aim of this chapter was to develop the potential for interpreting the link requirement by drawing on a combination of these two elements.

---

[140] Geiger et al. (n 132); Daniel Gervais, 'The Lisbon Agreement's Misunderstood Potential' (2009) 1 WIPO Journal 87.
[141] Geneva, 20 May 2015, PR/2015/779.
[142] For a detailed overview, see Chapter 5 by Geuze in this volume.

12. Geographical Indication protection in China
Haiyan Zheng

As one of the ancient civilizations, China has gained a reputation for its china, silk, herbal medicine and many other products. These globally well-known products are closely linked to specific climatic and geological conditions, as well as unique methods of production. However, compared to its long history and rich resources in GI products, China's Geographical Indication (GI) protection has a rather short history. It was not until the mid-1980s that China started its protection for GIs. The accession to international treaties prompted China to passively offer GI protection.[1] The beginning of Geographical Indication protection can be traced back to 1985 when China joined the Paris Convention for the Protection of Industrial Property.[2] Under the convention obligations, China started to protect indications of source and appellations of origin by way of administrative decrees. In 1987 the State Administration for Industry and Commerce (SAIC)[3] issued an administrative decree to stop a Beijing-based food company from using Danish Butter Cookies on its products.[4] In 1989 SAIC issued another administrative decree to protect Champagne from being misused as a generic term for a type of sparkling wine on Chinese markets, preserving it as an appellation of origin.[5] These probably are the first important events in China regarding GI-related

---

[1] See Min-Chiuan Wang, 'The Asian Consciousness and Interests in Geographical Indications' (2006) 96 TMR 906, 906.

[2] Paris Convention for the Protection of Industrial Property, 20 March 1883, 21 UST 1583, 828 UNTS 305 (hereafter Paris Convention).

[3] SAIC is a ministerial-level government agency which administers trade mark registration and protection, company name registration, anti-unfair competition, consumer protection and other market-related issues. The Trade Mark Office is under the umbrella of SAIC. See: http://www.saic.gov.cn

[4] See the Official Letter on the Protection of Appellations of Origin issued by the Trade Mark Office under SAIC on 29 October 1987, 《中华人民共和国商标法律法规汇编》 [*Compilation of Trade Mark Laws and Regulations of the People's Republic of China*] (1995), 311.

[5] See Notice of Stopping the Use of the Word 'Champagne' on Goods Classified as Alcoholic Drinks, Trade Mark Office, SAIC on 25 October 1989 available in the *Compilation of Laws* (n 4) 198.

administrative protection. After that, several legislative efforts were made and different government agencies have been involved in the realm of GI protection.

Currently China does not adopt a uniform approach in protecting GIs. Both trade mark and *sui generis* regimes are available for GI protection, which are complemented by laws on unfair competition, consumer protection and product quality. This chapter maps the different, often overlapping, options for GI protection and concludes with an analysis of the complexities arising from the concurrent operation of trade mark and *sui generis* models. It offers suggestions for improvements to the existing arrangement.

1. TRADE MARK REGIME

The philosophy of using trade marks to protect geographical indication is that GIs function quite similarly to trade marks. The primary purpose of the Trade Mark Law is to protect the interests of consumers by way of trade mark's source-identifying function and the quality guarantee function. Moreover, it protects the business interest of trade mark owners. Geographical indications designate geographical source, guarantee specific quality controlled by natural and human factors within the given geographical area and protect the investment made by generations of local producers on the reputation associated with the GIs. In that sense, GIs are understood as a subset of trade marks.[6] Generally speaking, a trade mark is an identifier of one single producer or service provider. A GI usually serves to identify a group of producers who share something in common, that is, the producers' products possess a certain quality, reputation or other characteristics which are essentially attributable to their geographical origin.[7] The availability of certification or collective marks makes it possible to accommodate GI protection within existing trade mark systems.[8] Both certification marks and collective marks identify groups of

---

[6] Lynne Beresford, 'Geographical Indications: The Current Landscape' (2007) 17 Fordham Intellectual Property Media & Entertainment Law Journal 979, 980–81.

[7] See Daniele Giovannucci, Elizabeth Barham and Richard Pirog, 'Defining and Marketing "Local" Foods: Geographical Indications for US Products' (2010) 13 JWIP 94, 104.

[8] See Irina Kireeva, Xiaobing Wang and Yumin Zhang, *Comprehensive Feasibility Study for Possible Negotiations on a Geographical Indications Agreement between China and the EU* (27 April 2009) 21, available at: http://www.ipkey.org/

users instead of one single business entity. The ownership of certification marks or collective marks is for applicants with a collective legal nature, usually associations of producers.[9] This corresponds to the collective nature of a GI registrant. Therefore, both certification marks and collective marks can be used in the trade mark system to protect GIs.

1.1 Legislative History

Before 1993 it was not possible to protect GIs within the trade mark system. The Implementing Rules of the Trade Mark Law of 1993 introduced provisions for the protection for certification marks and collective marks, thereby making this possible. In December 1994, based on the Trade Mark Law and its Implementing Rules, SAIC formulated and promulgated the Procedures for the Registration and Administration of Collective Marks and Certification Marks[10] which provided that certification marks could be used to certify the place of origin, raw materials, method of production, quality, accuracy or other characteristics of the said goods or services. This was the first administrative rule regarding GI protection in the national legal system.[11]

In its accession to the World Trade Organization, China made a commitment to introduce specific GI protection in its Trade Mark Law and the concept of Geographical Indication was officially accommodated in the revised Trade Mark Law of 2001.[12] This legislation elevated the legal basis for GI protection from administrative rule to national law.[13] But the

---

en/ip-law-document?filter_type=4&filter_category=0&filter_tag=0&searchword= GI+negotiations&filter_law_makingbody=0&filter_law_type=0&ordering=&is_ myresources=0&start_time1=&end_time1=

[9] In China, the government or its agencies cannot become trade mark registrants due to certain restrictions on the eligibility of ownership.

[10] Procedures for the Registration and Administration of Collective Marks and Certification Marks (issued by SAIC Dec. 30, 1994, effective 1 March 1995, repealed 1 June 2003), Chinese version available at http://vip.chinalawinfo.com/newlaw2002/SLC/slc.asp?db=chl&gid=18688

[11] Qinghu An, Speech at the International Symposium on Geographical Indications: Legal System on Geographical Indication Protection in China and Related Issues (25 June 2007) (Chinese version available at http://www.zjfw.org/xw-view.asp?bid=1&sid=2&id=504).

[12] Trade Mark Law (P.R.C.) (promulgated by the Standing Comm. Nat'l People's Cong., 23 August 1982, revised for the second time 27 October 2001, revised for the third time 30 August 2013) at Article 16 (2) [hereinafter Trade Mark Law], Chinese version available at http://sbj.saic.gov.cn/flfgl/flfg/201309/t20130903_137807.html

[13] See An (n 11).

Trade Mark Law itself does not regulate the procedures to register a GI. In 2002 the State Council amended the Implementing Rules into Regulations for the Implementation of the Trade Mark Law,[14] in which the practice of protecting GIs as certification or collective marks was further established. Art 6.1 of the Implementing Regulations stipulates, 'for geographical indications referred to in Article 16 of the Trade Mark Law, applications may be filed to register them as certification marks or collective marks under the provisions of the Trade Mark Law and these Regulations'. Definitions of collective marks and certification marks have been given in Art 3 of the Trade Mark Law: 'collective marks mean signs which are registered in the name of groups, associations or other organizations to be used by the members thereof in their commercial activities to indicate their membership of the organizations'; and 'certification marks mean signs which are controlled by organizations capable of supervising some goods or services and used by entities or individual persons outside the organization for their goods or services to certify the origin, material, mode of manufacture, quality or other characteristics of the goods or services'.

In 2003 SAIC issued the Measures for the Registration and Administration of Certification and Collective Marks[15] in response to the revised Trade Mark Law, making detailed provisions about the registration and administration of GIs. As a result, under the Trade Mark Law system, GIs can be registered and protected as collective marks or certification marks with the Trade Mark Office under SAIC. In order to facilitate the use of GI marks, SAIC issued the Measures on the Administration of Signs for Geographical Indication Products,[16] allowing the free downloading and use of the signs for GI products by qualified GI users. According to the Measures, the GI sign can be used together with the registered trade mark to indicate that the GI is registered with the Trade Mark Office and protected by Trade Mark Law.

---

[14] Regulations for the Implementation of the Trade Mark Law (promulgated by St. Council, 3 August 2002, effective 15 September 2002, revised 29 April 2014) [hereinafter Implementing Regulations], Chinese version available at http://sbj.saic.gov.cn/flfg1/flfg/201405/t20140522_145379.html

[15] Measures for the Registration and Administration of Collective Marks and Certification Marks (issued by State Administration for Industry and Commerce, 17 April 2003, effective 1 June 2003) [hereafter the SAIC Measures], Chinese version available at http://sbj.saic.gov.cn/flfg1/sbxzgz/200906/t20090603_60312.html

[16] Measures on Administration of Signs for Geographical Indication Products, Chinese version available at http://sbj.saic.gov.cn/flfg1/gfwj/200702/t20070206_54880.html

1.2 Definition of GI under the Trade Mark System

Art 16.2 of the Trade Mark Law provides that Geographic Indications 'refer to signs that identify a particular good as originating in a region, where a given quality, reputation or other characteristic of the goods is essentially attributable to its natural *or* human factors'. However, Article 7 of the SAIC Measures demands that GIs registered as collective or certification marks must demonstrate 'the relation between the given quality, reputation or any other characteristic of the goods and the natural *and* human factors of the region indicated by the geographic indication'. In practice it seems that both natural *and* human factors of the geographical region are indispensible for a product to gain the specific characteristics and the geographical term to qualify as a GI. Furthermore, the detailed provisions regarding GI registration contained in the SAIC Measures are the ones followed by the Trade Mark Office.

1.3 What can be Registered as a GI?

Under the trade mark system, a GI registered as a collective mark or a certification mark may be the name of the geographical region indicated or any other visual signs capable of indicating that a good originates from the region. The region is not required to be fully consistent with the name or boundary of the current administrative division.[17] The scope is much wider than that of the *sui generis* GIs in that the latter only allows for geographical names.

1.4 Special Registration Requirements

The general rule of ordinary trade mark registration has to be followed to register GIs as certification or collective marks. Moreover, there are additional requirements specified for GI trade mark applications. The applicant has to present the following additional evidence:

1. a document issued by the people's government which has jurisdiction over, or the competent authority of, the concerned industry approving the applicant's registration of the GI in question; for example, agricultural or fishery authorities;[18]
2. a description of the GI including: (i) the given quality, reputation or

---

[17] SAIC Measures (n 15), Art 8.
[18] Ibid., Art 6.1.

332 *Research handbook on intellectual property and GIs*

   any other characteristic of the goods indicated by the sign, (ii) the relation between the given quality, reputation or any other characteristic of the goods and the natural and human factors of the region indicated by the Geographic Indication, and (iii) the boundary of the region indicated by the GI;[19]
3. detailed information of the professionals and special testing equipment of the applicant or of any other organization authorized by the applicant to show its capability of supervising the particular quality of the goods indicated;[20]
4. the regulation governing the use of a collective mark or certification mark, including: (i) the purpose of using the mark, (ii) the quality of the goods to which the mark applies, (iii) the procedures for using the mark, (iv) the rights and obligations entailed in the use of the mark, (v) the liability the GI users shall bear for breaching the regulation, and (vi) the registrant's system for the inspection and supervision of the goods to which the mark applies.[21] The registrant of a GI imposes control over the use of the GI mainly through the implementation of this regulation.

For foreign applicants, they should appoint a trade mark agent to act for them[22] and further present documents certifying that the GI being applied for is protected in the country of origin, in addition to the documents mentioned in (2), (3) and (4) above.[23]

1.5 Special Provisions Regarding Wines and Spirits

There are special rules regarding protection on GIs for wines and spirits. Art 9 of the SAIC Measures provides that if several GIs for wines constitute homophonous or homographic names and these GIs can be distinguished from one another without misleading the public, each GI may be applied for as a collective mark or certification mark. Up to now there has been no case concerning such homophonous or homographic (that is homonymous) GIs for wines in China. The provision is formulated particularly to comply with the additional protection conferred upon GIs for wines and spirits by Art 23.3 of TRIPS.

Article 12 of the SAIC Measures further provides that one is prohibited

---

[19] Ibid., Art 7.
[20] Ibid., Arts 4, 5.
[21] Ibid., Arts 10, 11.
[22] Trade Mark Law, Art 18.2.
[23] SAIC Measures, Art 6.2.

from using another party's registered GI for wines or spirits to identify wines or spirits not originating in the place indicated by the GI in question, even if the true origin of the goods is indicated or the GI is used in translation or accompanied by expressions such as *kind, type, style, imitation* or the like.

1.6 Assignment of GIs as Certification or Collective Marks

In general, trade marks are transferable and can be sold to any other entity. A GI, by contrast, is linked to a certain geographical origin and should retain its local connection. Therefore GIs as certification or collective marks cannot be assigned as freely as ordinary trade marks. The Trade Mark Law only allows assignment of collective or certification marks subject to strict and specific preconditions. In general, the assignee of a GI should also comply with the same requirements that the original GI registrant needed to satisfy.[24] That is to say, under the Chinese trade mark legal system, the assignment or transfer of ownership of a GI certification or collective mark is strictly confined to the specified geographical region. In addition, the assignee must possess the same ability to exercise control over the quality of GI products as the assignor.

1.7 Use of GIs Registered as Certification or Collective Marks

There are concerns about whether small producers could be excluded from enjoying the benefits of GIs because the right holder may prefer large-scale producers which have standardized production processes and are easier to control. This does not constitute a problem under the trade mark system. By providing that anyone within the specified geographical area who satisfies the prescribed standard could ask for permission to use the GI and the owner could not refuse it, the trade mark system makes it possible that even small producers can share the benefits from GI protection.[25] If a qualified product meets the standards set by the owner of the certification mark, the producer must be permitted to use the mark

---

[24] Ibid., Art 16 (providing that when an application is filed for the assignment of a collective mark or a certification mark, the assignee shall have the relevant qualifications as a legal subject and shall comply with the provisions of the Trade Mark Law, the Implementing Regulations and these Measures. Where a transfer of a collective mark or a certification mark occurs, the transferee of the right shall have the relevant qualification as a subject and comply with the provisions of the Trade Mark Law, the Implementing Regulations and these Measures.).

[25] Implementing Regulations (n 14), Art 4.2.

fairly. The owner of the mark must exercise legitimate control over the use but may not discriminate against a producer who actually meets the standards. The requirement that owners of certification marks are not allowed to use the mark themselves in commercial or industrial activities warrants the impartiality.[26] Therefore, collective use is open to all producers in the specified region who comply with the rules or specifications. Likewise, anyone whose goods satisfy the conditions under which the GI is used may request membership from the collective mark registrant and the registrant must accede to this request in accordance with its articles of association. For those who do not request membership, fair use of the geographic name of the said GI must be allowed to describe the origin of their products.[27] This is contradictory to the Trade Mark Law, which states that the collective marks are provided to members of the trade mark owning group to indicate membership in the organization,[28] yet the current Implementing Regulation says that one needs not be a member of the registered collective to use a collective mark. This contradiction is the consequence of a drafting mistake in the Implementing Regulation. The 'GI' indicated here, which may be legitimately used by those who do not request membership of the trade mark owner, should read as 'geographical name'. That is why the SAIC made a correction in its issued Measures for the Registration and Administration of Collective Marks and Certification Marks.[29] If revised accordingly in the Implementing Regulation, it would constitute a fair use exemption of a geographical name under the trade mark system.

---

[26] A provision similar to US practice. See D Giovannucci, T Josling, W Kerr, B O'Connor and M Yeung, *Guide to Geographical Indications: Linking Products and their Origins* (International Trade Centre 2009) 66; see also the SAIC Measures (n 15), Art 20 (the registrant of a certification mark shall not use the certification mark on goods provided by himself or itself).

[27] Article 4.2 of the Implementing Regulations provides for fair use of the GI, and Article 18 of the SAIC Measures modifies it to be fair use of the geographical name.

[28] Trade Mark Law, Art 3(2) (providing the definition of collective marks as signs which are registered in the name of bodies, associations or other organizations to be used by the members thereof in their commercial activities to indicate their membership of the organizations).

[29] SAIC Measures (n 15), Art 18(2) (the fair use of a GI under Article 4.2 of the Implementing Regulations shall refer to the fair use of the geographic name of the said GI).

1.8 Control and Supervision Over the Use of GIs

The control of the use of GIs as certification or collective marks is exercised by the registrant itself according to the registrant's system for the control and supervision of the goods to which the collective or certification mark applies. The control and supervision system is clarified in the regulation governing the use of the said mark as an indispensable part of the application documents.[30] If the registrant of a collective mark or a certification mark fails to exercise effective governing of, or control over, the use of the mark and, as a result, the goods to which the said mark applies fail to meet the requirements of the regulation governing the use of the mark, causing damages to consumers, the administrative authority for industry and commerce can order it to rectify the situation within a time limit. If the registrant refuses to do so, a fine will be imposed.[31]

1.9 GI Protection and Enforcement

Protection of GIs as certification or collective marks follows the general rule of ordinary marks. Art 3.1 of the Trade Mark Law provides that 'the trade mark registrants shall enjoy the exclusive right to use the trade marks, and be protected by law'. In accordance with the general prescriptions, any of the following acts can constitute an infringement of the exclusive right to use a registered GI:[32]

- to use a mark that is identical with a registered GI in respect of the same goods without authorization of the proprietor of the registered GI;
- to use a mark similar to a registered GI in respect of the same goods or to use a mark identical with or similar to a registered GI in respect of similar goods, without authorization of the proprietor of the registered GI, where such use is likely to cause confusion;
- to sell goods that one knows bear a counterfeited registered GI;
- to counterfeit, or to make, without authorization, representations of a registered GI, or to sell such representations;
- to replace, without the consent of the GI registrant, its registered GI, and re-market the goods bearing the replaced GI;

---

[30] Ibid., Arts 10, 11.
[31] Ibid., Art 21.
[32] Trade Mark Law, Art 57.

336 *Research handbook on intellectual property and GIs*

- to intentionally provide a person with assistance for his infringement of a registered GI or facilitate such person's infringement of a registered GI; or
- to cause, in other respects, prejudice to the exclusive right of another person to use a registered GI.

In China there is a twin-track system to enforce one's exclusive trade mark rights. The right holder may either institute legal proceedings in the people's court or request the administrative authorities for industry and commerce (AICs) to take action. The AICs are empowered by the Trade Mark Law to investigate and handle trade mark infringement cases. They can order the infringer to immediately stop the infringing act, confiscate and destroy the infringing goods and tools specially used for their manufacture and impose a fine for counterfeiting the registered trade mark.[33] If the case is so serious as to constitute a crime, it shall be transferred by the AICs to the judicial authority for determination.[34] Alternatively, the interested party may directly bring a lawsuit to the people's court for a trade mark infringement case.

As of 30 June 2015, 2,790 GIs were registered as certification or collective marks with the Trade Mark Office under SAIC, among which 83 were foreign GIs. Parma Ham, Scotch Whisky, Thai Silk, Idaho Potatoes and Jamaica Blue Mountain are such examples.[35]

Meanwhile there are two independent *sui generis* systems available in China, implementing ministerial rules on GIs by the General Administration of Quality Supervision, Inspection and Quarantine (AQSIQ)[36] and the Ministry of Agriculture (MOA) respectively. Under a *sui generis* system, legal recognition and protection is of its own type or 'unique', that is specifically dedicated to this type of intellectual property. With their close links to the specific geographical area, GIs belong to the region itself and not to individual producers located therein. The collective nature of GIs makes public power deeply involved in their protection. Under a *sui generis* system, the rights to define a GI, to determine the eligible users and the enforcement of regulations are within the scope of public power. Governments intervene in terms of control and supervision on the

---

[33] Trade Mark Law, Art 53.
[34] Trade Mark Law, Art 61.
[35] The statistics are available at http://sbj.saic.gov.cn/dlbz/xwbd/201507/t20150723_159352.html
[36] Due to institutional reconstruction, AQSIQ was created in 2001 to incorporate the former State Bureau of Quality and Technical Supervision and the former State Administration for Entry-Exit Inspection and Quarantine.

quality and specific characteristics of the products marketed under GIs.[37] *Sui generis* protection puts more emphasis on the quality of GI products which requires strict controls over their production processes.

2. SUI GENERIS REGIME: AQSIQ PRACTICE

The AQSIQ rules have been heavily influenced by the AOC system of France. The former State Bureau of Quality and Technical Supervision, in close cooperation with the French Ministry of Agriculture, the Ministry of Finance and the Bureau National Interprofessionnel Du Cognac, promulgated the Provisions on Protection of Designations of Origin Products in 1999. This is the first administrative regime specifying protection for designations of origin.[38] After that the former State Administration for Entry-Exit Inspection and Quarantine promulgated the Provisions on the Administration of Marks of Origin in 2001. Following the incorporation of the two agencies into AQSIQ, a newly created government agency in 2001, AQSIQ promulgated the Provisions on the Protection of Geographical Indication Products (PPGIP)[39] in 2005, which replaced the above-mentioned two rules. All the rules mentioned above are administrative rules.

2.1 Scope of GI Products

According to PPGIP, geographical indication products are 'products designated by a geographical name after being examined and approved, which originate from a specific region, and whose quality, reputation or other characteristics are due to natural factors and human factors of the region of production'.[40] Products of GIs include: (i) those planted or cultivated in the said region; and (ii) those made, wholly or partially, of the raw materials from the region and produced or processed with the particular techniques in the region.[41] The coverage of GI products under PPGIP is

---

[37] See Kireeva, Wang and Zhang (n8), 16.
[38] Jing Dai, 《试论我国地理标志产品保护制度》 [On the Protection System of Geographical Indication Products in China] (2009) 3 Modern Business 277, 277.
[39] Provisions on the Protection of Geographical Indication Products (issued by General Administration of Quality Supervision, Inspection and Quarantine 16 May 2005, effective 15 July 2005); Chinese version available at http://www.gov.cn/gongbao/content/2006/content_292138.htm
[40] PPGIP, Art 2.1.
[41] PPGIP, Art 2.2.

rather wide, including agricultural products, handicraft or spirits, similar to those covered under the trade mark regime.

2.2 Registration Procedures

As a starting point there is a two-tier process for applicants to apply for registration of the GI products: first with provincial quality and inspection departments, then with the AQSIQ. After registration the producers intending to use the GIs for their products also have to go through a similar two-tier process to get the approval to use the GI.

2.2.1 Registration by the applicant

The applicants can either be entities designated by local governments, or enterprises or associations accredited by local governments.[42] For GI products of exporting enterprises, applications should be made to entry-exit inspection and quarantine departments of the prescribed area; other GI products should be applied for at the local (that is at or above county level) quality supervision departments.[43] According to PPGIP, the applicants should submit the following documents:[44]

1. the demarcation of the boundary of the GI products approved by competent local government;
2. the document approving the applicant's registration of the GI product by the competent local government;
3. evidence proving the qualifications of a GI product, including name and classification of the product; production area and geographical features; physical, chemical and organoleptic features, as well as their links to the natural and human factors; technical standards (including processing, safety and hygiene requirements, technical requirement for equipment); reputation of the products; production, sales of the product; and historical delineation of the product.
4. the technical standards for the GI product.

2.2.2 Examination and approval

The first level examination is conducted by provincial quality supervision departments or entry-exit inspection and quarantine departments. They draw up preliminary opinions on the application and then submit their

---

[42] PPGIP, Art 8.
[43] PPGIP, Art 11.
[44] PPGIP, Art 10.

report and application documents to AQSIQ. At the second level, AQSIQ first conducts formal checks on the application and will then publish a notice of acceptance in the AQSIQ Gazette as well as on its website if the application satisfies the formality requirement. If the application fails to meet the formality requirements, the AQSIQ will notify the applicant in writing.[45] Anyone who objects to the registration can file an opposition within two months after publication.[46] For an application without opposition or where the opposition is unsuccessful, the AQSIQ sets up an expert examination panel according to the features of the products in question. The expert panel then conducts a technical examination on the application and the AQSIQ will publish its approval of the application if it passes the technical review by the expert panel.[47]

2.2.3 Application by the producer to use a GI product

As a first step any producer within the geographical limits of the origin region who wants to use the GI sign on its product has to file an application before the local quality supervision department or entry-exit inspection and quarantine departments. The following documents should be submitted:[48]

1. an application to use the special sign on GI products;
2. a certificate issued by the competent authorities of the local government proving that the products concerned originate from the particular region; and
3. an inspection report issued by the relevant quality supervision department.

The successful application examined by provincial quality supervision department or entry-exit inspection and quarantine departments will then be subject to review by the AQSIQ as a second step. After the AQSIQ approves its application and publishes it in the AQSIQ Gazette, the producer will be eligible to use the sign in question.[49] This application process by users of GI products is quite similar to the registration system.[50]

---

[45] PPGIP, Art 13.
[46] PPGIP, Art 14.
[47] PPGIP, Arts 15, 16.
[48] PPGIP, Art 20.1.
[49] PPGIP, Art 20.2.
[50] See BM Bashaw, 'Geographical Indications in China: Why Protect GIs with Both Trade Mark Law and AOC-type Legislation?' (2008) 17 Pacific Rim Law & Policy Journal 73, 84.

2.3 Control and Supervision

Local quality inspection authorities exert routine control on the quality of the GI products and do so in a very detailed way, involving almost every aspect of production. This includes raw materials, production techniques, quality features, classifications of quality, quantity, packaging and labelling of GI products, as well as the printing, distribution, quantity and use of the special signs of the product, manufacturing environment, production equipment and conformity with standards of the product.[51]

2.4 Protection and Enforcement

The approved GI products are protected in accordance with PPGIP. There are three types of acts that can be categorized as infringing the legitimate rights of registrants:[52] (i) use without authorization or forging a GI and its specific marks; (ii) using names of GI products when they do not meet the quality standards or regulations, which involves unauthorized use by producers within the protected regions who cannot obtain approval because their products fail the requirements; and (iii) use of signs that are so similar to the protected signs that consumers will be misled into believing the products are protected GI products.

According to PPGIP, quality supervision and entry-exit inspection and quarantine departments are responsible for investigating the above-mentioned acts. Similar to the trade mark regime, interested parties can either lodge complaints with local quality supervision departments or bring lawsuits to the people's court. The quality supervision departments rely on the Law of the People's Republic of China on Product Quality, the Standardization Law of the People's Republic of China and the Law of the People's Republic of China on the Inspection of Import and Export Commodities to impose administrative penalties in dealing with cases of GI products.[53]

2.5 Protection of Foreign GI Products

PPGIP provides that separate provisions are to be formulated for the registration of foreign GIs.[54] Yet up to now there is no such provision

---

[51] PPGIP, Art 22.
[52] PPGIP, Art 21.
[53] PPGIP, Art 24.
[54] PPGIP, Art 26.

available. In 2007 the EU initiated a '10 plus 10' pilot project with AQSIQ, under which both sides presented a list of ten agricultural GIs respectively to seek protection in the each other's territories.[55] In addition, AQSIQ accepted an application from French Cognac in June 2009. And in December 2009 AQSIQ approved protection for Cognac pursuant to the Memorandum of Understanding on Geographical Indications signed by AQSIQ of China and the European Commission DG Trade, with reference to the PPGIP. It is the first foreign GI product protected by AQSIQ.[56]

3. SUI GENERIS REGIME: MOA PRACTICE

In addition to AQSIQ, the Ministry of Agriculture (MOA) also promulgated a set of administrative rules, namely the Measures for the Administration of Geographical Indications of Agricultural Products[57] in 2007 according to the Agriculture Law of the People's Republic of China and Law of the People's Republic of China on Agricultural Product Quality Safety. The MOA Measures entered into force in February 2008.

3.1 Scope of Protected Products

As defined in Article 2 of the MOA Measures, 'agricultural products' refer to primary agricultural products, including plants, animals, microorganisms and the processed products thereof obtained in agricultural activities.[58] Among the three types of protection regimes, the scope of

---

[55] The EU list comprises: Grana Padano; Prosciutto di Parma; Roquefort; Pruneaux d'Agen/Pruneaux d'Agen mi-cuits; Priego de Córdoba; Sierra Mágina; Comté; White Stilton Cheese/Blue Stilton Cheese; Scottish Farmed Salmon and West Country Farmhouse Cheddar. The Chinese AQSIQ list comprises: Dongshan Bai Lu Sun (asparagus), Guanxi Mi You (honey pomelo), Jinxiang Da Suan (garlic), Lixian Ma Shan Yao (yam), Longjing cha (tea), Pinggu Da Tao (peach), Shaanxi ping guo (apple), Yancheng Long Xia (crayfish), Zhenjiang Xiang Cu (vinegar) and Longkou Fen Si (vermicelli). See http://eeas.europa.eu/delegations/china/press_corner/all_news/news/2011/20110513_en.htm

[56] See Notice on Protection for Products of Cognac Geographical Indication issued by AQSIQ (No. 117 2009), available at http://kjs.aqsiq.gov.cn/dlbzcpbhwz/ggcx/201001/t20100106_134265.htm

[57] Measures for the Administration of Geographical Indications of Agricultural Product (issued by Ministry of Agriculture 25 December 2007, effective 1 February 2008) [hereinafter MOA Measures], Chinese version available at http://www.gov.cn/gongbao/content/2008/content_1071853.htm

[58] MOA Measures, Art 2.

protected products under MOA Measures is the narrowest, only covering agricultural products.

3.2 Registration Procedures

The MOA Measures is quite similar to PPGIP in that the registration procedures also involve a two-level process – provincial and national – for an applicant to obtain registration.

3.2.1 Registration by the applicant

Under the MOA Measures, agricultural product GIs are regarded as a collective right, so individuals or enterprises are not eligible for application. The eligible applicants should be professional cooperative organizations of farmers and industrial associations determined by governments at or above the county level. They should have the following capabilities: (i) to supervise and administer both GI labelling and the underlying products themselves; (ii) to provide guidance or service for the production, processing and marketing of GI agricultural products; and (iii) to independently bear civil liability.[59]

The applicant should prove that the following conditions have been met: (i) the name is composed of the name of the geographical area plus the generic name of the said agricultural product; (ii) the product has the requisite unique quality or special mode of production; (iii) the quality and characteristics of the product mainly rely on the unique natural and ecological environment as well as cultural and historical factors; (iv) the product has a specified production area; (v) the environment of the producing area and the product quality meet national compulsory technical standards.[60]

3.2.2 Examination and approval

Applicants may apply to provincial agricultural authorities and file the following documents: (i) a registration application form; (ii) a certificate on the qualifications of the applicant; (iii) a description of the unique characteristics of the product and the corresponding product quality appraisal reports; (iv) statements of the environment conditions of the producing area, technical standards for production and technical norms for product quality safety; (v) a document determining the territorial scope and distribution map of production area; (vi) a straight sample or a

---

[59] MOA Measures, Art 8.
[60] MOA Measures, Art 7.

sample picture of the product; and (vii) other necessary descriptive or evidentiary material.[61] After receiving the application, provincial agricultural authorities will conduct on-site verification and propose their preliminary examination opinion. For those that meet the conditions, they will send the filing documents and preliminary opinion to the Centre for Agri-food Quality and Safety (the Centre), operated under the MOA. For those that do not, they will notify the applicant of their opinion.[62] Within 20 working days after receiving the documents, the Centre examines the application and organizes expert examination. The expert committee should undertake the appraisal of the registration of GIs of agricultural products, work out appraisal conclusions independently and be responsible for the conclusions.[63] If the expert committee is in favour of the application after appraisal, the Centre will publish an announcement approving the application on behalf of the Ministry of Agriculture. Anyone who has an objection to the approval can file their opposition within 20 days with the Centre. For those without any objection, the Ministry of Agriculture makes an announcement and issues a Certificate of People's Republic of China on the Registration of Geographical Indications of Agricultural Products and publishes relevant technical regulations and standards. If the expert committee doesn't approve, the Ministry of Agriculture will make a decision not to register it and notify the applicant of the decision in writing.[64]

3.2.3 Application to use an agricultural product GI

Any user, that is, producer who satisfies the following conditions can apply to the registration certificate holder for using the GI on suitable agricultural products: (i) the agricultural product produced or traded by it originates from within the territory indicated in the registration certificate; (ii) the user has obtained the corresponding qualification for producing or trading the agricultural product concerned; (iii) the user is capable of producing or trading in strict accordance with the prescribed quality and technical norms; and (iv) it has the capacity for market development and production of the agricultural product concerned.[65]

---

[61] MOA Measures, Art 9.
[62] MOA Measures, Art 10.
[63] MOA Measures, Art 11.
[64] MOA Measures, Art 12.
[65] MOA Measures, Art 15.

3.3 Protection Term

Unlike the GIs registered as collective marks or certification marks, which have to be renewed every ten years, the MOA registration of a GI for an agricultural product will remain valid permanently.[66]

3.4 Control and Supervision

The MOA *sui generis* regime also emphasizes administrative supervision and control over the quality and source of products. Competent local agricultural authorities are responsible for conducting regular inspections and administering the use of GI signs, as well as evaluating the boundary requirement of geographical origin.[67] The producers of agricultural products with GIs also shoulder some responsibility by establishing a quality control tracing system.[68]

3.5 Protection and Enforcement

Forgery and use without authorization of GIs for agricultural products, or falsely claiming any registration certificates, are considered to violate the MOA Measures.[69] Like the PPGIP there is no direct provision in the MOA Measures concerning any administrative penalty. Rather, administrative punishment will be imposed according to the Law on Agricultural Product Quality Safety.[70]

3.6 Protection of Foreign GIs for Agricultural Products

Article 24 of the MOA Measures provides that the 'Ministry of Agriculture accepts applications for the registration of geographical indications of agricultural products from foreign countries, and protects them once they have been registered in China'. However, the specific measures on the application and registration of foreign GIs for agricultural products are yet to be formulated. As no such specific measures have been promulgated, no foreign GIs of agricultural products have been registered under the MOA regime so far.[71]

---

[66] MOA Measures, Art 13.
[67] MOA Measures, Art 18.
[68] MOA Measures, Art 19.
[69] MOA Measures, Art 20.
[70] MOA Measures, Art 23.
[71] See Kireeva, Wang and Zhang (n 8), 115.

4. GI PROTECTION UNDER OTHER LAWS

The Anti-Unfair Competition Law,[72] Product Quality Law[73] and Law on Protection of Consumer Rights and Interests[74] are enacted for the purposes of protecting producers and consumers. They only stipulate general rules, but can serve the purposes of GI protection.[75]

4.1 Unfair Competition

Under the Anti-Unfair Competition Law, falsely indicating the place of origin of commodities is categorized as an unfair competitive activity.[76] This law also prohibits business operators from using any false advertising or other means of false publicity in business activities regarding the origin of products.[77] It is obvious that the Law regulates place of origin from the perspective of consumer and producer protection, not from that of GI protection as such.[78]

4.2 Consumer Protection

The Law on Protection of Consumer Rights and Interests stipulates that consumers have the right to obtain genuine information on commodities or services, including information on place of origin, and business operators have the obligation to provide this information.[79] Thus, providing false information on place of origin constitutes an offence under the Law.

---

[72] Anti Unfair Competition Law (PRC) (promulgated by the Standing Committee, National People's Congress, 2 September 1993, effective 1 December 1993), English version available at http://www.lawinfochina.com/law/display.asp?id=648

[73] Product Quality Law (PRC) (promulgated by the Standing Committee, National People's Congress, 22 February 1993, effective 1 September 1993, revised for the first time 8 July 2000, revised for the second time 27 August 2009), English version available at www.lawinfochina.com/law/display.asp?id=615

[74] Law on Protection of Consumer Rights and Interests (PRC) (promulgated by Standing Comm. Nat'l People's Cong., 31 October 1993, effective 1 January 1994, revised 25 October 2013), Chinese version available at http://www.lawtime.cn/faguizt/117.html

[75] See Bashaw (n 50), 86.

[76] Anti Unfair Competition Law, Art 5(4).

[77] Ibid., Art 9.

[78] See Tian Furong,《地理标志法律保护制度研究》[*Study on the Legal System for Protection of Geographical Indications*] (2009) 272.

[79] Law on Protection of Consumer Rights and Interests, Arts 8, 20.

4.3 Product Quality Protection

The Product Quality Law forbids the inaccurate use of the place of origin of products.[80] However, the products mentioned in the Law refer to products processed and manufactured for the purpose of marketing.[81] A great number of GI products, namely primary agricultural products such as vegetables and fruits, are excluded from protection under this Law. Therefore, the scope of protection accorded by Product Quality Law to geographical indication is rather inadequate.[82]

In conclusion, there are three parallel ways available specifically addressing GI protection in China's legal system – within trade mark law, under the PPGIP regimes administered by AQSIQ and the MOA regime. Each of them is administered by different governmental agencies, with a distinct legal basis. There are also more general legal regimes (that is Unfair Competition, Consumer Protection and Product Quality regulation) targeting misleading conduct in the marketplace, which can also protect GIs.

5. MAJOR CHALLENGES FACED BY GI PROTECTION IN CHINA

GI protection in China faces many challenges. However, the most controversial issue is the concurrent operation of trade mark and *sui generis* models, and this is made more complicated by the fact that there are two parallel *sui generis* models operated by AQSIQ and the Ministry of Agriculture. As previously noted, trade mark protection falls within the competence of SAIC, while protection of GI products in general and protection of agricultural products in particular come under the administration of AQSIQ and the Ministry of Agriculture respectively. Since there are three independent and parallel systems of GI protection in China, the same GIs may be simultaneously protected, potentially obtaining three independent kinds of protection. Meanwhile, the possibility of registering place names qualified to be GIs as ordinary trade marks causes conflicts between such individual trade marks and GIs-as-trade marks within the trade mark regime. Therefore, there are two principal challenges: one is the conflict between ordinary trade marks containing geographical terms

---

[80] Product Quality Law, Art 5.
[81] Ibid., Art 2(1).
[82] See Tian Furong (n 78), 273.

and geographical indications by means of certification or collective marks; and the other is the overlap and ensuing conflict arising from the coexistence of trade mark and *sui generis* mechanisms.[83]

5.1 Conflicts between Trade Marks and GIs within the Trade Mark Regime

5.1.1 Causes of conflicts

In general, geographical terms which are descriptive cannot be registered as trade marks on the ground of lack of distinctiveness, that is, they are unable to distinguish the goods or services of one undertaking from those of others, a fundamental function of trade marks.[84] Meanwhile, this also prevents the possibility of the monopolization of geographical terms by a single entity.[85] Moreover, if the goods or services are not offered within the designated region, the geographically descriptive term could be misleading to consumers.

However, the Trade Mark Law before 1993 allowed registration of geographical names (even administrative units at or above the county level) as ordinary trade marks. Even misleading trade marks indicating a false place of origin continue to be valid if they were previously registered in good faith.[86] The 金华(JINHUA) trade mark for ham is one such example. The proprietor of JINHUA for hams is not established in Jinhua City. Rather it is located in another city within the same province, so it actually indicates a false origin. But it remains valid according to the law because it was a bona fide registration before 1993. Before China established its GI protection regime within the trade mark system, a huge number of trade marks containing geographical terms with GI significance were registered prior to 1993. Some of them have even become well-known marks thanks to the lasting and tremendous investment made by the owner. Furthermore the current Trade Mark Law only regulates the registration of geographical terms consisting of administrative regions as

---

[83] See Zhao Xiaoping,《地理标志的法律保护研究》[*Study on Legal Protection of Geographical Indications*] (2007) 285.

[84] WIPO, *Making a Mark: An Introduction to Trademarks for Small and Medium-sized Enterprises* (WIPO Publication No. 900, 2006) 4.

[85] See Tian Furong (n 78) 248.

[86] Bashaw (n 50) 79; see also Trade Mark Law, Art 16 (1) (where a trade mark contains a GI of the goods in respect of which the trade mark is used, the goods are not from the region indicated therein and it misleads the public, it shall be rejected for registration and prohibited from use; however, any trade mark that has been registered in good faith shall remain valid).

ordinary trade marks. However, geographical boundaries are determined by natural environment and human skills, which makes them not necessarily identical to administrative regions. The same is true with administrative place names under the county level. Such geographical terms cannot be prevented under Art 10.2 of the Trade Mark Law from being registered as ordinary trade marks. As a consequence, many geographical names below the county level, or non-administrative place names, have been registered as ordinary marks and these privately held rights are used to prohibit the use of such place names by local producers situated within the indicated place.[87] If there is a causal link between the geographical place and the quality, reputation or other characteristics of goods originated from this place, such a geographical name is eligible for GI protection. However, the registration of such a GI may be obstructed by the prior registered trade mark and this is when a controversy arises.

5.2 Proposed Solutions

The conflict between ordinary trade marks containing geographical terms and GIs applied for as certification or collective marks may be solved within the trade mark regime itself. The 'first in time first in right' principle is often proposed as an optimal solution to addressing the problem.[88] Between collective and certification marks on the one hand and ordinary trade marks on the other, simply following the principle of priority will be the general rule. But sometimes it is not so clear-cut an issue in practice.

The Jinhua Ham case is an exemplary case study of a situation where a rigid application of the priority principle would not produce satisfactory results. As mentioned above, JINHUA is a trade mark used on hams owned by Zhejiang Jinhua Ham Co. Ltd.[89] After becoming aware that Jinhua Ham could qualify as a protected GI, the Office for Protecting Jinhua Ham Certification Trade Mark filed an application for Jinhua City Jinhua Ham as a certification mark with the Trade Mark Office in 2003. Normally in accordance with the principle of priority, a subsequent confusingly similar sign (as is the case here) should be refused registration based on the prior registration of a mark which is used on identical or similar goods. However, sometimes one needs to settle such

---

[87] See Zhao Xiaoping (n 83) 281.
[88] D Gangjee, 'Quibbling Siblings: Conflicts between Trademarks and Geographical Indications' (2007) 82 Chicago-Kent Law Review 1253, 1263.
[89] See trade mark search result for 'Jinhua' at http://sbcx.saic.gov.cn:9080/tmois/wszhcx_pageZhcxMain.xhtml?type=reg&intcls=&regNum=130131&paiType=0

conflicts by taking historical factors as well as the interests of producers and consumers into consideration. While the proprietor of JINHUA is a company outside the boundary of Jinhua City, local Jinhua ham producers want Jinhua City Jinhua Ham to be registered as a GI so that their hams can be adequately protected. After difficult and lengthy negotiations and mediation, Jinhua City Jinhua Ham was published in the Trade mark Gazette in 2009 and registered as a certification mark GI.[90] This has created a de facto co-existence of quite similar marks under the trade mark system. Although coexistence is one solution for such conflicts, it should be subject to strict control and only allowed in exceptional cases. After all, intellectual property rights are generally subject to the principle of priority, which ensures exclusivity. Coexistence should be treated as an exception to this principle under very limited circumstances because it has certain adverse effects on the right holders. If the system allows coexistence, the holder of a GI sometimes has to tolerate the use of the GI by third parties provided that the parties use it in accordance with honest practices in industrial and commercial matters. Or in another scenario the owner of a prior trade mark has to take the risk of trade mark dilution.

There is a suggestion that the Trade Mark Law be amended to allow for the cancellation of ordinary trade marks containing GIs or to disallow renewal of such trade marks. As a consequence there would not be any prior trade marks that could potentially conflict with a subsequent GI application.[91] However, it seems unrealistic to revise the Trade Mark Law along these lines because it will result in legal uncertainty and deprive the trade mark holders of their investment and probably lead to more confusion among consumers. In particular, for those famous individually-owned geographical trade marks which are also eligible for GI protection, it is better to maintain the status quo because the trade mark has earned itself a good reputation in the marketplace. The owner has already considerably invested in the brand, so the potential cancellation of such famous regular trade marks or even coexistence between them and subsequent GIs would negatively impact upon the trade mark owner. Furthermore, in the minds of consumers, the geographical term points to a specific producer rather than a region after powerful presence of the brand in the marketplace for so many years. Therefore, continuing to recognise the

---

[90] See trade mark search result for 'Jinhua City Jinhua Ham' at http://sbcx.saic.gov.cn:9080/tmois/wszhcx_pageZhcxMain.xhtml?type=reg&intcls=&regNum=3779376&paiType=0

[91] See Bashaw (n 50) 90.

geographical term as an ordinary trade mark will better serve the interests of consumers.

One case decided by the Trade Mark Review and Adjudication Board (hereinafter TRAB)[92] reveals that the 'first in time first in right' principle may not apply if the registered trade mark has a GI nature and the use of the trade mark by the registrant might mislead the public as to the origin of its goods. In 2003 XIANG LIAN(literally meaning Hunan Lotus Seed) was registered by a Fujian-based company as an ordinary trade mark on lotus seeds and other products. The Hunan Xiangtan Xianglian Association filed an application with TRAB to cancel this registration on the ground that Xianglian was in fact a GI, referring to lotus seeds produced in Hunan Province. The trade mark owner defended itself by arguing that the disputed trade mark has distinctiveness as an ordinary trade mark. According to the evidence filed by the appellant, TRAB found that Xianglian is mainly produced in Hunan Province, and the lotus seeds have distinctive qualities which are essentially attributable to the local temperature, humidity, soil as well as the planting methods. Xianglian has been in use for over 1400 years to refer to the lotus seeds produced in Hunan Province. It satisfies the conditions established in Art 16.2 of the Trade Mark Law. Furthermore, TRAB found that the trade mark owner engaged in lotus seed trade with producers in Hunan Province before its registration. The trade mark owner knew that Xianglian referred to lotus seeds produced in Hunan Province but still applied for trade mark registration, which was liable to mislead the public as to the quality and origin of its product. This act was in violation of the provision of Art 16.1 of the Trade Mark Law (providing 'where a trade mark contains a geographic indication of the goods in respect of which the trade mark is used, the goods are not from the region indicated therein and it misleads the public, it shall be rejected for registration and prohibited from use; however, any trade mark that has been registered in good faith shall remain valid'). So TRAB cancelled the registration of the disputed trade mark. In this case TRAB established that unregistered GIs in China can also be protected.[93] This is the first case where TRAB recognised and protected an unregistered GI in a trade mark dispute case.

---

[92] As an agency in parallel to the Trade mark Office, TRAB is also under the administration of SAIC.

[93] Xinzhang Shi, 'Recognition and Protection of Geographical Indication in Trade Mark Dispute Cases for the First Time' (2009) Legal Affairs Correspondence, available at http://www.saic.gov.cn/spw/cwtx/200904/t20090409_55216.html

5.3 Priority Conflicts between Sui Generis GIs and Ordinary Trade Marks

5.3.1 Causes for conflicts

Conflicts will presumably not arise if the same GIs are owned by the same entities under different protection systems, though it might be regarded as a waste of time, money or energy to seek parallel avenues for protecting the same GI. However, if the same GIs are pursued by more than one unrelated entity via different protection systems, it is likely that conflicts will occur.[94] Moreover, the co-existence of the systems confuses applicants as to which avenue to take. In the absence of clarification regarding their differences, many of them have opted for cumulative registrations in all three of the relevant agencies. If conflicts arise out of different ownership decided by different authorities, the parties have to go to the court to resolve them, which could be expensive and time-consuming. This is very ineffective and uncertain from the perspective of GI stakeholders.

A given geographical name associated with a product may be protected as a GI under criteria set by one administration, but the same geographically significant term may be considered as having acquired distinctiveness under the Trade Mark Law and be eligible for registration as an ordinary trade mark.[95] At present there are no explicit rules in either the Trade Mark Law or its related regulations, or the *sui generis* administrative rules of the AQSIQ and Ministry of Agriculture which can resolve the conflict between rights granted under the trade mark system and those granted under the *sui generis* systems.

The case of 金华火腿 (Jinhua Ham) is a milestone regarding the resolution of the conflict between *sui generis* GIs and trade marks in China.[96] As noted above, the trade mark JINHUA and the GI 'Jinhua City Jinhua Ham' (registered as a certification mark) are concurrently valid under the trade mark system. Yet before the Jinhua City Jinhua Ham was able to obtain GI protection as a certification mark from SAIC, the Jinhua municipal authorities applied for a designation of origin (which later became a GI) for its hams and successfully obtained approval from the predecessor of AQSIQ. When the defendant, a company located in Jinhua

---

[94] Kireeva, Wang and Zhang (n 8) 147.
[95] WIPO, 'Geographical Indications and the Territoriality Principle' (SCT 9/5) (1 October 2002).
[96] *Zhejiang Food Co. Ltd. v Shanghai Taikang Food Co. Ltd.* (2003) Hu Erzhong Minwu (zhi) Chuzi No. 239 (Shanghai No.2 Interm. People's Ct., 25 December 2003) (unreported). See Tian Furong (n 78) 277–9.

352 *Research handbook on intellectual property and GIs*

which is authorized to use the designation of origin, put Jinhua Ham on the packages of its products, the individual trade mark proprietor sued him for infringement. So the dispute is in fact a conflict arising out of two protection systems, namely, trade mark protection versus designations of origin (*sui generis* GIs). After a trial, the court decided that the trade mark owner of Jinhua had the exclusive right, but was not entitled to prohibit the fair use of a third party. The fact that the defendant was authorized to use the designation of origin Jinhua Ham, which was approved by another government agency, granted the defendant the fair use exemption. Both parties enjoyed independent IP rights, namely a trade mark right and a *sui generis* GI right, both of which were protected by law. The court also ruled that in order to guarantee that their acts were legal and justifiable, both parties should respect each other's intellectual property rights and exercise their respective rights within the scope of protection, strictly following the relevant provisions.[97] Although the court decided the case by allowing coexistence of both sets of rights and the two parties accepted the judgment, the potential for clashes between the two systems remains the same.

5.4 Possible Solutions

From the JINHUA judgment, one can see that clashes between trade mark rights and GI rights could be settled based on principles of honest concurrent use. For historical reasons relating to permissive legislation, many place names with potential GI significance have been registered as individual trade marks. The trade mark owners have made tremendous efforts to enhance the reputation of the marks. On the other hand, the efforts and investment made by generations of local producers cannot be denied, either. Under such circumstances both rights should be protected by law provided that separate right holders use them fairly and honestly in the course of industrial and commercial activities.[98] However, co-existence is achieved at the price of compromises made by both trade mark and GI right holders.

---

[97] See Guoqiang Lv & Denglou Wu, 《我国地理标志法律制度的完善》 [Optimization of China's Legal System of Geographical Indications] (2006) 1 Legal Science 154, 158.
[98] Ibid., 159.

6. CONFLICTS BETWEEN SUI GENERIS GIs AND GIs REGISTERED AS TRADE MARKS

Different administrative authorities may confer the same GI right on different entities, define different geographical boundaries and enforce different quality standards. Take Shanxi Laochencu (literally meaning Old Vinegar in Shanxi) for example. AQSIQ recognized it as a protected product of designation of origin (now product of geographic indication) in 2004.[99] Then the Trade Mark Office registered it as a certification mark in 2010.[100] However, the production boundaries as well as the production standards determined by the two systems are so different that conflicts among the two right holders as well as producers are unavoidable. Under such circumstances, it is difficult to coordinate the systems to achieve efficient GI protection.

7. COMPARISON OF THE TRADE MARK REGIME AND SUI GENERIS REGIME OF GI PROTECTION

As mentioned before, the most prominent problem is the coexistence of the trade mark and *sui generis* systems for GI protection. By weighing the advantages and disadvantages of the two systems respectively and taking into consideration the conflicts precipitated by the co-existence of the two systems, it is suggested that in China, it would be better to maintain only one system. This should preferably be the trade mark mechanism because it accommodates the rationale of GI protection better than sui generis mechanisms, bearing legal and economic considerations in mind.

7.1 Cost

The trade mark system is already in place to conduct GI examination (as collective or certification marks). Whereas the *sui generis* system doesn't provide for protection for foreign GIs due to lack of procedural rules, it will be expensive to build up a comprehensive *sui generis* system to address international GI protection.

---

[99] Announcement No. 104 on Protection of 'Shanxi Laochencu' Product of Designation of Origin by AQSIQ (1994).
[100] See trade mark search result for 'Shanxi Laochencu' at http://sbcx.saic.gov.cn:9080/tmois/wszhcx_pageZhcxMain.xhtml?type=reg&intcls=&regNum=6173333&paiType=0

7.2 Obtaining Exclusive Rights

Despite the fact that in China it is usually government departments who set up an association as the applicant to initiate the registration of a certification or collective mark GI, under the trade mark regime it is the right holder who makes further decisions in seeking protection without government actions. On the international level, GIs as certification or collective marks can be applied for directly by the interested parties without an official government action being necessary, as is required under the Lisbon Agreement.[101] Trade mark registration also involves less government intervention in that the standards for inspection and verification are set by the certification or collective mark owner rather than the competent authorities. Those government agencies are not expected to take on many roles that are supposedly taken on by private parties, such as market functions like defining the production or operating standards, managing verification of compliance or controlling output.[102] Sometimes government agencies tend to be bureaucratic and demand complex procedures to be satisfied. The two-level examination and approval processes under both AQSIQ and MOA practices mirror the complexity of the procedures. From the producers' point of view, it is far more complicated to obtain *sui generis* GI registration.

7.3 Enforcement of Exclusive Rights

In cases where the use of a GI is not well regulated and monitored, or where misappropriation and abuses of GIs become rampant, GI protection can hardly achieve its goal of protecting producers, consumers and promoting local development. In countries like China, enforcement of GI protection is far more important in practical terms, as opposed to simply acquiring registered protection.

As far as the right holder is concerned, when facing misrepresentation or fraudulent use of GI, a private trade mark owner can take immediate legal action to reduce the negative effect to a minimum, while public authorities could be slow in reacting or responding due to bureaucratic procedures. Private owners will always try their best to maximize their profits and interests, whereas government agencies usually take other

---

[101] See WIPO, 'Document SCT/6/3 Rev. On Geographical Indications: Historical Background, Nature of Rights, Existing Systems for Protection and Obtaining Protection in Other Countries', 2 April 2002 (SCT/8/4) 22.

[102] Giovannucci et al. (n 26) 53.

factors, such as political ends, into consideration when dealing with GI protection, and do not always treat the interests of producers as a top priority.[103] From the perspective of administrative enforcement, an expeditious and efficient means is already in place for trade mark owners who want to lodge complaints against misuse and infringement of their exclusive right. Under the trade mark regime, the nationwide administrative forces for industry and commerce (AICs) guarantee rapid, convenient and effective enforcement against trade mark infringement. In fact the majority of trade mark infringement cases, including GI certification or collective mark cases are handled via these administrative means, namely by AICs at all levels throughout the country.[104] Comparatively, AQSIQ administrative forces put more emphasis on product quality supervision in the production channels and inspection of imported or exported goods according to their defined functions by the State Council, mostly dealing with what happens in the workshops where processing and production take place, not in the circulation channel (market supervision is within the competence of SAIC).[105] Neither does the Ministry of Agriculture have experienced forces to handle GI infringement issues.

7.4 Legal Considerations

7.4.1 Trade mark law ranks higher than AQSIQ and MOA administrative regimes

The hierarchy of Chinese national legal system starts from the Constitution, as the highest level, then come national laws made by the National People's Congress or its Standing Committee, after which are administrative regulations made by the State Council based on the Constitution and national laws, followed by regional laws made by the provincial people's congress and then to administrative rules made by ministries or provincial governments for the purpose of implementing national laws, administrative regulations or regional laws.[106] According to the classification, the Trade Mark Law is a national law promulgated by the Standing Committee of the National People's Congress. The PPGIP and MOA Measures are

---

[103] Ibid., at 53.
[104] See 《中国商标专用权的行政保护》 [Administrative Protection for Trademark Exclusive Rights in China], http://lunwen.5151doc.com/Article/HTML/184113.html (only Chinese version available).
[105] See Yumin Zhang, 《地理标志的性质和保护模式选择》 [On Its Character & Mode Choice of Geographical Indications] (2007) 6 Law Science Magazine 6, 11.
[106] See An (n 11).

administrative rules promulgated and implemented by ministerial agencies, which are lower in the legal hierarchy than national laws.

7.4.2 International protection

(i) Protection pursued by domestic GI right holders Trade mark systems have been established for an extended period of time and are well accepted in most countries. GI right holders have the option to apply for trade mark registration in those countries where protection for GIs as collective marks or certification marks is available.[107] Even absent the possibility of collective or certification marks, they can always count on regular trade marks to protect GIs. By contrast, there has been no harmonized international system regarding GI protection up to now. The Lisbon Agreement has only 28 contracting parties and China is not one of them.[108]

Overseas GI protection can also be achieved by way of bilateral trade agreements or specific GI protection agreements via the provision of lists of protected GI terms in the annex, but it relies heavily on the initiatives and efforts of governments and these take time to negotiate and conclude. So far few such bilateral agreements have been reached between China and other countries.[109] By comparison, trade mark protection is always available in other jurisdictions where the GI producers have a market interest. While they can file the applications individually in these countries, the Madrid international registration system makes it much easier and cheaper to achieve the same goal. In particular, Rule 9(4)(x) of the Common Regulations stipulates that, where the basic application or the basic registration relates to a collective or certification mark, the international application should contain an indication to that effect.[110] Successful examples of Chinese GIs registered as certification trade marks include Zhangqiu Scallion and Guanxi Sweet Shaddock (also known as a pomelo). After commercial success on domestic markets, the trade mark

---

[107] Paris Convention, *supra* note at Article 7bis. (It sets a legal basis to make it possible to register GIs as collective marks in other countries and requires Member States to provide protection for collective marks. The obligation is incorporated into TRIPS Agreement by Article 2.1.)

[108] http://wipo.int/treaties/en/ShowResults.jsp?lang=en&treaty_id=10. For recent initiatives to expand membership, see Chapter 5 by Geuze in this volume.

[109] One such bilateral agreement is the Free Trade Agreement between China and Peru. Meanwhile China and the EU are still in the process of negotiating a possible agreement on GIs.

[110] WIPO, *Addendum to Document SCT/6/3 Rev. (Geographical Indications: Historical Background, Nature of Rights, Existing Systems for Protection and Obtaining Protection in Other Countries)* at 10, WIPO Doc. SCT 8/5 (2 April 2002).

owners explore the commercial potential internationally by using the Madrid system to seek GI protection in overseas markets.[111]

(ii) Protection pursued by foreign GI right holders   At present it is difficult to register foreign GIs in China under the *sui generis* systems. Lacking in procedural rules, neither the PPGIP nor the MOA Measures can be used to register such foreign GIs. So far, AQSIQ has put several foreign GIs in its recording list, but they are based on the conclusion of bilateral agreements which are time-consuming and costly. The easiest way available is to register them as certification or collective marks under the Trade Mark Law.[112] Therefore, only the trade mark mechanism fully complies with the TRIPS requirement in terms of national treatment and enforcement.

7.4.3 Administrative appeal and judicial review

The trade mark regime provides for administrative appeal exercised by TRAB for decisions of the Trade Mark Office in terms of refusal or opposition to registration. At the same time, the Law allows judicial review of administrative decisions made by the TRAB on refusal of registration, opposition, cancellation, revocation, as well as the administrative penalties such as fines made by AICs.[113] This is particularly important for the fulfilment of the TRIPS Agreement. By contrast, neither the *sui generis* AQSIQ nor the MOA rules explicitly provide for the rights of administrative appeal or judicial review.[114] If the application fails to be accepted by the administrative agency, there is no remedy to correct the application form or any other administrative appeal procedures. Nor is there any judicial review for opposition, cancellation or revocation decisions. The TRIPS Agreement requires Member States to provide for judicial remedies for any intellectual property in their legislation.[115] From this aspect, only the trade mark regime enables China to fulfil its WTO obligations.

At present there are some Chinese scholars who advocate protecting GIs as an independent commercial sign in parallel with trade marks,

---

[111] See Shiping Chen,《平和琯溪蜜柚品牌效益凸显》[Brand Effects Obvious of Pinghe 'Guanxi Sweet Shaddock'] (2009) 2 China Fruit News 37, 37–8.
[112] Ibid., at 109.
[113] See Trade Mark Law, Arts 34, 35, 45, 54.
[114] See Bashaw (n 50) 100.
[115] TRIPS Agreement, Art 41.4. (Parties to a proceeding shall have an opportunity for review by a judicial authority of final administrative decisions and, subject to jurisdictional provisions in a Member's law concerning the importance of a case, of at least the legal aspects of initial judicial decisions on the merits of a case. However, there shall be no obligation to provide an opportunity for review of acquittals in criminal cases.)

certification marks and collective marks under the Trade Mark Law.[116] This is also a common legislative practice adopted in some countries, such as Indonesia (Law on Marks) and Russia (Federal Law on Trademarks, Service Marks and Appellations of Origin of Goods).[117] The advantages of such a model include granting a rather straightforward and definite protection on GIs without the necessity of framing a separate new law. In addition it facilitates the determination of priority and classification of products if both trade marks and GIs are subject to examination under the Trade Mark Office,[118] which is not available under the current dual system. The separation of GIs from certification or collective trade marks is quite a big change, which needs discrete analysis and scrutiny. In view of the fact that a *sui generis* system under AQSIQ and the Ministry of Agriculture are still in operation, it seems too early to categorize geographical indication as an independent sign under the trade mark system. Otherwise it will make GI protection more complicated and confusing. Nevertheless, it is a direction for Chinese legislators to undertake further consideration on the method of GI protection in China.

---

[116] See Xiaoxia Li, 《地理标志商标法保护模式的重构》 [On the Reconstruction of Protecting Model of Geographical Indications by Trade Mark Law] (2009) 1 Journal of Xinyang Normal University (Philosophy and Social Sciences Edition) 95, 97.
[117] See O'Connor & Co, *Geographical Indications and TRIPs: 10 Years Later... Part II – Protection of Geographical Indications in 160 Countries around the World* (Report commissioned for EC (DG Trade) 2007).
[118] See Li (n 116), 97.

PART IV

CRITICAL ISSUES

13. Learning to love my PET – the long road to resolving conflicts between trade marks and Geographical Indications
Burkhart Goebel and Manuela Groeschl

1. INTRODUCTION

The conflict between trade marks and Geographical Indications (GIs) has been a prominent, if not *the* most prominent, topic in the legislative, forensic and academic debate on GIs during the last 25 years.[1] Unsurprisingly, it evolved largely in parallel to the rise of protection systems for GIs. The debate has been intense and often emotional. For the proponents of GI protection the mere possibility of a trade mark protected in one part of the world being identical or similar to a GI protected in another part of the world amounted to an onslaught on the cultural heritage of GI users, which undoubtedly deserved universal protection. For trade mark owners the hypothesis that a prior trade mark could be deprived of its exclusivity or – worse – be completely expropriated on the basis of a GI existing in another part of the world was seen as nothing but a blunt confiscation of

---

[1] A selection of instructive articles on the issue includes Florent Gevers, 'Conflicts between Trade Marks and Geographical Indications – The Point of View of the International Association for the Protection of Industrial Property (AIPPI)' in WIPO, *Symposium on the International Protection of Geographical Indications* (WIPO 1995) 143; Henning Harte-Bavendamm, 'Geographical Indications and Trade Marks: Harmony or Conflict?' in WIPO, *Symposium on the International Protection of Geographical Indications* (WIPO 1999) 61; Dev Gangjee, 'Quibbling Siblings: Conflicts between Trade Marks and Geographical Indications' (2007) 82 Chicago-Kent Law Review 1253; Justin Hughes, 'Champagne, Feta, and Bourbon: The Spirited Debate About Geographical Indications' (2006) 58 Hastings L.J. 299; Annette Kur and Sam Cocks, 'Nothing but a GI Thing: Geographical Indications under EU Law' (2007) 17 Fordham Intell. Prop. Media & Ent. L. J. 999; Ruth L Okediji, 'The International Intellectual Property Roots of Geographical Indications' (2007) 82 Chicago-Kent Law Review 1329; Burkhart Goebel, 'Geographical Indication and Trade Marks in Europe' (2005) 95 TMR 1165; Burkhart Goebel, 'Geographical Indications and Trademarks – The Road from Doha' (2003) 93 TMR 964; Tim Josling, 'The War on *Terroir*: Geographical Indications as Transatlantic Trade Conflict' (2006) 57 Journal of Agricultural Economics 337.

valuable private property and an equally despicable onslaught on one of the major achievements of the French revolution, namely the protection of private property against arbitrary confiscation by the state.[2]

Beneath the politically charged debate over public versus private property and cultural heritage versus entrepreneurial achievements, there was a profoundly legal debate. Trade marks and GIs could certainly be in conflict with each other. They are both signs that designate the origin of a product, in one case the geographical origin of a product and in the other case the commercial origin of a product. Any given word could theoretically serve both functions. Capri is an island in Italy and was the protected trade mark of a Ford car in the 1970s. Sapporo is not only the name of a city in Japan, but was also adopted as the company name and then brand of the local brewery which developed into a major international brewery, protecting Sapporo as a trade mark for the beer it brewed and not as an indication for beer from Sapporo around the world. Quite often entrepreneurs used the names of the places where their companies came from in their brands or as their brands. Other roots for conflicts between trade marks and GIs were laid by emigration, where emigrants called their new places and their foodstuffs or local and artisanal products in the countries they emigrated to by the same names they had originally called them. Additional conflicts arose from the fact that names sometimes simply coincide. This is true for names of human beings, names of places and brands. The profoundly legal debate was now about the mechanism by which these conflicts should be resolved. Which conflict rules should be applied? Should the established and indeed almost universal rules existing under trade mark laws built on the principles of priority, exclusivity and territoriality be applied or should there be special rules for conflicts between trade marks and GIs deriving from *sui generis* laws on the protection of the latter?

In our opinion – after numerous national court decisions and, in particular, two milestone decisions under public international law – fairly clear guiding principles for resolving conflicts between trade marks and geographical indication have emerged, in particular over the last decade. The following chapter will discuss how these rules emerged and what they are. The focus will be on public international law and European Union law. It will conclude with a discussion of current public international law initiatives and identify a few issues which remain unresolved.

---

[2] See L Bergeron, 'Biens Nationaux' in F Furet and M Ozouf (eds), *Dictionnaire critique de la Révolution française* (Flammarion 1992) 77 *et seq.*

2. SETTING THE SCENE

Principles for conflict resolution between a trade mark and an identical or similar sign used to designate identical or similar goods or services are based on three widely accepted and rather straightforward principles: priority,[3] exclusivity,[4] and territoriality.[5] When a conflict arises, the prior right will prevail. It will enjoy exclusivity for the territory in which it is protected with the better priority. As there are millions of trade marks

---

[3] *Anheuser-Busch Inc. v Budějovický Budvar* (C-245/02) [2004] ECR I-10989, [2005] ETMR 27 at [98] ('Finally, "priority" of the right in question for the purposes of [Art 16(1) of TRIPS] means that the basis for the right concerned must have arisen at a time prior to the grant of the trade mark with which it is alleged to conflict ... that requirement is an expression of the principle of the *primacy of the prior exclusive right*, which is one of the basic principles of trade-mark law and, more generally, of all industrial-property law' (emphasis added)).

[4] See e.g. Art 6*quinquies*(B)(i) of the Paris Convention ('Trade marks covered by this Art may be neither denied registration nor invalidated except in the following cases: (i) when they are of such a nature as *to infringe rights acquired by third parties* in the country where protection is claimed'); Art 9 of the Community Trade Mark Regulation No 207/2009 (CTMR) ('A Community trade mark shall confer on the proprietor *exclusive rights* therein'); WTO Panel Report *European Communities – Protection of Trade Marks and Geographical Indications for Agricultural Products and Foodstuffs*, complaint by the United States (WT/DS174/R) 15 March 2005, at [7.602 *et seq*] ('The text of Art 16.1 [TRIPS] stipulates that the right for which it provides is an "exclusive" right. This must signify more than the fact that it is a right to "exclude" others, since that notion is already captured in the use of the word "prevent". Rather, it indicates that this right belongs to the owner of the registered trade mark alone, who may exercise it to prevent certain uses by "all third parties" not having the owner's consent. The last sentence provides for an exception to that right, which is that it shall not prejudice any existing prior rights. Otherwise, the text of Art 16.1 is unqualified. Other exceptions to the right under Art 16.1 are provided for in Art 17 and possibly elsewhere in the TRIPS Agreement. However, there is no implied limitation vis-à-vis GIs in the text of Art 16.1 on the exclusive right which Members must make available to the owner of a registered trade mark. That right may be exercised against a third party not having the owner's consent on the same terms, whether or not the third party uses the sign in accordance with GI protection, subject to any applicable exception.'). See also WTO Panel Report, complaint by Australia (WT/D290/R) 15 March 2005, at [7.602 *et seq*.]. For a discussion of exclusivity with a focus on GIs see Gangjee (n 1) 1254–1261.

[5] *Koninklijke Philips Electronics NV v Lucheng Meijing Industrial Company; Nokia Corporation v HMRC* (Joined Cases C-446/09 and C-495/09) [2011] ECR I-12435 at [AG65] ('Furthermore, it is to be borne in mind that protection of intellectual property rights is based on the *principle of territoriality*. By virtue of that principle, holders may prohibit the unauthorised use of their right only in those States in which it enjoys protection') (emphasis added).

protected around the globe, it is not surprising that courts in almost every country have relevant experience with adjudicating trade mark cases and can rely on a well-established body of case law built on these three basic principles. It is important to mention at this early stage that the three principles of priority, exclusivity and territoriality (PET) are subject to a number of qualifications in trade mark law. PET is the point of departure, not the end of the journey, as every trade mark practitioner will appreciate. Exceptions to a simple registration priority rule are made on the basis of prior use rules, novelty rules or bad faith concepts. Bad faith or lack of novelty is also a concept that helps mitigate against the adverse effects of a too-strict territoriality rule. Finally, even exclusivity is subject to a number of exceptions, such as fair descriptive use or use of one's own name or address.[6] However, the existence of these exceptions underlines the importance of the basic rules of priority, exclusivity and territoriality. Under trade mark law all rights are equal and, as a matter of fact, no right is more equal than others.

The setting is somewhat different for GIs. To begin with, there is a wealth of protection systems ranging from unfair competition, to collective and certification marks, to *sui generis* systems. These in turn are protected under national laws, regional laws, bilateral agreements or multilateral agreements.[7] These protection systems may cover a wide range of types of

---

[6] On limitations, see Max Planck Institute for IP and Competition Law, *Study on the Overall Functioning of the European Trade Mark System* (15 February 2011) at [2.237]–[2.241] ('Art 6 TMD and Art 12 CTMR provide for three exceptions to the exclusive rights of trade mark proprietors: use of one's own name or address (lit. a), descriptive use (lit. b), indication of intended purpose of product (lit. c) . . . Indications of product characteristics may also be used in a trade mark sense; e.g., an indication of geographical origin was allowed even though it had the same appearance as a trade mark . . . The provision is primarily designed to prevent a proprietor from prohibiting third parties from using descriptive terms which are within the scope of the mark, i.e. because they are similar to the mark or are the same as elements of a composite mark that are themselves descriptive. It may also authorise a third party to use a trade mark if such use consists in giving an indication concerning the kind, quality or other characteristics of products marketed by that third party'); US WTO Panel Report (n 4) at [7.647] ('Art 17 expressly permits Members to provide limited exceptions to the rights conferred by a trade mark, which include the right provided for in Art 16.1 . . . The Panel has already found that the [EU GI] Regulation limits the availability of the right provided for in Art 16.1. Therefore, to the extent that it satisfies the conditions in Art 17, this limitation will be permitted under the TRIPS Agreement.').

[7] WIPO Secretariat, 'Document SCT/6/3 Rev. on Geographical Indications: Historical Background, Nature of Rights, Existing Systems for Protection and Obtaining Protection in Other Countries' 2 April 2002 (SCT/8/4); with Addendum 2 April 2002 (SCT/8/5); OECD, Appellations of origin and geographical indications

GIs from simple geographical indications (for example Made in Macau) to qualified geographical indications (for example Swiss watches) and appellations of origin (for example Cognac). For the purpose of this chapter we shall ignore unfair competition protection or protection through the trade mark system (certification or collective marks). By and large, in these categories conflict rules coincide with those under regular trade mark laws. This is not the case for *sui generis* protection systems which are typically modelled after the ancestor of all *sui generis* systems, the French laws for the protection of appellations of origin.[8] Under these regimes there is often a perception that a GI is an expression of a regional or national cultural heritage and a common good that is somehow superior to the individual private property right contained in a brand protected through a registered trade mark. This could have the effect that a prior trade mark could be invalidated (expropriated) on a basis of an appellation of origin protected with a more recent priority date, or could at least no longer be renewed once an identical or similar designation had been protected as a GI.

A case in point is the (in)famous Art 5(6) of the Lisbon Agreement:

> If an appellation which has been granted protection in a given country pursuant to notification of its international registration has already been used by third parties in that country from a date prior to such notification, the competent Authority of the said country shall have the right to grant to such third parties a period not exceeding two years to terminate such use, on condition that it advise the International Bureau accordingly during the three months following the expiration of the period of one year provided for in paragraph (3), above.

This provision had been interpreted as requiring the phase-out of conflicting trade marks after a two-year grace period irrespective of whether this trade mark registration enjoyed an earlier priority date.[9] Conceptually, this rule implies that a GI is somehow superior to a trade mark. This assumption was so widespread in Europe that the rule was proposed as a European-Community-wide rule in a draft of EC Regulation 2081/92.[10]

---

in OECD member countries: Economic and legal implications, Working Paper, COM/AGR/APM/TD/WP/(2000)15/final (2 April 2002).

[8] On French laws, see Norbert Olszak, *Droit des appellations d'origine et indications de provenance* (Tec & Doc – Lavoisier, 2001); see also Chapter 2 by Stanziani and Chapter 11 by Marie-Vivien in this volume.

[9] See Ludwig Baeumer, 'Protection of Geographical Indications under WIPO Treaties and Questions Concerning the Relationship between Those Treaties and the TRIPS Agreement' in WIPO, *Symposium on the Protection of Geographical Indications in the Worldwide Context; Eger, Hungary 1997* (WIPO 1999) 9, 19 & 23.

[10] Commission Proposal [1990] OJ C 30/91 P 9, art 14(1)(4); reprinted in GRUR Int 1991, 277 (in German).

The draft was changed only at a very late stage upon intervention of the Council. It was replaced by Articles 14(3) and 14(2) of that Regulation, which gave exclusivity to prior well-known marks and provided for coexistence between a – then not clearly defined – group of other marks that were prior, but not well known in Europe with a later GI.[11] A prior GI would always serve to block a later trade mark according to Articles 14(1) and 13 of the Regulation, and this appeared to be a watered down expression of the superiority concept. That concept had already made it into Community law, in its more absolute form, under the former EC Wine Regulation. Until 2008 this Regulation provided that the prior mark conflicting with a later GI for wine had to be invalidated unless it had been registered for more than 25 years.[12] This rule was also adopted in some bilateral agreements such as the Wine Agreement between the European Communities and Australia which led to the phase-out of a number of prior marks used for wines and spirits in Australia.[13]

The real conflict between trade marks and GIs was therefore the conflict of two conflict resolution mechanisms. The issue was which one would prevail: the trade mark rules firmly built on priority, exclusivity and territoriality or the *sui generis* rules built on the assumption that the 'common good' of GIs was somehow superior to the private property right of trade

---

[11] See Arts 14(2) & (3) of Council Regulation (EEC) No 2081/92 of 14 July 1992 on the Protection of Geographical Indications and Designations of Origin for Agricultural Products and Foodstuffs [1992] OJ L208/1. This Regulation was repealed and replaced by Council Regulation (EC) No 510/2006 and this in turn was replaced by Regulation No 1151/2012 of the European Parliament and of the Council of 21 November 2012 on quality schemes for agricultural products and foodstuffs [2012] OJ L323/1.

[12] Council Regulation (EC) No 1493/1999 of 17 May 1999 on the Common Organisation of the Market in Wine [1999] OJ L179/1, Annex VII, Part F, No 2 ('Moreover, the holder of a well-known registered brand name for a wine or grape must which contains wording that is identical to the name of a specified region or the name of a geographical unit smaller than a specified region may, even if he is not entitled to use such a name . . . , continue to use that brand name where it corresponds to the identity of its original holder or of the original provider of the name, provided that the brand name was registered at least 25 years before the official recognition of the geographical name in question by the producer Member State in accordance with the relevant Community provisions. . . and that the brand name has actually been used without interruption.').

[13] Agreement between the European Communities and Australia on trade in wine [1994] OJ L86/31; repealed by Agreement between the European Communities and Australia on trade in wine [2009] OJ L28/3; see further Chapter 10 by Stern in this volume; The Australian Government FAQ (2010) at: http://www.agriculture.gov.au/SiteCollectionDocuments/ag-food/wine/agreement-faqs.pdf

marks and could therefore destroy its existence, or at least its exclusivity, irrespective of priority and territoriality.

The debate was completely open in the early 1990s when the major global players negotiated the TRIPS Agreement, which was the prime opportunity to agree on a conflict resolution mechanism. It is presumably fair to say that some negotiators may not have fully appreciated the significant conceptual differences that existed on this issue. Others may have appreciated them and may have preferred not to force the negotiators to come up with a clear solution, while a third group may simply have thought that they had successfully negotiated their preferred conflict resolution mechanism into the TRIPS Agreement. Whatever the thinking of the trade negotiators was, the outcome in the TRIPS Agreement was certainly not clear enough to provide a straightforward answer to the question how such a conflict should be resolved. What followed was more than a decade of uncertainty, debate and extended litigation. All of this eventually clarified the conflict rules.

Today we can say that there are no special rules when it comes to the conflict between a prior trade mark and a GI as compared to conflict between a prior trade mark and another sign. The superiority-based rule has not become the conflict resolution mechanism at global level. On the contrary, conflicts are resolved on a basis of priority, exclusivity and territoriality, and this is due mostly to two landmark decisions, considered below: the WTO Panel decision in *European Communities – Protection of trade marks and geographical indications for agricultural products and foodstuffs*, and the judgment of the Grand Chamber of the European Court of Human Rights (ECHR) in *Anheuser-Busch v Portugal*. These two decisions so fully endorse the exclusivity (the Panel Report) and the priority (the ECHR judgment) principles that they leave virtually no room for a superiority-based *sui generis* conflict resolution mechanism.

3. TWO DECISIVE MILESTONES: THE WTO PANEL REPORT AND THE ECHR JUDGMENT

3.1 The First Milestone: The WTO Panel Report

The relationship between GIs and trade marks was at the core of a widely acclaimed WTO Dispute Settlement decision on GI legislation in Europe. On 15 March 2005, the WTO published a Panel Report[14] on the dispute

---

[14] US Panel Report (n 4). The Australian Panel Report (WT/D290/R) includes identical considerations.

between the US and Australia, on the one hand, and the European Communities (EC), on the other hand, over the compatibility of various aspects of Regulation No 2081/92 with the TRIPS Agreement and GATT 1994. The decision was not appealed as both sides claimed victory of their case. The Panel Report was eventually adopted by the WTO Dispute Settlement Body on 20 April 2005. The Panel Report is *the* fundamental decision holding that the conflict between a prior trade mark and a later GI be resolved on the basis of the conflict rules established under trade mark law, with exclusivity and priority being the main applicable principles. It thereby effectively shut the door to a 'superior right' approach among the Members of the WTO.

With their claim, Australia and the US *inter alia* contested the legality of Art 14(2) of Regulation No 2081/92. According to the pre-WTO dispute settlement proceedings interpretation given to this provision by the EC it provided for a broad rule of co-existence between a prior trade mark and a later GI that was confusingly similar to that prior mark. According to the EC, a conflict between a trade mark and a GI would generally be resolved in favor of the GI pursuant to Arts 14(1) and 13 of the Regulation. Where the trade mark had the better priority it might, however, continue to be used provided that the trade mark was registered in good faith before the date on which application for registration of a GI was lodged and there were no grounds for invalidity or revocation of the trade mark as provided for by Art 3(1)(c) and (g) and Art 12(2)(b) of the EU Trade Mark Directive (TMD) respectively.[15] Only under very exceptional circumstances – basically the prior mark being well known – would the trade mark prevail over the GI. These rules are – at least *prima facie* – an expression of a *sui generis* approach and not an endorsement of trade mark law conflict rules.

The claim was raised as an 'as such' claim. No specific adverse decisions based on Art 14(2) were challenged, but the legality of the very structure of the EU GI regime was. Australia and the US argued that Art 14(2) of the Regulation (providing for co-existence between a prior trade mark and a later GI that was identical or confusingly similar) was in breach of Art 16.1 of the TRIPS Agreement in that it prejudiced the exclusivity of trade mark rights warranted under that provision. The US pleaded that by imposing co-existence upon the trade mark owner rather than adhering to the priority principle, the trade mark owner could no longer exercise 'the

---

[15] First Council Directive 89/104/EEC of 21 December 1988 to approximate the laws of the Member States relating to trade marks [1989] OJ L40/1; replaced by Directive 2008/95/EC of the European Parliament and of the Council of 22 October 2008 to approximate the laws of the Member States relating to trade marks [2008] OJ L299/25 (hereafter, the TMD).

exclusive right that lies at the heart of his trade mark right'.[16] According to the United States, exclusivity was indispensable given the essential function of trade marks to identify the commercial source of a product,[17] and exceptions to the exclusive nature of a trade mark were only allowed where explicitly provided for in the TRIPS Agreement.[18] There could not be a general exception for GIs from the exclusivity rule since, due to their nature as designations, these had to be considered conflicting signs within the meaning of Art 16.1.[19]

The EC raised three defenses. It argued that under Art 24.5 of the TRIPS Agreement co-existence between the earlier trade mark and the later GI was actually mandatory. In any event, the Regulation had been grandfathered by Art 24.3 of the TRIPS Agreement pursuant to which the level of protection of GIs that existed in a Member State immediately prior to the date of entry into force of the TRIPS Agreement shall not be diminished. Finally, the EC pleaded that the co-existence provided for under the Regulation was so limited that it was justified under Art 17 of the TRIPS Agreement, which permitted a limited exception for fair use of descriptive terms.[20]

It is important to understand the conceptual differences between the three lines of defense. The first two are *sui generis* defenses. The conflict between trade marks and GIs was to be resolved under a set of special rules in the TRIPS chapter on GIs. The third line of defense, in contrast, moved the conflict resolution mechanism back to the world of trade mark law, arguing that the provision of Art 14(2) of the Regulation was compatible with the limited exceptions trade mark law itself provided to the priority, exclusivity and territoriality rule.

From the three defenses raised by the EC, the Panel only found the limited exception argument under Art 17 of the TRIPS Agreement to have some merit. The first defense, mandatory co-existence under Art 24.5, was rejected in its entirety. Art 24.5 of the TRIPS Agreement stipulates that the validity or the use of a prior trade mark that was acquired in good faith may not be compromised by a measure of a Member applying the rules on the protection of GIs under the TRIPS Agreement. The EC concluded from the existence of this provision that it implicitly provided for a mandatory rule of co-existence between a prior trade mark and a later GI. The

---

[16] Submissions by the United States, WT/DS/174/R/Add.1 (23 April 2004) Annex A-2 at [134].
[17] Ibid., at [145 *et seq*].
[18] Ibid., at [142 *et seq*].
[19] Ibid., at [139].
[20] US Panel Report (n 4), at [7.513].

right to use would not be affected by co-existence, whilst the exclusivity of the prior mark would be. They further argued that Art 24.5 provided for an implicit exemption of the exclusive right to use a prior trade mark guaranteed by Art 16.1. The Panel dismissed these arguments finding that there was insufficient support to 'imply in Art 24.5 either the right to prevent confusing uses or a limitation on the right to prevent confusing uses'.[21] It emphasized that within the WTO Member States, a trade mark owner's exclusive rights entitled him to enforce it against any third party using the mark without his consent, whether or not the third party's mark enjoyed protection as a GI.[22] An exemption to the exclusivity rule would require that Art 24.5 TRIPS expressly stipulated such an exemption:

> The Panel considers that it is difficult to sustain an argument that a limitation which is allegedly implied can prevail over an obligation in a WTO covered agreement which is express. It is evidently the position under the European Communities' domestic law that an implied positive right to use a registered geographical indication prevails over the negative right of a prior trade mark holder to prevent confusing uses. However, such an interpretation of the TRIPS Agreement is not possible without a suitable basis in the text.[23]

The Panel concluded from this that under Art 16.1 of the TRIPS Agreement WTO Member States were required to enable trade mark owners to prevent certain third-party uses of conflicting signs, including uses of conflicting signs as GIs. The Panel also dismissed the EC's defense that it had grandfathered its regime on GIs that predated the TRIPS Agreement as provided for in Art 24.3 of the TRIPS Agreement. The Panel held that the stand-still provision of Art 24.3 preserving existing levels of protection may only apply to individual GIs, but does not protect an entire legal regime.[24]

The Panel then turned to an internal trade mark law defense and considered whether Art 14(2) of Regulation No 2081/92 could be considered a limited exception under Art 17 of the TRIPS Agreement. Art 17 provides for limited exceptions to the exclusive rights granted to the trade mark owner under Art 16.1. As an example, Art 17 names fair descriptive use. For determining whether co-existence provided for under Art 14(2) of Regulation No 2081/92 could be considered a limited exception under Art 17 of the TRIPS Agreement, the Panel had to establish the scope of Art 14(2). In their pleadings, the EC had construed Art 14(2) narrowly as

---

[21] Ibid., at [7.619].
[22] Ibid., at [7.603].
[23] Ibid., at [7.618].
[24] Ibid., at [7.513].

the exception rather than the rule. They had explained that in most cases where there was likelihood of confusion between a prior trade mark and a later GI, the GI could be refused under a broadly construed Art 14(3) of the Regulation, making the rule one where the prior trade mark prevailed over the GI. Art 14(3) of the Regulation stipulated that a prior trade mark could give cause for refusing the registration of a GI where, in the light of that trade mark's reputation, renown and the length of time it had been used, registration was liable to mislead the consumer as to the true identity of the product. The EC had stated in support of their defense:

> The complainant's claim is based on the unwarrantedly narrow interpretation of Art 14 (3) of Regulation 2081/92. When properly interpreted, that provision allows the registering authorities to refuse the registration of any confusing geographical indication.[25]

In other words, the EC deflated the scope of application of the co-existence rule provided for in Art 14(2) of the Regulation in order to squeeze the provision through the needle-eye of the narrow descriptive use exemption of Art 17 TRIPS. The Panel accepted this interpretation and was 'satisfied that where the likelihood of confusion was relatively high, the exception in Art 14(2) will not apply'.[26] In the Panel's view, this constituted a limited exception under the TRIPS Agreement since co-existence would only be at issue where likelihood of confusion with a prior trade mark was remote. The Panel noted that co-existence was limited to GIs subject to registration under Regulation No 2081/92.[27] It emphasized that co-existence could only be invoked for the indications as registered and that the co-existence rule under Art 14(2) could not justify a trade mark infringing use of an alleged translation of the registered GI.[28]

The Panel thus affirmed the application of the general rules of conflict – priority, exclusivity and territoriality – in situations of conflict between trade marks and GIs. It did not endorse any special regime for the conflict between trade marks and GIs, giving full consideration to the priority and exclusivity guarantees of Art 16.1 TRIPS Agreement. On the other hand, it found limited co-existence to be consistent with Art 17 of the TRIPS Agreement provided that likelihood of confusion with the prior trade mark was remote. So, while the Panel clearly shut the door to a 'GIs are

---

[25] Submissions by the European Communities, WT/DS174/R/Add.2 (24 February 2004) Annex B-1 para. 286.
[26] US Panel Report (n 4) at [7.679].
[27] Ibid., at [7.656].
[28] Ibid., at [7.657]; see also Goebel (2005) (n 1), 1186–1193.

superior' rule, it recognized the existing trade mark law qualifications of the priority and exclusivity rule.[29] But where exceptions are possible they are to be found in the trade mark rules and not elsewhere.

3.2 The Second Milestone: ECHR in *Anheuser-Busch v Portugal*

The second milestone is more limited in its geographical reach, but still important enough to deserve a mention directly after the WTO Panel Report. It is the *Anheuser-Busch v Portugal* decision of the Grand Chamber of the European Court of Human Rights (ECHR),[30] which was of fundamental importance for firmly putting the priority principle on the European map. The geographical reach should not be underestimated either: It goes well beyond the European Union and covers all 45 Member States of the Protocol No 1 to the European Convention on Human Rights under which the claim was advanced.[31]

For our purposes, the core finding of the judgment of the Grand Chamber of the ECHR is its confirmation that even a trade mark application enjoys the guarantees for the protection of possessions under Art 1 of Protocol No 1. This is so, because the application priority creates a vested right and makes the trade mark application an object of private property rather than merely a hope of obtaining a property right at some stage in the future. The background of the case was the following. With its complaint to the ECHR, Anheuser-Busch challenged a judgment of the Portuguese Supreme Court. The Portuguese Supreme Court had confirmed in its judgment the refusal of registration of Anheuser-Busch's trade mark BUDWEISER on the ground that it conflicted with the GIs Českobudějovický pivo and Českobudějovický Budvar, which were protected under a Bilateral Agreement concluded between the Czech Republic and the Republic of Portugal. The Court held that the term BUDWEISER and the terms Českobudějovický pivo and Českobudějovický Budvar were neither similar nor could they be confused by the average consumer in Portugal, but that the word Budweiser could be understood as a German translation of the two terms. Given that the Bilateral Agreement protected

---

[29] For an analysis of the potential for co-existence created by this panel report see also Gangjee (n 1); Hughes, (n 1), 327 *et seq*.
[30] *Anheuser-Busch, Inc. v Portugal*, Case No 73049/01 [2007] ETMR 24 (ECHR Grand Chamber, 11 January 2007).
[31] Protocol to the Convention for the Protection of Human Rights and Fundamental Freedoms, ETS No 9 (20 March 1952). The list of Member States to the Protocol is available at: http://conventions.coe.int/Treaty/Commun/ ChercheSig.asp?NT=009&CM=8&DF=&CL=ENG

GIs against use in translation, the Court decided to refuse the trade mark. This gave rise to an expropriation issue: the trade mark application predated the conclusion of the Bilateral Agreement by six years. This is why the interpretation of the priority principle was so important in this case.

As already mentioned, the complaint to the ECHR was based on Art 1 of Protocol No 1 to the European Convention on Human Rights. The provision reads:

> Every natural or legal person is entitled to the peaceful enjoyment of his possessions. No one shall be deprived of his possessions except in the public interest and subject to the conditions provided for by law and by the general principles of international law.
>
> The preceding provisions shall not, however, in any way impair the right of a State to enforce such laws as it deems necessary to control the use of property in accordance with the general interest or to secure the payment of taxes or other contributions or penalties.

Explicit reference regarding intellectual property is not provided for in the Convention. However, the ECHR had already previously held that the concept of 'possession' under the Convention had an autonomous meaning which was not confined to ownership of physical goods and was independent from the formal classification in domestic law. Therefore, rights and interests other than physical goods could also be regarded as 'possessions' for the purpose of Art 1 provided that they constituted assets.[32]

In the first judgment by the Second Section of the ECHR in 2005, while it was affirmed that registered trade marks amounted to 'possessions', the court denied that the right to possession extended to trade mark applications.[33] The Section held that prior to registration, even though the applicant would own 'an undeniable ownership interest' enjoying a 'certain legal protection', a trade mark application lacked the 'legitimate expectation' that it will actually materialize, which was necessary for constituting a 'possession'.[34] Anheuser-Busch appealed the case to the Grand Chamber, where contrary to the Second Section, the appellate

---

[32] *Beyeler v Italy*, Case No 33202/96 (ECHR, 5 January 2000) at [100]; *cfr* also *SmithKline & French Labs. Ltd. v The Netherlands*, Case No 12633/87 (ECHR, 4 October 1990) at [70]; *British-American Tobacco. Ltd. v The Netherlands*, Cae No 19589/92 (ECHR, 20 November 1995) at [91]; *Melnychuk v Ukraine*, Case No 28743/03 (ECHR, 5 July 2005).

[33] *Anheuser-Busch, Inc. v Portugal*, Case No 73049/01 (ECHR, 2nd Section, 11 October 2005).

[34] See further, B Goebel, 'Trade Marks are "Possessions" – but Applications are Not (Yet)' (2006) 1 JIPLP 240.

court endorsed the principle of trade mark applications being protected as possessions. The Chamber found 'that such applications may give rise to a variety of legal transactions, such as a sale or licence agreement for consideration, and possess – or are capable of possessing – substantial financial value'.[35] It concluded that 'these elements taken as a whole suggest that the applicant company's legal position as an applicant for the registration of a trade mark came within Art 1 of Protocol No 1, as it gave rise to interests of a proprietary nature'.[36] The Grand Chamber thereby led the priority principle into the European Convention on Human Rights, providing it with a legal importance that could no longer be ignored in any European legislation. The complaint itself was eventually dismissed as the Grand Chamber scrutinized the Portuguese judgment only against an 'arbitrariness' test. Given that there were some doubts as to priority in light of conflicting appellations of origin for the term 'Budweiser Beer' that existed at the time of the application for the trade mark and had been cancelled subsequently but not at the time when the Bilateral Agreement was concluded, the Grand Chamber did not consider the judgment of the Portuguese Supreme Court to be arbitrary.[37]

4. FURTHER DEVELOPMENTS ENDORSING PRIORITY, EXCLUSIVITY AND TERRITORIALITY

4.1 The Lisbon Agreement

The changing international environment as regards the protection of GIs has not left the Lisbon system untouched. Discussions arose on the interpretation of the Lisbon Agreement concerning the way the Lisbon system can fit in with, or be adapted to, the TRIPS Agreement. This, in particular, concerned situations of conflict between GIs and prior trade marks of third parties, an area that is only partly addressed in Art 5(6) of the Lisbon Agreement.

Art 5(6) stipulates that if an appellation which has been granted protection in a given country pursuant to notification of its international registration but was already in use by third parties in that country, and assuming that no refusal is notified, such third party may be granted a

---

[35] *Anheuser-Busch, Inc. v Portugal* (n 30) at [76].
[36] Ibid., at [78].
[37] Ibid., at [84].

phase-out period of two years. As mentioned above, this provision had been interpreted by some authors as requiring the phase-out of conflicting trade marks after a two-year grace period irrespective of whether that trade mark enjoyed a better priority.[38] Such an approach would, however, have been incompatible with the principles of priority, exclusivity and territoriality and would have raised serious concerns, for instance, as to its compatibility with the TRIPS Agreement. WIPO has therefore attempted to mitigate these concerns.

In a document prepared for the Working Group on the Development of the Lisbon System, the WIPO Secretariat clarified that Art 5(6) only applies in case no refusal was notified under Art 5(3).[39] Art 5(3) of the Lisbon Agreement provides for refusal of an international registration within a one-year period from the receipt of the notification of the registration by the Member State. Elsewhere the WIPO Secretariat observed that the Lisbon Agreement is neutral on the grounds on which protection of an international registration may be refused. It stated that it is left to the domestic forum to decide for the grounds on which a refusal can be based and that such grounds may encompass the conflict of an international registration with earlier trade marks.[40] No limits on the grounds that may be invoked in support of a refusal under Art 5(3) hence exist. Any challenge of the grounds of refusal must be before the courts of the refusing country.[41] By now it has also been clarified through Rule 16 of

---

[38] Cf. Baeumer (n 9) at [37], [56].

[39] WIPO, LI/WG/DEV/1/4 Prov (8 January 2010) at [144] ('If a country had not notified a declaration of refusal, then there was a provision under Art 5(6) of the Agreement which stipulated that prior use had to be terminated and that the country had the right to postpone the elimination of that prior use until two years after the refusal period had ended, at the latest. That provision only appeared to apply at that point in time. Indeed, if a country had notified a declaration of refusal, Art 5(6) and the corresponding Rule 12, according to their wording, would not appear to have application any longer').

[40] WIPO, LI/WG/DEV/1/2 Rev (10 February 2009), Annex II at [16b] ('For instance, a contracting country may refuse to protect an appellation of origin because it considers that the appellation has already acquired a generic character in its territory in relation to the product to which it refers or because it considers that the geographical designation does not conform to the definition of an appellation of origin in the Lisbon Agreement or because the *appellation would conflict with a trade mark or other right already protected in the country concerned*' (emphasis added)).

[41] Daniel Gervais, 'The Lisbon Agreement's Misunderstood Potential' (2009) 1 WIPO Journal 87, 98; see also Mihály Ficsor, 'Challenges to the Lisbon System' 31 October 2008 (WIPO/GEO/LIS/08/4) at [19] ('Art 5(3) and Rule 9(2)(ii) require an indication of the grounds for a declaration of refusal if the competent authority of

the Implementing Rules that a registered application of origin can be invalidated even where there had been no refusal under Art 5(3) in the first place. The application can be invalidated for any grounds the national legislator deems appropriate, including prior trade marks.[42]

This endorsement of priority, exclusivity and territoriality under the Lisbon Agreement through interpretation and an amendment of the Implementing Rules has led to a practice which shows that refusal of protection on grounds that the appellations conflict with prior trade marks seems broadly accepted.[43] For instance, the Czech appellation Bud was refused by a number of Lisbon Member States, such as Haiti, Macedonia, Peru and Nicaragua, on the ground that it conflicted with Anheuser-Busch's trade mark registrations for BUD enjoying priority in those countries.[44] Similarly, the Georgian appellation KHVANCHKARA was refused by Bulgaria based on an earlier national trade mark.[45]

Yet, if a country does not refuse protection of the appellation of origin registered under the Lisbon Agreement within the one-year period provided for under Art 5(3), Art 5(6) comes into play. Then a subsequent registration of an appellation of origin might still trump a prior trade mark which will have to be phased out after two years. This leaves a one year window for priority. Interestingly, the WIPO Secretariat stated that the mandatory phase-out period of two years provided for in Art 5(6) does not apply where a refusal is, in whole or in part, withdrawn under Rule 11. If a refusal is withdrawn, a longer period for co-existence is allowed, possibly for an indefinite period.[46] The WIPO Secretariat explained that such a longer period might be the result of an understanding reached between the country of origin and the country that had notified the refusal.[47]

---

the contracting country declares that it cannot ensure the protection of an appellation of origin whose international registration has been notified to it, but neither the Agreement, nor the Regulations specify the grounds on which a declaration of refusal can be based. *It happens in practice, and thus it seems broadly accepted, that an internationally registered appellation of origin is, and can be, denied protection in a contracting country to the Lisbon Agreement because existing prior rights would conflict with that appellation*' (emphasis added)).

[42]   Daniel Gervais, 'Reinventing Lisbon: The Case for a Protocol to the Lisbon Agreement (Geographical Indications)' [2010–2011] 11 Chicago Journal of International Law 67, 96.
[43]   Ficsor (n 41) at [19].
[44]   See Lisbon Registration No. 598 for BUD; available via the Lisbon Express database at: http://www.wipo.int/lisbon/en/
[45]   See Lisbon Registration No. 862 for KHVANCHKARA.
[46]   Gervais (n 42), 95.
[47]   WIPO, LI/WG/DEV/1/4 Prov (8 January 2009) at [161]–[162] (The 'Secretariat indicated that its understanding was that that provision only applied in case no

Just to be clear, Art 5(6) is anything but perfect. It still provides theoretically for the possibility of the phase-out of a prior right in situations where the owner of the prior right failed to oppose the appellation of origin. Appropriately interpreted in accordance with public international law rules, such a possibility should not be enforceable for being incompatible with both the TRIPS Agreement and fundamental rights guarantees. At the same time any adverse effects of Art 5(6) have so far been mitigated by accepting refusals to recognize Lisbon appellations on the basis of prior trade marks.

4.2 EC Regulation on Wines

The recent developments at the international level endorsing priority, exclusivity and territoriality in situations of conflict between GIs and trade marks continued with a reform of the EC Regulation on wines. Upon entering into force of (the now repealed) Regulation No 479/2008,[48] the

---

refusal was notified and not in case of the withdrawal of a refusal. This view was based on the fact that Art 5(6) itself, as well as Rule 12 of the Lisbon Regulations, laid down a procedure for notification at the end of the one-year period in which a refusal could be issued and specified that the two-year period for phasing out prior uses of the denomination concerned in a given country that did not refuse to protect the international registration in question would have to start at the end of that one-year period and had to be notified to the International Bureau within three months after the end of the one-year period . . . it was, of course, a question of interpretation whether or not Member States were also bound to eliminate prior use in case of the withdrawal of a refusal within two years from the date of such a withdrawal. Although this question had never arisen in practice under the procedures of the Lisbon system, the Secretariat had once been asked by a Member State for its understanding in this regard, as the Member State in question was, at that time, preparing the withdrawal of a refusal and was wondering whether it could grant a period to prior users that was longer than two years from the date of the withdrawal. After ample consideration, the Secretariat had indicated to the Member State in question that it was its understanding that, in case of the withdrawal of a refusal, a longer period was allowed. In this respect, the Secretariat referred to [(LI/WG/DEV/1/2 Rev.) at [18]] and said that such a longer period might be the result of an understanding reached between the country of origin and the country that had notified the refusal. If a longer period would not be allowed in the given situation, the country that had refused would simply wait until the longer period had passed before withdrawing the refusal. The Secretariat was of the view that this would not be in the interest of right holders, nor of the Lisbon system. Allowing a longer transitional period in the case of the withdrawal of a refusal would, after all, have the advantage that during the transitional period the international registration would benefit from protection against other third parties.').

[48] Council Regulation (EC) No 479/2008 of 29 April 2008 on the Common Organisation of the Market in Wine [2008] OJ L148/1, repealed by Council

absolute prevalence of GIs over trade marks adopted under the former Wine Regulation was abandoned. The former approach was clearly inconsistent with Arts 16.1 and 24.5 of the TRIPS Agreement and likely to be in violation of the fundamental guarantee of private property rights as protected under the EU Treaty and the European Convention on Human Rights. Pursuant to former Regulations No 1493/1999[49] and No 2392/89,[50] a later-filed GI for wine in principle prevailed over a prior trade mark. Only well-known trade marks that were registered at least 25 years before the official recognition of the geographical name in question and which had been used since then without interruption had been allowed to co-exist. Other trade marks were subject to cancellation.[51] This latter limited exemption was even not provided for in the original Regulation No 2392/89. It was included after an amendment of the Regulation following the famous TORRES case. In this case, trade mark registrations for TORRES had been in use for wine for many years. The prior trade mark registrations were put at risk upon the registration of the name Torres Vedras as a GI for a valley north of Lisbon which the government claimed was a traditional wine growing region under Regulation No 2392/89. The protection of the GI would have had as a consequence the invalidation of the TORRES trade marks as they conflicted with the later-filed geographical indication.[52] Regulation No 479/2008 (and now Regulation No 1234/2007) opted for a more balanced approach allowing refusal of protection of a later-filed GI where in light of a prior trade mark's reputation and renown it is liable to mislead the public and providing for co-existence for all other prior trade marks.[53] This approach is consistent

---

Regulation (EC) No 491/2009 amending Council Regulation (EC) No 1234/2007 on a Common Organisation of Agricultural Markets and on Specific Provisions for Certain Agricultural Products (Single CMO Regulation) [2007] OJ L 299/1.

[49] Council Regulation (EC) No 1493/1999 (n 12).

[50] Council Regulation (EEC) No 2392/89 Laying Down General Rules for the Description and Presentation of Wines and Grape Must [1989] OJ L232/13.

[51] Annex VII, Part F, No 2(2) of Regulation No 1493/1999 (n 12).

[52] See AG Skol, 'Geographical Indications and International Trade' 20 June 2003 (WIPO/GEO/SFO/03/15) 2; see also N Resinek, 'Geographical Indications and Trade Marks: Coexistence or "First in Time, First in Right" Principle?' [2007] EIPR 446, 449 (for the approach taken under Regulation No 2392/89).

[53] Art 43(2) of Regulation No 479/2008 (now Art 118k(2) of Regulation No 1234/2007) reads: 'A name shall not be protected as a designation of origin or geographical indication where, in the light of a trade mark's reputation and renown, protection is liable to mislead the consumer as to the true identity of the wine.' Art 44 of Regulation No 479/2008 (now Art 118l of Regulation No 1234/2007) produces a similar effect. A prior GI trumps a subsequent and 'infringing' trade mark. However, a prior trade mark (whether by registration or by use)

with the approach taken under Regulation No 2081/92 for agricultural products and foodstuffs which was found compatible with the TRIPS Agreement by the WTO Panel Report.[54]

In addition, Regulation No 479/2008 (and now Regulation No 1234/2007) adopted as uniform priority date the date on which the application for protection of the GI was submitted to the Commission, thereby endorsing both territoriality and priority.[55]

4.3 EC Regulation No 1151/2012 on Agricultural Products and Foodstuffs

The shift in the law of the European Union on the protection of GIs towards a greater respect of the principles of priority and exclusivity of trade marks is also visible in certain provisions of new Regulation No 1151/2012 on quality schemes for agricultural products and foodstuffs, repealing and replacing Regulation No 510/2006 which itself repealed and replaced Regulation No 2081/92.

Former Regulation 510/2006 was considered the cornerstone of the European system for the protection of GIs. Its relevance had even increased when the Court of Justice of the European Union (CJEU) held in *Anheuser-Busch v Rudolf Ammersin*[56] that the Community GI system established under it was exhaustive, thereby preempting any Member State level protection for GIs which would otherwise qualify for EU protection, that is, those that establish a link between specific characteristics of the product and the particular region where it was produced. Consequently, qualifying GIs that are not protected at Community level enjoy protection neither in the EU nor in the Member States. This also means that it should be impossible to invalidate a conflicting trade mark on the basis of, for instance, a national French appellation of origin that could have been protected under the Community system but was not.[57] The finding of

---

shall continue to exist notwithstanding the subsequent GI registration, provided no grounds exist to invalidate or revoke the mark under trade mark law. Art 44 concludes: 'In such cases the use of the designation of origin or geographical indication shall be permitted alongside the relevant trade marks.'

[54] See US Panel Report (n 4).
[55] B Goebel, 'Trade Marks as Fundamental Rights – Europe' (2009) 99 TMR 931, 949.
[56] *Budějovický Budvar v Rudolf Ammersin GmbH* (C-478/07) [2009] ECR I-7721, [2009] ETMR 65.
[57] Ibid., at [129], [170]–[173].

the CJEU has now been expressly endorsed by Regulation No 1151/2012; exception is only made for GIs of third countries.[58]

That the rules under Regulation No 510/2006 and now Regulation No 1151/2012 have shifted towards an endorsement of priority also becomes clear from their further provisions. With regard to refusal of a trade mark application because of a GI, Regulation No 510/2006 allowed for such refusal only where the trade mark application was submitted after the date of submission of the GI application to the Commission (Art 14(1)). Previously, former Regulation No 2081/92 had provided for the refusal of the registration of a trade mark application if the trade mark application was lodged prior to the GI but registered after the publication of the protection of the GI under EU law.[59] Hence, the registration date was decisive and not the application priority. The latter approach was clearly incompatible with the principle of priority and the nature of trade mark applications as possessions under the European Convention of Human Rights. The approach taken in Regulation No 510/2006 has now also been implemented in Regulation No 1151/2012 (new Art 14(1)).

The issue of appropriate reference date for priority and protection of GIs prior to their registration arose in the recent case *Bayerisches Bier II*.[60] The case was special insofar as, under Regulation No 2081/92, the reference date for priority was publication in the EC's Official Journal according to Art 6(2), but the GI at issue had never been published in this way.[61] It had been protected under the simplified procedure of

---

[58] Recital 24 reads:

To qualify for protection in the territories of Member States, designations of origin and geographical indications should be registered only at Union level. With effect from the date of application for such registration at Union level, Member States should be able to grant transitional protection at national level without affecting intra-Union or international trade. The protection afforded by this Regulation upon registration, should be equally available to designations of origin and geographical indications of third countries that meet the corresponding criteria and that are protected in their country of origin.

[59] Art 14(1)(3) of Regulation No 2081/92 reads: 'This paragraph shall also apply where the application for registration of a trade mark was lodged before the date of publication of the application for registration provided for in Art 6(2), provided that that publication occurred before the trade mark was registered.'

[60] *Bavaria NV v Bayerischer Brauerbund eV* (C-120/08) [2010] ECR I-13393, [2011] ETMR 11.

[61] Art 14(1)(1) of Regulation No 2081/92 reads:

Where a designation of origin or geographical indication is registered in accordance with this Regulation, the application for registration of a trade

Art 17 of Regulation No 2081/92. Under that procedure, no assessment at Community level or publication in the Official Journal was required. Protection was acquired by being listed in the Annex to Regulation No 1107/96[62] based on the information of the Member State about the national protection of the GI.[63] Art 17 aimed at preventing established GIs which were already protected under the legal system of the Member States from being subject to the same requirements and time limits as new GIs.[64] The Court of Justice clarified that protection of a GI required transparency, for example by publication of the registration, in order to satisfy the principle of legal certainty. The Court considered that the publication of Regulation No 1107/96 containing the claimed GI actually satisfied the transparency requirement.[65]

While having dropped the above-discussed problematic language of former Regulation No 2081/92 as regards priority, Regulation No 510/2006 did not yet fully endorse the priority principle. It established priority for a possible co-existence between a prior trade mark and later GIs by referring to two dates, the priority of the GI in its country of origin or – reflecting Art 24(5) TRIPS Agreement[66] – 1 January 1996[67] (Art 14(2)). By determining the date of protection in the country of origin as priority date, a trade

---

mark corresponding to one of the situations referred to in Art 13 and relating to the same type of product shall be refused, provided that the application for registration of the trade mark was submitted after the date of the publication provided for in Art 6(2).

[62] Commission Regulation (EC) No 1107/96 of 12 June 1996 on the Registration of Geographical Indications and Designations of Origin Under the Procedure Laid Down in Art 17 of Council Regulation (EEC) No 2081/92 [1996] OJ L148/1.

[63] Art 17(1)(1) of Regulation No 2081/92 provided: 'Within six months of the entry into force of the Regulation, Member States shall inform the Commission which of their legally protected names, or, in those Member States where there is no protection system, which of their names established by usage they wish to register pursuant to this Regulation.' Art 17(1)(2) required the Commission to register such GIs after a light touch review and Art 17(1)(3) allowed for transitional national protection during this process. Art 1 of Regulation No 1107/96 also refers to this fast track or simplified procedure.

[64] *Federal Republic of Germany and Kingdom of Denmark v Commission of the European Communities* (Joined Cases C-465/02 and C-466/02) [2005] ECR I-9115 at [AG55]-[AG56].

[65] See *Bavaria NV v Bayerischer Brauerbund* (n 60) at [60]–[65].

[66] Arts 24.5 (a), 65.1 of TRIPS; see also Nuno Pires de Carvalho, *The TRIPS Regime of Trade Marks and Designs* (Kluwer Law, 2006) at [24.5].

[67] For the basis for 1 January 1996 being incorporated into Regulation No 510/2006 see Roland Knaak, 'Die EG-Verordnung Nr. 510/2006 zum Schutz

mark that was registered in another Member State of the European Union was hence theoretically at risk of being phased out in case a conflicting GI was registered at Community level, which would without doubt have given rise to a challenge under the European Convention on Human Rights considering the ECHR's judgment in *Anheuser-Busch v Portugal*.[68] This has therefore now been corrected by new Regulation No 1151/2012, which provides for the date of submission of the application for the protection of the GI to the Commission as the uniform priority date that will apply to all products for which GIs can be registered.[69] Hence, the new priority rule is the well-established rule of application priority for the territory for which protection is sought. This is compliant with both the principles of priority and territoriality.

5. AVOIDING UNDUE MONOPOLIES – EXPLAINING THE EXEMPTIONS INHERENT WITHIN TM LAW

We have seen that the rules for resolving conflicts between trade marks and GIs have shifted towards priority, exclusivity and territoriality. That raises the question whether these principles are indeed adequate for resolving such conflicts. There must be a reason why contrary to, for instance, the conflict between a trade mark and a copyright or a trade mark and a design right, there was, and partly still is, such a passionate defense of *sui generis* rules giving superiority to GIs. One reason lies certainly in the perceived conceptual differences between a common good and private property, but there had also been the notion of unfairness; the perception of the poor rural community that is deprived of its cultural heritage by clever private entrepreneurs who know how to play the trade mark

---

von geographischen Angaben und Ursprungsbezeichnungen' GRUR Int. 2006, 893, 898.

[68] See the *Anheuser Busch* ECHR decisions at (n 30) and (n 33).
[69] Art 14(2) of Regulation No 1151/2012. See also Recital 31 that provides:

The scope of the protection granted under this Regulation should be clarified, in particular with regard to those limitations on registration of new trade marks set out in [the TMD 2008/95/EC] relating to trade marks that conflict with the registration of [PDOs] and [PGIs] as is already the case for the registration of new trade marks at Union level. Such clarification is also necessary with regard to the holders of prior rights in intellectual property, in particular those concerning trade marks and homonymous names registered as [PDOs] or as [PGIs].

system. Assuming – for the sake of argument – that such situations may occur we need to examine whether an internal trade mark law conflict solution mechanism based on priority, exclusivity and territoriality has the tools for resolving them. In our opinion this is the case. The monopoly right granted under trade mark law has never been unlimited.[70] Trade mark law has always known internal exemptions to the rules of priority, exclusivity and territoriality, for example as reflected in Art 17 of the TRIPS Agreement.[71] Employing the tools furnished by trade mark law in situations of conflict between trade marks and GIs allows for just results on both sides – avoiding undue monopolies but, at the same time, honoring the private property rights acquired by the respective rights holder.[72]

The main mechanisms are refusing trade marks as geographically descriptive or misleading and thereby avoiding a monopoly right in the first place; refusing a trade mark because of a conflict with an existing prior GI (priority and exclusivity rule) and finally limiting the monopoly right of a registered mark by providing for coexistence with geographically descriptive terms. We will draw on European trade mark law to provide illustrations of these mechanisms, although they generally exist in trade mark regimes around the world.

5.1 Refusing TMs as Geographically Descriptive

Trade mark law has long acknowledged the need to keep certain geographical signs free, providing for a possibility to refuse trade mark application if they either consist of a geographically descriptive term or if they are deceiving as to the geographical origin of the products for which the trade mark shall be registered. Art 3(1)(c) and (g) of the EU Trade mark Directive stipulates:

> The following shall not be registered or, if registered, shall be liable to be declared invalid:
> (c) trade marks which consist exclusively of signs or indications which may serve, in trade, to designate the kind, the quality, quantity, intended purpose, value, geographical origin, or the time of production of the goods or of rendering of the service, or other characteristics of the goods;
> . . .
> (g) trade marks which are of such a nature as to deceive the public, for instance as to the nature, quality or geographical origin of the goods or services.

---

[70] Okediji (n 1) 1364 *et seq.*
[71] Kur and Cocks (n 1) 1013.
[72] See also Hughes (n 1) 386.

The Community Trade Mark Regulation (CTMR)[73] contains identical provisions in Art 7(1)(c) and (g).

The standard test for determining whether a GI needs to be kept free was developed by the Court of Justice in its *Chiemsee* decision.[74] Interestingly, the case did not concern a trade mark application but the judgment was based on a reference from the Regional Court of Munich in a trade mark infringement case where the use of the word CHIEMSEE in a logo was challenged. Since CHIEMSEE also was the name of a German lake, the validity of the allegedly infringed trade mark had first to be established. In this regard, the Regional Court of Munich sought guidance from the Court of Justice. The Court of Justice observed (emphasis added):

> 31. Thus, under Art 3(1)(c) of the Directive, the competent authority must assess whether a geographical name in respect of which application for registration as a trade mark is made designates a place which is *currently associated in the mind of the relevant class of persons with the category of goods concerned, or whether it is reasonable to assume that such an association may be established in the future.*
>
> 32. In the latter case, when assessing whether the geographical name is capable, in the mind of the relevant class of persons, of designating the origin of the category of goods in question, regard must be had more particularly to the *degree of familiarity* amongst such persons with that name, with the characteristics of the place designated by the name, and with the category of goods concerned.
>
> 33. In that connection, Art 3(1)(c) of the Directive does not in principle preclude the registration of geographical names which are *unknown* to the relevant class of persons – or at least unknown as the designation of a geographical location – or of names in respect of which, because of the type of place they designate (say, a mountain or lake), such persons are unlikely to believe that the category of goods concerned originates there.[75]

A trade mark application may thus be refused where there is current association or where there is tangible evidence to assume that such an association may be established in the future. The latter has to be measured against the degree of familiarity of the term in question, that is whether it is likely that consumers will come to believe that the designated products originate from the relevant geographical location. Where there is no association with the geographical name or where no such association is likely to be developed in the future for the goods or services concerned by the public in the relevant territory, there is no reason to protect the name from being

---

[73] Council Regulation (EC) No 207/2009 of 26 February 2009 on the Community Trade Mark [2009] OJ L 78/42.

[74] *Windsurfing Chiemsee Produktions-und Vertriebs GmbH (WSC) v Boots- und Segelzubehör Walter Huber and Franz Attenberger* (Joined Cases C-108/97 and C-109/97) [1999] ECR I-2779, [1999] ETMR 585.

[75] Ibid., at [31]–[34].

monopolized. In particular, there is no goodwill or value endorsed in the designation other than that created by the use of the trade mark that could be prejudiced. Besides, as will be shown below, registering a geographical name as a trade mark does not automatically bar any use of the geographical name (here: Chiemsee) in connection with products similar to those for which protection is claimed under the trade mark that come from the said geographical region. Fair descriptive use remains possible.

The *Chiemsee* doctrine limits the registrability of geographically descriptive names as trade marks. At the same time, it does not exclude the possibility of words that are also geographical names being registered as trade marks. For example, the trade mark FUERTEVENTURA for clothing was found eligible for registration by the Office of Harmonization for the Internal Market (OHIM) as the Office found it unlikely that the average Community consumer would consider the Spanish island with the name Fuerteventura as an indication of the geographical source for certain clothing. According to OHIM, even though consumers in the EU might know the island of Fuerteventura, there was no reason to assume that consumers might think that the goods were made or designed on the island, given that Fuerteventura was not famous or on the verge of becoming famous for its clothing industry. In such a situation, there was no reason to keep a geographical name free.[76] Based on similar considerations, OHIM confirmed registration of the trade mark GREAT CHINA WALL for clothing. It observed that, although the Great Wall of China was one of the greatest wonders of the world and an appealing tourist attraction, it was not known in the Community for the presence of any facilities for the production of clothing textiles.[77] The rules for Europe were further clarified in the ALASKA litigation. The trade mark ALASKA for mineral waters was disputed and while the invalidation division of OHIM found the trade mark to be geographically descriptive, the Board of Appeals and the General Court considered the trade mark to be fanciful and distinctive, given that consumer had not become used to identify waters from the Americas by their geographical names. According to the General Court there was also no serious possibility that water from Alaska would be imported into the EU.[78]

---

[76] *Fuerte Ventura*, OHIM Opposition Division Decision No 1130/1999, 12 November 1999 (unreported).

[77] GREAT CHINA WALL (R 281/2009-1) OHIM First Board of Appeal, 26 January 2010 at [29].

[78] ALASKA (329 C 505503-1) OHIM Cancellation Division, 30 September 2004, at [21]–[25]; (T-225/08) General Court, 8 July 2009. The somewhat doubtful decision was appealed to the CJEU as (C-365/09) but has now been settled and withdrawn.

5.2 Refusing a Trade Mark Because of a Conflict with a Community GI

Moreover, the EU GI legislation itself relies on the principles of priority and exclusivity for resolving situations of conflicts between prior GIs and later trade marks. Art 14(1) of Regulation No 1151/2012 (similarly numbered in both predecessor Regulations) stipulates:

> Where a designation of origin or a geographical indication is registered under this Regulation, the registration of a trade mark the use of which would contravene Art 13(1) and which relates to a product of the same type shall be refused if the application for registration of the trade mark is submitted after the date of submission of the registration application in respect of the designation of origin or the geographical indication to the Commission.
>
> Trade marks registered in breach of the first subparagraph shall be invalidated.
>
> The provisions of this paragraph shall apply notwithstanding the provisions of [the TMD].

According to Art 14(1), a protected GI may constitute a ground for refusal of a later trade mark if the trade mark triggers any of the GI infringement provisions in Art 13(1)(a). This includes use of a GI for similar, though unprotected, products or exploitation of the reputation of that GI. Art 13(1) (b) further bars 'any misuse, imitation or evocation, even if the true origin of the products or services is indicated or if the protected name is translated or accompanied by an expression such as "style", "type", "method", "as produced in", "imitation" or similar'. Finally, and most broadly, the use of a GI is prohibited under Art 13(1)(c) and (d) against 'any other false or misleading indication as to the provenance, origin, nature or essential qualities of the product' and 'any other practice liable to mislead the consumer as to the true origin of the product'. Similar provisions are contained in the EC Regulations on wines and spirits.[79] In a departure from the former Regulations, the new Art 14(1) further clarifies the relationship between GI protection and trade mark protection providing that a later trade mark triggering one of the prohibitions in Art 13 may be refused under Regulation No 1151/2012 notwithstanding the provisions of the TMD.

The priority principle endorsed in the EU legislation on GIs is complemented by Art 14(2) of Regulation No 1151/2012 (similarly numbered in both predecessor Regulations) shielding rights to GIs against any prior registration of a trade mark in bad faith. Art 14(2) states:

---

[79] Priority of a Community GI was under scrutiny in *Bureau national interprofessionnel du Cognac v Gust Ranin Oy* (Joined Cases C-4/10 and C-27/10) [2011] OJ C269/8, [2011] ETMR 53 (relating to Cognac).

Without prejudice to Art 6(4), a trade mark the use of which contravenes Art 13(1) which has been applied for, registered, or established by use if that possibility is provided for by the legislation concerned, in good faith within the territory of the Union, before the date on which the application for protection of the [GI] is submitted to the Commission, may continue to be used and renewed for that product notwithstanding the registration of a [GI], provided that no grounds for its invalidity or revocation exist under [the CTMR or the TMD]. In such cases, the use of the [GI] shall be permitted as well as use of the relevant trade marks.

By clarifying that the situation under Art 14(2) is without prejudice to Art 6(4), the EU legislator emphasizes the exceptional character of coexistence between GIs and trade marks, thereby honoring the WTO Panel's finding that the provision needs to be construed narrowly in order to constitute a limited exception under Art 17 TRIPS.[80] Art 6(4) stipulates that a later-filed GI shall not be protected where, in the light of a trade mark's reputation and renown and the length of time it has been used, registration of the GI would mislead the consumer as to the true identity of the product. The grounds for invalidity or revocation under the CTMR referred to in Art 14(2) include trade marks which consist exclusively of indications which may serve, in trade, to designate the geographical origin of goods or to deceive the public as to the geographical origin of the goods.[81] Similar provisions can also be found in the EC Regulations on wines and spirits.

The question of bad faith was at the heart of the *Gorgonzola/Cambozola* case[82] where the *Handelsgericht Wien* (Commercial Court of Vienna) sought guidance from the CJEU regarding the use of the trade mark CAMBOZOLA when confronted by the (later) GI Gorgonzola. The CJEU specified that the concept of good faith must be viewed in the light of the entire body of legislation, both national and international, in force at the time when the application for registration of the trade mark was lodged. Most importantly, the proprietor of the trade mark may benefit from a presumption of good faith if the legislation in force at the date of filing did not protect the GI at issue.[83] As regards an analysis of the particular circumstances under which the application was submitted, the Court

---

[80] US Panel Report (n 4) at [7.679].

[81] Arts 3(1)(c), (g) and 12(2)(b) of the TMD. See also Section 5.1 above (refusing TMs as geographically descriptive).

[82] *Consorzio per la tutela del formaggio Gorgonzola v Käserei Champignon Hofmeister GmbH et al.* (C-87/97) [1999] ECR I-1301, [1999] ETMR 454; see also Roland Knaak, 'Die Rechtsprechung des Europäischen Gerichtshofs zum Schutz geographischer Angaben und Ursprungsbezeichnungen nach der EG- Verordnung Nr. 2081/92' GRUR Int 2000, 401, 404 *et seq*.

[83] Ibid., at [35].

found that only the domestic court before which the case was pending was in a position to carry out an analysis of that nature and determine whether the application was lodged in good faith.[84] The Court thus honored the principle of territoriality acknowledging that while it may be justified to assume bad faith in one country due to the specific situation found in that country this may not be the case in another. Applying the interpretative criteria developed by the Court of Justice, the Austrian Supreme Court (*Oberster Gerichtshof*) dismissed the case, finding that the application for CAMBOZOLA was in good faith as it was in accordance with the Austrian law applicable at the date when the trade mark application was lodged. At that time, the GI Gorgonzola was not protected under Austrian law, since the Stresa Convention had then not yet been ratified by Austria.[85]

Similarly, the *Corte d'Appello di Torino* dismissed Bayerischer Brauerbund's complaint for the invalidation of the prior BAVARIA trade marks of the Dutch brewery Bavaria based on the subsequent Community GI Bayerisches Bier.[86] The Dutch brewery had registered several trade marks containing the element BAVARIA, including a word mark BAVARIA. Bayerischer Brauerbund had sought a ruling declaring the trade marks invalid on the ground that they conflicted with the (later) Community GI Bayerisches Bier. The *Corte* reasoned that the trade marks were registered in good faith since BAVARIA had been used for beer by the Dutch brewery since the 1930s and Bayerisches Bier had not been known in Italy at the time of their registration. In light of this, the *Corte* granted the complaint only with regard to the word mark BAVARIA which it found to be geographically descriptive, hence subject to invalidation under the Trade Mark Directive. According to Art 14(2), a trade mark that may be invalidated under trade mark law must give way to a geographical indication despite having been registered in good faith. EU GI legislation and trade mark law therefore go hand in hand. With regard to the other trade marks, the Corte dismissed the claim, reasoning that the additional figurative elements and/or the additional word 'Holland Beer' ruled out any geographically descriptive or deceptive character.[87]

---

[84] Ibid., at [36].
[85] Case 4 Ob 25/01, Oberster Gerichtshof, 10 July 2001, published in GRUR Int 2002, 445, 447.
[86] Case No 299/07 RG, Corte d'Appello di Torino, 2 February 2012. The similar German case involving the trade mark 'BAVARIA HOLLAND BEER' of the BAVARIA brewery has recently been referred back to the Court of Appeals Munich (*Bayerisches Bier II*, Case I ZR 69/04, BGH, 22 September 2011).
[87] See Volker Schoene, 'Corte d'Appello di Torino: "Bavaria" – Marken koexistieren mit "Bayerisches Bier"' GRUR-Prax 2011, 322.

By this, the Corte d'Appello applied the interpretative guidelines given by the CJEU in its preliminary ruling on the case.[88] It had referred the question of whether the recognition of a GI was to have no adverse effects on the validity or usability of prior trade marks that incorporated the GI or a confusingly similar indication. The CJEU answered this by stating that the registration as GI must be interpreted as allowing co-existence of the prior trade mark and the later GI under Art 14(2), unless the trade mark was registered in bad faith or was affected by the grounds for invalidity or revocation set out in the TMD, thereby not affecting the validity or usability of a prior trade mark.[89] As already mentioned above, this has now been explicitly recognized under Regulation No 1151/2012.

In the context of the case, the CJEU also elaborated on the relationship between Art 14(2) and Art 14(3) of former Regulation No 2081/92 (now Art 14(2) and Art 6(4) of Regulation No 1151/2012). According to Art 14(3), a GI shall not be registered where, in the light of a trade mark's reputation and renown and the length of time it has been used, registration is liable to mislead the consumer as to the true identity of the product.[90] The Court held that Art 14(3) did not have any effect on the interpretation of Art 14(2). While both provisions cover the situation of conflict of a prior trade mark with a later GI, the provisions were independent, having separate objectives and functions and being subject to different conditions (scope, consequences and addressees are different).[91] Art 14(3) governs the registrability of a GI in a situation of a conflict with a prior well-known trade mark if the use of such a GI was misleading.[92] By contrast, Art 14(2) involves ascertaining whether a competing trade mark may continue to be used (as in the present case), if it was registered in good faith and not subject to the grounds for invalidity or revocation provided for by Art 3(1)

---

[88] *Bavaria NV and Bavaria Italia Srl v Bayerischer Brauerbund eV* (C-343/07) [2009] ECR I-5491, [2009] ETMR 61.
[89] Ibid., at [125].
[90] Art 14(3) of Regulation No 2081/92 reads: 'A designation of origin or geographical indication shall not be registered where, in the light of a trade mark's reputation and renown and the length of time it has been used, registration is liable to mislead the consumer as to the true identity of the product'; Art 6(4) of Regulation No 1151/2012 reads: 'A name proposed for registration as a designation of origin or geographical indication shall not be registered where, in the light of a trade mark's reputation and renown and the length of time it has been used, registration of the name proposed as the designation of origin or geographical indication would be liable to mislead the consumer as to the true identity of the product.'
[91] *Bavaria* (n 88) at [117].
[92] Ibid., at [118], [120].

(c) and (g) and Art 12(2)(b) of the Trade Mark Directive.[93] Given the different scope of application of the provisions, the fact that there was no likelihood of confusion on the part of the consumer, for the purposes of Art 14(3), between the later GI and the prior trade mark did not exclude the use of the mark from being covered by Art 14(2) or being considered a situation referred to in Art 13(1) (misleading use) or being subject to one of the grounds for invalidity or revocation as provided for by Art 3(1)(c) and (g) and Art 12(2)(b) respectively of First Directive 89/104.[94]

This standard provided for in situations of conflict between GIs and trade marks gives adequate consideration to the interests of the holders of both rights and is also internal to trade mark law, as laid out above.[95]

5.3 Refusing TMs Because of a Conflict with a National GI

The conflict between a national GI and a Community trade mark application was at issue in the *Anheuser-Busch (BUD)* judgment of the CJEU of 29 March 2011.[96] The judgment concerned an opposition of a trade mark application based on a prior GI protected under the Lisbon Agreement in France, Italy and Portugal, and under a bilateral treaty in Austria. The opposition was brought under Art 8(4) of the Community Trade Mark Regulation. Art 8(4) reads:

> Upon opposition by the proprietor of a non-registered trade mark or of another sign used in the course of trade of more than mere local significance, the trade mark applied for shall not be registered where and to the extent that, pursuant to the Community legislation or the law of the Member State governing that sign:
> (a) rights to that sign were acquired prior to the date of application for registration of the Community trade mark, or the date of the priority claimed for the application for registration of the Community trade mark;
> (b) that sign confers on its proprietor the right to prohibit the use of a subsequent trade mark.

The CJEU affirmed that a GI could be considered a 'sign used in the course of trade' for the purposes of trade mark law, thereby giving

---

[93] Ibid., at [119], [121].
[94] Ibid., at [124].
[95] See Section 5.2 above (refusing TMs as geographically descriptive). For a proposal to resolve the conflict between trade marks and geographical indications by importing the reasoning underlying the doctrine of 'honest concurrent use' see Gangjee (n 1) 1283; for a different approach considering GIs as a form of prior art limiting the trade mark owner's use of the priority club see Okediji (n 1) 1365.
[96] *Budějovický Budvar v OHIM* (C-96/09) [2011] ECR I-2131, [2011] ETMR 31.

national GIs their place as a possible prior right. Its qualification for trumping a later trade mark was only depending on whether the indication had a real presence on the relevant market in the European Union, because:

> The common purpose of the two conditions laid down in Art 8(4) of Regulation No 40/94 is to *limit conflicts between signs by preventing an earlier right which is not sufficiently definite* – that is to say, important and significant in the course of trade – *from preventing registration of a new Community trade mark*. A right of opposition of that kind must be reserved to signs which actually have a real presence on their relevant market. [emphasis added][97]

The Court specified that for a GI to block a trade mark application this required that the indication had been used in a substantial part of its area of protection,[98] in particular in the territory in which it was protected (here: France, Italy, Portugal and Austria).[99] Genuine use is required neither for owners of trade mark rights nor for holders of geographical indications opposing a later trade mark.[100]

This case is – under the auspices of trade mark law – again an important expression of what has been analyzed throughout this chapter: when it comes to the conflict between trade marks and geographical indications there are no special 'superior right' rules applying in favor of GIs. With its decision the CJEU reversed a very comprehensive, but erroneous judgment of the General Court. The General Court had considered that use of the designation outside the territory for which the GI was protected could be taken into account; it also considered it possible to take into account use made after the date of application for the trade mark. Most likely these findings were driven by an approach favoring GIs and it is exactly this 'special rule approach' which the CJEU dismissed when reversing the judgment of the General Court.

5.4 Co-existence of TMs with a Geographically Descriptive Term

A further exemption of the priority and exclusivity rule embodied in trade mark law that contributes to an adequate balance between trade marks and GIs is fair descriptive use. An example is Art 6(1)(b) of the Trade Mark Directive which stipulates:

---

[97] Ibid., at [157].
[98] Ibid., at [159].
[99] Ibid., at [162].
[100] Ibid., at [142]–[146].

The trade mark shall not entitle the proprietor to prohibit a third party from using, in the course of trade, . . . indications concerning the kind, quality, quantity, intended purpose, value, geographical origin, the time of production of goods or of rendering of the services, or other characteristics of goods or services.

This limitation on the scope of trade mark rights was the subject of a preliminary ruling of the CJEU in *Gerri/Kerry*.[101] Gerolsteiner Brunnen, a German company, claimed that the marketing of mineral water in Germany bearing a label including the name Kerry Spring prejudiced their prior trade mark rights in the term GERRI. The contested mineral water originated from a spring called Kerry Spring in Ireland. The Court held that fair use was not ruled out by the mere fact that a sign was used as a trade mark[102] and the signs at issue were confusingly similar.[103] It pointed out that the purpose of Art 6(1)(b) was to reconcile the interests of the parties.[104] Therefore, 'an overall assessment of all the relevant circumstances' of the case had to be carried out for establishing whether there was fair use of the younger designation.[105] The assessment was left for the national court which the CJEU – rightly – found best suited to examine the specific situation in the relevant country of conflict. The CJEU refined its approach to descriptive fair use in two later cases, *Anheuser-Busch*[106] and *Céline*,[107] which involved the question of fair use of confusingly similar trade names. The CJEU identified the following factors for assessing fair use, that is, whether the use was in accordance with honest practices in industrial or commercial matters:

- the extent to which the use of the third party's sign is understood by the relevant public, or at least a significant part of the public, as indicating a link between the third party's goods or services and the trade mark proprietor;
- the extent to which the third party ought to have been aware of that;
- the trade mark's reputation from which the third party might profit in marketing his goods or services.[108]

---

[101] *Gerolsteiner Brunnen GmbH & Co. v Putsch GmbH* (C-100/02) [2004] ECR I-691, [2004] ETMR 40; see also Goebel (2005) (n 1) 1181 *et seq.*
[102] Ibid., at [19].
[103] Ibid., at [25].
[104] Ibid., at [16].
[105] Ibid., at [26].
[106] *Anheuser-Busch Inc. v Budějovický Budvar* (C-245/02) [2004] ECR I-10989, [2005] ETMR 27.
[107] *Céline SARL v Céline SA* (C-17/06) [2007] ECR I-07041, [2007] ETMR 80.
[108] *Anheuser-Busch* (C-245/02) (n 106) at [83]; *Céline* (C-17/06) (n 107) at [34].

Applying the fair use exemption as understood by the CJEU thus provides for an adequate reconciliation of the interests embodied in GIs and trade marks.

6. TO BE CONTINUED

Our analysis has shown that priority, exclusivity and territoriality are increasingly accepted as the most relevant principles for resolving the dispute between trade marks and GI and that trade mark law's intrinsic limitations of the scope of the monopoly right ensure that equitable solutions can be crafted in cases of conflicts.

It would, however, be too early to proclaim an 'end of history' under the assumption that priority, exclusivity and territoriality are now the firmly established conflict resolution principles. The consensus is fragile, as the continued debate in international fora reveals. It is important to stay vigilant and actively contribute to the currently ongoing reform processes on the various systems of GI protection in order to ensure that the achievements of the last decade are not casually abandoned. The most important such forums are the WTO and WIPO; more specifically the Doha Round negotiations and the Lisbon Agreement reform process.

Two contentious issues which are under negotiations at the WTO Doha Round are: (1) the establishment of a system of notification and registration of GIs, and (2) the extension to other products than wines and spirits of the higher scope of protection afforded to wines and spirits by the TRIPS Agreement.[109] Two main sets of proposals on the creation of a multilateral register for GIs – one led by the EU, the other led by the US – have been submitted over the years. Binding wider protection for geographical indications for all WTO Member States is a key goal for the European Union in the negotiations, whereas the United States favors a database with merely informational character and to which participation is strictly voluntary.[110] However, nothing in the current deliberations at WTO level indicates that the protection of GIs – whether or not expanded to all products other than wines and spirits – is supposed to prejudice prior trade mark rights that exist in the individual WTO Member States. With their proposal of 2005, the European Union has committed itself

---

[109] See Report by the Chairman, 21 April 2011 (TN/IP/21). On these two issues see Chapter 7 by Handler and Chapter 8 by Martín in this volume.

[110] For the US proposal see WTO, Proposed draft TRIPS Council decision on the establishment of a multilateral system of notification and registration of geographical indications for wines and spirits, 31 March 2011 (TN/IP/W/10/Rev. 4).

to the principle of territoriality. Also, taking into account the outcome of the WTO Panel Report in the dispute between the EU and the US as well as Australia on the relation between GIs and trade marks, the EU's approach of mandatory co-existence that would have allowed GIs to be protected and, more importantly, used in any WTO Member State regardless of possibly conflicting prior trade marks, was abandoned. The EU acknowledged that:

> [There] are instances in which, under the current EU proposal, countries could give the appearance that a certain GI would be protectable in their territory, where, in fact, conflicting trade marks could in practice result in a GI not being allowed to be used in the course of trade.[111]

Under the proposals of both the EU and the US, it would thus be left to the national legislator and the national courts to decide whether a GI should be refused protection in the specific country where there are conflicting trade marks. Clarifying and confirming that rule in the respective proposals would be helpful.

Within the framework of the WIPO, a working group was set up in 2008 to discuss a possible development of the Lisbon System.[112] The stated aim of the development was to make the Lisbon system more attractive to Members and prospective Members, while preserving the principles and objectives of the Lisbon Agreement. The process culminated in the adoption of the Geneva Act of the Lisbon Agreement on Appellations of Origin and Geographical Indications on 20 May 2015 in Geneva. The conflict between trademarks and appellations of origin or GIs featured prominently in the course of the negotiations of the Geneva Act.[113] A first

---

[111] WTO, Communication from the European Communities 14 June 2005 (WT/GC/W/547), 4; see also the EU's modified proposal of 2008 WTO, Draft modalities for TRIPS related issues, 19 July 2008 (TN/C/W/52). For the problems resulting from a wider protection of GIs, see Hughes (n 1) 373 *et seq*.

[112] See for more information: http://www.wipo.int/meetings/en/topic.jsp?group_id=45

[113] For instance, in the session of December 2011, the WIPO Secretariat presented a proposal providing for a co-existence of prior trade marks with which it intended to 'safeguard prior rights to use a trade mark for a sign that corresponds to an internationally registered geographical indication, except whether the earlier trade mark has been acquired in bad faith'; WIPO, 'Notes on the draft new instrument concerning the international registration of geographical indications and appellations of origin' 7 October 2011 (LI/WG/DEV/4/4), Notes on Art 12, at [12.02]. In the session of December 2012, the presented proposal simply referred to the minimum standard of the TRIPS Agreement; WIPO, 'Notes on the draft new instruments', 28 September 2012 (LI/WG/DEV/6/4), Notes on Art 13 at [13.01].

assessment of the 2015 Geneva Act is mildly positive with a view to the conflicts between trademarks and GIs. First, the Geneva Act does not contain a provision like Art. 5(6) of the Lisbon Agreement. Secondly, Art. 11(3) of the Geneva Act which provides for the refusal or invalidation of a conflicting trademark limits such a refusal or invalidation to later trademarks. Thirdly, Arts 15 and 19 of the Geneva Act allow for the refusal or invalidation of an appellation of origin or geographical indication. It is understood that a prior trademark may constitute a ground for such a refusal or invalidation. Much will depend on how the Geneva Act will be implemented in practice. Still, it provides a good basis for resolving the conflicts between trademarks and GIs according to the well-established principles of priority, territoriality and exclusivity and this is a positive development.

7. CONCLUSION

If we look closely at the debate over the solution of conflicts between trade marks and GIs and analyze the legislation as well as the case law of the last decade it cannot go unnoticed that the almost universally accepted basic principles of priority, exclusivity and territoriality have become the guiding principles for these conflicts. Most stakeholders have learned to love their PET. As with any living creature, this pet changes with time and according to circumstances. Sometimes there is greater emphasis on exclusivity; sometimes the emphasis is on fair descriptive use. But the basics stay the same: priority, exclusivity and territoriality are the guiding principles for the conflict between trade marks and GIs, as opposed to any special rules based on an assumption of inherent superiority of one type of rights over the other. This is a hugely important achievement that will help in building further bridges between trade mark owners and GI rights holders.

14. The Budweiser cases: Geographical Indications v trade marks
*Christopher Heath**

It would be an interesting but very substantial exercise to collect all the decisions of all the courts and all the trade mark registries of all the countries where the battles have taken place. Many lawyers and their families in many places must be grateful that these two parties apparently cannot produce a once and for all world-wide settlement.
(Robin Jacob, LJ, in *Budejovicky Budvar Narodni Podnik v Anheuser-Busch Inc* [2009] EWCA Civ 1022 at [6])

1. INTRODUCTION

The beer-brewing tradition in the Bohemian town of Budweis/Ceske Budejovice dates back to the 13th century. The conflict that has been brewing between beer originating from this town and beer originating from the United States sold under the trade marks 'Budweiser' and 'Bud' since 1895 is not quite as old as that, yet dates back to the time before the First World War. Conflicts have arisen in about 30 countries and have raised a number of most interesting legal problems. But contrary to patent lawsuits worldwide that are decided on more or less the same facts, where the same patents and the same parties are concerned, the issue of the Budweiser cases is more complex. These cases are based on contract, registered marks, unregistered (but perhaps well-known) marks, trade names, on geographical indications protected by bilateral treaties or international

---

* This chapter is limited to those aspects of the Budweiser disputes that concern Geographical Indications. A complete overview of the disputes including questions related to contracts between the parties, and issues under trade mark law, has been published as: Christopher Heath, 'The Budweiser Cases – A Brewing Conflict', in C Heath and A Kamperman Sanders (eds), *Landmark IP Cases and Their Legacy, Macao Conference Papers vol. 6* (Kluwer Law International, 2010) 181. The book chapter also contains a table listing all Budweiser disputes worldwide to the extent known to the author. Consent by Kluwer for an amended republication is gratefully acknowledged. Most of the research for this article was done up to 2010. Further developments up to June 2015 were included, where possible.

agreements and on conflicts between trade marks and Geographical Indications. Due to differing marketing strategies, differing dates of market entry and the use of a variety of different indications, the facts to be decided by each national court were different, and so were the results.

2. THE PARTIES

2.1 Anheuser-Busch (AB)

The US company Anheuser-Busch (AB) has brewed and sold beer in the United States since 1876, as of 1895 under the name 'Budweiser'.[1] According to a statement by Adolphus Busch, before a district court in Eastern Wisconsin on 26 April 1894, the name 'Budweiser' was chosen because the beer brewed by AB could be likened to the beer brewed in Budweis, or Bohemia at least, in terms of quality, colour, taste and fragrance. AB, now the largest beer brewery in the world, sells beer to the entire American continent, as well as many Asian and some European countries, and advertises under the slogan 'the King of beers'. The trade mark 'Budweiser' was registered in the US on behalf of AB on 23 July 1907.

In Europe, the US beer-brewing company Anheuser-Busch sells its beer

---

[1] The following facts were compiled from the different lawsuits as mentioned below. Comprehensive case histories were given by the UK, Japanese, the Australian courts and the Advocate General in *Budějovický Budvar v Rudolf Ammersin GmbH* (C-478/07) [2009] ECR I-7721. As the author has not been personally involved in any of the Budweiser cases, the sources for the most part have been in publicly available information. In some cases the author could rely on the Max-Planck network of friends, and, for Italy, some information was kindly supplied by Stefano de Bosio who represented BB in Italy. The article in part draws upon previous studies: Christopher Heath, 'Il caso Budweiser', 2004 Rivista di Diritto Industriale, part II, 82 (in Italian), and Christopher Heath, 'A Hungarian Chapter to the Budweiser Saga' (2009) 40 IIC 328. For the cases in Portugal, attention is drawn to Antonio Corte-Real, 'The Conflict between Trade Marks and Geographical Indications – The Budweiser Case in Portugal', in C Heath and A Kamperman Sanders (eds), *New Frontiers of Intellectual Property Law*, 25 IIC Studies (Hart, 2005) 149. The early history of the Budweiser trade mark in the US can be traced from a number of cases that AB had initiated against US competitors: *Anheuser-Busch Brewing Ass'n v Fred Miller Brewing Co.*, 87 F.864 (CCD Wis 1898); *Anheuser-Busch v Budweiser Malt Products Corp.*, 295 F.306, 309 (2d Cir. 1923); *Anheuser-Busch v Cohen*, 37 F.2d 393, 397 (D Md 1930); *Anheuser-Busch v Power City Brewing Co.*, 28 F.Supp. 740 (WDNY 1939); *Anheuser-Busch Inc. v DuBois Brewing Co.*, 175 F.2d 370 (3d Cir).

under the following brands: 'Bud' in Greece and the Benelux countries; 'Budweiser' in Italy, Ireland, the United Kingdom, Spain, Portugal, Finland, Sweden and Denmark; 'Anheuser-Busch B' in Germany and Switzerland; and, finally, 'American Bud' in Austria.[2]

2.2 Budweiser Budvar (BB)

In 1795 in the Bohemian town of Budweis, German-speaking inhabitants founded an undertaking called 'Die Budweiser Brauberechtigten Bürgerliches Brauhaus' (BBBB). In 1895, the Czech-speaking inhabitants established another brewery under the name 'Cesky Akciovy Pivovar'.

In January 1946, the Cesky brewery was nationalised and carried on business as a National Enterprise. The German-speaking brewery, now denounced as a centre for 'traitors and collaborators', was confiscated and transferred to the national administration. By Ministerial Decree of 10 July 1948, retroactive as of 1 January 1947, BBBB's previously confiscated property was incorporated into the National Enterprise. Beer exported by BB between 1920 and 1960 largely concentrated on the trade name 'Budvar' (which is an abbreviation of Budejovice Pivovar and means Budweiser Brew), at least outside the German and Austrian markets. It was only at the end of the 1960s that BB harmonised its labels and relied on the strength of the indication 'Budweiser'. BB sells under the 'Budweiser Budvar' indication in about 40 countries.[3]

In countries in which employment of the 'Budweiser' designation is prohibited, the beer is sold under the label 'Czecvar', for example in the United States. Other labels include 'Budejovicky Budvar' (used in about ten countries),[4] 'Budweiser Budbräu', 'Budweiser', 'Bud' or 'Budvar'.

The beer that is sold by BB is described as follows:

> Budweiser Budvar is a typical Budweiser beer characterised by its slightly sweet, delicate and delicious taste with a pleasant aroma of hops and a slight

---

[2] This information was obtained from the site: www.bradfordlicensing.com in 2010. Yet as Anheuser-Busch is no longer a client of this firm, the information has meanwhile been deleted.

[3] Germany, Austria, Switzerland, Spain, Norway, Poland, Azerbaijan, Georgia, Vietnam, Morocco, Kyrgyzstan, Nigeria, Benin, Togo, France, Croatia, Latvia, Lithuania, Estonia, Kazakhstan, Mongolia, Serbia, England, Hungary, Israel, Ireland, Russian, Ukraine, Bosnia and Herzegovina, Albania, Moldova, Romania, Bulgaria, Luxembourg, Belgium, the Netherlands, Turkey, Macedonia, Armenia, Tajikistan, Portugal, Slovakia and Pakistan.

[4] Italy, Finland, Malta, New Zealand, Cyprus, Japan, Thailand, Sweden, Denmark, the Maldives and Hong Kong.

bitterness. In it are wedded many years of care, the know-how of generations of brewers and the special taste of ingredients which are all Czech.[5]

In 1967, a number of geographical indications were registered in (then) Czechoslovakia and subsequently abroad in the member states of the Lisbon Agreement. Based on domestic registrations effected on 22 November 1967, four of these appellations of origin (AOs) (Nos. 1–4) were subsequently published as international registrations under the Lisbon Agreement in March 1968.[6] All Lisbon AOs containing 'Budweiser' are:

1. 'Ceskobudejovicke Pivo/Budweiser Bier/Biere de Ceske Budejovice/ Budweis Beer' (No. 49)
2. 'Budejovicke Pivo-Budvar/Budweiser Bier-Budvar/Biere de Budweis-Budvar/Budweis Beer-Budvar' (No. 50)
3. 'Budejovicky Budvar/Budweiser Budvar' (No. 51)
4. 'Budejovicke Pivo/Budweiser Bier/Biere de Budweis/Budweis Beer' (No. 52)
5. BUDĚJOVICKÝ MĚŠŤANSKÝ VAR/ BUDWEISER BÜRGERBRÄU (No. 837).[7]

In every case, the area of production was identified as 'Ville du Ceske Budejovicky (Budweis)', and the owner as 'Organisations qui, dans la Region, s'occupent de la production des produits mentionnés'. All these appellations were designated for beer. All four original appellations were rejected by two Lisbon Agreement member states: by Haiti on 18 April 1969 and by Mexico on 11 June 1969. For a complete overview, see Section 4.2.3 below.

Under the Athens Accession Treaty to the European Union, Art 20, the following is mentioned:

---

[5] See: http://www.budvar.cz/html/en/home.html

[6] Information concerning all registered AOs under Lisbon are available on the 'Lisbon Express' database, at: www.wipo.int/ipdl/en/search/lisbon/search-struct.jsp

[7] Prior to the amendments of the Rules governing the Lisbon Agreement, appellations could be registered including their translations. This possibility has now been curbed (see below), yet this has no bearing on appellations already registered, as is made clear in the register itself: 'With regard to international registrations effected from April 1, 2002 onwards, the translation of an appellation of origin and, where relevant, its transliteration, will appear under a heading separate from that of the appellation of origin itself.' A further registration (No. 598) was made on 10 March 1975 for: Bud.

The names 'Budějovické pivo', 'Českobudějovické pivo' and 'Budějovický měšťanský var' shall be registered as protected geographical indications (PGI) and listed in the Annex in accordance with specifications submitted to the Commission. This is without prejudice to any beer trademark or other rights existing in the European Union on the date of accession.

3. 'BUDWEISER' LAWSUITS AROUND THE WORLD

For the last 35 years, Budweiser lawsuits have spanned five continents. The litigation of these cases has not been confined merely to Europe, but has also taken place in New Zealand, Australia, Japan, Korea, South Africa and Argentina. The highest density, however, is in Europe, and cases increasingly come before the Court of Justice of the European Union (CJEU).[8] The only major country in which, as yet, no litigation seems to have taken place is France.[9] The suits would appear to be a consequence of AB's determined efforts to push BB off the world's markets (at least under the indication 'Budweiser') since the mid-1970s. The struggle has as much to do with a commercial conflict as with history:

> The recognition of [BB's] part in the tradition of beer production in this part of their country informs and informed their actions and leads to a

---

[8] See the (subsequently elaborated) decisions: *Budějovický Budvar, národní podnik v Rudolf Ammersin GmbH* (C-216/01) [2003] ECR I-13617; *Anheuser Busch Inc. v Budějovický* (C-245/02) [2004] ECR I-10989; *Budějovický Budvar, národní podnik v OHIM* (Joined Cases T-53/04 to T-56/04, T-58/04 and T-59/04) [2007] ECR II-57 (GC); *Budějovický Budvar, národní podnik v OHIM* (Joined cases T-225/06, T-255/06, T-257/06 & T-309/06) [2008] ECR II-3555 (GC); *Anheuser-Busch v OHIM* (T-191/07) [2009] ECR II-703; *Budějovický Budvar National Corporation v Rudolf Ammersin GmbH* (C-478/07) [2009] ECR I-7721; [2009] ETMR 65 (CJEU Grand Chamber); *Budvar národní podnik; Budějovický Budvar v Anheuser-Busch* (C-482/09) [2011] ECR I-8701; *Anheuser-Busch v Budějovický Budvar národní podnik* (C-96/09) [2011] ECR I-2131, overturning decision (T-309/06) and referring back.

[9] However see the references in *Budějovický Budvar* (joined cases T-225/06 etc) (n 8), at [57]: First, that on 30 June 2004, the TGI Strasbourg held that 'BUD' was not an indication that could be protected under the Lisbon Agreement (to which France is a member), and that this decision was under appeal before the Appeal Court Colmar, and, second, that the French Patent Office on 19 May 2005 had refused registration of the trade mark BUD for beer on behalf of AB, and that this decision had become final. The CFI decision that had based its reasoning on BB's appellations of origin has subsequently been overturned by the CJEU (see (n 8)), as for a successful opposition the CJEU required an actual use of the opposing indication that was of more than local significance.

perception in them ... of a commercial struggle with an American company which has 'only' been making and selling beer since the last quarter of the nineteenth century, and which used the name of their town to promote its own beer.[10]

It is not without irony that the head of AB's intellectual property department could state: 'We are forced to fight country-by-country. It would be wise to have one body [determine this], but we are happy to take each case one by one. Most of our problems have come in a handful of countries mostly confined to Europe.'[11]

3.1 The Trade Mark Cases

Cases based on trade mark law solely were litigated in about fifteen countries, plus both instances of the CJEU, and the European Court of Human Rights. As these cases have no direct bearing on the issue of Geographical Indications, they are not further elaborated here.[12]

3.2 Cases Involving the Lisbon Agreement

Although the Lisbon Agreement is supposed to provide absolute protection for registered appellations of origin, BB's success has been mixed. It has prevailed in Israel on several occasions,[13] as well as in Bulgaria,[14] had mixed success in Hungary,[15] but (for the moment) has failed in

---

[10] Quoted from the Australian decision *Anheuser-Busch v Budějovický Budvar* (2002) 56 Intellectual Property Reports 182, 229 (Federal Court of Australia, 5 April 2002).
[11] Cited in (2003) 127 Managing Intellectual Property 14.
[12] The reader is thus directed to the chapter by Heath in (n *) above.
[13] Israeli Supreme Court, 10 January 1990, 22 IIC 255 – 'Budweiser', and 13 December 1992, 25 IIC 589 [1994] – 'Budweiser II'. The court held that appellations of origin as registered under the Lisbon Agreement could not be contested except in the country where the appellation was registered (here: Czechoslovakia); Israeli Supreme Court, 13 December 1992, 25 IIC 589 [1994] – 'Budweiser II'; also a third request by A.B. to register its marks failed: Israeli Supreme Court, 28 IIC 596 [1997] – 'Budweiser III'.
[14] Sofia City Court, 11 February 2000, Civil Case 950/1999; Sofia Appeal Court, 16 May 2000, Civil Case 7325/2000; Bulgarian Supreme Court, 7 February 2001, Civil Case 333/2000.
[15] Hungarian Supreme Court, decision of 21 March 2007, 40 IIC 353 [2009] – 'Bud', and Hungarian Supreme Court, 28 March 2007, 40 IIC 357 [2009] – 'Budweis Beer'.

Italy[16] and Portugal.[17] In Italy, protection was denied, *inter alia*, because beer as such could be produced anywhere and was not linked to a certain milieu, as was required under the Lisbon Agreement.

3.3 Cases Involving Bilateral Treaties on Geographical Indications

BB was relatively successful in invoking bilateral treaties in order to counter AB's efforts to register trade marks. In Portugal, AB managed

---

[16] Italian Supreme Court, 21 May 2002, 34 IIC 676 [2003] – 'Budweiser'. Subsequent to the Italian Supreme Court decision, the following developments have occurred: Decision of Rome District Court, 17 May 2005, and Appeal Court Rome, decisions of 5 March 2007 and 10 May 2010, all holding that as a consequence of the 2002 Supreme Court decision, 'Budweiser' could not be protected as denomination of origin under the Lisbon Arrangement 1958, allegedly because a 'Budweiser' beer could be produced anywhere, and that 'Budweiser' cannot be protected nor legitimately used as a geographical indication, not even under European law, although the beer brewed in Budweis/Ceske Budejovice – under the Accession Treaty of Athens of 2003 – has been acknowledged to have special characteristics due to its geographical origin and thus to enjoy protection as a PGI under Regulations 2081/1992 and 510/2006; Milan District Court, 30 July 2008 denying protection for the denomination 'Budweiser' even before the declaration of nullity by the Supreme Court in 2002; this decision was reversed by Appeal Court Milan, 16 June 2011 acknowledging that BB's trade mark registration corresponding to a well-known and reputed denomination of origin, namely 'Budweiser' beer in relation to the traditional Czech beer brewed in Budweis-Ceske Budejovice, is unlawful because of the inherent risk of misleading consumers; as such the relevant registration is null and void irrespective of the priority in the registration thereof. In a decision of the Italian Supreme Court of 13 September 2013 (case no. 21023/13, [2015] 46 IIC 888), the above-mentioned decisions of the Rome courts were overturned and remanded. The Supreme Court held that a 'geographical indication' need not necessarily be the official name of a place, but could also be an indication in a different language or a term used in the past. The court therefore annulled the revocation of BB's marks as ordered by the Rome courts. The case is now (May 2015) again pending before the Appeal Court Rome with the order to evaluate BB's argument that the Accession Treaty explicitly protects the beer brewed in Budweis as a Geographical Indication. If this was affirmed, it would also remove the basis for the Supreme Court's 2002 decision and thereby establish the validity of the AO under the Lisbon Agreement for Italy. In a complementary decision of 19 September 2013 (case no. 21472/13, [2015] 46 IIC 891), the decision of the Appeal Court Milan of 2011 was affirmed, and the Supreme Court held: 'Even geographical names used in the past, but not anymore, can continue to function as geographical indications when their reputation continues, although they are no longer used.'

[17] In a default judgement before the Lisbon First Instance Court, civil case 7906, 3rd division of the 13th circuit, 8 March 1995.

to register 'Bud', but not 'Budweiser'.[18] In Switzerland,[19] however, AB was precluded from using the mark 'Bud', as it was deemed similar to BB's protected indication 'Budweiser'. In Austria, both 'Budweiser' and 'Bud' were indications protected under an agreement between Austria and the Czech Republic.[20] The CJEU held that Austria was bound by the agreement.[21] BB had already prevailed in the interim action,[22] yet lost its

---

[18] Portuguese Supreme Court, 23 January 2001, 34 IIC 682 [2003] – 'Budweiser III'. This ruling was not based on the Lisbon Agreement, where the default judgement ordered invalidation of the appellation of origin, but rather on the bilateral agreement. AB appealed to the European Court of Human Rights against this alleged expropriation. The court ultimately rejected the case: *Anheuser-Busch, Inc. v Portugal*, Case No 73049/01 [2007] ETMR 24 (ECHR Grand Chamber, 11 January 2007). See K Beiter, The Right to Property and the Protection of Interests in Intellectual Property [2008] 39 IIC 714; see also Chapter 13 by Goebel and Groeschl in this volume.

[19] Swiss Supreme Court, 15 February 1999, GRUR Int. 1999, 1072, 1073 – 'BUD'.

[20] Agreement between the Republic of Austria and the Socialist Republic of Czechoslovakia on the Protection of Indications of Origin, Appellations of Origin and Other Indications of Origin for Agricultural products, Austrian Bundesgesetzblatt 1981/75, and Executive Provisions, Austrian Bundesgesetzblatt 1981/76.

[21] *Budejovicky Budvar v Rudolf Ammersin* (C-216/01) (n 8). This decision has now been qualified by the subsequent CJEU decision, *Budejovicky Budvar v Rudolf Ammersin* (C-478/07) (n 8). The Court concluded that after the accession of the Czech Republic to the EU, only simple (rather than qualified) Geographical Indications could be protected by bilateral agreements between two member states. Where an indication qualified for protection under Regulation 510/2006, but had not been registered, no protection could be afforded under national law, which included bilateral agreements between member states. This leaves open two questions. First, whether this ruling also applies to bilateral agreements between a member and a non-member state (given the fact that non-European indications can also be protected under Regulation 510/2006), and to what extent this ruling affects obligations under the Lisbon Agreement, where the protected indication originates in another member state, or in a non-Member State (see below). The ruling is highly problematic in that it amounts to an expropriation of previously protected Geographical Indications and therefore hardly stands up to the criteria as established by the European Court of Human Rights in *Anheuser-Busch, Inc. v Portugal*, Case No 73049/01 [2007] ETMR 24 (ECHR Grand Chamber, 11 January 2007). Criticism has been voiced by Knaak, 'Der Fall "Bud" – Schutzentziehung geistigen Eigentums durch Unionsrecht', GRUR 2012, 705.

[22] Austrian Supreme Court, 1 February 2000, Case No. 4 Ob 13/00s. The issue had become a brawl between the Austrian Supreme Court and the Commercial Court in Vienna. After the CJEU had decided the case, the Austrian Supreme Court referred the case back with the order to clarify whether 'Bud' could be considered a simple or indirect geographical indication in the Czech Republic. The latter was denied by the Vienna Commercial Court on 23 March 2006,

registered trade mark 'Bud' due to non-use, and further failed to stop AB from using 'Bud' based on the bilateral agreement.[23]

4. ANALYSIS, SUGGESTIONS AND CONCLUSIONS

The above decisions dealt with a great variety of legal aspects. These can be divided into those made under contract law (not further elaborated here), under trade mark law (mentioned to the extent that Geographical Indications are involved), and under the law of Geographical Indications.

4.1 The Angle of Trade Marks and Trade Names

As mentioned above, most disputes concerned questions of the law on trade marks, trade names or geographical indications. A number of different issues relating to the relationship between geographical indications and trade marks were argued in the manifold disputes.

---

a decision overturned and remanded back by the Vienna Appeal Court. The Vienna Commercial Court referred new questions to the CJEU that decided (in *Budejovicky Budvar v Rudolf Ammersin* (C-478/07) (n 8)) that should 'Bud' be considered a qualified (rather than a simple) Geographical Indication, protection under a bilateral treaty between two member states was pre-empted by Regulation 510/2006 on the protection of Geographical Indications. And as the Czech Republic seemed to regard 'Bud' as a qualifying geographical indication (quality and characteristics of the product must essentially be due to the geographical origin, including natural and human influences), protection could not be afforded thereto. The above decision of the Vienna Commercial Court was overturned by the Vienna Appeal Court that found the first instance court to be in breach of the order rendered by the previous Supreme Court decision, namely to conduct a survey of how 'Bud' was perceived by consumers in the Czech Republic. Upon further appeal, the Austrian Supreme Court, decision of 9 August 2011, held that an indication could not be at the same time a 'simple' and a 'qualified' geographical indication, and if the Czech Republic treated 'Bud' as a qualified geographical indication, failure to register it under the Council Regulations 2081/1992 and 510/2006 meant that it could no longer be protected. As this decision seems to put an end to the litigation in Austria, AB is not infringing any of BB's rights by using 'Bud' in commerce in Austria.

[23] Austrian Patent Court, 12 December 2001, [2001] Österreichisches Patentblatt 128. In May 2005, it was reported that the Salzburg District Court had held against BB in the latter's attempt to enforce the Geographical Indication 'Bud' against AB, or to prevent the latter from registering 'Bud'. The case has been appealed, yet its relationship with the cases mentioned in (n 22) is not clear.

4.1.1 Registrability

(a) General considerations   The first issue is to what extent would a trade mark containing 'Budweis' be registrable as such? After all, there is no shortage of registered trade marks for the denomination 'Budweiser' in Europe and elsewhere, be it for AB or BB.[24] To the extent that these are mere word marks, they should not have been registered, according to the standards of European trade mark law at least, if 'it can be reasonably expected that such an indication can be used as a geographical indication of those goods according to the understanding of these relevant circles',[25]

---

[24] In the Community trade mark register alone there are 16 trade marks listed containing 'Budweiser' (some also by third parties). AB held 16 marks containing the element 'Bud', BB three. For the register, see: http://oami.europa.eu/eSearch/#advanced/trademarks

[25] The 'Gold Standard' in this respect is *Windsurfing Chiemsee Produktions- und Vertriebs GmbH v Boots- und Segelzubehör Walter Huber* (Joined Cases C-108/97 & 109/97) [1999] ECR I-2779 at [25]–[27], [31]–[35]: Under Art 3(1)(c) of the EU Trade Marks Directive, which prohibits the registration of descriptive marks in the absence of evidence of acquired distinctiveness, 'the competent authority must assess whether a geographical name [being applied for] designates a place which is currently associated in the mind of the relevant class of persons with the category of goods concerned, or whether it is reasonable to assume that such an association may be established in the future'. In the latter case,

> when assessing whether the geographical name is capable ... of designating the origin of the category of goods in question, regard must be had more particularly to the degree of familiarity amongst such persons with that name, with the characteristics of the place designated by the name, and with the category of goods concerned. However, Article 3(1)(c) ... pursues an aim which is in the public interest, namely that descriptive signs or indications relating to the categories of goods or services in respect of which registration is applied for may be freely used by all, including as collective marks or as part of complex or graphic marks. Article 3(1)(c) therefore prevents such signs and indications from being reserved to one undertaking alone because they have been registered as trade marks.

It would significantly help the tenuous relationship between trade marks and Geographical Indications if the above standard was consistently applied. Yet it appears that even WIPO advocated the registration of geographical terms as individual trade marks, as was the case for 'Sidamo', a geographical area in Ethiopia well known for coffee:

> The government of Ethiopia decided that instead of trying to protect Ethiopian coffee's geographical origin, it would be better to protect its commercial origin, which it would do through registering trade marks. This was seen as a more direct route of protection because it would grant the government of Ethiopia the legal right to exploit, license and use the trade marked names in relation to coffee goods to the exclusion of all other traders. Unlike a GI, a trademark

which is certainly the case for beer. Such registrations make it difficult for BB, or for other breweries in Budweis, to market their beer under the correct denomination of origin (but also see below for possible defences to this argument). The European cases before the GC and CJEU did not concern this point, however, but rather dealt with the prior trade mark rights of BB, or prior conflicting registrations under the Lisbon Agreement, while the Appeal Court in Paris stressed that BB itself was not entitled to register those appellations protected under the Lisbon Agreement also as trade marks.[26] And while the Italian Supreme Court in 2002 declared the appellations of origin under the Lisbon Agreement related to 'Budweiser' null and void,[27] the same court in 2013 held in respect of the term 'Budweiser' that 'a trade mark consisting of a geographical name – albeit not constituting a geographical indication – may be used and registered only where there is no link between the product and the geographical location, or the products typical thereof'. 'Budweiser', when used by AB, was thus geographically misdescriptive to Italian consumers. Also for this point, the issue of whether only 'official' place names should give rise to misconceptions was argued before the court, but denied: after all, Italian consumers refer to 'Ragusa', 'Fiume' or 'Nizza' as place names, although these towns are now officially called 'Dubrovnik', 'Rijeka' and 'Nice'.

While the CJEU finally rejected AB's application for 'Budweiser' for beer (case C-214/09), it did so on the basis of conflicting prior trade marks of BB rather than the geographical nature of the denomination.

---

registration does not require a specific coffee to be produced in a specific region or have a particular quality in connection with that region. Using trade mark registrations, the government of Ethiopia could then produce greater quantities of specialty coffees from all over the country. (http://www.wipo.int/ipadvantage/en/details.jsp?id=2621)

A previous draft of the Lisbon Revision 2015 (LI/WG/DEV/9/2 of 16 April 2014) in Art 11 (2)(b) contained a similarly contorted notion that a trade mark conflicting with a Geographical Indication could co-exist 'if the goods in respect of which the trade mark is registered originate in the geographical area of origin and comply with any other applicable requirements for using the application of origin or the geographical indication'. This disregards the fact that a trade mark does not and cannot guarantee a geographical origin, as the mark can be freely transferred and the goods be replaced (even though the holder of the trade mark 'Jaffa Oranges' may start selling oranges from Jaffa, there is nothing under trade mark law to prevent him from also selling oranges from Valencia or Sicily under that mark). Fortunately, this provision has been excluded from the final Lisbon text of 2015.

[26] Paris Appeal Court, 1 October 1986, (1986) Propriété Industrielle Bulletin Documentaire (PIBD) III, 314.
[27] See (n 16).

For 'Bud', things were complicated, as 'Bud' is not a place name (although airlines use it as an abbreviation for 'Budapest'), but is protected under the Lisbon Agreement, and AB's trade mark application concerned both figurative and word marks. The GC in cases T-225/06 et al. analysed AB's prior rights under the Lisbon Agreement in the following terms:

> The contested decisions state that the Board of Appeal examined whether the sign BUD 'is an [AO] at all'. In that connection, the Board of Appeal held that the opposition at issue could not succeed under Article 8(4) of [the CTMR] on the basis of a right 'presented as an [AO]', but which in fact 'is not . . . an [AO] at all'. Further, the Board of Appeal held that the question whether the sign BUD was treated as a protected [AO], inter alia in France pursuant to the Lisbon Agreement, 'is of secondary importance'. It follows that the Board of Appeal ruled on the very classification as an [AO], without examining the extent of the protection of the [AO] at issue in the light of the claimed national rights.
>
> First – without needing to examine, in connection with the first part of the plea, the effects of the Lisbon Agreement on the protection of the claimed earlier right under French law – the Court points out that the registration of [AOs] under that agreement is made at the request of the authorities of the contracting countries, in the name of natural or legal persons, public or private, having, according to their national legislation, a right to use such appellations. In that context, the authorities of the contracting countries may, within a period of one year from the receipt of the notification of registration, make a declaration, with an indication of the grounds therefor, that they cannot ensure the protection of an [AO] (Article 5(1) and (3) of the Lisbon Agreement).
>
> Secondly, the [AO] registered under the Lisbon Agreement cannot be deemed to have become generic for as long as it is protected as an [AO] in the country of origin. Where that is the case, the protection conferred on the [AO] is ensured, and it is unnecessary to seek any renewal (Article 6 and Article 7(1) of the Lisbon Agreement).
>
> Thirdly, under Rule 16 of the Regulations under the Lisbon Agreement, when the effects of an international registration are invalidated in a contracting country and the invalidation is no longer subject to appeal, that invalidation must be notified to [WIPO's] International Bureau by the competent authority of the contracting country. In that event, the notification is to indicate the authority which pronounced the invalidation. It follows that, under the Lisbon Agreement, the effects of a registered [AO] can be declared invalid only by an authority in one of the contracting countries party to that agreement.
>
> In the present case, the [AO] 'bud' (No 598) was registered on 10 March 1975. France did not declare, within the period of one year from the date of receipt of notification of the registration, that it could not ensure the protection of that [AO]. . .
>
> Since the effects of the [AO] 'bud' have not been declared definitively to be invalid in France, the Board of Appeal ought, under Article 8(4) of [the CTMR], to have taken account of the relevant national law and the registration

made under the Lisbon Agreement, and did not have the power to call in question the fact that the claimed earlier right was an '[AO]'.[28]

4.1.2 Defences

In general, BB relied on two different defences; firstly, the use of its own commercial or trade name (not further analysed here), and/or secondly, the use of a Geographical Indication.

(a) Right to use? Use of a Geographical Indication as a defence is essentially the mirror image of relying on a Geographical Indication in order to prevent registration. BB's defence here could be twofold. Either to rely on a right of use conferred under the GI regime established under EU Regulation 2081/92,[29] or under Art 6 of the Trade Mark Directive which establishes a 'descriptive use' defence to a claim to trade mark infringement. In this respect, attention should be drawn to the WTO Panel decisions on GIs according to which there is no qualitative difference between trade marks and geographical indications in the sense that the latter could take precedence over the former. Rather, an earlier trade mark should prevail over a subsequent Geographical Indication as a rule, yet limited exemptions may be provided.[30] The WTO panel ruled that the former

---

[28] *Budějovický Budvar v OHIM* (Joined cases T-225/06 etc.) (n 8) at [83]–[87], [90]. Although this decision has been overturned by the CJEU decision (C-96/09) (n 8), it in fact confirmed the GC's findings on the legal nature of the Lisbon Agreement and its relationship to EU law.

[29] Council Regulation (EC) No 2081/92 of 14 July 1992 [1992] OJ L208/1. This has been successively replaced by Council Regulation (EC) No 510/2006 of 20 March 2006 on the Protection of Geographical Indications and Designations of Origin for Agricultural Products and Foodstuffs [2006] OJ L93/12; Regulation No 1151/2012 of the European Parliament and of the Council of 21 November 2012 on Quality Schemes for Agricultural Products and Foodstuffs [2012] OJ L323/1.

[30] WTO Panel Report *European Communities – Protection of Trade Marks and Geographical Indications for Agricultural Products and Foodstuffs*, complaint by the United States, (WT/DS174/R) 15 March 2005, and WTO Panel Report, Complaint by Australia (WT/D290/R) 15 March 2005. The limited exemptions analysed by the panel concern Art 14 of Regulation 2081/92 which, somewhat in line with the Lisbon Agreement, assumes that (even subsequently) registered GIs take precedence over previously registered marks, a position that can still be found in Art 14.2 of the successor Regulations. For a contrasting perspective, see Chapter 13 by Goebel and Groeschl in this volume. The view as voiced by the WTO panel is based on the assumption that Geographical Indications, just like trade marks, only come into existence once they are applied for registration. However, registration of a Geographical Indication is preceded by often decades of actual use, particularly in countries that in the past did not provide for a GI register. In other words, priority in respect of Geographical Indications must be

Regulation 2081/92 amounted to a limitation of a trade mark owner's rights under Art 16 TRIPS, yet qualified as a permissible limited exception under Art 17 of TRIPS. The defence of relying on a Geographical Indication would thus be limited to cases where the trade mark application precedes the existence or recognition of the Geographical Indication in that country.[31] This may be of interest in the case of recent bilateral agreements on Geographical Indications that contain the name 'Budweiser', or in light of the Accession Treaty between the Czech Republic and the European Union (Section Agriculture), which states the following:

> The names 'Budějovické pivo', 'Českobudějovické pivo' and 'Budějovický měšťanský var' shall be registered as protected geographical indications (PGI) and listed in the Annex according to the specifications submitted to the Commission. This is without prejudice to any beer trade mark or other rights existing in the European Union on the date of accession.

The question is thus whether the above registration as a PGI would give BB a better position in marketing its beer in those countries where BB's use of 'Budweiser' is currently impermissible (Spain, Denmark, Sweden and Italy). Without the specific addition on existing trade marks, registration under Regulation 2081/92 and its successors could give owners of a registered PGI the right to prohibit use of the indication for comparable products not originating from the region, under Art 13.1. While the scope of protection for GIs is further elaborated below, it should be sufficient to say at this stage that Art 13.1 of the current EU Regulation 1151/2012

---

assessed based on custom and actual use rather than registration (a position that has been implicitly called into question by the CJEU in the decision *Budejovicky Budvar v Rudolf Ammersin* (C-478/07) (n 8). This decision is questionable, though (see above n 22). The definition of an AO under the Lisbon Agreement takes the 'reputation' of products as a precondition to protection, confirming that this reputation is established prior to registration.

[31] As otherwise, either the trade mark should not have been registered in the first place, or is subject to subsequent invalidation (the latter was established for 'Budweiser' by the Italian Supreme Court, 19 September 2013 (n 16). The court held that even if 'Budweiser' had become well known for AB in the 1930s in Italy (an untenable finding in the first place), the indication was misleading when used for beer not originating from Budweis). Even if there is no formal recognition of an indication, the trade mark may still be subject to invalidation in cases where the mark gives rise to misconceptions of the geographical origin: *Bavaria NV and Bavaria Italia Srl v Bayerischer Brauerbund eV* (C-343/07) [2009] ECR I-5491, [2009] ETMR 61. On the other hand, the case makes clear that a subsequently-registered Geographical Indication does not give a right to enjoin the use of a conflicting prior mark unless such a conflicting mark should not have been registered in the first place due to its geographical or misleading nature.

(as well as Art 3 of the Lisbon Agreement) does not extend protection to active use as a translation for the GI registrant, yet does give a right to prevent third parties from using such a translation, that is, a right to exclude. In other words, the protected indication 'Budejovicke Pivo' does not give BB the right to use a translation thereof, and thereby to force a concurrent use in countries where BB is, for whatever reason, prohibited from using the indication 'Budweiser'.[32]

(b) Defence   Although the above registration does not confer a right on BB to use of the indication 'Budweiser', BB may still be able to rely on Art 6(1)(b) of the Trade Mark Directive 89/104. Under this provision, the trade mark owner may not prohibit third parties from using indications on the geographical origin of goods. Whether this would be of use to BB

---

[32] The WTO panel in DS 290 (n 30) at [7.574] states: 'Australia also alleges that three Czech beer GIs, 'Budějovické pivo', 'Českobudějovické pivo' and 'Budějovický měšťanský var' could be used in a manner that would result in a likelihood of confusion with the prior trademarks BUDWEISER and BUD, registered in respect of beer. In response to a direct question from the Panel, the European Communities did not deny that these GIs could be used in a manner that would result in a likelihood of confusion with these prior trademarks. Instead, it pointed to an endorsement on the three GI registrations that they apply 'without prejudice to any beer trademark or other rights existing in the European Union on the date of accession'. This might imply that it accepts a likelihood of confusion, but considers that there are other means besides Art 14(3) to deal with that. And, further, at [7.657], the Panel notes:

> The Regulation curtails the trade mark owner's right in respect of certain signs but not all signs identical or similar to the one protected as a trade mark. It prevents the trade mark owner from exercising its right to prevent use of an indication registered as a GI in accordance with its registration. The Panel recalls its finding earlier *that the GI registration does not confer a positive right to use any other signs or combinations of signs nor to use the name in any linguistic versions not entered in the register*. The trade mark owner's right is not curtailed against any such uses. If the GI registration prevented the trade mark owner from exercising its rights against these signs, combinations of signs or linguistic versions, which do not appear expressly in the GI registration, it would seriously expand the exception and undermine the limitations on its scope. (emphasis added)

While the CJEU has consistently denied the direct applicability of TRIPS on Community legislation, it is likely that an interpretation of TRIPS by a WTO panel related to a Community Regulation on geographical indications will be taken into account by the court. The Lisbon Revision 2015 resolves the whole issue of official or unofficial names, translations or not, by allowing for the registration of 'the name of a geographical area, or another denomination known as referring to such area'. The latter could thus also include names that do not correspond to the official names in a given country, e.g. 'Ceylon' tea from Sri Lanka.

depends on the answers to two questions. Firstly, to what extent does Art 6(1)b) of the TMD, with 'indications on the geographical origin', make provision for translations of such indications? This is clear where such translations are indeed understood in a geographical sense. However, according to many decisions (for example in Denmark),[33] this was not the case for Budweis (but, one could add, the relevant public might have made even less sense of the name 'Ceske Budejovice'). At the end of the day, public recognition cannot be a limiting factor, as, otherwise, most geographical origins abroad would not qualify for such a defence in the absence of sufficient public knowledge. Rather, the provision is intended to allow goods to be correctly labelled as to their geographical origin, and 'Budweiser Beer' that originates in Ceske Budejovice (or its translated equivalent) certainly qualifies no less than Ceylon Tea (where Sri Lanka rather than Ceylon is the official denomination of the country). In this respect, Art 6(1)(b) should be the mirror image of Art 3(1)(c) that according to the CJEU,

> pursues an aim which is in the public interest, namely that descriptive signs or indications relating to the categories of goods or services in respect of which registration is applied for may be freely used by all, including as collective marks or as part of complex or graphic marks. Article3(1)(c) therefore prevents such signs and indications from being reserved to one undertaking alone because they have been registered as trade marks.[34]

As the following paragraph explains, the CJEU has not clearly endorsed this view, though.

BB, although relying on the geographical origin of its products, also uses 'Budweiser' as a trade mark. To what extent, then, is Art 6(1)(b) of the Trade Mark Directive also applicable if a third party uses a descriptive indication as a trade mark? Precisely this was the question the German Supreme Court referred to the CJEU.[35] In that particular case, the designation 'Kerry Spring' was also used to distinguish the defendant's products from other goods (here, mineral waters) and to indicate the source of the goods, and not merely describe the place of origin. The designation was used, therefore, as an actual trade mark, and this argument could also apply to the products marketed by BB. The CJEU has not yet given a clear answer to this question.[36] While the Advocate

---

[33] Copenhagen District Court, 16 April 1998, 31 IIC 104 [2000] – 'Budweiser'.
[34] *Windsurfing Chiemsee* (n 25).
[35] Federal Supreme Court, 7 February 2002, English version in 34 IIC 555 [2003] – 'Gerri/Kerry Spring'.
[36] *Gerolsteiner Brunnen GmbH & Co. v Putsch GmbH* (C-100/02) [2004] ECR I-691, [2004] ETMR 40.

General and the European Commission would have answered the question in the affirmative, the CJEU made the defence under Art 6(1)(b) conditional upon use of the mark in accordance with honest practices in trade, which it subsequently further clarified in case (C-245/02): 'The condition of "honest practices" is, in essence, an expression of the duty to act fairly in relation to the legitimate interests of the trade-mark proprietor.' It should be a matter for the national courts to address this in each individual case, yet it has never been alleged (not even by AB) that BB, in using its indications, has the actual intention of causing confusion with AB's products or misappropriating reputation. As of yet, BB has invoked only the defence of using 'Budweiser Budvar' as the name of its enterprise (that is corporate or trading name, as part of an 'own name' defence), while the potential for use as a geographical origin under Art 6 remains unexplored.

4.2 Geographical Indications[37]

4.2.1 The Paris Convention and Madrid Agreement

(a) The Paris Convention The protection of Geographical Indications started with the notion of protection against misleading use. In order to determine what exactly may be considered misleading, the general principles under trade mark law serve as a useful guideline. However, some differences should be noted. For one, trade mark protection of Geographical Indications is not only denied in cases concerning misleading use, but also in those which concern descriptive use. This is understandable, as a descriptive trade mark cannot confer a distinctive commercial origin. On the other hand, the use (rather than the registration) of a descriptive indication does not cause confusion, and nor is it misleading in any way. Further, trade mark law is not only concerned with actual, but also with potential conflicts. Registration can be denied not only if the place is currently not associated with certain goods or services, but also if this could be the case in the future. Finally, public interest might require denying proprietary protection by registration for one single enterprise, where (non-misleading) use in commerce is completely permissible. In fact, the argument that registration of a Geographical Indication should be denied because other undertakings might have a legitimate interest in using

---

[37] The following explanations draw upon Christopher Heath, 'Geographical Indications: International, Bilateral and Regional Agreements', in Heath and Kamperman Sanders (eds) (n 1) 97.

this indication already presumes that such use by other undertakings is lawful.[38]

This concept of preventing misconceptions in trade was adopted by the Paris Convention (1883) and the Madrid Agreement for the Repression of False or Deceptive Indications of Source (1891). In particular, Art 10 of the Paris Convention prevents the use of false (but not misleading) indications of source:

> False Indications: Seizure, on Importation, etc., of Goods Bearing False Indications as to their Source or the Identity of the Producer
> (1) The provisions of the preceding Article shall apply in cases of direct or indirect use of a false indication of the source of the goods or the identity of the producer, manufacturer, or merchant.
> (2) Any producer, manufacturer, or merchant, whether a natural person or a legal entity, engaged in the production or manufacture of or trade in such goods and established either in the locality falsely indicated as the source, or in the region where such locality is situated, or in the country falsely indicated, or in the country where the false indication of source is used, shall in any case be deemed an interested party.

Of course, this provision is rather rudimentary, not only in its remedies (which refer to Art 9) but also in its approach. First, it does not address the question of generic indications – on the face of it, the sale of 'Hamburgers' under such indication would amount to a 'false indication of the source', unless the hamburgers came from Hamburg[39] – and, second, a number of Old-World place names have found their way to the New World and could therefore be legitimately used under Art 10, but still cause confusion.

---

[38] I have further elaborated these principles in the 'Brewing Conflict' book chapter see (n *) above, 195–201.

[39] The history of this provision demonstrates that the initial objective was to prevent only the 'fraudulent' use of an indication, for which the original provision required the additional use of a fictitious trade name in conjunction with the false indication. This was only changed at the 1958 Lisbon Conference where there were still widely different views on the issue of whether courts in any given country should be qualified to find an indication generic or descriptive (as 'Hamburger' no doubt would be). The French delegation, based on its understanding of the Madrid Arrangement and the draft Lisbon Agreement (in preparation at the time) took the view that only the country of origin (here: Germany) could determine whether 'Hamburg' was descriptive or generic. See S Ladas, *Patents, Trademarks and Related Rights, Vol. III* (Cambridge, MA: Harvard University Press, 1975) 1577–1580; D Gangjee, *Relocating the Law of Geographical Indications* (CUP, 2012) 41–73.

(b) The Madrid Agreement 1891 The Madrid Arrangement, a special arrangement permitted under Art 15 of the Paris Convention, takes this rather rudimentary approach to protection under Art 10 of the Paris Convention further in that:

- It protects against false or 'misleading' direct or 'indirect' indications (Art 1),[40] or all indications that are capable of deceiving the public as to the source of goods (Art 3*bis*).
- It allows national courts to decide what applications are considered generic, but excludes regional wine indications from such discretionary authority (Art 4). In other words, national courts of member states are not entitled to find an indication such as 'Champagne' generic, although such indication might be regarded as generic by the public in this member state.

There was an intense discussion during the Washington Revision Conference as to whether Art 4 should be extended to all indications for products which derive their natural qualities from the soil and climate (rather than only wine products), yet this was rejected.[41]

The latter point is vital for understanding the Lisbon Agreement, as it limits the competence of national courts. National courts ('tribunals') should neither be entitled to consider an indication generic, nor should they be entitled to question the validity of an indication once that indication has been protected in the country of origin, communicated to the International Bureau and examined by the other countries.[42]

---

[40] An example would be the decision of the Japanese Patent Office to refuse registration of the mark 'Loreley' for wine products which bore no relation to Germany: Japanese Patent Office, 23 October 1991, (1993) 24 IIC 409.

[41] Ladas (n 39), 1592.

[42] In this respect, the Lisbon Revision 2015 adds a somewhat obscure footnote to Art 12 ('Protection against becoming generic') that reads:

> For the purposes of this Act, it is understood that Article 12 is without prejudice to the application of the provisions of this Act concerning prior use, as, prior to international registration, the denomination or indication constituting the appellation of origin or geographical indication may already, in whole or in part, be generic in a Contracting Party other than the Contracting Party of Origin, for example, because the denomination or indication, or part of it, is identical with a term customary in common language as the common name of a good or service in such Contracting Party, or is identical with the customary name of a grape variety in such Contracting Party.

To the extent that grape varieties are concerned, the footnote is difficult to reconcile with Art 4 Madrid Agreement. See below for further comments on this issue.

(c) The Versailles Peace Treaty 1919 as a multilateral agreement The first agreement where an absolute form of protection was put into practice was the treaty of Versailles, which in Art 275 obliged Germany, Austria, Hungary and Bulgaria:

> On condition of reciprocal treatment on the part of the Allied and Associated powers, to give binding force to the law or decisions of the country to which a regional appellation of products of the vine belongs. The tribunals are no longer free to decide that a regional appellation of spirits, for instance the name 'Cognac', is a generic or descriptive term, if by the law of the country of origin (in the instance cited, France), or by its court's decisions, the appellation in question is a true appellation of origin belonging to the products of a specified region.[43]

With the attempts to revise the Madrid Arrangement, as well as the Versailles Treaty, the protection of Geographical Indications was taken beyond the ambit of misconception and already some way towards absolute proprietary protection. Such absolute proprietary protection is also sought by bilateral treaties for the protection of Geographical Indications.

4.2.2 Bilateral treaties

In bilateral treaties the principle was established that the law of the country of origin should govern the question of whether a designation was a true indication of origin belonging to the producers in a specified region, or whether it could be treated as generic or descriptive.

In recent years, the EC has concluded a number of bilateral agreements on the protection of appellations of origin with, *inter alia*, Bulgaria, Hungary and Romania. These agreements became obsolete once these countries joined the European Community. Outside Europe, the EC has concluded agreements with Morocco, Tunisia, Canada, Chile and South Africa, as well as with Australia.[44]

In these agreements, both parties must agree in advance on a list of mutually protected Geographical Indications. Accordingly, subsequent opposition by private parties may not be raised, and the number of protected Geographical Indications may only be broadened by mutual

---

[43] Ladas (n 39), 1597.
[44] For details, see http://ec.europa.eu/trade/policy/countries-and-regions/agreements/; The Wine Agreements are summarised in O'Connor & Co, *Geographical Indications and TRIPs: 10 Years Later... Part I – A Roadmap for EU GI Holders to Get Protection in other WTO Members* (Report commissioned for EC (DG Trade) 2007). The same pattern can now be found for bilateral Free Trade Agreements, an instrument whereby the EU tries to export its standards for protecting Geographical Indications to countries such as Korea.

consent and a subsequent amendment. Transition periods facilitate the phasing out of indications incorrectly used by the contracting member state. In almost all bilateral agreements on the protection of Geographical Indications, provisions can be found indicating that 'diluting' an indication by additions such as 'type', 'method', and so on, is not permissible.

Over the course of the European Budweiser litigation, the CJEU dealt with the relationship between bilateral agreements between member states, and protection of Geographical Indications under Regulation 510/2006.[45] At issue was the indication 'Bud', protected under the bilateral agreement between the Czech Republic and Austria, considered within the context of EC Regulation 510/2006. According to the court:

> [Regulation 510/2006] confirms, in respect of the 10 new Member States, what was already the case for the 15 old Member States, that is to say, that the national protection of existing qualified geographical indications is permitted only if the conditions in the rules of the transitional legislation specifically laid down in respect of such indications are met, including the condition that an application for registration be submitted within six months, which the Czech authorities neglected to do with regard to the designation 'Bud' at issue in the main proceedings.

Those particular systems and, especially, the express authorisation granted, on certain conditions, to the member states to maintain, on a transitional basis, the national protection of existing qualified Geographical Indications would be difficult to understand if the Community system of protection of such indications were not exhaustive in nature, implying that the Member States retained in any event an unlimited capacity to maintain such national rights.

> In the light of the foregoing . . . the Community system of protection laid down by Regulation No 510/2006 is exhaustive in nature, with the result that that regulation precludes the application of a system of protection laid down by agreements between two Member States, such as the bilateral instruments at issue, which confers on a designation, recognised under the law of a Member State as constituting a designation of origin, protection in another Member State where that protection is actually claimed despite the fact that no application for registration of that designation of origin has been made in accordance with that regulation.[46]

Thus, bilateral agreements between member states can no longer be invoked for those indications that would otherwise be registrable under

---

[45] *Budejovicky Budvar v Rudolf Ammersin* (C-478/07) (n 8).
[46] Ibid., at [127]–[129].

Regulation 510/2006 (or by extension, its successor Regulation 1151/2012), in which respect the court notes the following:

> In that regard, it must be borne in mind that . . . [the] Court observed that the first question referred in that case dealt with the hypothesis that the name 'Bud' constitutes a simple and indirect indication of geographical provenance, that is to say, a name in respect of which there is no direct link between a specific quality, reputation or other characteristic of the product and its specific geographical origin, so that it does not [qualify for protection under the EU regime] and which, moreover, is not in itself a geographical name but is at least capable of informing the consumer that the product bearing that indication comes from a particular place, region or country . . .[47]

In view of the registrations made under the Czech Accession Treaty (see above) the inconsistency is evident. One would of course have to observe that there is a link between the specific quality of the product and its geographical origin, a point made by the Austrian Supreme Court in its second 'Bud' decision on 9 August 2011. However, the Austrian Supreme Court could also have drawn the opposite conclusion, namely that as simple Geographical Indication, 'Bud' should not have been registered under the Lisbon Agreement. After all, the indication 'Bud' would probably not be registrable under the Regulation, as it does not qualify as the direct name of a region or specific place (and, *nota bene*, neither is this the case for 'Feta', 'Cava' or 'Vinho Verde'). According to the CJEU, 'a simple designation of geographical provenance is indirect if it is not in itself a geographical name but is at least capable of informing the consumer that the product bearing that indication comes from a particular place, region or country'.[48]

---

[47] Ibid., at [73].

[48] Ibid., at [82]. The issue of European pre-emption is complicated by the fact that there are various reasons why an indication may not have been registered under the Community Regulations. The one reason the CJEU has in mind is simple inertia. But what if an application for 'Pilsener' was refused on a Community level because it was deemed generic in a number of EC countries? Should the indication then also lose protection in those countries where the indication is considered geographical rather than generic, or where it is protected under bilateral agreements? The issue is different for 'Budweiser', as prior trade mark rights cannot as such prevent registration of an indication. The whole concept of precedence of the Community registration system has been questioned by R Knaak, 'Der Fall "Bud" – Schutzentziehung geistigen Eigentums durch Unionsrecht', GRUR 2012, 705. Knaak highlights that under the Community system of IP enforcement, the courts of one country are not entitled to revoke or invalidate a registered IP right registered in another country. For this reason, according to Knaak, only the courts in the Czech Republic would have been entitled to rule the validity of the registered

4.2.3 The Lisbon Agreement for the Protection of Appellations of Origin

(a) History The Lisbon Agreement concluded in 1958 is limited in its application to AOs that are more narrowly defined than Geographical Indications in general. Protection is based on registration irrespective of confusion:

> Every member state of the Paris Convention that also adheres to the agreement undertakes to protect in its own territory all appellations of origin of other member states for those products registered on the express condition that protection is also afforded in the home countries. The expression 'qualified' means that the right of an appellation of origin first of all needs to be recognised in the country of origin. The agreement thereby imposes on all member states a uniform set of rules, yet without separating this from national rules . . . Registration of an appellation of origin under the agreement can only be demanded by the country of origin. . . . Protection must thus be granted against all attacks of the exclusive rights given to those entitled to use the appellation, be it against the unlawful use . . ., be it against the fraudulent imitation of an appellation.[49]

The Lisbon Agreement thus regulates two particular issues: the procedure for obtaining protection, and the rights conferred. It came into force in 1966 with the original member states of Cuba, Czechoslovakia, France, Haiti, Israel, Mexico and Portugal. Subsequently, the following countries acceded to the Agreement (at the time of writing): Hungary (1967), Italy (1968), Algeria (1972), Tunisia (1973), Bulgaria (1975), Burkina Faso (1975), Gabun (1975), Togo (1975), Congo (1977), the Czech Republic and Slovakia (1993), Costa Rica (1997), Yugoslavia (1999 – now Serbia (1999), Montenegro (2006), Macedonia (2010), Moldova (2001), Georgia

---

indication 'Bud'. Just as AB took its case on expropriation of a trade mark application to the ECHR, BB could do likewise. See also n 21 above. It is also pretty clear that pre-emption can only relate to the issue of property rights and not to protection against misleading use: *Fage UK Ltd v Chobani UK Ltd* [2014] EWCA Civ 5 at [173]: 'It would be very surprising if the effect of EU law were to prevent member states from taking action against deceptive marketing.'

[49] *Actes de la Conference du Lisbonne 1958* (Geneva 1963), 814–815. The Lisbon Revision 2015 in Art 2 now specifically refers to appellations of origin *and* geographical indications. In the context of the Lisbon Agreement, adding GIs is not a 'significantly new subject matter', as has been alleged by the US Ambassador in an opening statement to the Diplomatic Conference on 30 October 2014, as many of those appellations already protected under the Lisbon Agreement should properly be classified as Geographical Indications, because there is no inseparable link between soil and product. This is primarily so for non-food items such as Czech porcelain or Iranian carpets.

(2004), North Korea (2005), Peru (2005), Nicaragua (2006) and Iran (2006). Five countries have signed, but not ratified the Agreement: Greece, Romania, Spain, Morocco and Turkey. The agreement has been substantially revised,[50] and, in the following, reference is also made to the revisions as adopted on 21 May 2015 (Lisbon Revision 2015).[51] However, even once the Revision comes into force (three months after five member states of the Paris Convention (including the EU as an intergovernmental organisation) have deposited their instruments of ratification or accession (Arts 28, 29), AOs in the current member states of the Lisbon Agreement that have already ratified the Lisbon Revision 2015 will be afforded 'no lower protection than is required by the Lisbon Agreement or the 1967 Act' (Art 31). The current legal framework and the following analysis thereof will thus remain applicable to those appellations of origin currently registered and for those countries that are currently member states to the Lisbon Agreement.

(b) Registration of appellations The registration procedure is regulated in great detail in Art 5, the core provision of the agreement. An application can only be made by a member state of the agreement, not by any individual or organisation:

> It is indeed for each contracting country to decide, on grounds specific to it, which national authorities are to be involved in the application of the provisions of the Lisbon system. The experience of the International Bureau is that there are generally a number of such authorities. In France, for example, the authority in power to request an international registration under the Lisbon Agreement is the General Director for Competition, Consumers and Prevention of Fraud (DGCCRF), whereas the authority competent to receive notifications from the International Bureau is the National Institute of Industrial Property (INPI) and the authority able to grant to third parties established on its territory a maximum period of two years in accordance with Article 5(6) is the National Institute of Appellations of Origin (INAO).[52]

An application by a member state has to be addressed to the International Bureau of WIPO in Geneva and meet with certain formality requirements.

---

[50] For the revision process in detail, see Chapter 5 by Geuze in this volume.

[51] http://www.wipo.int/meetings/en/doc_details.jsp?doc_id=304144. The final version was based on the draft prepared for the Working Group of the Development of the Lisbon System (Appellations of Origin), Ninth Session, Geneva, 23 to 27 June 2014, document LI/WG/DEV/9/2 (16 April 2014).

[52] WIPO Secretariat, 'Questions to be Examined With a View to the Modification of the Regulations under the Lisbon Agreement' 10 May 2000 (LI/GT/1/2) at [15].

The application has to specify the applying country, the application date, who is entitled to use the appellation, the product and product classification for which the appellation is used, the area of production and the provisions or decisions on which domestic registration is based. The application is then published by the International Bureau, and the other member states are duly notified of the same. The question to what extent other member states could refuse the appellation in their respective countries was intensely discussed in the sessions chaired by the Italian Tullio Ascarelli.[53] While some countries, France and Italy in particular, wanted the right of refusal to be limited to certain conditions, for example the fact that the appellation had become generic in a country, the original proposal had not envisaged this, and, ultimately, no such limitation was introduced. Therefore, all member states under the agreement have a right to refuse to protect another country's appellation of origin, yet are obliged to give specific reasons for such rejection. The country of origin has a right to judicial or administrative review of such a decision. A rejection, however, is limited by two conditions. Firstly, it can only be declared by 'the administrations' of the receiving country, not by any court of law. Secondly, the rejection can only be declared within one year from the date of notification.[54] In the latter respect, Art 5(4) is very clear: 'Such declaration (of protection) can no longer be opposed by the administrations of the member states after the expiration of one year from [the receipt of the notification of registration].'[55]

It is thus very clear that in the absence of a notification to the contrary, an appellation of origin duly registered with the International Bureau is afforded protection in all other member states without any limitations. And if an appellation afforded protection in the other member states has already been used by third parties in a member state prior to the date of international registration, the administration in this member state has the

---

[53] *Actes de Lisbonne* (n 49), 832.

[54] It is this part where the Lisbon Revision 2015 differs most from the original Agreement. The Revision 2015 in Art 9 mentions 'refusal, renunciation, invalidation or cancellation'. Initial refusal and cancellation for lack of protection in the country of origin were already stipulated in the original Lisbon Agreement, while renunciation and invalidation were not. Invalidation is now dealt with in Art 19, though it is not specified on which grounds this may occur, or by which authority.

[55] The Lisbon Working Group in 2002 amended the Regulations to allow for (national) revocations and Rule 16 was modified accordingly in 2002. See Lisbon Working Group, 'Report Adopted by the Working Group' 12 July 2000 (LI/GT/1/3) at [70], [83]. This is both inconsistent with the Agreement itself, and was procedurally flawed.

possibility of granting these third parties a period of up to two years to discontinue use, under Art 5(6).[56]

WIPO mentions the following regarding refusals:

> [Sixty-two] refusals of protection, concerning 51 international registrations, have been entered in the national register. The grounds most frequently given for refusal by the authorities of the contracting countries are that the appellation of origin for which registration is sought conflicts with an earlier mark that is protected in the country concerned.[57]

(c) Scope and contents of Lisbon rights The scope of protection under Lisbon is generous. Art 3 provides that protection is not only afforded to the appellation as such, but also against the use of translations, or the use of the appellation in combination with additions such as 'type', 'manner', 'imitation', or the like. It is interesting to note that the Czechoslovakian delegation was particularly interested in having the appellation 'Pils' further protected against variations such as 'Pilsner' or 'Pilsen', and was assured that the provision did indeed extend to such variance.[58] Usually the international registration would already indicate common translations of the appellation so as to facilitate the application of Art 3.

No less important is the provision of Art 6, which takes Art 4 of the Madrid Arrangement one step further. Not only wine products, but all appellations of origin duly registered with the International Bureau and not opposed within one year 'may not be considered as having become generic for as long as they are protected as an appellation of origin in the country of origin'. The Protocols of the Lisbon Agreement give the following explanation for this provision:

> The General Assembly holds it necessary to regulate this case explicitly. After all, there could be opportunities where member states might wish to find exceptions to this fundamental rule that an appellation of origin once registered could never again be considered as generic.[59]

---

[56] The Lisbon Revision 2015 now has certain provisions on prior use (Art 13), namely in regard of trade marks. 'Prior' refers to the date of international registration of the Geographical Indication. The trade mark in such case prevails when registered 'in good faith', a term that thereby becomes of vital importance in solving this conflict. If 'good faith' is interpreted as excluding 'causing confusion at the time of application', prior trade mark rights may fare better then when interpreted as 'upon knowledge that such term was used in other countries as a geographical indication'.

[57] WIPO (LI/GT/1/2) (n 52) at [7].

[58] *Actes de Lisbonne* (n 49), 834.

[59] Ibid., 838.

In other words, because member states might well be tempted to regard appellations of origin as generic even after the one-year period for objection has expired, Art 6 had to be drafted in the most explicit terms possible. In this respect, attention should be drawn to an Italian Supreme Court decision that had to interpret this provision.[60] The case concerned the appellation 'Pilsen', registered as an international appellation of origin under the Lisbon Agreement on 22 November 1967 and published in March 1968. The exact appellation is 'Plzen', and the given translations are 'Pilsen Pils', 'Pilsener' and 'Pilsner'. An Italian brewery had inserted the word 'Pilsener' on its product labels, and was sued by the Czech undertaking Pilsen Urquell, an enterprise entitled to use the appellation of origin as specified under the Lisbon Agreement. While the Italian Supreme Court took note of Art 6, it interpreted the provision as a presumption only, and not an unchallengeable right:

> The owner of such a right, while not being in possession of the unchallengeable right which the appellant claims (against the literal meaning of the words of the above-cited Article 6 of the agreement) can nonetheless rely, as a consequence of the registration, on a presumption of legitimacy in its use.

The Italian Supreme Court has reiterated this position in the 2002 Budweiser decision.[61]

(d) Specific issues related to the Budweiser cases

(I) THE LISBON REGISTER   The Lisbon Register features five registrations related to Budweiser, as mentioned above. As to the first four, the status of these indications as reported by WIPO's Lisbon Express database is the following:

---

[60] 'Pilsen Urquell', Italian Supreme Court, 3 April 1996; [1998] ETMR 169.

[61] In both cases, as well as in the case of 'Bavaria' (Italian Supreme Court, 27 September 2012, forthcoming in 46 IIC (2015)), the reporting judge was Giuseppe Maria Berruti. In all three cases, international or bilateral agreements were arguably misinterpreted. This is lamentable also for the fact that it may undermine Italy's position in having its numerous Geographical Indications protected abroad. The example of 'Parmesan' shows how critical particularly the issue of generic use is to Italy. The 'Pilsen' case was incorrectly decided because Art 6 is not a mere presumption, but meant to give legal certainty. This is also clear when compared to the Lisbon Revision 2015 that now seems to allow the authorities to hold an indication generic even beyond the period in which an application can be refused. Even read in this way, the Italian Court should then have examined whether at the time of registration, 'Pilsener' was considered generic in Italy. Under the Lisbon Revision 2015, the provision (Art 12 of the revised text) cannot be read as a 'presumption'.

ČESKOBUDĚJOVICKÉ PIVO / BUDWEISER BIER / BIÈRE DE ČESKÉ BUDĚJOVICE / BUDWEIS BEER

Number	49
Date	22.11.1967
Holder	The organisations that produce, in this region, the said products
Appellation	ČESKOBUDĚJOVICKÉ PIVO BUDWEISER BIER BIÈRE DE ČESKÉ BUDĚJOVICE BUDWEIS BEER
Publication	N° 1: 03/1968
Country of Origin	CZ
Nice Classification	32
Product	Beer
Area of Production	City of České Budějovic (Budweis)
Refusal	HT – 18.04.1969 MX – 11.06.1969 PE – 16.06.2006 NI – 07.06.2007 IR – 10.12.2007
Invalidation	IT – 19.09.2003 PT – 13.03.2006
Period Granted	CU – 27.05.1969–20.05.1971
Withdrawal	GE – 11.01.2006 (Date of the initial refusal: 21.12.2005) NI – 07.06.2007 (Date of the initial refusal: 29.11.2006)
Legal Basis	Executive Order N° 12.594/66-01/34 of November 3, 1966 of the Ministry of Food Processing Industry
Notification Article 5(2):	No longer applicable
Notification Article 14:	Currently not applicable
Note:	With regard to international registrations effected from April 1, 2002 onwards, the translation of an appellation of origin and, where relevant, its transliteration, will appear under a heading separate from that of the appellation of origin itself.

The last indication (Budweiser Bürgerbräu) has been refused by many more countries:

IL – 30.05.2001; CR – 03.09.2001; MX – 08.01.2002; IT – 18.01.2002; HU – 22.01.2002; HT – 23.01.2002; PT – 06.02.2002; PE – 16.06.2006; NI – 07.06.2007; IT – 17.01.2007; IR – 10.12.2007

(II) NATIONAL COURTS   The Budweiser cases in Israel, Italy and Hungary are essentially the only court cases in which details of the Lisbon Agreement were discussed in some depth. After all, the Lisbon Agreement presents several problems regarding its interpretation that were also discussed in the consultations of the working group for amendment of the implementing regulations for the Lisbon Agreement held on 10–13 July 2000.

Products eligible for protection – Art 2   As opposed to EU law on geographical indications of origin, the designations eligible for registration are not restricted to wines and foodstuffs. For example, the register – comprising over 900 designations – also includes musical instruments and porcelain.[62] One question disputed in the Budweiser cases was whether beer can bear an indication of origin at all, since beer can be produced in any quality anywhere. Although there are certainly many beers to which this statement applies, this is by no means the case regarding all beers. Under EU law, 18 beers from three different countries have been registered as protected GIs, not least beer from Budweis under the designations 'Budejovicke Pivo', 'Ceskobudejovicke Pivo' and 'Budejovicky Mestansky var'. For this reason, the Supreme Court of the Republic of Hungary rightly rejected the argument that beer could not be the subject-matter of an AO, whereas the Italian Supreme Court wrongly took precisely the opposite view.[63]

---

[62]   As mentioned above, the Lisbon Revision 2015 now clearly endorses this broad interpretation of appellations of origin by adding the category of 'geographical indications'.

[63]   The Lisbon First Instance Court in a default judgement had also taken this view:

> Neither the beer manufactured by the defendant nor any other beer have characteristics or qualities that are exclusively or essentially connected to natural factors (soil, climate, etc. . . .) or to human factors existing only in the locality where they are manufactured or any other specific locality. The characteristics and properties of any beer are determined by the raw materials used in the manufacture and in the manufacturing method itself. Neither the raw materials nor the manufacturing method are influenced by natural or human factors

Indications eligible for protection – Arts 2 and 3   Another aspect called into question in various Budweiser proceedings was whether names other than official place names (so-called indirect appellations or GIs) could be eligible for registration at all. Since the year 1945 the official name of the place of origin of the beer is no longer Budweis but, rather, Ceske Budejovicke. Art 2 of the Lisbon Agreement only mentions the name of a country, region or locality.[64] Often these are not even translations, but the customary names of places or regions in former times, for example Saigon (now Ho Chi Minh City), Ceylon (now Sri Lanka), Madras (now Chennai) or Formosa (now officially ChineseTaipei, and unofficially Taiwan). The French town of Nice is called 'Nizza' not only in Italy, but also in Germany although it has been French since 1793. The language of commerce may thus be different from the official one. Depending on the country involved, it does not appear sensible to sell the product under a name that is not common as a geographical name in that country. For Italians the 'Münchner Oktoberfest' takes place in Monaco (which for the rest of the world is a principality near Nice (or Nizza, see above) in France), for Americans in Munich, for the French in Munique, and, last but not least, for the Bavarians in their local dialect as Minga. The value of geographical indications would diminish considerably if they were only eligible to be registered in the national language of the respective country, resulting in terms which might be almost completely unknown abroad. Bangkok, officially called Krung Teph, comes to mind. Since Art 2 of the Lisbon Agreement does not refer to the official name, it should be reasonable to enter the name in all languages that are in common international use. This was the common practice until the year 2002, as reflected in the entries for Budweis beer in various languages. After the amendment of the implementing regulations, translations are now only included in the explanatory information and no longer in the indication as such. This can

---

existing only in a determined place or are exclusively or essentially related with that place or area. Therefore it is perfectly possible to manufacture beer with the same qualities and characteristics in different geographical places or areas.

There is no doubt that beer can be produced in such a manner, and no doubt that AB is one of the largest producers of such industrially manufactured beer. However, Budweiser or Pilsener Urquell in the Czech Republic are 'artisan' beers produced with local ingredients and for that reason qualify for protection under the Czech Act on Appellations of Origin of 12 December 1973.

[64]   The Lisbon Revision 2015 has taken this issue into account and clarifies that registration is also available for names 'known to refer' to a certain place (Art 2). It would thus appear that this makes the discussion about 'official' place names moot and allows the registration of an indication under a number of names, be they official or otherwise known.

have consequences for the scope of protection (see below). There are also certain reservations regarding the amendment procedure itself. Art 3 of Commission Regulation No. 1898/2006 implementing Regulation No. 510/2006 permits registration 'only in the languages which are or were historically used to describe the said product in the defined geographical area', which would be relevant at least in the case of historical place names such as Budweis, Carlsbad or Ragusa. However, the designation 'Bud' is certainly not eligible for registration and should be cancelled by the Czech Office, something that would significantly help to defend the Lisbon register.

The Hungarian decision also stresses that invented names such as 'Budvar', when applied on their own, are not eligible for registration. What is remarkable as regards the Czech entries is indeed that they do not merely indicate the locality and the product, but also contain indications otherwise to be expected in trade marks, for example 'mestansky var/Bürgerbräu' (AO No. 837) or 'Budvar/Budbräu'. The trade-mark-like nature of the indications is underscored by the fact that they only accrue to a single enterprise and not to a group. However this aspect of individual ownership is also a common feature in registrations of Geographical Indications for mineral water.

Scope of protection   The scope of protection arises from Art 3 of the Lisbon Agreement and, according to its wording, also extends to 'any usurpation or imitation, even if the true origin of the product is indicated or if the appellation is used in translated form or accompanied by terms such as "kind", "type", "make", "imitation", or the like'. The provision goes beyond Art 10 of the Paris Convention and Art 3(2) of the Madrid Arrangement, which render the scope of protection dependent solely on whether an indication of source is false or misleading. Regulation 510/2006 contains an almost identical definition of the scope of protection in its Art 13(1)(b). As has been pointed out above, it is difficult to read anything but a purely defensive right into Art 3 of the Lisbon Agreement or Art 13(1)(b) of the EC Regulation. This is particularly unsatisfactory in the case of the Lisbon Agreement, because existing rights of third parties can only be called into doubt with regard to the appellation of origin itself (Art 5(6)), and not with regard to the scope of protection defined in Art 3. It is for this reason as well that the amendment of the implementation Regulation dating from 2002 is questionable.[65]

---

[65]   According to the explanation furnished by WIPO prior to the amendment,

Requesting authorities frequently give the name of the appellation of origin in the national language together with its translation into a certain number of

Protection of third parties   Possibly the most difficult questions of the Lisbon Agreement concern existing rights accruing to third parties, their possibility of raising objections and the jurisdiction of national courts in countries other than the country of origin to declare the indication invalid. There are two provisions in this respect in the Agreement.

1. According to Art 5(4), objection to an international registration may be declared by the authorities of other countries within a period of one year only. For example, the Patent Office of Haiti filed an objection to the international registration of the designation 'Bud' effected on 10 March 1975, on 11 February 1976. Objections were also filed against the entry of 'Pilsener Bier' (e.g. in France) and 'Budweiser'

---

other languages. The practice . . . seems pointless, however, in view of Article 3 of the Lisbon Agreement which stipulates that 'protection shall be ensured against any usurpation or imitation, even if (. . .) the appellation is used in translated form (. . .).' In other words, Article 3 of the Agreement means that an AO contained in an international registration is protected against any use in translation, even if that translation is not referred to in the international registration. Consequently, it is suggested that it should only be possible for the indication of the AO, as referred to in Rule 1(2)(iv) of the Regulations to be provided in the official language or languages of the country of origin. Nevertheless, it is undeniable that the translation of the name of the AO may constitute useful information for users of the Lisbon system and for third parties in general, particularly where the translation differs considerably from one language to another. In order to maintain such information in the framework of an international registration, the Regulations could provide the possibility for requesting authorities to furnish one or more translations of the AO, not as part of the indication of the AO as referred to in Rule 1(2)(iv), but as additional (optional) information. Such translations would appear under a separate heading on the application form. They would in no way be checked by the International Bureau. (WIPO (LI/GT/1/2) (n 48) at [29]–[31])

This suggestion has been taken up by the Regulation that in Rule 5 now distinguishes between mandatory contents (Rule 5(2)) and optional contents (Rule 5(3)). Rule 5(2)(iii) now mentions 'the appellation of origin for which registration is sought, in the official language of the country of origin or, where the country of origin has more than one official language, in one or more of those official languages'. Under Rule 5(3)(ii), it is optional to furnish 'one or more translations of the appellation of origin, in as many languages as the competent authority of the country of origin wishes'. However, it has to be pointed out in view of the above analysis that the registration of translation is not 'pointless' at all, as Art 3 Lisbon Agreement is related to the scope of the appellation as regards possible infringements, but not as regards a right to use. The Lisbon Revision 2015 clarifies the extent of protection in Art 11. Here, protection is provided against the use of the indication in translation, and, for a well-reputed indication also against the use for dissimilar goods.

(e.g. in Mexico and Haiti). In the case of 'Budweiser', Cuba used the grace period specified in Art 5(4), but did not refuse protection.[66]

2. If a country does not lodge an objection, this too has an effect upon the existing rights of third parties according to Art 5(6), a provision which should be quoted in full:

> If an appellation which has been granted protection in a given country pursuant to notification of its international registration has already been used by third parties in that country from a date prior to such notification, the competent office of the said country shall have the right to grant to third parties a period not exceeding two years to terminate such use, on condition that it advises the International Bureau accordingly during the three months following the expiration of the period of one year provided for in paragraph (3) above.

In fact, the provision does little to clarify the conflict between a mark and an appellation of origin, because it never refers to a registered mark, and it does not clearly indicate how to solve the conflict if the prior user has already obtained some protection due to such use. Ladas interprets the provision as follows:

> Persons who have been using an appellation of origin as a generic term may ask for a term of two years at the end of which they must discontinue such use. This tends to imply that if such private persons do not ask for the term of two years they would have to discontinue immediately any generic use of the appellation of origin. This result would not be consistent with the administration's right of refusal within one year. Certainly this does not apply in the case where the alleged appellation of origin does not comply with the definition of Art 2 or in the case where such appellation infringes vested trade mark rights. Furthermore, in many countries a treaty itself cannot affect private rights, and in any case, it would be up to the courts rather than to the administration to resolve a conflict between the attempted international registration of an appellation of origin and pre-existing private rights.[67]

If the appellation has been used by third parties prior to the product's international registration, such use may continue for a transitional period of two years at the most, provided the Patent Office of the relevant country grants its approval and notifies WIPO accordingly. The Lisbon Agreement thus envisages no co-existence between the indication of origin and a regis-

---

[66] It is thus interesting to speculate what will happen once AB starts selling its beer in Cuba.

[67] Ladas (n 39) 1604–1605. Ladas's point of view confirms the interpretation of generic use as mentioned earlier. The Lisbon Revision 2015 is much more explicit on prior user rights that according to Art 13 take precedence when used or registered 'in good faith'.

tered or unregistered trade mark. This is hardly in line with TRIPS, yet the Lisbon Agreement falls under the exemption provision in Art 5 of TRIPS as a special Agreement under Art 19 of the Paris Convention.

Judicial review   The Lisbon Agreement is primarily addressed to national administrations. For example, Art 5(4) of the Lisbon Agreement refers only to the national authorities. This begs the question of the role of the courts in interpreting the agreement. Can the courts of other member states scrutinise the international registration of an appellation of origin and refuse protection if they believe that the appellation failed to meet the requirements for registrability, has subsequently become generic, conflicts with other marks, and so on? The text of the agreement does not necessarily give a definite answer to this question. Rather, a solution must be based on the structure of the whole Agreement, the historical context between the Madrid Arrangement and the Lisbon Agreement, the Protocols of the Preparatory Conference of the Agreement and the revised Rules. In this respect the following should be distinguished.

(1) The question to what extent the courts can find that an internationally registered appellation of origin can become generic after registration has already been addressed above. A comparison between the corresponding provision of the Madrid Arrangement and Art 6 of the Lisbon Agreement makes this unequivocally clear. While under the Madrid Arrangement national courts have the power to refuse protection to Geographical Indications if considered generic, the Arrangement makes an exemption in this respect for wine products. It is already stated under the Madrid Arrangement that the courts may not hold Geographical Indications for wine products to be generic, regardless of public perception. The Lisbon Agreement carries this exemption further to all those appellations duly registered under the Lisbon Agreement. Thus, even though the public in a particular country might regard an appellation as a generic term, the courts of this country cannot deny protection to this appellation. Here, the concept of appellations of origin as proprietary rights clearly takes precedence over the concept of protection by way of unfair competition principles. Art 6 of the Lisbon Agreement only concerns those cases where appellations have become generic over time. If an appellation is considered generic from the beginning, the national authorities of that country must deny protection within one year. It is submitted that the courts are not empowered to subsequently invalidate an appellation due to its being generic. But see also point (5) below.

(2) It is equally clear that it is up to the courts to decide to what extent indications similar to those of the registered appellations of origin constitute an infringement under Art 3 of the Lisbon Agreement. This is an issue

that may well be raised in infringement procedures, and the courts are called upon to settle such matters.

(3) Art 7 of the Lisbon Agreement obliges every member to protect an internationally registered Geographical Indication by domestic law as long as that indication is protected as an appellation of origin in the country from where it originates. This provision seems to indicate that both the administration as well as the courts of any given member state are competent to inquire to what extent an internationally registered appellation is still protected in its country of origin. In doing so, the legal and factual situation in the country of origin rather than the country in which the court or administration is situated would have to be examined.

(4) This leaves the question to what degree the courts in any given member state may determine not only whether the appellation of origin is still protected in its home country – and if so, to what extent, but also whether the appellation of origin indeed satisfies the definitional criteria set out under Art 2 and thus qualifies for protection as an appellation of origin under the agreement. That the courts are competent to determine these issues has been held by the following decisions: by the Appeal Court Douai in 1976,[68] and over the course of the Budweiser cases before the Lisbon District Court, the Italian Supreme Court, and the Hungarian Supreme Court. That the courts are not competent in this respect has been held by two decisions of the Israel Supreme Court,[69] the Bulgarian Supreme Court[70] and argued by two French academics.[71] Other academics who have written on the Lisbon Agreement do not mention the problem at all.[72] Given the importance of this question, this is somewhat surprising.

---

[68] Appeal Court Douai, 30 June 1976, Gazette du Palais 1976, 648 – 'Pilsheim'.
[69] Israel Supreme Court, 10 January 1990, 22 IIC 255 [1991] – 'Budweiser'; Israel Supreme Court, 13 September 1992, 25 IIC 589 – 'Budweiser II' [1994].
[70] Bulgarian Supreme Court, 7 January 2001, Civil Case 333/2000.
[71] A Devlétian, Case Comment on Appeal Court Douai, (1976) 76 Revue Internationale de la Propriété Industrielle 178; M Plaisant, Case Comment on Appeal Court Douai, (1977) Gazette du Palais 233.
[72] E.g. Vivez, 'L'arrangement de Lisbonne du 31 Octobre 1958' (1968) La Semaine Juridique 2198; G Ronga, 'L'arrangement de Lisbonne du 13 Octobre 1958' (1967) Revue Trimestrielle 425; FK Beier, 'Geographische Herkunftsbezeichnungen und Ursprungsangaben' [1968] GRUR Int. 69; R Knaak, 'The Protection of Geographical Indications According to the TRIPS Agreement' in F-K Beier and G Schricker (eds), *From GATT to TRIPS – The Agreement on Trade-Related Aspects of Intellectual Property Rights* IIC Studies, Vol. 18 (Weinheim, 1996) 117; S Escudero, 'International Protection of Geographical Indications and Developing Countries' Working Paper No. 10, South Centre (July 2001); E Faure, R Plaisant and JM Auby, *Le droit des appellations d'origine* (Litec, 1974); L Bäumer, 'The International Protection of Geographical Indications' in

Both the Lisbon and Italian Supreme Court denied protection to the appellation 'Budweiser' because beer was not a product that could qualify for an appellation of origin. The Lisbon court[73] in this respect merely states that:

> the expressions 'Budweiser' or 'Bud' do not consist of a geographic denomination of a region or locality that can be used to designate any good originating from that place ... Accordingly, when the registrations of the appellations of origin ... were granted to the defendant, something was registered that did not meet the minimum legal requirements to be protected as appellations of origin and consequently should never be registered as such ...

The court does not examine the question to what extent it is entitled to decide this matter. The Milan Appeal Court (the second instance court for the initial Budweiser dispute that subsequently went up to the Supreme Court)[74] declared that the appellations of origin 'Budweiser' and 'Bud' could not have been registered as they did not qualify as appellations of origin, and because the plaintiff AB (Anheuser-Busch) already enjoyed prior conflicting rights based on an unregistered but well-known trade mark. While the court examines at length to what extent beer can qualify as an appellation of origin and why the circle of those entitled to use the appellation was amended at a certain time, it never questions its own competence to deny protection to a valid appellation of origin.[75] The one court that did examine this question at length was the Israel Supreme Court in two decisions. In the first, it held as follows:

1. An appellation of origin which is internationally registered pursuant to the Lisbon Agreement can only be denied protection in Israel

---

WIPO, *Symposium on the International Protection of Geographical Indications, Weisbaden Oct. 1991* (WIPO publication No 713(E), 1992) 21.

[73] Lisbon Court of First Instance, 8 March 1995, Civil Case 7906, 3rd division of the 13th circuit, on file with the author. See also Antonio Corte-Real, 'The Conflict Between Trade Marks and Geographical Indications – The Budweiser Case in Portugal', in C Heath and A Kamperman Sanders (eds), *New Frontiers of Intellectual Property Law*, 25 IIC Studies (Hart, 2005) 149.

[74] Milan Appeal Court, 1 December 2000, (2000) 350 Rivista di Diritto Industriale II, 112 – 'Budweiser II'.

[75] In its decision of 16 June 2011, the Milan Appeal Court reached the opposite conclusion as to whether Budweiser could be protected as a Geographical Indication, though not based on the Lisbon Agreement, but in the context of European law (where the power of the courts to determine these issues has never been questioned). This point was not part of the subsequent decision by the Italian Supreme Court of 13 September 2013 (above n 16), but may well play a role in the pending proceedings before the Rome Appeal Court.

after the conclusion of opposition proceedings and registration in the appellation of origin register if the appellation is no longer an appellation of origin in the country of origin.
2. This connection to the legal situation in the country of origin applies to the cancellation proceedings before the trade mark office as well as to proceedings before a civil court.
3. Within the framework of an action for breach of an appellation of origin the defence is not admissible that the appellation of origin is not valid because it should not have been registered at all as an appellation of origin, pursuant to the Appellations of Origin (Protection) Law, 5725-1965.

In the findings, which are worth reproducing, it is stated:

> The objective of the Appellations of Origin Law – against the background of the Lisbon Agreement – was to give a foreign appellation in Israel the same degree of protection given to it in the country of origin. If we allow the courts in Israel to revoke a foreign appellation of origin protected in the country of origin, we shall deviate sharply from the purpose of the law and from Israel's international commitment. This purpose and commitment will be met if a court in Israel will only cancel a foreign appellation of origin or make a declaration as to the revocation thereof on those grounds upon which the Registrar himself could do so, that is to say, only if the foreign appellation of origin is not protected in the country of origin . . . In this way, expression is given to the basic principle of the Lisbon Agreement, pursuant whereto once a foreign appellation is registered, [it cannot be regarded as generic according to Art 6]. This principle appears expressly in the Appellations of Origin Law, which provides that 'the validity of the registration of an appellation of origin which is made pursuant to a notice received pursuant to section 17 is the same as its validity in the country of origin'. It follows from this that if in the country of origin the law is – like the law in Israel – that after registration the appellation of origin may not be opposed on the ground that the registration from the start was not lawful, because the appellation is not an appellation of origin but a mark of provenance or type only, then also in Israel there is no longer any possibility of opposing the appellation of origin on this ground . . . [The AO in dispute was not proved to be challenged or vulnerable in Czechoslovakia and . . .] it is clear that by way of direct opposition, the claim may not be raised before the Registrar that the registration of the foreign appellation should be struck out on the ground only that, at the time of registration thereof, it was not an appellation of origin. This result is called for by the policy [foundations of Lisbon, which] was the giving of comprehensive protection to a foreign appellation of origin and giving it the status in the foreign state which it enjoyed in the original state . . . (see Tilmann, 'Die Geographische Herkunftsangabe' 415 (1976)). Only if in the original state the appellation is no longer protected will it cease to be protected in the foreign state. In this matter, it is not made clear, either in the Appellations of Origin Law or in the Lisbon Agreement, whether the decision in the foreign country relating to the absence of protection in the original country must be based on the decision of the Registrar or the courts in the country of origin

that the appellation is not protected therein (direct opposition in the country of origin) or whether it is sufficient that, in the opinion of the foreign state, there is a ground on the strength of which the appellation of origin may be deleted in the country of origin (indirect opposition in the foreign state). We have no need to resolve this question in the matter before us, because the claim that the appellation of origin of the appellant was not protected in the country of origin, because there is a ground for opposing the same in the country of origin, was not raised before us. Indeed, if the foreign appellation is protected in the country of origin, it must be protected also in the foreign country.

...

[The court concluded that] if it is not possible to oppose the actual validity of a registered appellation of origin directly before the Registrar, it is not possible to oppose the validity of the registered appellation of origin indirectly in the court. The Israeli legislature wished to protect a registered appellation of origin and opened the door to direct opposition thereof on limited grounds [which did not include] ... indirect opposition ... thus facilitating damage to the registered appellation of origin on additional grounds ... It should not be assumed that, in the case before us, that which cannot be obtained directly can be obtained indirectly. This matter is particularly prominent in the case of a foreign appellation of origin. The objective of the Appellations of Origin Law against the background of the Lisbon Agreement – was to give a foreign appellation in Israel the same degree of protection given to it in the country of origin. If we allow the courts in Israel to revoke a foreign appellation of origin protected in the country of origin, we shall deviate sharply from the purpose of the law and from Israel's international commitment. This purpose and commitment will be met if a court in Israel will only cancel a foreign appellation of origin or make a declaration as to the revocation thereof on those grounds upon which the Registrar himself could do so, that is to say, only if the foreign appellation of origin is not protected in the country of origin. Thus, we give protection in Israel to appellations of origin which arise outside Israel. Thus also, we ensure the protection outside Israel of Israeli appellations of origin. The district court, as we have seen, reached the opposite conclusion. It seems to me that the mistake is to be found in the fact that it regarded the case before us as a case which turned on a question of the jurisdiction of the civil court ... the civil court has jurisdiction to consider the validity of the appellation of Origin. The question is not one of jurisdiction, but of a ground. The question is, on what grounds may the competent court invalidate an appellation of origin? The existence of general jurisdiction does not confer permission to invalidate appellations of origin on any ground whatsoever ... We have already indicated the clear policy of the legislature – following the Lisbon Agreement – to give protection in Israel to foreign appellations of origin to the same extent as that foreign appellation of origin is protected in the country of origin. The giving of the possibility to the court in Israel to decide that a foreign appellation of origin, valid in the country of origin, is not valid in Israel will be contrary to the policy of the legislature and the Lisbon Agreement, the fulfilment of which is the objective which every commentator must gain in interpreting the Appellations of Origin Law.[76]

---

[76] *Budweiser I* (1991) 22 IIC 255, 261–263 (Israel SC, 1990). What is puzzling in this otherwise laudable decision is the fact that Tilmann's book, cited in

In its second decision in the matter, the Supreme Court stated in further detail:[77]

1. In the Lisbon Agreement, the state of Israel assumed international obligations to the agreement's member states, so that it must be assumed that, when the law on appellations of origin was enacted, the legislature sought to give validity to such obligations. Any legal interpretation must therefore be in accordance with the Lisbon Agreement.
2. It is the principle of the Lisbon Agreement that its members mutually recognise and honour the property rights reserved, provided that such rights have been recognised by the states and are duly registered. Such protection in favour of a foreign state is absolute and exclusive with the only exception provided for in Sec. 5(3) of the Agreement, which allows the declaration that a specific appellation of origin cannot be protected. The reasons for this can be that the alleged appellation is only an indication of provenance or that the person claiming protection is not entitled to use it.
3. The only reason to strike out the registration of a foreign appellation of origin is that the appellation is no longer protected in the country of origin or has ceased to be protected there.

Hence, the Supreme Court in Israel reiterated that the matter did not concern the question whether the court lacked jurisdiction to decide on the validity of a national registration under the Lisbon Agreement, but, rather, the question as to how far the jurisdiction of the court to review the matter goes. Thus, the third parties mentioned in Art 5(6) cannot obtain a declaration of invalidity for a registered indication from the courts after the fact, unless this takes place in the AO's country of origin. Otherwise, Art 5(6) would be meaningless. Art 8, referring to court proceedings, only relates to the enforcement of protection but not to its revocation. A comparison of the provisions in the Paris Convention, the Madrid Arrangement and the Lisbon Agreement shows that the objective of the latter Agreement was specifically to increase legal certainty for appellations in the other member countries by creating a unitary right amongst all member states that could and should only be attacked in the home country. Whereas protection under Art 10 of the Paris Convention and Art 4 of the Madrid Arrangement is limited to the risk of false or misleading indications, this was to change under the Lisbon Agreement, which was based specifically on the registration system. The legal certainty thus achieved was to be obtained through the procedure according to Art 5(3) and (4), and would be undermined if third parties were then permitted

---

the decision, does not contain such passage on p. 415 (or anywhere else, for that matter).

[77] *Budweiser II* (1994) 24 IIC 589 (Israel SC, 1992).

to challenge the indication outside the country of origin. Rather, courts outside the country of origin are only entitled to verify whether protection in the country of origin has lapsed, and not to question the prerequisites of registration.

This is no different than in the case of a bilateral agreement protecting certain geographical indications. The French would certainly not be amused if the Hungarian courts undertook to examine whether 'Champagne' was considered generic in Hungary despite the fact that Hungary is meant to protect this indication based on the Versailles Peace Treaty. Or if the US courts decided that 'Cognac' was devoid of protection for one reason or another, despite the bilateral agreement between France and the US (1971) to protect 'Cognac' and 'Armagnac' as French indications, and 'Bourbon' and 'Bourbon Whisky' as US-made ones: 'The tribunals are no longer free to decide that a regional appellation of spirits, for instance the name "Cognac", is a generic or descriptive term.'[78] In the author's opinion, there are therefore sound arguments for interpreting the Lisbon Agreement in line with the interpretation adopted by the Israeli courts.[79]

(5) The new implementing regulations for the Lisbon Agreement in force since 1 April 2002 now provide for the possibility in Rule 16 for individual states to deposit a declaration of invalidity of an indication of origin at WIPO. However, if the Lisbon Agreement itself does not provide for such a possibility (which it does not appear to do), then this cannot be correct. If the member countries were in agreement about such a possibility, it would have been necessary to amend the Agreement rather than only the implementing regulations.

(III) RELATIONSHIP TO COMMUNITY LAW   As a number of countries are members of the European Union as well as being signatories to the Lisbon Agreement (Portugal, France, Italy, the Czech Republic, Hungary, Bulgaria), the relationship between Community protection under Regulation 1151/2012 (and its predecessors) and the Lisbon Agreement should be clarified. Contrary to bilateral agreements (see

---

[78] Ladas (n 39), 1597. Bilateral agreements often go one step further than the Lisbon Agreement in that they do not allow the national courts to even ascertain whether an indication is still protected in the country of origin.

[79] A further argument in favour of this view is the fact that all bilateral agreements concluded by the EU (as mentioned above) contain long lists of geographical names the other contracting party is meant to protect. It would be odd if such protection could be undermined by the courts considering certain indications to be generic, or not protectable for other reasons.

436   *Research handbook on intellectual property and GIs*

above), both the CJEU and the GC have shown a certain deference to the Lisbon Agreement. The CJEU has noted that obligations under the Lisbon Agreement are covered by Art 307 of the EC-Treaty (prevalence of international obligations over Community law), and that it did not have to directly address the question of whether protection under Lisbon is pre-empted by the Community-wide exhaustive protection for GIs.[80]

Also, the GC in case *Budejovicky Budvar narodni podnik v OHIM* (T-225/06 et al.)[81] gave significant weight to the registrations under the Lisbon Agreement.[82] It may thus be safe to say that obligations under the Lisbon Agreement (and the appellations protected thereunder) can be invoked both against registrability of conflicting trade marks, and as rights of use, even if such would not be in accordance with Community law.

5. COMMENTS AND CONCLUSIONS

The issues mentioned above are complicated by the sheer number of different factual and legal frameworks, and by the way certain issues, such as protection under the Lisbon Agreement, feature in different contexts. The author would like to briefly (1) comment on some issues of interest, and (2) summarise his position, before (3) speculating as to the future outlook.

5.1 Comments

It is of course understandable that different opinions should arise regarding most of the legal issues discussed above, and indeed such diversity is reflected in the court decisions listed and analysed above. As is often the case with legal matters, views depend on upbringing, education and preconception. This author, for example, when visiting the US for the first time thought that 'Budweiser' was a beer from, well, Budweis. American tourists visiting Germany or the Czech Republic might err the other way round thinking that the Budweiser beer they can purchase originates from the US.

But be this as it may, when reading the extensive case history and decisions, two jurisdictions do stand out as rather odd. One is Austria, where

---

[80]   *Budejovicky Budvar v Rudolf Ammersin* (C-478/07) (n 8) at [100]–[102].
[81]   General Court of the European Union, Cases T-225/06, T-255/06, T-257/06 and T-309/06 decided on 16 December 2008, *Budejovicky Budvar narodni podnik v OHIM*.
[82]   See the text accompanying (n 28) above.

even the Advocate General had reason to make some uncomplimentary remarks about the Austrian judiciary in relation to the Budweiser disputes,[83] and the second is Italy. Here:

1. The courts took the view that beer as such could not qualify for protection as a GI despite several such beer registrations under European GI law.
2. The courts further held that AB had obtained a reputation in the sense of Art 6*bis* of the Paris Convention in Italy in the 1930s solely on the basis of a handful of advertisements in foreign language magazines that had made their way to Italy, and those advertisements actually on file with the courts date from the 1950s. One could add that none of AB's beers could be sold in Italy until the 1960s, and that BB was running a huge refrigerated warehouse in Trieste that was operative until the Second World War.
3. The Italian Public Ministry, comparable to the Advocate General, declared, in oral proceedings before the 2002 Italian Supreme Court, that it supported AB's position, but declined to give reasons 'in view of the fact that it is already lunchtime'.
4. The representative of AB in Italy in an open letter to the editor of the *Rivista di Diritto Industriale*,[84] just as the representative of AB in Germany had done in a letter to the Max Planck Institute, called into question the author's academic motives for publishing the Italian case comment.

Finally, there are reservations regarding the nature in which the amendments to the implementing regulations of the Lisbon Agreement in 2000 were brought about. They were prepared by a working group consisting of representatives of the member countries as well as of three NGOs, namely the European trade mark association ECTA, AIPPI and the international trade mark association INTA, which according to the working report participated in the discussions although they merely have observer status under Art 9 (viii) of the Lisbon Agreement. Furthermore, ECTA and INTA are associations that represent the interests of trade mark proprietors, and the conflict between the Lisbon Agreement and trade mark rights is probably the most explosive one in this area, not least in view of the Budweiser cases. What is even more problematic, therefore, is

---

[83] Opinion of the Advocate General Colomer in (C-478/07) (n 8) at [53]–[55] et seq.
[84] A Vanzetti, Lettere al Direttore, [2004] Rivista di Diritto Industriale II, 349.

438 *Research handbook on intellectual property and GIs*

that one of the ECTA representatives, DC Ohlgart, had represented the interests of Anheuser-Busch in Germany and before the CJEU. During the consultations, the ECTA representative spoke out in favour of allowing the member countries to register the invalidity of a deposited indication of origin,[85] a point of view which coincides with the motions filed by Anheuser-Busch in the court proceedings in Israel, Italy (pending at that time), Portugal (pending at that time) and Hungary, and which may therefore appear to be a less than disinterested proposal.

5.2 Conclusions

1. Both for the issues under trade mark law, and the law on Geographical Indications, it should be acknowledged that 'Budweis' and 'Budweiser' are Geographical Indications for beer, regardless of the fact that the place is now called Czeske Budejovice.
2. To the extent that beer is produced with local ingredients, as is the case for BB's beer, such beer is eligible for protection as a Geographical Indication.
3. 'Budweiser' as a geographical indication for beer has significance for conflicting trade mark applications in two ways. To the extent that it is a protected indication, it counts as a prior right. As a geographical name, it does not allow for the registration of marks for those goods or services that currently or in the future are or may be produced in this place.
4. The indication also has significance for a potential 'descriptive use' defence under trade mark law where the indication is used in a geographical sense.
5. The prerequisites for registrability of appellations of origin under the Lisbon Agreement can only be examined by the authorities of the country of origin.
6. Appellations of origin registered under the Lisbon Agreement cannot be deemed generic, and cannot be invalidated due to prior rights beyond the one-year period specified by the Agreement.
7. The appellation 'Bud', registered under the Lisbon Agreement and several bilateral agreements, may be useful (or, according to the Austrian Supreme Court, not really), but should be deleted by the Czech authorities, because it does not indicate a place.
8. Geographical indications protected under Community Regulations or

---

[85] Lisbon Working Group, 'Report Adopted by the Working Group' 12 July 2000 (LI/GT/1/3) at [80]–[85].

the Lisbon Agreement give a right of use only as registered. The right to exclude others, however, also extends to translations commonly used to denominate the geographical area abroad.

5.3 Outlook

I add only this: is it too much to hope that before they embark on this next round of litigation, these warring parties can be brought to understand the merit of mediation and the advantages of a lasting truce to hostilities? I recommend it to them.[86]

It is indeed difficult to see how the conflict between AB and BB can be solved but for an outright purchase of the Czech enterprise. AB's first attempt at a takeover in the 1990s was ultimately vetoed by the then Economic Minister Vaclav Klaus, later President of the Czech Republic. Rumours of a planned takeover resurface every now and then.[87]

---

[86] *Budejovicky Budvar Narodni Podnik v Anheuser-Busch Inc* [2009] EWCA Civ 1022 at [65] (Ward LJ).

[87] E.g. Süddeutsche Zeitung of 22 July 2003. The value of BB is an estimated €400 million. The acquisition by Budejovcy Mestansky Pivovar, also a Czech brewery in Budweis, by AB made headlines in January 2012:

> THE US-BASED brewer of the Budweiser beer has moved a step closer to ending a century-old legal dispute over the right to use the name, after buying a Czech brewery which also uses the brand. Anheuser-Busch – which markets the better-known 'Budweiser' beer, which is also brewed in Kilkenny – has bought one of two breweries in the Czech town of Ceske Budejovice, which in the German language was known as 'Budweis'. Having bought Budejovicky Mestansky Pivovar (BMP), and its own 'Budweiser Beer' brand, Anheuser-Busch has tightened its grip on the name – potentially bringing the dispute to an end. Another brewery in Ceske Budejovice – the better-known 'Budvar' brewery, whose beers are also available in Ireland – also still claims use of the name, but may also itself be put up for sale in the coming years. The Budejovicky Budvar brewery was one of the prized assets of the former Czechoslovakian government, Czech Position reports, with the trademark dispute becoming more pronounced in the 1970s when the government opted to increase its exports. The Czech government has long been considering spinning off assets like Budvar, however – and it's thought that whenever the brewery is sold, most likely through a public flotation, that Anheuser-Busch could step in and buy a significant stake. Around 120 legal battles have been fought in various countries between Anheuster-Busch and Budvar – with Budvar usually coming out on top – and another 40 cases are still alive in courts around the world. (http://businessetc.thejournal.ie/budweiser-buys-czech-brewer-in-bid-to-seal-control-of-budweiser-brand-327520-Jan2012/)

15. Geographical Indications and protected designations of origin: intellectual property tools for rural development objectives
Dominique Barjolle

1. INTRODUCTION

This chapter aims to develop a theoretical approach identifying the main factors which contribute to successful Geographical Indications (GIs). To achieve this objective, we highlight the interactions between the mechanisms of intellectual property rights protection, those linked to economic transactions and those linked to sustainable rural development. Consequently, our theory will take into account three levels of interactions: markets, organisations (or institutions) and territory. As far as GIs are concerned, we refer to the special case of traditional products whose names can be potentially protected under international treaties, such as those qualifying as GIs under Art 22.1 of TRIPS, or subject-matter-specific protection regimes, such as Protected Designations of Origin (PDOs) or Protected Geographical Indication (PGIs) recognised in Regulation (EU) No 1151/2012 on Quality Schemes for Agricultural Products and Foodstuffs. The intellectual property protection of GIs, as defined in TRIPS, must be implemented by national regulations in each member state. The foundation of the actual legal basis of protection is the 'unfair competition' framework, which regulates the conduct of firms in the marketplace: they obtain the means to fight against unfair imitations, produced outside the geographical area of the traditional product, in almost all the cases at lower prices because of less demanding processes, and by abusing the trust of the consumers about the genuine provenance of the product.[1]

In order to set up a coherent theoretical framework, it is necessary to analyse a complex set of interactions at three different levels of functioning:

---

[1] See generally, E Thévenod-Mottet, 'Legal and Institutional Issues Related to GIs' (WP1 report, SINER-GI, 2006).

1. Market: it refers to the level where economic value is generated when transactions occur;
2. Organisation: a core element which enables the expression of economic value is the collective (producer) organisation and governance of the supply chain; and
3. Territory: the benefits from direct and indirect effects of the production and sales of the products, including the non-trade or 'public goods' concerns operate at the territorial level.

We will now provide an overview of the system functioning as a conceptual framework, to understand how the three levels interact. This is the principal contribution to the state of the art in the field of knowledge regarding the impact of GIs on territorial development. This scheme takes place in the fourth part of the paper, after the presentation of the theoretical backgrounds (section 2) and the empirical results based on case studies, collected during a previous European research programme[2] (section 3).

2. THEORETICAL ASPECTS

Our purpose is to explore the interactions between the market, organisation and territorial impacts of the functioning of a set of actors involved in the implementation of a GI.

2.1 Market as the Place Where Economic Value is Generated

The economic value linked to qualities (physical characteristics of the products and associated processes of production) which leads to the differentiation of the product can be distinguished from the added value itself. Making a distinction between these two notions leads to a better understanding of the effect of the protection due to the intellectual property right.

Value added is a concept developed in the macro-economy. It refers to the additional value of a commodity over the cost of commodities used to produce it from the previous stage of production. The value added

---

[2] 'PDO-PGI: Markets, Supply Chains and Institutions' FAIR contract no. CT95–0306. The project participants were: A Fearne and N Wilson, *Wye College* (GB), K De Roest et al., *CRPA* (IT), K Galanopoulos and K Mattas, *University of Thessaloniki*, C Fotopoulos, A Vakrou et al., *NAGREF* (GR), B Sylvander and B Lassaut, *INRA-UREQUA*, M Leusie, *Chrysalide* (FR), K Van Ittersum et al., *Wageningen* (NL), D Barjolle, J-M Chappuis and M Dufour, *IER-EPFZ* (CH).

corresponds to the remuneration of the factors of production, which are the labour, the capital and the risks.

Economic value is a broader concept. It is obtained at the moment of the transaction (sale), and it remunerates beyond the factors of physical production as well as additional characteristics such as the image or reputation acquired over time. The more subjective or socially constructed product qualities and intellectual property rights are the two main facilitators of the generation of economic value at the producer level. Without intellectual property rights, the efforts of the producers/sellers may be in vain, due to the claims of product equivalence/similarity or even sometimes deceptive assertions (parasitism) of competitors willing to benefit without due cause.

Product differentiation is a classical strategy in the economic value creation process, well-known in marketing theories,[3] where the main focus is on the price setting at a market equilibrium removed from the initial market to a niche market because the product has a specific and unique quality which makes it different from the standard product.[4] Quality can refer to physical characteristics as well as to characteristics related to the production processes. These characteristics can be known more or less precisely at the moment of the transaction. The consumer preference varies according to the combination between all kind of characteristics and prices, as well as to the willingness to pay a premium for this product instead of for another with a different combination of quality/price. The nutrition content, the freshness, the colour and general aspect of a food product are the main physical characteristics known at the moment of the transaction. The taste and the organoleptic characteristics are generally unknown at the moment of the transaction. Related to the process of production, both the region of origin (at least the country of origin) and the manner of production (organic, integrated production, and so on) are generally known at the moment of the purchase. Consumer trust towards the product and the process of production can usually be enhanced by diverse certificates of guarantee.

The basis for establishing a specific type of juridical regime for intellectual property rights protection (trade marks or GIs) can be traced to the lack of a common understanding to what quality refers exactly to, as well as the difficulty faced by consumers when assessing quality and its

---

[3] See generally, M Porter, *Competitive Advantage* (The Free Press, 1985).
[4] See generally, EH Chamberlin, *The Theory of Monopolistic Competition* (Harvard University Press, 1933); Pour la traduction Française: *La Théorie de la Concurrence Monopolistique* (Presses Universitaires de France, 1953).

appropriate price at the moment of purchasing a product. Several neoclassical economists have highlighted the kind of market failure linked to information asymmetry between the producer/seller and the consumer/buyer. Akerlof has studied the link between the quality and the uncertainty.[5] His approach to uncertainty is based on the assumption that, in certain trade contexts, such as second-hand car sales, uncertainty is inherent in the nature of the transaction itself. Asymmetry of information about products is essentially linked (i) to experience goods,[6] for which quality, especially sensory quality, can be assessed by the consumer only at the time of actual consumption, and (ii) to trust goods,[7] for which the quality can never be assessed by the consumer him/herself (production methods or nutrition content for example). Shapiro[8] demonstrates that premium prices are not in contradiction to the general welfare theory, because they are necessary to cover reputation costs (certification and control costs can also be added to this head of costs), in which any firm has to invest in order to gain consumers' trust.

Several recent research findings highlight the trend that quality products are increasingly in demand by consumers.[9] In the context of the current pressure of globalisation of trade and pressure imposed by retailers, these new demands from the consumer side allow producers to develop interesting strategies. Producers can more easily create a niche position on the market when the consumers can express their preferences and willingness to pay a premium for products that have a strong identity and use good quality signals. Here intellectual property rights tools constitute the legal basis to preserve the identity values of traditional products and allow this mechanism of generating economic value for the producers when responding to the consumer demand, in spite of high competition and the market

---

[5] GA Akerlof, 'The Market for "Lemons": Quality Uncertainty and the Market Mechanism' (1970) 84 Quarterly Journal of Economics 488.

[6] P Nelson, 'Information and Consumer Behaviour' (1970) 78 Journal of Political Economy 311.

[7] M Darby and E Karni, 'Free Competition and the Optimal Amount of Fraud' (1973) 16 Journal of Law and Economics 67.

[8] C Shapiro, 'Premiums for High Quality Products as a Return to Reputations' (1983) 97 The Quarterly Journal of Economics 659.

[9] D Barjolle, P Damary and B Schaer, 'Certification Schemes and Sustainable Rural Development: Analytical Framework for Assessment of Impacts' in U Rosenberg and A Sellier (eds) (2011–2012) 37–38 *Revue: Sviluppo Locale* anno XV; For the motivations behind consumer preferences in Europe, see also D Giovannucci, T Josling, W Kerr, B O'Connor and M Yeung, *Guide to Geographical Indications: Linking Products and their Origins* (International Trade Centre, Geneva, 2009) 7–8.

power of the retailers. This very high concentration of retail is one of the major obstacles of fairness in the negotiations over agriculture. The position of retailers is very dominant, and abuses are frequent. In effect, retailers have plenty of alternative suppliers and this pool will always increase with the further globalisation of markets. This is unfortunately not counterbalanced by a sufficient degree of concentration on the producers' side. On the contrary, the latter group is very fragmented and diverse. In this situation, the development of collective labelling strategies for traditional products has seen dramatic advances over the past few decades.[10]

2.2 Organisation as Instrumental in Retaining Economic Value

An easy way to assess economic performance is to compare costs and benefits. For a certain level of consumer willingness to pay, the benefit is higher at a minimal level of costs. To rationalise the processes is the most classical source of savings. In the particular case of products differentiated by their geographical origin, savings on production costs are limited by some constraints described in a list of requirements which have to be respected by the producers and processors, that is, a product specification or 'code of practices'. When formally embodied as rules in a code of practice, these constraints are linked to a qualification initiative led by the producers while interacting with the agency responsible for the implementation of the legal basis for the protection of registered GIs in a given country.[11] In any case, for a product differentiated by its origin and benefiting from IP protection, a constraint in the process of production leads to an additional cost which will affect the final consumer price. While this logic can be perceived as paradoxical, it can be explained in the following manner. In fact, the actors can choose to fix some rules even if they impose higher production costs, because the method of production adopted corresponds to specific conditions which – although they would be perceived as 'constraints' elsewhere – in the case of GIs are just the natural conditions

---

[10] B Sylvander, G Allaire, G Belletti, A Marescotti, E Thévenod-Mottet, D Barjolle and A Tregear, 'Les Dispositifs Français et Européens de Protection de la Qualité et de l'Origine dans le Contexte de l'OMC: Justifications Générales et Contextes Nationaux', Symposium international à Lyon, 9–11 March 2005, *Territoires et Enjeux du Développement Régional*.

[11] These juridical bases vary considerably between countries in their institutional natures, their regulatory authorities, their qualification procedures and degrees of expertise required. See Thévenod-Mottet (n 1). These legal frameworks also vary depending on the grounds and motives of their acceptance by the national legislatives authorities themselves. See Sylvander et al. (n 10).

of the particular socio-technical context in the region of origin.[12] The costs of production are consequently always relative. The logic of the reduction of production costs and the law of comparative advantages are not sufficient to simultaneously maintain the fairness of competition and retain the production in the region of origin, which also ensures that consumers are not misled.

It is interesting to consider, as a second source of costs savings, the opportunity to bundle interests and establish common rules which facilitate repetitive individual decisions. In the case of regionally specific or traditional products, collective organisations are the 'strategic alliances' between groups of farmers, producers and processors to coordinate the production and sales of these products. The institutional form of such organisations varies, from loose networks to more formal associations, as does the extent of the supply chain which is incorporated.[13] Certain collective organisations are able to generate those so-called 'transaction costs' savings. This aspect helps to explain the interest of producers in adopting a collective strategy of differentiation for a traditional product. Transaction costs cover usually both the costs of 'discovering the adequate prices' and the 'costs to negotiate and conclude separate contracts for each transaction'.[14] The search for information on the product and its selling conditions, the contract negotiations, the possible intervention of lawyers, the breach or non-carrying out of contracts, the associated legal proceedings and claims to enforce a contract: all these common operations have to be taken into account. In the case of a collective strategy of differentiation, the major decisions, both at technical and commercial levels, are facilitated thanks to collective rules. Contract negotiations are therefore limited, because the quality is assessed through a procedure which is defined for the whole supply chain by an agreed set of standards, and the prices are very often known through a common system of market information.

However, in order to achieve those kinds of transaction costs savings, the ability of the collective organisation to function effectively is of crucial importance. The efficiency and effectiveness of strategic choices

---

[12] For example, the production of cheese such as L'Etivaz (a Swiss AOC) from raw milk, which is heated in traditional copper cauldrons over open wood fires and therefore a more labour-intensive process.

[13] For further details, see S Réviron and J-M Chappuis, 'Geographical Indications: Collective Organisation and Management' in E Barham and B Sylvander (eds), *Labels of Origin for Food* (CABI, 2011) 45.

[14] RH Coase, 'The Nature of the Firm' (1937) 4 Economica 386. Traduction française: 'La Nature de la Firme' (1987) 2(1) Revue Française d'économie 133.

are core elements, and the trust between the actors plays a crucial role in the collective decision-making process. Strategy is the result of the ability to take decisions in relation to the stakes of the context, while seeking coherence with a vision of the goals to be reached. In the case of products with a geographical name, the organisation is necessarily collective, which is inevitably more difficult in terms of strategic management. The decision-making processes are more onerous because of the additional negotiation processes. As the interests of the different actors (many of whom may compete within a certain level of the supply chain), may be contradictory, the strategic capacity is potentially less efficient. In terms of the product, the collective management concerns the definition of the product itself, identifying its specific qualities and the related production processes. Previous research[15] has identified the main characteristics of the economic value creation process within the framework of a supply chain for a differentiated product. The pertinence and adequacy of the technical definitions in its code of practice are clearly linked to the motivations of the actors and initiators of the procedure, as well as to the coherence between these motivations and the current market and general economic stakes.

In terms of place and price, the collective organisations responsible for GIs have limited power because they are limited by the public competition policies.[16] The regulation of the quantities is allowed only in cases where increased production would impact on quality (for example, the well-known case of vineyard yields which influence the wine qualities). This regulation of production volumes is the main tool used by collective organisations. It has an effect on quantity, quality and price.

---

[15] This capacity of collective organisation has been highlighted as a crucial factor of success in the approaches of differentiation based on the geographical origin for agro-food products. See D Barjolle, J-M Chappuis and M Dufour, 'Key Success Factors of Competitive Position for Some Protected Designation of Origin (PDO) Cheeses' in J-F Hocquette and S Gigli (eds), *Indicators of Milk and Beef Quality* (EAAP Publ. 112, Wageningen Academic Publishers, 2005) 245; D Barjolle and B Sylvander, 'Fondements Théoriques et études Empiriques de la Protection Européenne de l'Origine des Produits Agroalimentaires', in *Actes du colloque SFER-Enita Clermont*, 5–6 October 2005. It has been studied in-depth in several previous publications. See D Barjolle, J-M Chappuis and S Réviron, 'Organisation and Performance of the Origin Labelled Food Alliances', in AR Bellows, *Focus on Agricultural Economics* (Nova Science, 2005) 91; D Barjolle, S Réviron, B Sylvander and J-M Chappuis, 'Fromages d'origine: dispositifs de gestion collective', Colloque INRA-INAO, Paris 2005.

[16] For a more detailed analysis, see E Raynaud and E Valeschini, 'Collectif ou Collusif? A Propos de l'Application de Droit des Ententes aux Certifications Officielles de Qualité' [2005] Revue Internationale de Droit Economique 165.

2.3 Territory as Beneficiary of Impacts

The collective negotiation mechanisms which relate to the internal redistribution, between the actors in the supply chain, of the economic value generated on the market, are really the key factors in the social construction of a commercial performance. In turn, this can have a major impact on sustainable rural development at the territorial level.

The capacity of the collective organisation to coordinate and channel the diverse interests of the participating firms, which are otherwise competitors, and to develop synergies with other actors in the region is a crucial factor, enabling the efficient use of the financial resources devoted to the promotion of the product. According to Pecqueur,[17] the bundling of local resources around a basket of goods leads to an increase in the consumer willingness to pay, especially since the resources are interdependent as well as complementary and set up a coherent image for the territory.

According to the established definition of sustainable development and the analysis of the recent evolution of the multifunctionality of agriculture in western European countries,[18] the impact on the territory can be divided into several categories:

- Economic effects, essentially local direct and indirect jobs being generated or sustained;
- Social effects, often indirect, but directly linked to the economic effects, and categorised as the protection of culture or heritage, social cohesion and inclusion, identity, and pride;
- External environmental effects, often not rewarded or penalised directly through the market, which can be: (a) positive effects, such as landscape protection, or local resources protection (animal and vegetal biodiversity); as well as (b) negative effects, such as a load on natural resources (water, air, soil) or the welfare of the animals (conditions of detention of the cattle);
- External effects on human health, which are not penalised if negative (food risk factors in non-hereditary diseases), but which can be rewarded if positive in so far as they are known by consumers.

---

[17] B Pecqueur, 'Qualité et Développement Territorial: l'Hypothèse du Panier de Biens et de Services Territorialisés' (2001) 261 Économie rurale 37.

[18] See the following special issue for an overview: H Renting, H Oostindie, C Laurent, G Brunori, D Barjolle, A Jervell, L Granberg and M Heinonen, *Multifunctionality of Agricultural Activities, Changing Rural Identities and New Institutional Arrangements*, International Journal of Agricultural Resources, Governance and Ecology (2007) (Special Issue).

Several case studies conducted during a European research programme (SUS-CHAIN)[19] show that the main and constant factor explaining the impacts of an initiative for the valorisation of a local traditional product is the capacity of the concerned actors within a territory (beyond the collective organisation managing the product) to mobilise their own as well as some external resources, to build up local as well as broader networks and to gain some useful social and commercial competencies.

2.4 Hypothesis

We want to assess the links between: (i) the market as the place where economic value creation process takes place; (ii) the organisation, as a condition for the effectiveness of collective strategies; and (iii) the territory, as the beneficiary of potential impacts on sustainable rural development.

Insights from previous research allow us to formulate the following hypothesis:

The process and interactions:
within the supply chain,
between the actors in the chain and the buyers and consumers,
between the chain and other actors involved in the territory,
influence the commercial, organisational and territorial performances.

This hypothesis will be discussed in light of studies of concrete cases in several European countries.

3. CASE STUDIES

3.1 Commercial Performances are Directly Linked to the Organisational Performance

We have studied 21 cases of legally recognised GIs, Protected Designations of Origin (PDOs) as well as Protected Geographical Indications (PGIs),[20] within the framework of the European research programme: 'PDO-PGI, Market, Supply Chain & Institutions' (see Table 15.1).

---

[19] D Roep and H Wiskerke (eds), *Nourishing Networks – Fourteen Lessons About Creating Sustainable Food Supply Chains* (2006), at: www.agriboek.nl/

[20] PDOs and PGIs were the legal categories of GIs used in the Regulations which preceded the present Regulation 1151/2012; i.e. Regulation 510/2006 and Regulation 2081/92.

Table 15.1   The 21 concerned PDO-PGI Product Supply Chains (as described in their original language)

Country	Products
France	Cantal (cheese), Agneau du Quercy (sheep meat), Comté (cheese), Pommes de terre de Merville (potatoes), Huile d'olives de Nyons (olive oil)
Greece	Feta (cheese), Zagora apples (fruit), Peza Olive Oil (olive oil)
Italy	Prosciutto di Parma (ham), Parmigiano Reggiano (cheese), Fontina (cheese)
The Netherlands	Noord-Hollandse Goudse (cheese), Noord-Hollandse Edammer (cheese), Boeren Leidsche met Sleutels (cheese), Opperdoeze ronde (potatoes)
Great Britain	West Country Farmhouse Cheddar Cheese, Scottish Lamb (meat), Jersey Potatoes
Spain	Jamón de Terruel (ham), Ternasco de Aragon (lamb meat)
Switzerland	Gruyère (cheese), Abricot Luizet du Valais (apricots)

We should clarify at this stage that the cases presented in this chapter were carefully selected within the framework of this European research programme, according to neutral criteria which ensure that they are representative of the general situation of the GIs in Europe.[21]

3.1.1 Organisational performance as the determining factor for commercial performance

Research indicates that the specificity of the product (specific quality) determines the commercial potential upon which the product can rely, and this specificity is mainly due to two factors: one is an invariable factor, which is linked to the product's own nature, the second is a socially constructed factor, which is the result of the management of this nature over time, to make it permanently fit with its natural and human context. The collective management rules and the modalities of their enforcement over a period of time have an influence on the nature of the product itself, and permit the organisation to express the potential of the product in a variable manner towards consumers. Three main modalities of the collective management of the product have been identified (cf. Table 15.2).

In previously published papers, we have already discussed the key

---

[21] The word limits of this chapter do not permit us to present the case studies in detail. We refer the interested reader to the online versions of the case studies, available at: www.origin-food.org.

Table 15.2 *Coordination and cooperation at the collective management level*

Adaptation of the raw material to the final product	No management	Selection, raw material sorting	Orientation according to the final product qualities
Payment depending on the raw material quality	Yes, but according to inadequate criteria	No	Yes, on the final quality of the product
Definition of the product (code of practice) adapted to the demand	Inflexible, opportunistic, impervious	Moderated	Flexible: every single actor can develop his own quality profile while respecting the code of practice
Code of practice control	Uneven, biased	Limited or missing	Effective and unbiased
Final product taxation (based on quality grading)	Incomplete or biased	Limited or missing	With class change
Volumes management	Yes, inflexible (quotas) or not legitimated	No	Yes, flexible (class change, area, etc.)
Type of management	Inflexible management	Neutral management	Efficient management

Source: Adapted from D Barjolle, J-M Chappuis & B Sylvander, 'From Individual Competitiveness to Collective Effectiveness: A Study on Cheese with Protected Designation of Origin', Proceedings of the 59th EAAE (European Association of Agricultural Economists) Appeldoorn (NL), 22–24 April 1998.

elements determining the commercial performance according to the classical approach of market and competition analysis in marketing management sciences, based on several PDO or PGI case studies in Europe.[22] We have identified the following main factors as key:

- The attractiveness of the reference market itself;
- The public support policies;

[22] D Barjolle et al., 'Key Success Factors' (n 15).

- The effectiveness of the internal coordination between the actors in the chain;
- The product's specificity, which is simultaneously a causal feature of and influenced by the collective management of quality over time.

We also concluded that the protection and registration of GIs (here, PDOs) as such is not sufficient to ensure a positive economic result. Additionally, for supply chain actors, the room for manoeuvre depends on the effectiveness of the institutional coordination mechanisms.[23]

3.1.2 Protection of commercial value through intellectual property rights

It is also interesting to examine the effectiveness of legal protection for the commercial potential of the product. We have previously explored this through an examination of the European regulation 510/2006,[24] according its own stated objectives regarding market aspects, which are the following:

- Protection of the name, against both misleading as well as evocative, that is 'free riding' uses. This also allows the name to be protected against degenerating into generic status through continuous misuse;[25]
- Clear information provided for consumers, while diminishing the risk of confusion with similarly named substitute products;
- Diversification of offers in the marketplace, since GI protection supports the diversity of products, including artisanal and non-standardised products, as opposed to an industrial or large-scale production of undifferentiated products.

---

[23] D Barjolle, S Réviron and B Sylvander, 'Création et distribution de valeur économique dans les filières de fromages AOP' (2007) Economies et Sociétés, Série 'Systèmes agroalimentaires', AG, n°29, 9/2007, 1507–1524. This engages in comparative analysis between Beaufort, Comté, Cantal, Mont d'Or (FR), Vacherin Mont d'Or, Gruyère, and Emmentaler (CH). These cases illustrate the manner in which the commercial potential of an origin product is mobilised by the collective organisations and the negotiation processes which determine the distribution of the value added along the chain (in the case of cheeses between milk producers, cheese makers and ripeners).

[24] Council Regulation (EC) No 510/2006 of 20 March 2006 on the protection of GIs and designations of origin for agricultural products and foodstuffs [2006] OJ L93/12. This replaced its similarly titled predecessor, Council Regulation (EEC) 2081/92 [1992] OJ L208/1. Regulation 510/2006 has subsequently been replaced by Regulation 1151/2012.

[25] On genericide, see Chapter 18 by Gangjee in this volume.

Table 15.3 Creation of commercial value for the nine PDO cheeses studied

	Name protection	Clear information to the consumer	Diversification of the offer
Parmigiano Reggiano	xxx	xxx other parmesans	xxx
Comté	xx	xx other hard cheeses	xxx
Cantal	xx	xx Salers	x
Gruyère	xxx	x	xxx
Fontina	xx	xx Fontal	xxx
Feta	xxx	xxx Feta from abroad	xxx
Boeren Leidse met Sleutels	x	xxx Pan Pan	xxx
NH Edamer	x	xx other Edams	x
WCF Cheddar	x	xx (other Cheddar)	xx

Note: 'x' indicates the importance of each dimension for a given product: x = small; xx = medium; xxx = large.

Source: D Barjolle, J-M Chappuis & M Dufour, *Protection des Appellations d'Origine et des Indications Géographiques en Suisse, Recommandations Finales* (Projet européen FAIR CT 95-306, PDO-PGI Products: Market, Supply Chain and Institutions, Institut d'Economie Rurale de l'EFPZ, 1999).

We have selected these three axes as good indicators of protection and analyse their level of relevance for the nine PDO cheeses studied in the frame of the European programme mentioned above.

The importance of the name protection depends on the attractiveness of the name outside the traditional original area (cf. Table 15.3, first column). In that regard, a group of high-profile products like Parmigiano Reggiano, Feta or Gruyère are running very high risks of degeneration into generic expressions. Indeed, because of their high visibility and high household penetration rate, including outside the production area and even abroad, those products' names are misappropriated by unscrupulous firms. For a second group of products, the notoriety is mainly regional or national (for example Fontina, Cantal, Comté). The risk for these products is the emergence of marginal counterfeits in their production zone (for example modification of the traditional recipes in order to reduce the costs or the abusive use of a name for products made in an adjacent region). For this second group of products the code of practice is useful, even if risks and stakes are not as important as for the first group of products. Finally, a third group of products appears far less, or even not at all, threatened by a degeneration risk. Often, those products are semi-generic (a name which has become generic associated with a geographic localiser). For these products, the objective of name protection is not an issue.

The more these products have numerous close substitutes, the more consumer protection (confusion risks due to these close substitutes) is at stake (cf. Table 15.3, second column). The precise identification of those products allows consumers to be certain of the nature and the exact provenance of the product. The need for this regulation is also very important for the products whose names are a mix between a generic name and a geographic localiser: they will gain a differentiation power in the minds of consumers, who could otherwise be indifferent to or ignorant of the exact origin of the product.

Diversification implies a varied offer of artisanal products, as opposed to large-scale production of standardised products produced through industrial processes (cf. Table 15.3, third column). Regional products contribute to the realisation of this objective only if an artisanal structure of milk or cheese production actually exists on the ground. Moreover, Regulation 510/2006 (or its successor Regulation 1151/2012) does not explicitly aim to support artisanal production structures. In this regard, previous research on cheese suggests that only three products do not contribute to it in a decisive way. The Cantal has a not very detailed or elaborate code of practice, which has allowed a very high concentration of cheese-making units. This situation has led to a very low milk price and therefore to the decline of the farm enterprises in marginal areas. The Noord-Hollandse Edammer is made of milk produced in intensive milk factories and is processed in one unique industrial cheese dairy. As for the Boeren-Leidse met Sleutels, the number of firms, for which production alternatives (that is other non-GI cheese made by the same factories) are almost equivalent, is too low.

3.1.3 Effect on agricultural income and support to local employment

Several of the products studied are situated in disadvantaged areas and contribute, because of the relatively higher selling price to consumers, to a better remuneration of the producers (the price paid for milk is higher) and the cheese makers (cf. Table 15.4). Those products (Parmigiano Reggiano, Comté, Gruyère, Fontina, Feta) allow an economic efficiency based on a

Table 15.4 Qualitative measurement of the support to agricultural income effect

	Parmigiano Reggiano	Comté	Cantal	Gruyère	Fontina	Feta	B. Leidse met S.	NH Edamer	WCF Cheddar
Supporting agricultural income (Keeping jobs in marginal areas)	xxx	xxx	x	xxx	xxx	xxx	x	x	x

Note: 'x' indicates the importance of each dimension for a given product: x = small; xx = medium; xxx = large.

Source: D Barjolle, J-M Chappuis & M Dufour (1999), Table 15.2 above.

higher workforce intensity[26] and a higher employment rate per product. The most artisanal products are the ones for which the contribution to the expected effects is the highest. This is linked to the higher workforce intensity in artisanal units. The remuneration linked to commercial valorisation allows economic returns to reinforce farms' viability and to protect a heritage and a landscape threatened by a marketplace which is otherwise dominated by price competition.

The rest of the products present contrasting situations. It is therefore more difficult to draw conclusions which can be of general application. Cantal suffers from a low price on the market because of its decline in many of the products produced by individual producers, regarding aspect and taste, and because the constraints in the code of practices do not generate any increase in the qualitative value. The British and Dutch cheeses have not been incorporated under the Regulation for enough time to assess them. They are located in areas where alternatives exist for equal work remuneration (high workforce cost). They are not pillars of the local agricultural economy, as the products from the first group are.

According to the examples studied, the success on the market of some products from marginal areas allows the market remuneration of an artisanal product which is workforce intensive. This is especially the case when the workforce opportunity costs are low. The farms' viability in these zones would be lower without this remuneration. It is, however, difficult to measure how far intellectual property protection does or does not directly cause such effects, as expected by the European regulation on PDOs and PGIs, due to the causal complexity of the factors influencing these effects. For example, the agricultural income is reinforced by targeted agricultural policy measures (such as support to farms in mountainous areas) before being directly influenced by the protection of the geographic name of a cheese produced with milk from the agricultural farms in the areas. The landscape protection can also be supported by specific measures (regional parks, the creation of territorial zones, protection of buildings via heritage laws), without any direct link with GI protection. According to the examples studied, the success on the market of certain products from less favoured regions allows market remuneration for a production craft character that is more extensive in the local workforce. This is especially since the opportunity cost of the labour is weak. The viability of the exploitations in these zones would be less feasible without this remuneration.

---

[26] This refers to the intensity of labour necessary for producing each product, i.e. the number of working days per unit of product.

3.1.4 Territorial performance

The indirect effects on territory are difficult to isolate and to measure. We estimated the contribution of these same PDO cheese makers to the fulfilment of the purposes of the European regulation concerning expected indirect effects (cf. Table 15.5).

In the case of products where the specifications are the strictest as regards the manufacturing process conserving living micro-organisms, the sensorial variety will be the highest. These results are confirmed by the scholarship of researchers in sensory quality and microbiology of cheeses.[27] The management of a high sensorial quality correlates to and keeps pace with an effective management of the quality by the collective organisation (cf. Table 15.5).

The protection of a threatened local heritage is stronger in the marginal zones where the collective management of the product is effective, with an interesting result for the creation of economic value. The links with the other economic activities of the region, in particular catering (promoting local cuisine) and tourism, are potentially significant. The export of the product outside the region and its contribution to the attendance of visitors to the region and to the attractiveness of the region in general are indicators to be considered.

The contribution of the PDO to the protection of local landscapes is a criterion more difficult to estimate. The impact of certain practices which are required by the product specifications, such as the use of fodder consisting of unrefined feeds that helps to maintain pastures and grasslands in a general way, prohibiting silage, and the outdoors grazing for cattle are the criteria which were considered here. Generally speaking, these kinds of evaluations are made by comparisons either by the researchers or based on experts' statements.

In conclusion, even if it is difficult to formulate general lessons as regards the territorial impacts of the mobilisation of a GI's potential for added value, we can conclude that the capacity of the local actors to mobilise tools and support policies around such regional products is the determining element of territorial performance. Indeed, the identification of rural development stakes in the concerned zone and their consideration in the definition of specifications allows, within the context of other more general frameworks (such as the agricultural or regional policy), legal protection to help in the realisation of a territorial performance with

---

[27] B Martin, I Verdier-Metz, S Buchin, C Hurtaud and J-B Coulon, 'Facteurs de Production et Qualité Sensorielle des Produits Laitiers d'Origine' (Colloque INRA-INAO, Paris, 2005).

Table 15.5 Indirect effects of some PDO cheeses

Expected indirect effects:
Is the product one of reasons for which the situation was consolidated or improved from the point of view of the expected effect?

	Parmigiano Reggiano	Comté	Cantal	Gruyère	Fontina	Feta	B. Leidse met S.	NH Edamer	WCF Cheddar
Satisfaction of the need for variety of the consumers (Organoleptic variety within the same product)	xxx	xxx	xx	xxx	xx	xxx	xxx	xx	xx
Protection of a heritage (The PDO allows the continuation of endangered assets and liabilities)	xxx	xxx	x	xx	xx	xx	xx	x	x
Protection of the landscape (Contributes to the preservation of threatened or typical cultivated landscapes)	xxx	xxx	xxx	xxx	xxx	xxx	xx	x	x

Note: 'x' indicates the importance of each dimension for a given product: x = small; xx = medium; xxx = large.

Source: D Barjolle, J-M Chappuis & M Dufour (1999), Table 15.2 above.

non-market effects. These conclusions are supported by recent research, which attempts to develop a nuanced and sophisticated methodology for assessing the direct and indirect impacts of GI protection.[28]

4. DISCUSSION

Concluding with a synthesis of the empirical results presented above, we want to underline the following elements which apply to the products which can guarantee, based on their geographical origin, a specificity that helps justify their protection as GIs:

- **At the market level:** The creation of economic value (surplus and added value) is all the more substantial and more long-lasting if strict specifications defining conditions of production are aligned with local agricultural projects and the production rules are defined in a participative manner, including the actors concerned within the chain.
- **At the collective organisation level:** The modalities for the collective management of quality are crucial for the evolution of the product. The quality of a product at any given moment determines the potential for creation of economic value. This quality is also the result of another potential, inherent to the nature of the product, which is then influenced by the social construction of the collective organisation which subsequently regulates the product. The history and natural conditions of the product shape the rules, but the rules subsequently shape the product.
- **At the territorial level:** The economic value of the product's image is tethered to landscape-related, cultural and symbolic attributes, which must be communicated to the consumers. It is all the more important as the effect of rarity (that is exclusivity or niche image) is considerable, and as the attributes are interlinked, being also affected by other actors relating to local economic promotion (in the sense of the joined theory of the basket of the goods developed by Pequeur, 2001).[29]

---

[28] Giovanni Belletti et al., *The Effects of Protecting Geographical Indications: Ways and Means of their Evaluation*, 2nd edn (Swiss Federal Institute of Intellectual Property, Publication No 7 (09.11) 2011).
[29] B Pecqueur, 'Qualité et Développement Territorial: l'Hypothèse du Panier de Biens et de Services Territorialisés' (2001) 261 Économie Rurale 37.

The capture and the distribution of the added value along the chain depends on links between the actors of the chain which are both reliable and formally recognised, as well as on their capacity to be negotiated with the external buyers in the region (wholesalers, important distributors).[30] This tendency asserts itself in the context of globalisation of markets which evolve in their nature, as well as the framing of domestic agreements within the constraints of international (trade) agreements.[31] To reconsider and to redefine the frames of negotiations between the actors within the sector then becomes crucial for the survival of the organisation and for preserving the nature of the product. This is as least as important as the legal framework for intellectual property protection of geographical names.

For GIs, the conceptualisation of the effects of intellectual property protection on the territory from which the product arises requires the application of multiple theoretical approaches which interact in a systematic manner (see Figure 15.1).

The diagram in Figure 15.1 represents three levels of analysis to be taken into account and their interactions at two different levels:

- At the top of the diagram, the elements are derived from strategic choices and tactics which can be influenced by both (i) the supply chain actors, at the individual level of the agricultural companies or the actors downstream in the chain (network), and (ii) at the collective level of the organisation which manages the product as well as actors at the territorial level.
- Towards the bottom, elements on which these same actors cannot intervene directly are presented. At the level of the territory, it is always influenced by the evolution of agricultural policy or the policies of the territory according to national level planning, but as part of a long-term initiative and with sometimes durable external alliances to be created.

The three levels of analysis are as follows:

1. **Market and consumers:** This is a context in which actors have more direct agency and the manner of intervention can be formalised in the product specifications, but also by means of product definition and the implementation of marketing strategies, by appropriate choices in

---

[30] Barjolle and Sylvander, 'Fondements Théoriques' (n 15).
[31] L Boltanski and L Thévenot, *Les Economies de la Grandeur* (PUF, 1987).

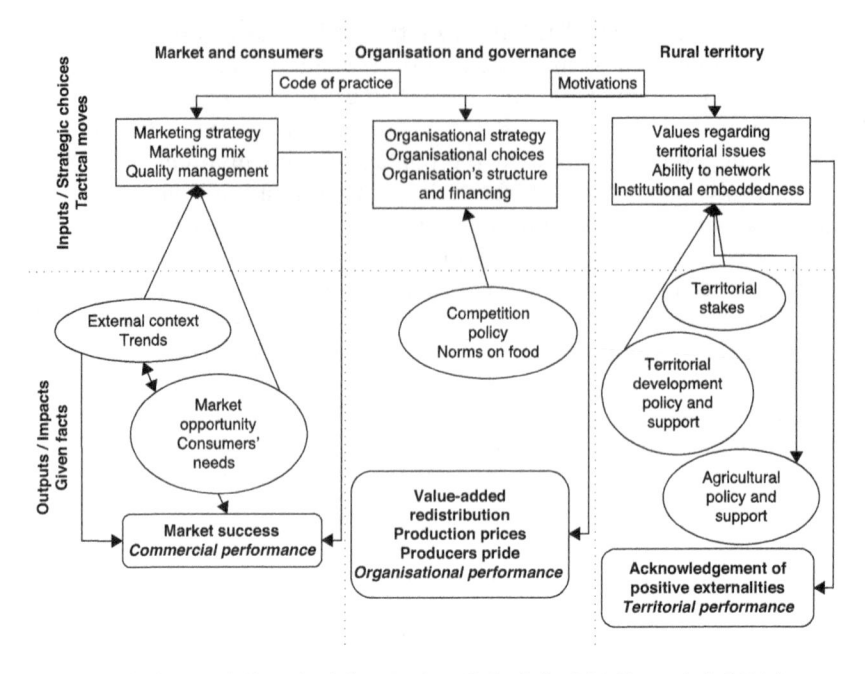

460 *Research handbook on intellectual property and GIs*

Source: This is a partially revised diagram from D Barjolle, J-M Chappuis & S Réviron, 'Organisation and Performance of the Origin Labelled Food Alliances', in AR Bellows, *Focus on Agricultural Economics* (Nova Science, 2005) 91.

Figure 15.1 Conceptual framework of the interactions within the system of market-organisation-territory

the marketing mix, and by the crucial element of the management of quality. The main elements which are going to influence the strategy and the final performance are very evidently the elements of the external context (such as the broader economic climate), trends in consumption, but also opportunities found in the market (the absence of product substitutes, weaknesses of the competitors, and so on) or the needs of the consumers (such as their inclination to pay a premium for the product). These various elements are going to influence the success of the strategy. A strong interaction exists between the organisational strategy and the marketing strategy, and via this link, the organisational strategy determines largely the commercial performance.

2. **Collective organisation and governance:** This addresses the regulation of the product and production methods. The collective organisation acts in accordance with a defined strategy and makes organisational choices, within a structure of rules and with limited financial means.

The main external factors which influence the frame of its strategic freedom are the policies of the competition on the one hand, and the quality standards of the food supply on the other hand. The main result of the organisational strategy can be measured according to the redistribution of economic value created by consumers (expressions of the commercial performance for each of the levels of the sector). Production prices are an integral part of this organisational performance and deserve to be clarified because they will largely determine the territorial performance of the system. A qualitative element of the organisational performance is the pride of the producers, whether farmers or downstream actors, which expresses the linkages established between them and products beyond mere monetary values.
3. **Territory:** At the level of the territory, the capacity for the collective organisation and the other actors from the territory to create the non-market or public goods for the territory (for example by redrafting the specifications to reduce the intensity of chemical treatments or making it obligatory to use local plant or animal species), to create networks and more generally a 'fit' or integration with other local institutions, is going to determine the territorial performance. The efficiency of these actions which affect territorial performance can be measured by the acknowledgement of positive externalities on the territory by the stakeholders themselves, such as by surveys and other forms of qualitative empirical evidence.

5. CONCLUSION

The theoretical approach adopted here reveals the processes and the interactions (within the sector; between the actors of the sector and the buyers as well as consumers; between the sector and other stakeholders within the territory) that influence the commercial, organisational and territorial performances, as they evolve through a complex game of interactions. While quantitative research remains important, the research and approaches described in this chapter remain indispensable for allowing us to get an accurate sense of the effects of GI protection and the importance of organisational structures. It must also be remembered that the difficulties associated with quantification are significant: the concerned sectors are not equipped with systematic statistics for the relevant indicators that measure performances in various dimensions. However, quantification is necessary to pursue systematic reflection in this domain. The examples of failures of the collective organisation in the construction or in the preservation of a product and its label of origin serve as reminders that the balances between the product, its qualities and its promise to the consumers are never guaranteed or settled. Changes in the socio-economic context or in the needs of the consumers, the appearance of new competitors as well as the loss of economic efficiency within the supply chain all call for a permanent attentiveness.

The mobilisation of the skills necessary for the functioning of the system represents an unavoidable variable, which is rarely taken into account because of its intangible nature but which it is necessary to recognise as important. In the context of globalisation of trade and the dominant position of large distributors, the survival of numerous regions depends on the capacity of such collective organisations to define and to implement organisational strategies as well as marketing strategies, in order to create positive externalities for these territories. If appropriately accommodated within the IP protection of geographical indications and supported by the stakeholders (in the broad sense) of the territory concerned, these collective organisations, if they have the chance to manage a product to its potential economic value, can be the mainspring of important profits and benefits.

16. Social gains from the GI for Feni: will market size or concentration dominate outcomes?
*Dwijen Rangnekar and Pranab Mukhopadhyay**

1. INTRODUCTION

Increasing social welfare is an objective of public policy. However, it is well recognized that interest groups influence public policy in a manner that best suits their group interests.[1] In this paper we study the expected role of Geographical Indications (GI) as a welfare-enhancing policy intervention. When conceived as a protective measure, GI recognition was expected to provide protection to products that have traditional knowledge embodied in them. There are conflicting views on the ability of GIs to increase aggregate welfare. The institutions that are required to garner rents created by GIs may just not exist in developing countries.[2] Further, being a protective measure, it stands in contrast to the overarching logic of a globalized agri-food regime.[3]

We explore the possible impacts of the GI on the stakeholders of Feni, an alcoholic drink distilled in Goa, a small state on the western coast of India.[4] The Government of Goa and the Goa Cashew Feni Manufacturers and Bottlers Association, in a joint application, successfully registered a

---

* Dwijen's battle with cancer ended before this work went to print. This chapter is dedicated to the memory of our time together, tasting Feni and the many intellectual discussions around it. – PM

[1] See for example Andrea Prat, 'Rational Voters and Political Advertising' in Barry R Weingast and Donald A Wittman (eds), *The Oxford Handbook of Political Economy* (Oxford University Press 2006) 52; Mancur Olson, *The Logic of Collective Action: Public Goods and the Theory of Groups* (Harvard University Press 1965) 15.

[2] J Hughes, 'Coffee and Chocolate – Can We Help Developing Country Farmers Through Geographical Indications?' International Intellectual Property Institute (Washington, DC 2009).

[3] K Raustiala and SR Munzer, 'The Global Struggle over Geographic Indications' (2007) 18 European Journal of International Law 337, 342.

[4] Feni is made from cashew juice as well as coconut sap. A detailed discussion on these follows later in the chapter.

464 *Research handbook on intellectual property and GIs*

GI for Feni with the Indian GI registry on 27 February 2009.[5] Feni is the first Indian alcoholic beverage (under class 33 of the GI Act) to be registered as a GI in India. This chapter uses both secondary data as well as the findings of a field survey for the analysis.[6]

Goa's economy like most former colonial economies was largely dependent on the primary sector at the time of independence when it joined the Indian union in 1961. The economy rapidly transformed, with the share of the primary sector in the Gross State Domestic Product (GSDP) dropping from 31 per cent in 1970–71 to 12 per cent in 2000–2001.[7] However, the sectoral distribution of employment did not follow this trend and this has created matching problems for the state. Consequently, attempts to bolster income in the primary sector have significance for economic welfare. It is in this context that some of the potential benefits of a GI for Feni should be analysed.

In the last four decades, the reported production of Feni in Goa declined (from more than 1 million litres to 0.87 million litres, see Table 16.1), which presumably had adverse welfare implications for the distillers. This general decline should be placed in the wider context of increased production and availability of a range of alcohols that have emerged, clubbed together as 'Indian Made Foreign Liquors' (IMFL),[8] with production increasing from 0.2 million litres in 1971 to 19 million litres in 2004; see Table 16.1. Thus, we weigh different arguments for the relative decline

---

[5] GI Application No. 120, published in (3 October 2008) 27 Indian Geographical Indications Journal 37. The role of government here needs to be noted: (a) many GIs in India and elsewhere have government agencies and departments as co-applicants; (b) many countries in the South have depended on some form of public support to acquire GIs; and (c) GIs in the North, at least in Europe, have invariably had similar public institutional support. For an overview of state involvement in GI systems, see D Gangjee, 'Proving Provenance? Geographical Indications Certification and its Ambiguities' (2015) World Development [forthcoming].

[6] The baseline survey covered 429 distillers in different cashew zones (out of 1532 zones), nine bottlers (out of a total 19), 59 retailers (out of 9145 licensees) and 24 wholesalers (out of 94 licensees). The survey was conducted between March and October 2007 and covered the whole state of Goa. See Pranab Mukhopadhyay, Santosh Maurya and Suryabhan Mourya, 'Baseline Survey of the Feni Industry' (SHODH, Nagpur 2008) 21–22 for details.

[7] *Statistical Handbook of Goa* (Directorate of Planning, Statistics and Evaluation, Government of Goa (GoG); Government Printing Press, Panaji).

[8] IMFL 'means brandy, whisky, gin, rum, vodka, milk punch, wines or beer manufactured in India and such other liquor as may be declared by the Government as Indian made foreign liquor', see Goa Excise Duty Act & Rules 1964, Rule 2.kk.

Table 16.1 Production of Feni, IMFL and beer (all figures in millions)

Year	Production of cashew liquor (litres)	Production of IMFL other than beer (litres)	Production of beer (litres)	Excise from country liquor (Rs)	State excise revenue (Rs)	Total tax revenue (Rs)
1971	1.08	0.2	1.96	1.93	13.5	48.2
1986	0.98	1.617	10.34	9.46	68.95	605.91
1996	0.73	12.88	23.75	9.25	290.3	3245.09
2000	0.48	14.35	21.30	5.99	390	5382.33
2004	0.87	18.9	22.8	10.7	550	7657.3

Source: Government of Goa (GoG) (various years).

of Feni in comparison to newer beverages. Following this analysis, the chapter anticipates that the structure of the industry may change in a manner that may disenfranchise traditional small distillers and lead to their displacement, following entry of large companies with wider marketing channels, thus defeating a primary objective of introducing GIs – to protect traditional local producers.[9]

The rest of the chapter is organized as follows: Section 2 provides a brief discussion on Geographical Indications and is followed by Section 3 which traces the evolution of Feni as a product and its specification for registered GI status. Section 4 examines the role of GIs in changing the market size and production structure, and the chapter concludes in Section 5 with a note of caution that unless appropriate public policy is devised, the GI could have an outcome that is contrary to the one envisaged by its proponents. Public intervention to promote Feni could take the form of either fiscal concessions or marketing support.

2. GEOGRAPHICAL INDICATIONS: AN OVERVIEW

The specifications that constitute a GI present opportunities to closely tie a commodity to a demarcated geographical area, thus presenting opportunities for economic gains to be accumulated in the identified

---

[9] There have been reports in the media of the expression of interest from companies like United Breweries, Reliance, etc. to enter the Feni business. See S Prabhudesai, 'Mallya also Plans Feni Export' *Goa News* (20 August 2002); and R Roy, 'Feni Now Seeks Global Flavour' *The Times of India* (3 May 2003).

region. The antecedents to GIs can be found in various historical practices of trading bodies and guilds of merchants and craftsmen where either marks or seals of quality indicating provenance were used to distinguish goods.[10] Evidence is also available of Egyptians stamping wine jars to indicate vintage and provenance as did the Romans thereafter.[11] The contemporary legal emergence of GIs should also be placed in the context of the diversity of interrelated notions indicating geographical origin, such as those available in the Paris Convention, the Madrid Agreement and the Lisbon Agreement.[12]

Among the remarkable features of the global and multilateral debate concerning GIs is this: unlike any other instrument of intellectual property, a large and growing number of countries from the Global South campaign for stronger IP in this area. This is evident at the World Trade Organization (WTO), where negotiating stronger protection for GIs is now part of the Doha Round negotiations.[13] This positive characterization of, and interest in, GIs from the Global South, including India, follows the way in which GIs have been identified as a policy intervention for development across numerous multilateral fora. Thus, for example, at the Food and Agriculture Organization (FAO), GIs are presented as enabling better terms for commodity prices for small-scale agriculture producers, whereas at the World Intellectual Property Organization (WIPO) GIs are favourably considered as a tool for protecting traditional knowledge, and at the United Nations Environmental Programme (UNEP) deliberations they are thought of as forestalling the potential for biopiracy.[14]

Tying together these various policy concerns is the idea that GIs

---

[10] See FI Schechter, *The Historical Foundations of the Law Relating to Trade Marks* (Columbia University Press 1925).

[11] See M Maher, 'On Vino Veritas? Clarifying the Use of Geographic References on American Wine Labels' (2001) 89 California Law Review 1881, 1883–4.

[12] For details on these three treaties administered by WIPO, see Chapter 5 by Geuze in this volume.

[13] For an early discussion of the WTO negotiations concerning strengthening GI provisions, see D Rangnekar, 'The Socio-Economics of Geographical Indications: A Review of the Empirical Evidence from Europe' UNCTAD/ICTSD Issue Paper No. 4 (May 2004). For more recent analysis, see Raustiala and Munzer (n 3); see also Chapter 7 by Handler and Chapter 8 by Martín in this volume.

[14] For a brief discussion of the policy concerns that underlie the manner in which GIs are being framed, see D Rangnekar, 'Expert Opinion: Geographical Indications' in A Najam et al. (eds), *Trade and Environment: A Resource Book* (ICTSD; International Institute for Sustainable Development, the Regional and International Networking Group, 2007). See also Chapter 15 by Barjolle in this volume.

can promote endogenous development based primarily on endogenous resources.[15] Briefly, the specifications that constitute a GI present opportunities to closely link a commodity to a demarcated geographical area. Thus, it is potentially beneficial for rural products, handicrafts and artisanal goods or agricultural commodities, as evidenced by the promotion of GIs by a number of development agencies.[16] In the next section we examine the product Feni, and its characteristics, laying the ground for critically discussing the GI and its expected outcomes in Section 4.

3. FENI – EVOLUTION AND SPECIFICATION

Feni is variously written in historical and contemporary documents as well as labels – 'Feni', 'Fenny', 'Fennim' and 'Fenni' – and these various spellings appear to be culturally acceptable. The word has its origins in the local language, *Konkani*, and its local use suggests it means 'froth', probably referring to the frothing of the fermentation process and also to the bubbles that form when Feni is poured into a glass.[17] The term tends to be used for two different liquors: one that is distilled from coconut toddy (*maad*) and the other from the juice of cashew apples (*caju*). Even while coconut Feni has a much longer history in Goa, the Feni GI is exclusively reserved for cashew liquor and the application is acutely silent on the former.[18] This study focuses on cashew Feni which has obtained a GI.

There is a rich and enduring diversity to Feni distilling processes, represented by the various materials and techniques used and also the shared aspects to the cultural repertoire.[19] The process begins with the harvesting

---

[15] See TN Jenkins, 'Putting Postmodernity into Practice: Endogenous Development and the Role of Traditional Cultures in the Rural Development of Marginal Regions' (2000) 34 Ecological Economics 301.

[16] Here consider the United Nations Conference on Trade and Development programme in India, conducted collaboratively with the Government of India, 'Enabling Poor Farmers and Artisans to Protect their Unique Products through GIs', that has run from 2003 to 2009. This has been funded by the UK government's Department for International Development.

[17] See Mukhopadhyay, Maurya and Mourya, 'Baseline Survey' (n 6) 14; and D Rangnekar, 'Geographical Indications and Localisation: A Case Study of Feni' (ESRC Report 24 September 2009) 5.

[18] See Rangnekar (n 17) 20–32 (for a discussion on the specifications and the process leading to the exclusion of coconut Feni) and 49–50 (for resolving this exclusion).

[19] See Mukhopadhyay, Maurya and Mourya, 'Baseline Survey' (n 6) 15; and Rangnekar (n 17) 28.

Table 16.2   Official description of Feni

Item	Description
Physical	Colourless clear liquid which when matured in wooden barrels acquires a golden brown tint
Aroma	Fruity pleasant, strong cashew apple
Ethanol content	v/v 37.1 to 42.8
Tartaric acid ppm	Maximum 200
Acetic Acid ppm	Maximum 60
Esters ppm	Minimum 50
Aldehydes ppm	Minimum 25
Ammonium Chloride	Nil
Copper ppm	Maximum 10 ppm
Methanol ppm	To pass the test
Furfural	5g per 100l of absolute alcohols
Residue on evaporation	Not more than 0.5g/100ml

Source: GI Application No. 120, published in the Indian Geographical Indications Journal (3 October 2008) 41; at: http://ipindia.nic.in/girindia/journal/27.pdf

of fallen cashew apples which are then crushed and the juice collected and fermented. The fermented cashew juice, with no other additives or ingredients, is placed for distillation in a cauldron (locally called a *Bhann*), earlier made of clay but now largely made of copper. The vapour escapes through a pipe (locally called a *Nalli*) at the top of the *Bhann*, which extends to another pot, called a *Lavni*, that rests diagonally to the *Bhann*. The *Lavni* is cooled by the constant and manual pouring of water and this allows the vapours to condense. This first distillation produces a liquor called *Urrack*, that is also consumed during the distilling season as a 'refreshing' and mildly alcoholic beverage. For the second distillation, *Urrack* is mixed with a batch of fermented cashew juice and produces the second distillate that is called *Cazulo*.[20] Naturally, this has a higher alcoholic strength. Oral traditions and residual practices indicate that Feni was triple distilled; however, presently few distillers appear to practise a third distillation and this change appears to be culturally acceptable.[21] For that matter, the GI specification is for a double-distilled liquor (see Table 16.2).

Over time, like with any other cultural artefact, there have been changes

---

[20]   See Mukhopadhyay, Maurya and Mourya, 'Baseline Survey' (n 6) 15.
[21]   See Rangnekar (n 17) 33. Interestingly, the research suggests that the change to the number of distillations was driven by changing market (i.e. consumer) preferences.

in practice, alongside the introduction of new materials and techniques. For instance, cashew apples used to be crushed by feet. This would involve various members of the family, including children, thus generating a wider social context to Feni distilling and also generating opportunities for tacit learning. While some distillers continue with this practice, a number of different materials and techniques, such as a stone-mortar that is powered by draught animals and also mechanical crushers are currently in use.[22] Even materials and techniques in distilling have witnessed changes, one of which is the notable replacement of the *Nalli* and the *Lavni* by a coil, either aluminium or copper, that lies submerged in a water-tank made from cement, and the use of either glass containers (food grade) plastic jerry cans, or even metal cans, to collect the condensed vapours. The baseline report suggests that almost 78 per cent of distillers use semi-traditional techniques.[23]

A key distinguishing feature of Feni – apart from aspects of the distilling technique, such as whether it is double or triple distilled – would be its alcoholic strength. This is measured by a hydrometer and the readings, called *grau*, give a measure of the alcoholic strength of the liquor. For instance, 18 *grau* would translate into a 42.8 per cent v/v.[24] The survey finds wide variation in *grau*. For instance, most distillers (88 per cent) reported a range between 19 and 22 and most bottlers (86 per cent) a range between 18 and 25.[25] The most popular level was 21 among the distillers (see Figure 16.1).

It is this history and diversity of cultural practices that the GI addresses. The specifications of the GI that were eventually registered are that it is a double-distilled alcoholic beverage made only from the fermented juice of fallen cashew apples produced through the months of March–May with a *grau* of 19–20.[26] The following details are noteworthy:

- Fallen and ripe apples are 'normally' used.
- The liquor is distilled in a traditional pot-still with the *Bhann* being made of copper and using an earthen *Lavni*.

---

[22] Ibid., 24–26.
[23] By semi-traditional we imply the use of copper pots (instead of traditional clay pots) along with barrels filled with water for condensation. It includes cases where a coil is used for condensation and a plastic barrel or receptor is used to collect the distillate (Mukhopadhyay, Maurya and Mourya, 'Baseline Survey' (n 6) 25).
[24] See Rangnekar (n 17) 28.
[25] See Mukhopadhyay, Maurya and Mourya, 'Baseline Survey' (n 6) 29.
[26] GI Application No. 120, published in the Indian GI Journal (3 October 2008).

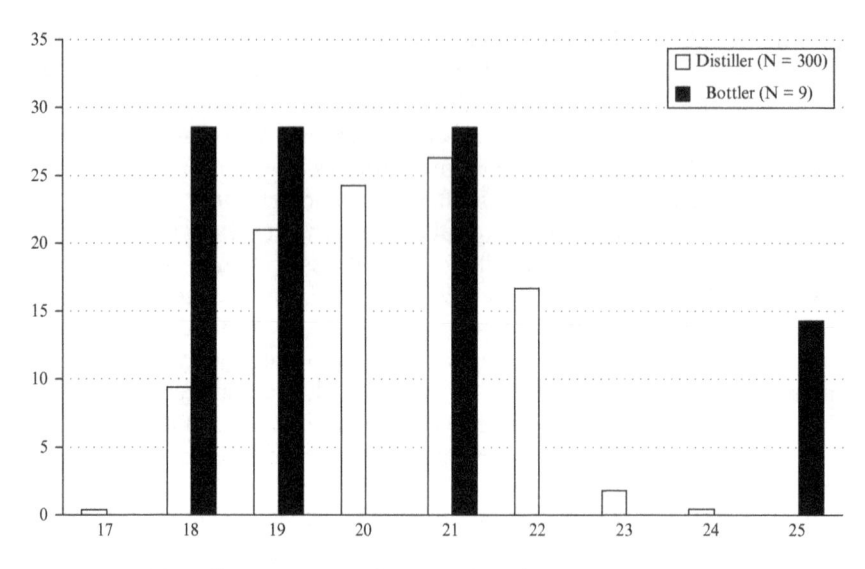

Source: Mukhopadhyay, Maurya and Mourya, 'Baseline Survey' (n 6), 30.

Figure 16.1 Proportion of distillers and bottlers of Feni, and their alcoholic concentration (measured in grau, horizontal axis)

- The first distillation produces a liquor of 16–17 *grau*.
- The second distillation begins with a mix of 130 litres of *Todap*[27] with 110 litres of fermented cashew juice, to produce Feni of 19–20 *grau*.

These specifications are remarkable for a variety of reasons. On the one hand, they are carefully expressive of cultural practices such as prescribing that only fallen and ripe cashew apples are used. Equally, the specifications suggest an awareness of changing practices and the cultural acceptance of particular changes, as in setting out that Feni is double-distilled.[28] Yet, we argue that the specifications are also too prescriptive and could potentially exclude a large segment of distillers.[29]

This specification is narrow and would easily be violated in practice on the following grounds – the instruments of distillation have changed

---

[27] *Todap* is the liquor that is produced from the first distillation and it is also consumed as *Urrack*.
[28] Rangnekar (n 17) 28.
[29] This is assuming that the specification is enforced. There is a telling absence of research on the enforcement of product specifications within the various GI producer groups in India.

over the years as distillers have adapted newer implements. The *Lavni* has changed from being earthen to accommodating other materials which are less fragile, including large plastic cans. The condensation unit has transformed from bamboo, and stems to coils in water tank, etc. Finally, the *grau* of Feni itself varies widely. Less than 45 per cent of the distillers claimed their Feni to have a concentration between 19 and 20 *grau*.[30] This implies that a majority of distillers cannot legally call their product Feni, as per the GI specification. It also implies that the GI applicants in their attempt to specify a product have potentially reduced product diversity, thus effectively reducing the number of distillers who would legally qualify for use of the Feni label for their liquor.

Legally speaking, apart from the possibilities of common law user rights, any non-adherence to the articulated specifications would lead to a prohibition on using Feni as a label for distilled liquor.[31]

4. GIs, PRODUCT DIFFERENTIATION, MARKET SIZE AND CONCENTRATION

The current concern of Feni bottlers is the market limitation that they face both within and more importantly outside Goa.[32] Feni cannot be sold legally in any other state because it is classified as a 'country liquor'. The advantage of being classified a country liquor is the levy of a lower excise duty. However, there are restrictions on its sale outside the state (or even its districts in some cases) of its production, which presents an institutional bottleneck for expanding sales.[33] This is part of the colonial legacy

---

[30] Mukhopadhyay, Maurya and Mourya, 'Baseline Survey' (n 6) 30.
[31] Other features of potential exclusions are discussed in Rangnekar (n 17) 32–44.
[32] See Times News Network (TNN), 'Say Cheers to Feni!' *The Times of India* (1 June 2010).
[33] 'Country Liquor' in Goa is described as 'liquor manufactured in any part of India other than Indian made foreign liquor'. See Goa Excise Duty Act & Rules 1964, s 2c, and liquor is further described as 'spirits of wine, methylated or denatured spirits, spirits, wines, toddy, beer, feny and all liquids consisting of or containing alcohol, wash, other than medicinal, and toilet preparations'. Different states have different rules for the transport of country liquor. In Goa, the Goa Excise Duty Act & Rules 1964, Chapter IX in conjunction with Rule 19B deals with the 'Import, Export, Transport and Possession of Country Liquor'. It permits the import and export of country liquor with other states (i.e. federal units) in India. However, while Goa permits the export of country liquor, its entry into other states is governed by the rules of that state and in most other states there are restrictions on the movement of country liquor. See for example Punjab Excise Act

and post-Independence evolution of excise laws. This can be overcome by amending the excise laws and re-classifying Feni. An idea being pursued by the Goa Cashew Feni Manufacturers and Bottlers Association is to re-classify Feni as a 'Heritage' liquor in order to avoid the market limitation faced as a 'country liquor'.[34]

A more important part of the problem of market limitation is the battle to reverse the declining share of Feni in the alcoholic beverages market in Goa itself (see Table 16.1). This decline cannot be attributed exclusively to the workings of excise laws but has to be explained by local factors. We now turn to a discussion of these issues as they are critical in anticipating the impacts of legal recognition as a GI.

4.1 Institutional Issues and Market Segmentation

The cashew plant was introduced in India by the Portuguese, bringing it from Brazil, in the 16th century.[35] While cashew plantations spread across the sub-continent and the nut has become a prized product, the use of the apple for alcoholic distillation remained exclusively in India-Portuguesa, specifically Goa.[36] The reason for this 'localization' is not clear. The oddity is that the raw material (cashew apple) existed elsewhere in India as did wine and alcohol distillation technologies. To speculate, there might be an institutional and cultural dimension, assuming consumers' tastes and preferences across the Indian sub-continent were not widely different. This curious and interesting puzzle we leave for another project.

Goa became independent from Portuguese colonization in 1961 and joined the Indian union, thus enabling increased trade, investment and social mobility between the two regions. In the alcohol segment, while Indian liquor (non-country liquor) made its way to Goa, the local produce of Goa, Feni, was legally prohibited from trade and sales in the rest of the country.

1914, and Punjab Liquor Import, Export, Transport and Possession Order, 1932, Rules 5 & 13 respectively: 'The import of country fermented liquor is prohibited' and 'Country fermented liquor shall not be transported except within the district in which it is made'.

[34] TNN, 'Say Cheers to Feni!' *The Times of India* (1 June 2010).

[35] Directorate of Cashewnut & Cocoa Development, 'About Cashew', http://dccd.gov.in/cashew.htm; and AR Desai, SP Singh, JR Faleiro, M Thangam, S Priya Devi, SA Safeena and NP Singh, 'Techniques and Practices for Cashew Production' (2010), Technical Bulletin No. 21, ICAR Research Complex for Goa, Ela, Old Goa-403402, Goa, India, http://www.icargoa.res.in/TB%20No.21.pdf

[36] RC Mandal, DG Dhangar and CV Dhume, 'Cashew Apple Liquor Industry in Goa' (1985) 108 Acta Horticulturae (ISHS) 172, 173.

Indian-Made Foreign Liquor (IMFL) production in Goa grew by over 94 times and Beer by 11 times between 1971 and 2004. In the same period, the volume of Feni production declined by nearly 20 per cent (as per official figures). The relative share of government's revenue from country liquor also declined substantially (see Table 16.1). Evidently, the Feni industry has not seen growth comparable to IMFL and has been losing its market share to competing products. The decline in aggregate production is matched by a decline in per capita consumption since both the size of the resident population as well as tourist inflow into Goa has increased in this period.[37]

4.2 Causes for Decline in Demand

This leads us to ask: why has Feni production not grown over the years but actually declined? The possible answers anticipate the impact of GI and the role public policy may need to play in securing the social welfare of traditional small distillers. We discuss the following possible explanations: (i) supply-side issues which deal with reported production, including (a) smaller apple production (the botanical argument) and (b) unreported production (the political economy argument); and (ii) demand-side issues which deal with consumer preferences. We critically examine these arguments in view of two developments in this sector: First, an increase in reported area under cashew plantation by over 68 per cent in the period 1960–2004 and second, an increase in efficiency of juice extraction technology.[38]

4.2.1 Botanical argument
Cashew plantation owners obtain two products each year – the apple and the nut. Of the two, evidently, it is the nut which is the prized product. Therefore, cashew planters are focused on maximizing the nut yield either through greater nut clusters or the size of the nut.[39] In this sense, cashew juice extraction for Feni production is a subsidiary industry to the cashew nut industry. Reportedly the new varieties of cashew being planted by

---

[37] Goa's population has increased from 0.79 to 1.34 million between 1971 and 2001. Similarly the tourist inflow has increased from 0.7 million to 2.3 million in 2006. See *Statistical Handbook of Goa* (n 7).

[38] Mandal, Dhangar and Dhume, 'Cashew Apple Liquor Industry in Goa' (n 36) 175.

[39] Cashew was the second most valuable item among all the agri-horticultural export commodities from India, selling 97,550 MT of cashew kernels valued at Rs. 17.8 billion during the year 2001–2002. MOFPI, 'Cashew/Coconut Feni' (MOFPI).

Table 16.3   Area under cashew plantation (in hectares)

Year	Total area under cashew nut (in hectares)	Total area sown under crops (in hectares)	Cashew as a proportion of total area sown (%)	Number of licences issued for stills
1986	46 888	130 575	36	3037
1996	51 360	165 506	31	3991
2000	53 767	171 356	31	3217
2004	54 858	169 200	32	3095
2009	55 732	160 320	35	2851

Source: GoG (various years).

farmers do not have a large, juicy apple like the traditional varieties.[40] The smaller size of the apple and its lower juice content could be a possible explanation for the reduction in Feni production.

This argument has certain limitations. It is not clear why despite adoption of new cashew varieties that yield low juice content there should be a reduction in output when there has been an increase in the area under cashew cultivation (see Table 16.3). Further, it is not feasible that there is a shortage of cashew juice since Goan distillers could receive cashew apples from the bordering areas of Maharashtra and Karnataka as Feni distillation is not permitted there.[41] So even though there could be a decline in local supply, this could be supplemented from across the state border (see Table 16.4). Therefore, the decline in Feni production cannot be entirely on account of botanical factors.[42]

4.2.2   Political economy

A second possible explanation for the reported decline in output could be the under-reporting of output to avoid payment of excise tax. This is

---

[40] Some of the respondents mentioned this during the field survey in 2007–8.

[41] News reports suggest this is the case – at least for illicit distilling HBD, 'Maharashtra Plans to Make Feni Too' *O Hearldo* (Panaji, 22 June 2009); Rangnekar, 'Geographical Indications and Localisation: A Case Study of Feni' (n 17) 39; TNN, 'Say Cheers to Feni!' (n 34).

[42] In 2010–11, Goa accounted for only 6 per cent of the national area under cashew plantation and less than 0.1 per cent of the global cashew production area. Neighbouring Maharashtra, which accounts for almost one-third of the national cashew production, has been contemplating starting production of cashew-based alcohol. Goan producers felt that this would further erode their shrinking market. See 'Maharashtra Plans to Make Feni Too' *O Heraldo* (22 June 2009).

Table 16.4   Cashew production in India (2004–5)

State	Area in '000 hectares	Production in '000 tonnes	Yield in kilogramme per hectare
Maharashtra	181	208	1231
Andhra Pradesh	183	107	588
Orissa	149	91	669
Kerala	78	71	947
Tamil Nadu	135	65	507
Karnataka	119	57	491
Goa	56	24	436
West Bengal	11	11	1000
Others	33	19	576
All India:	945	653	720

Source: Agricultural Statistics at a Glance 2011, Government of India, New Delhi, Table 5.6b, http://eands.dacnet.nic.in/latest_20011.htm

a well-known world strategy that producers and other individuals resort to, in order to avoid full payment of their taxes, by under-reporting their true output and/or income. Economy-wide studies in India on the size of the grey economy suggest that this can be fairly large.[43] Would the Feni industry exhibit similar grey markets?

If we combine the positive output effect of an increased area under cashew plantations with the improvement in juice extraction technology, the official data on Feni production possibly underestimates actual production by a large percentage. Under *ceteris paribus* conditions, the cashew liquor production in the state should have been much higher than the reported output in 2004 (a back of the envelope calculation suggests this to have been in the range of 1.2–1.6 million litres, approximately).[44] However, this would still mean that the output has at best doubled while that of IMFL products has grown 16–20 times over. Furthermore, if Feni has large unreported production, the same may be expected of the IMFL segment for the same reasons. Irrespective of the above hypothesizing, the reasons for relative decline in Feni production in comparison to IMFL production cannot fully be explained by this

---

[43]   Arun Kumar, 'India's Black Economy: The Macroeconomic Implications' (2005) 28 South Asia: Journal of South Asian Studies 249, 257.

[44]   In terms of state revenue, the share of excise revenue from country spirits too has dropped from 20 per cent to 7 per cent which may have led to a decline in the state's interest in promoting cashew liquor (see Table 16.1).

476 *Research handbook on intellectual property and GIs*

Table 16.5 Price of cashew liquor and price index (in Rupees)

Year	Price of cashew liquor in Panaji/Mapusa		GDP deflator (1999–2000 base)
	Retail (750ml)	Wholesale (18 bottles 750ml)	
1965	5	74.17	6.8
1970	7.72	106.53	8.9
1986	24	360	33.7
1995	36	634	77.9
2000	62	873	103.5
2004	70	1029	121.1

Source: GoG (various years) and authors' calculations.

under-reporting argument.[45] We now turn from supply-side factors to demand-side factors.

4.2.3 Demand-side factors

The third possibility is a decline in demand for Feni over a period of time due to four possible reasons: (i) a decline in incomes; (ii) a rise in relative prices of Feni; (iii) availability of new competing substitute products at competitive prices; and (iv) a change in consumer tastes and preferences. Let us consider each of these in turn.

(i) **Income:** Goa has witnessed a rapid increase in per capita incomes of over 6 per cent per annum (on average in real terms) during 1980–2005. Therefore, the drop in demand could not be due to a decline in income.

(ii) **Relative prices:** The relative price of Feni does not seem to have increased but actually declined[46] (see Table 16.5).

(iii) **Competing products range:** Once Goa joined the Indian union in 1961, there was an expansion of the alcoholic product range given

---

[45] Also, there is no *a priori* reason to expect that evasion in Feni is greater than the IMFL segment.

[46] The GDP deflator (which is a proxy for the average price of aggregate output in the economy) increased by about 17 times while Feni prices increased by about 13 times in the same period (1965–2004). This effectively means that the price of Feni (relative to a composite goods basket) actually declined over this period. The GDP deflator was obtained by dividing the GDP at current prices by GDP at constant prices.

that beer and IMFL production both started in the state and IMFL could be easily imported from other states.

(iv) **Tastes and preferences:** Therefore, the massive increase in consumption (and reported production) of IMFL and beer simultaneous with a decline in Feni is probably indicative of a shift in consumer tastes caused by an expansion of the alcoholic product range.[47] This pertains to not only the resident Goan population but the increased inflow of tourists who are responsible for sustaining alcohol production in the state.

Given that Feni is steadily losing market share to IMFL products, the question that arises is what prompted the registration of a GI for Feni? After all, the small producer does not have any individual incentive to seek a GI as he is a relatively minor player in the alcohol market. So who are the people who would benefit the most from this legal strategy? This leads us to a discussion of the theory of the firm and the implications of GI registration (and its accompanying specifications) on the Feni distillers.

Firms typically prefer an imperfect market structure since it allows them to earn 'super normal' profits. However, there are two necessary conditions for monopoly rents to accrue – entry deterrence and the lack of a near substitute. Monopoly rents get dissipated as the number of firms increase. Super normal profits or monopoly rents become zero under perfectly competitive conditions when there is free entry or exit of firms.[48] It is, therefore, the aspiration of every firm to curb competition in order to increase its monopoly rent by creating either entry deterrence[49] or product differentiation.[50]

---

[47] Feni has a strong cashew aroma which may not be appreciated by consumers, particularly those who are used to the odourless liquors found in the IMFL range. Confirmation of the shift in consumer taste would have to wait for consumer surveys, but if our conjecture is correct, this would not bode well for stakeholders in the Feni industry.

[48] The theory of Contested Markets suggests that even in the presence of market concentration (few producers), firms may not earn super normal profits as long as there is free entry in the industry, see WJ Baumol, JC Panzar and RD Willig, *Contestable Markets and the Theory of Industry Structure* (Harcourt, Brace Jovanovich, Inc. 1982); Makoto Yano, 'A Price Competition Game under Free Entry' (2006) 29 Economic Theory 395 (Symposium in Honour of Mukul Majumdar).

[49] Entry deterrence could occur due to existence of natural monopolies (mines, oil firms), due to firm behaviour (e.g. excess capacity creation, price cutting) or state initiated policies (e.g. defence, health, pension funds).

[50] Lack of near substitutes could be natural (e.g. salt) or created by altering consumer tastes and preferences by advertising and creating awareness

4.3 Entry Deterrence and Market Concentration

A GI provides the possibility of both entry deterrence (geographical protection to local producers) as well as market differentiation. Entry deterrence – which implies that only those who satisfy the product specification can legitimately use the GI – can be ensured both by private legal actions taken by producers and through the agencies of the state (legal and executive bodies). Additionally, when a product has a GI label, in the eyes of the consumer, it will be distinguished from other substitutes. However, market differentiation will depend largely on producers being able to ward off competition from near substitutes and change consumer tastes and preferences. It has been argued that 'the institutional setting for GIs is that of competitive markets' because GIs (unlike trade marks)[51] are a quasi-public good.[52] If there are rents accruing to firms in this industry, it would attract fresh investment either from existing firms or new firms that would geographically locate in the region to take advantage of the GI till rents are dissipated.

By definition, GIs have a degree of locational specificity which requires either raw materials or processing and production to take place in a demarcated geographical territory.[53] To the extent that there are such specifications in the GI, there will be particular requirements to be adhered to by incumbents and new entrants. The economic consequence of this feature is to generate additional costs upon those seeking to enter the market segment of a GI. This disallows any possibilities of adopting alternative materials and techniques and also of locating production and processing according to a purely 'least cost' parameter. The potential

---

(e.g. toothpaste, cars, perfumes, clothes, eco-label products, etc.) or state initiated (speed limits, seat belts, helmets).

[51] See GianCarlo Moschini, Luisa Menapace and Daniel Pick, 'Geographical Indications and the Competitive Provision of Quality in Agricultural Markets' (2008) 90 American Journal of Agricultural Economics 794, 795.

[52] See Rangnekar, 'The Socio-Economics of Geographical Indications' (n 17) 19.

[53] Consider the predicament confronting Newcastle Brown Ale when seeking to expand production at a new location across the river Tyne and leaving its site where it operated for 100 years, it was thus required to rescind the GI as the specifications were being contravened. Conversely, the producers of Blue de Bresse, a French cheese, decided to rescind their GI so that they could license the use of the name to cheese produced in New Zealand. See Rangnekar, 'The Socio-Economics of Geographical Indications' (n 17) 37.

entrant will have to locate processing and/or production in the geographical area demarcated in the GI.[54]

There are numerous qualifications to the above. To begin with, the protection conferred by a GI is such that the knowledge related to production and processing remains in the public domain. Thus, potential entrants can adopt and adapt them either in the same or at other locations – and enter the market with their substitutes to the GI. Naturally, these alternatives will necessarily be prohibited from using the GI.[55] There are numerous examples of such 'substitutes': Cava (from Spain) for Champagne and Agava (from South Africa) for Tequila and Mescal.

The current GI specification for Feni will create a market segmentation of two types of cashew liquor – one that meets GI certification standards and one that does not. For Feni producers who do not label their product and market exclusively through traditional networks, the presence or absence of a GI label on their product may not affect their markets so much, at least in the short term. However, for those who want to access the bottled Feni segment, this would be an important factor for market access. Other states like Maharashtra who have indicated their intentions of allowing distillation of cashew liquor will not be able to label the product as Feni.[56]

4.4 Product Differentiation and GIs

Product differentiation for a GI good occurs at a number of levels – one is the product guarantee offered by the label indicating that the product adheres to GI specifications. This distinguishes the product from others within the product category that fails to satisfy the GI specification. A GI for a product increases market visibility and its label is read as a proxy for quality.[57] A second layer of product differentiation is that which occurs

---

[54] For the possible sustainable development benefits of this location-specific (i.e. territorial) requirement, see Chapter 15 by Barjolle in this volume.

[55] For a discussion of the rules prohibiting misleading or misappropriating uses, see D Rangnekar, 'Geographical Indications: A Review of Proposals at the TRIPS Council – Extending Article 23 to Products other than Wines and Spirits' (UNCTAD/ICTSD Capacity Building Project on IPRs and Sustainable Development, Issue Paper No. 4, 2003).

[56] TNN, 'Say Cheers to Feni!' (n 34).

[57] Presumably a product labelled with a GI certification would expect to distinguish itself in the eyes of the consumer as a seal of authenticity or quality. Studies on food product labelling suggest that consumer awareness makes a significant difference to demand, see for example K Kiesel, D Buschena and V Smith, 'Do Voluntary Biotechnology Labels Matter to the Consumer? Evidence from the

through the firms' brand itself; thus differentiating between firms that offer the GI product.

Larger firms will overlay this with individual branding efforts to signal quality to the consumer and therefore distinguish themselves from others. Quality will be endogenously determined by firms conforming to the GI certification.[58] The GI, therefore, is a stepping stone for further quality differentiation where firms with a good reputation would also be able to charge a higher price.[59] Some price dispersion may not be due to product heterogeneity, that is, quality, add-on features, after-sales service, but a manifestation of the level of ignorance in the market.[60] Consequently, consumers tend to be willing to pay a higher price, a premium, for those products that are *reputable* in delivering a certain level of quality. Firms with a well-known existing brand-name are likely to have higher sales than their counter-parts because of brand recognition.[61] This apart, in certain GIs, there is the possibility of proliferating 'types' within the GI-specifications. An exemplar of this is the different types of Scotch whiskey that have been fostered under the umbrella of Scotch.[62]

4.5 Market Size and Concentration

The entry of larger investment and brand creation would be in anticipation of market expansion (volume or value) for higher profits. The existing firms and distillers have a first mover advantage while new firms may encounter large transaction costs in establishing their product. There is persistence of market concentration, and hence monopoly rents, in the 'ready to eat' breakfast cereal sector. This sector is not characterized by a

---

Fluid Milk Market' (2005) 87 American Journal of Agricultural Economics 378, 390.

[58] See Moschini, Menapace and Pick (n 51) 795.

[59] See R Rafael and T Sekiguchi, 'Reputation and Turnover' (2006) 37 The RAND Journal of Economics 341, 345.

[60] GJ Stigler, 'The Economics of Information' (1961) 69 Journal of Political Economy 213, 225.

[61] C Bonnet and M Simioni, 'Assessing Consumer Response to Protected Designation of Origin Labelling: A Mixed Multinomial Logit Approach' (2001) 28 European Review of Agriculture Economics 433, found that in the case of French Camembert cheese, some consumers were willing to buy more of the PDO labelling if it was available at the same price as its competitor but were found to respond more effectively to brand names. In a choice experiment it was found that consumers are willing to pay a higher price for a quality assurance label from a known brand than for a lesser known brand with the same quality label.

[62] Rangnekar, 'The Socio-Economics of Geographical Indications' (n 17) 32.

complicated technology which would otherwise deter market entry, except that entry deterrence occurs on account of the costs to launch a product and gain shelf-space. Companies in the alcoholic beverage business who have an established presence and a strong market infiltration are likely to have lower transaction costs in brand creation.[63] They might also be better able to create brands and establish product differentiation in a commodity which currently has no 'standard' except what the GI specification suggests.[64]

If quality is directly associated with fixed costs, higher quality firms are expected to edge out worse performers and there could be an increase in concentration even with an increase in market size.[65] However, brand-building involves advertising expenses that are fixed in nature.[66] If distillers were to aspire to creating a brand and a label with GI certification then they would experience an increase in fixed costs. Given imperfect capital markets, this would imply that only those with collateral to raise credit in the market would be able to access the advantages of GI certification, which would typically be large producers. The scale of production would have to be sufficiently large to justify the incurring of such expenses.

So even if the GI does increase market size (in volume terms), there would be an unequal access to the GI label. Whether the market will actually expand or not, is crucially dependent on one factor – consumer acceptance of Feni in preference to other competitors like IMFL. As we have discussed earlier, a plausible way of explaining the decline in Feni production over the years, could be linked to consumer tastes and preferences.

---

[63] This was evident in the case of the German meat industry, see U Enneking, 'Willingness-to-Pay for Safety Improvements in the German Meat Sector: The Case of the Q&S Label' (2004) 31 European Review of Agriculture Economics 205, 218.

[64] Currently there are no national-level alcohol producers who are involved in Feni distillation or bottling but there have been occasional reports that some have evinced interest; see Prabhudesai, 'Mallya also Plans Feni Export' (n 9).

[65] See for example WJ Baumol and RD Willig, 'Fixed Costs, Sunk Costs, Entry Barriers, and Sustainability of Monopoly' (1981) 96 The Quarterly Journal of Economics 405, 416. However, this result may not hold if quality is positively correlated to variable costs. For example, S Berry and J Waldfogel, 'Product Quality and Market Size' (2010) 58 The Journal of Industrial Economics 1, 12, found that while the restaurant industry with quality linked to variable costs displayed low concentration, even with a large market size, the reverse was true for the newspaper industry where the concentration increased with market size as quality here was linked to fixed costs.

[66] See for example J Sutton, *Sunk Costs and Market Structure: Price Competition, Advertising and the Evolution of Concentration* (MIT Press 1991) 230.

The survey found that the bulk of Feni sales by distillers are made directly to retailers (including bars and restaurants) and not to bottlers, suggesting that the sale of Feni is largely of the unbranded variety. Local customers who are traditional Feni drinkers procure unbottled Feni from known and trusted distillers, middlemen or local *tavernas*.[67] The branded and bottled Feni is only bought by those who are carrying the product out of the state and at bars where customers have no information on the quality of unbranded Feni. The need for branding arises because of anonymous sales requiring product certification. The tourist typically would not have any idea whether an unbranded bottle of Feni is trustworthy or not – a classic case of information asymmetry. As we are well aware, information asymmetry can lead to adverse selection and market distortions which can reduce people's welfare.[68] Evidently, there is no consensus on whether GIs would also lead to increased market concentration since the local net social welfare benefits of GIs will depend on the combined effect of size and distributional impacts. It is quite possible that even with an increase in market size, there could be a deterioration in welfare if the market concentration increases within the Feni industry.

The worst case scenario for the traditional Feni distiller would be the entrance of a large number of firms either in Goa or in other cashew growing states. We conclude this chapter with a discussion of different scenarios that might emerge hereafter.

5. DISCUSSION

The Feni industry therefore faces multiple uncertainties, due to the possibility of: (a) the entry of large distillers in Goa who could displace existing small producers; (b) the production of cashew liquor in other states called by names other than Feni; and (c) a further decline in relative demand for Feni due to increased competition from the IMFL segment of alcoholic beverages.

Legal recognition as a GI would not be able to stall any of these eventualities, as they do not violate the exclusive rights granted by this branch of IP law. The first eventuality could be avoided by reserving distilling licences to traditional or small brewers, thereby restricting capital flows

---

[67] See Mukhopadhyay, Maurya and Mourya, 'Baseline Survey' (n 6) 35.
[68] See George A Akerlof, 'The Market for "Lemons": Quality Uncertainty and the Market Mechanism' (1970) 84 The Quarterly Journal of Economics 488; and; Moschini, Menapace and Pick (n 51).

into this industry. The second possibility is unfortunately beyond the jurisdiction of the Goa state to regulate. The best that local producers can hope for is that the popularity of the name Feni will give them an advantage in the market. The third factor stated above has as much to do with tastes and preferences as it has to do with pricing structure. A careful brand building by one or more firms, the Feni Association, or the government could woo consumers, causing them to change their taste preferences in favour of Feni and reverse the falling demand. This is feasible given that Feni is seen as an identity drink for Goans and export of Feni to countries with large Goan expatriate populations has been encouraging.[69] If Feni receives preferential tax treatment, this measure may help the sector beat the competition from the IMFL segment. This intervention would lie in the domain of tax policy rather than IP law.

In an industry dominated by small distillers who are not part of any producer group or association it is the more organized segment of this industry that is likely to reap the benefits of protection unless there is active intervention by the state or civil society. Even the long-run benefits of local large producers is not guaranteed as they could be edged out by new entrants with a larger production base, or by mergers and acquisitions (as has happened in India in the soft drinks segment, among others). Apart from formal recognition as a registered GI, the absence of continued state intervention may defeat the primary objective of GI protection, that is, protecting the welfare of traditional distillers.

---

[69] See, for example Anon, 'Feni Goa-ing Global' *Ambrosia* (May 2007); P Kamat, 'Goan Distillery to Patent Feni, Tap Potential' *Times News Network* (Panaji, 16 May 2002); MOFPI, 'Cashew/Coconut Feni' (n 39) 41; and Roy, 'Feni Now Seeks Global Flavour' (n 9).

17. From *terroir* to pangkarra: Geographical Indications of origin and Indigenous knowledge*

Brad Sherman and Leanne Wiseman

There has been a marked increase over the last few decades or so in the number of countries which recognise and protect Geographical Indications of origin (GIs). There has also been a steady expansion in the types of things that are protected. While this is not that surprising given the growing interest in slow food and traditional products,[1] what is more surprising, at least at first glance, is the increased attention that has been given to the potential use of GIs to support and promote Indigenous interests. GIs have been associated with Indigenous traditional knowledge in two ways. Firstly, it has been suggested that they could be used as a mechanism to protect and sustain Indigenous interests.[2] This is because,

---

* The Kauma people from the Adelaide Plains in South Australia use the term '*pangkarra*' to describe the characteristics of a particular place and how it shapes things from that place. See Max Allen, 'Terroir Australia', The Weekend Australian Magazine (13 Dec 2003) 48. We would like to thank Dev Gangjee for his comments on the chapter.

[1] On this see C Bramley and J Kirsten, 'Exploring the Economic Rationale for Protecting Geographical Indicators in Agriculture' (2007) 46(1) Agrekon 47; B Ilbery and M Kneafsey, 'Registering Regional Speciality Food and Drink Products in the United Kingdom: The Case of PDOs and PGIs' (2000) 32(3) Area 317; Elizabeth Barham, 'Translating Terroir: The Global Challenge of French AOC Labeling' (2003) 19 Journal of Rural Studies 127; see also Chapter 3 by Barham in this volume.

[2] See Teshager Worku Dagne, 'Harnessing the Development Potential of Geographical Indications for Traditional Knowledge-based Agricultural Products' (2010) 5(6) JIPLP 441; Commission on Intellectual Property Rights, *Integrating Intellectual Property Rights and Development Policy* (2002), Chapter 4; David Downes, 'How Intellectual Property Could be a Tool to Protect Traditional Knowledge' (2000) 25 Columbia Journal of Environmental Law 253, 281: Ranjay K Singh, RC Srivastava, Adi Community and Monpa Community, 'Biological Geographical Indicators of Traditional Knowledge Based Products and Green Technology from Arunachal Pradesh: An Initiative for Safeguarding IPR of Communities' (2010) 9(4) Indian Journal of Traditional Knowledge 689. Terri Janke, *Minding Culture: Case Studies on Intellectual Property and Traditional Cultural Expression* (Geneva, WIPO, 2003), 36 ('given that indigenous peoples'

as the WIPO Intergovernmental Committee on Intellectual Property and Genetic Resources, Traditional Knowledge and Folklore noted, some traditional cultural expressions may qualify as goods which could be protected by geographical indications.[3] Secondly, and more ambitiously, it has also been suggested that the regimes used to regulate GIs might be used as a template on which *sui generis* schemes to protect Indigenous knowledge might be modelled.

It is the aim of this chapter to look at the suggestion that geographical indications of origin offer an effective and appropriate way of protecting Indigenous traditional knowledge. In particular, we ask: is it realistic to expect that GIs can be used to protect Indigenous knowledge? While it is possible to imagine the law of GIs protecting aspects of traditional knowledge, we argue that beyond this it is unlikely that they would be ever be modified in a way that would allow Indigenous knowledge to be accommodated on its own terms.

There are a number of questions that need to be addressed before we are able to explore this issue in more detail: one of the most important being what do we mean by 'traditional knowledge' and 'geographical indication of origin'? This is especially important given the uncertainty that surrounds these broad terms.[4] In part these problems are made worse by the growing demand, which has been prompted by the push for international protection, for globally relevant definitions that can be applied equally in Switzerland and Sudan. This is particularly the case with the term 'traditional knowledge', which has found favour at the international level and amongst intellectual property scholars. While WIPO's decision to pursue a top-down international treaty to regulate traditional knowledge means that a broad-based definition of traditional knowledge is almost inevitable, for others, however, the attempt to develop a functional, universal definition is a folly, not least because 'the term traditional knowledge is so broad that

---

cultural expression reflects their belonging to land and territories, this may allow some scope for indigenous people to use geographic indications for their clan names, and language words for regions'). There are, however, critics. See, for example, Madhavi Sunder, 'IP3' (2006) 59 Stanford Law Review 257, 302; Shivani Singhal, 'Geographical Indications and Traditional Knowledge' (2008) 3(11) JIPLP 732, 738.

[3] WIPO, 'Consolidated Analysis of the Legal Protection of Traditional Cultural Expressions', 2 May 2003 (WIPO/GRTKF/IC/5/3) 52.

[4] Marion Panizzon, 'Traditional Knowledge and Geographical Indications: Foundations, Interests and Negotiating Positions' (NCCR Trade Working Paper No 2005/01, Oct 2006) 6 ('Apart from politics, a main reason for deferring [discussions at the International level] at this stage is the fact that sufficiently clear concepts of scope and level of protection for TK and expanded GI's are still lacking').

any attempt to define it is necessarily incomplete'.[5] This concern is borne out by experience in Australia, where the decision as to whether a person is classified as 'Aboriginal' or 'Torres Strait Islander' has, at times, been controversial and problematic. Questions such as who decides whether a person is Indigenous and the criteria used to make this decision are not only highly politicised, they also carry significant cultural, social, spiritual, legal and economic consequences. These problems have been compounded by the fact that many Indigenous peoples have forcibly been removed from the country that forms such a strong connection to tradition. This is not to suggest that removal from country necessarily equates to a loss of tradition so much as that the nature of the connection to country and, with it, what is meant by traditional knowledge changes; a change that makes the task of developing a global definition all the more problematic. While we will, where possible, try to speak generally about Indigenous traditional knowledge, we will draw our examples from Australia.

There is also some uncertainty about what is meant by a 'geographical indication of origin'. In part this arises because there is no international consensus about the minimum standards for protection, nor about what the criteria for protection should be. Instead, there are a range of national and regional laws suited to local interests that have been used to regulate the names of products (and indirectly also the product itself) including GIs, designations of origin, indications of source and appellations of origin. The fact that the 'manner in which the protection of GIs has been conceptualised in different countries at different times has varied considerably' has created an ambiguity in basic terminology.[6] It has also meant that with 'the exception of design law, there is probably no category of intellectual property law where there exists such a variety of concepts of protection as in the field on geographical indications'.[7] These problems are exacerbated by the fact that, like many things today, GIs are 'no longer what they used to be'.[8] This is particularly notable in Europe, which is the spiritual home of GIs, where there has not only been a splintering in the

---

[5] Shivani Singhal, 'Geographical Indications and Traditional Knowledge' (2008) 3(11) JIPLP 732, 738.

[6] Oskari Rovamo, 'Monopolising Names? The Protection of Geographical Indications in the European Community' (2006) Helsinki University, Pro Gradu Thesis, 5. See also Bernard O'Connor, *The Law of Geographical Indications* (Cameron May, 2004) 23.

[7] S Escudero, 'International Protection of Geographical Indications and Developing Countries' Trade Working Paper 10 (South Centre, Geneva: 2001); cited in Bramley and Kirsten (n 1) 49.

[8] Rovamo (n 6) 5.

types of protection, but also a change in the standards for protection: the EC no longer requires that a product should be endowed with exclusive characteristics by its geographic origin in order to qualify for protection.[9]

While GIs take a number of different forms, it is possible to characterise them in terms of how the connection between the named product and its geographical origin is determined.[10] At one extreme, some types of GIs – which we will refer to as *natural geographical indications of origin*[11] – are only granted where an applicant can show that the characteristics of the named product are a direct result of the place where the product was produced or, as the Lisbon Agreement of 1958 states, the 'quality and characteristics' of a product are 'due exclusively or essentially to the geographic environment, with its inherent natural and human factors'.[12] From this perspective, a natural geographical indication is a form of collective intellectual property in which 'the protection is related to the product itself and is neither dependant on a specific right holder nor on consumer deception or confusion'.[13] For our purposes, the defining feature of this form of GI is the role that 'nature' plays in shaping the named product. At the other extreme, and in contrast to natural geographical indications, there are a range of GIs which are much more dependent on consumer perceptions about the product in question and its competitors, rather than on the role that nature plays in shaping the product. While with this type of GI there is a connection between product and place, the status of this connection is not something that needs to be proved as an end in itself: instead, what is important is the way that consumers view this connection. To avoid confusion, whenever we speak of GIs in this chapter we will refer to natural geographical indications of origin.

***

---

[9] This is reflected in the definition of a GI, where a reputational link to the region of origin is sufficient. See Art 5(2) of Regulation (EU) No 1151/2012 on Quality Schemes for Agricultural Products and Foodstuffs.

[10] While this distinction could also be seen in terms of stronger and weaker GIs, we wish to highlight the connection between product and the environment in which it is grown. Thus, instead of speaking of the relative level of protection, we wish to focus on the type of connection between nature and right.

[11] For an interesting account of the important role that natural geographical indications of origin play as ideal types and rhetorical device in the law see Dev Gangjee, *Relocating the Law of Geographical Indications* (Cambridge University Press, 2012) 77–126.

[12] See Art 2(1) of the Lisbon Agreement for the Protection of Appellations of Origin and their International Registration, 1958.

[13] Rovamo (n 6) 1–2.

There are a number of similarities and points of connection between geographical indications of origin and Indigenous traditional knowledge. One of these relates to the way that GIs and traditional knowledge are perceived. More specifically it arises because GIs and traditional knowledge are both seen as outsiders – as maverick, *sui generis*, dangerous, ill-fitting and different. While there has been a lot written about the image of Indigenous peoples as dangerous outsiders,[14] there has been less consideration given to the status and standing of the laws that regulate GIs. Despite this, however, it is clear that GIs share many of the attributes that are often (pejoratively) associated with traditional knowledge. The outsider status of geographical indications is reflected, for example, in the debates about whether they properly belong within intellectual property law and in their characterisation as a *sui generis* form of protection.[15] The outsider status is also reflected in the fact that GIs run against the grain of mainstream intellectual property law. This can be seen, for example, in the remark that 'geographical indications as a form of intellectual property challenges the law, culture and economic logic of American business, orientated as it is towards liberal economic theory based in individual ownership'.[16]

The outsider status of GIs is reinforced by the fact that while intellectual property law is often portrayed as modern and progressive, this is not the case with geographical indications of origin, which are seen to be pre-modern and traditional. While patents, design, and copyright are concerned with advancement, novelty, progress and the protection of the new, in contrast GIs are more concerned with preserving and sustaining localised traditional (almost indigenous) practices. Moreover, while modern intellectual property law is seen to be abstract, agnostic, standardised and global, GIs, like the laws that regulate them, are still rooted in the local: they are parochial, judgemental and disparate. Although neither of these caricatures stands up to scrutiny, nonetheless they still underpin the way that intellectual property law generally and GIs more specifically are often viewed.

For many this maverick status is a problem that needs to be rectified. For others, however, it is a cause for celebration in so far as it provides new ways of thinking about intellectual property law. To this end, we have recently seen a small but growing number of commentators who have turned to the law of geographical indications of origin on the basis

---

[14] See e.g. Amanda Nettelbeek and Robert Foster, *In the Name of the Law* (Wakefield Press, 2007).

[15] Stephen Stern 'Are GIs IP?' (2007) 29 EIPR 39. See also D Rangnekar, 'The Intellectual Properties of Geography' (2009) 31 EIPR 537.

[16] Barham (n 1) 129.

that it offers an alternative (conceptual) framework for thinking through some of the issues facing intellectual property law.[17] Geographical indications of origin are also being celebrated because they take account of broader social and cultural interests, which is one of the reasons for the growing interest in geographical indications of origin and traditional knowledge.

There are a number of other reasons why Geographical Indicators are seen to be 'especially suitable' for use by indigenous and local communities.[18] In addition to the general economic arguments made in favour of using GIs to protect Indigenous traditional knowledge,[19] there have also been a number of more specific explanations in favour of protection. One of these is that Geographical Indicators provide a means of protecting *traditional* knowledge (in the strict sense of the word). Indeed, one of the features of many of the regimes used to protect GIs is that they reward traditional cultural values and knowledge rather than promoting innovation per se, as is the case with most of the other forms of intellectual property.[20] Given the emphasis on the preservation and protection of localised traditional values, there is an obvious synergy between the law of geographical indications and traditional knowledge. As one commentator noted, geographical indications accord with the 'emphasis that indigenous communities typically place upon their traditional ways of life, including their relationship with their ancestors land, waters, and living ecosystems'.[21] There is also a belief that Indigenous peoples and the owners of geographical indications share a common concern with the protection of the environment. Another related reason for the growing interest in GIs

---

[17] See e.g. Mario Biagioli, 'Nature and the Commons: The Vegetable Roots of Intellectual Property' in Jean-Paul Gaudilliere, Daniel J Kevles and Hans-Jorg Rheinberger (eds), *Living Properties: Making Knowledge and Controlling Ownership in the History of Biology* (Max Plank Institute for History of Science Preprint 382, 2009) 241–250; Rosemary Coombe, 'Cultural Agencies: "Constructing" Community Subjects and their Rights' in Peter Jaszi and Martha Woodmansee (eds), *Making and Unmaking Intellectual Property* (Chicago University Press, 2011) 79.

[18] Downes (n 2), 269; Kal Raustiala and Stephen Munzer, 'The Global Struggle over Geographical Indications' (2007) 18(2) The European Journal of International Law 337, 345 ('GIs and traditional knowledge share several attributes'); Bramley and Kirsten (n 1) 47.

[19] Downes (n 2) 268; Bramley and Kirsten (n 1).

[20] Bramley and Kirsten (n 1). It is important to note that GIs do not offer direct content (i.e. product or process) protection. See S Frankel, 'The Mismatch of Geographical Indications and Innovative Traditional Knowledge' (2011) 29(3) Prometheus 253.

[21] Downes (n 2) 272.

as a means of protecting traditional knowledge flows from the role that they both play in symbolising quality and guaranteeing authenticity (and all the problems that accompany it).[22]

An additional reason why geographical indications are thought to be particularly well suited to Indigenous knowledge is because there are no limitations on the period of protection. The fact that the legal rights remain in force as long as the collective tradition is maintained has obvious benefits for traditional knowledge.[23] Another argument in favour of using GIs to protect traditional knowledge is that unlike most other intellectual property rights, geographical indications of origin are not freely transferable:[24] the rights remain connected to the group or collective that initiated the rights in the first place. This helps to ensure that knowledge, practices and rituals remain with and under the control of the community. A further reason why geographical indications are thought to be particularly well suited to protecting Indigenous knowledge is because they are able to accommodate group rights.[25] In so doing they help to overcome one of the shortfalls of copyright and patent protection; namely, that they focus on the individual at the expense of broader collective interests. This is not the case, however, with geographical indications of origin, which tend to prioritise collective interests over those of the individual.[26]

Yet another advantage of using GIs to protect traditional knowledge is that they offer a way of ensuring that Indigenous law is incorporated into the legal regimes used to regulate Indigenous culture.[27] One of the consequences of the decentralised nature of the geographical indications system is that questions such as who is to count as an Indigenous person, what type of subject matter ought to be protected, and how and when that subject matter might be used by third parties do not have to be resolved in advance, in the abstract or at a distance. Rather than being answered by centralised agencies, these questions are answered by the community or group that is seeking protection. In the same way in which collectives

---

[22] Raustiala and Munzer (n 18) 346; Rovamo (n 6) 14; Downes (n 2) 269 (both 'protect and reward traditions while allowing evolution').

[23] Singhal (n 5) 733; Downes (n 2) 269; Raustiala and Munzer (n 18) 345.

[24] Downes (n 2) 269.

[25] While community or group interests play a prominent role in many Indigenous Communities, this does not mean that decisions within Indigenous communities are made collectively. Cf. Singhal (n 5) 733.

[26] Raustiala and Munzer (n 18) 345.

[27] See E Daes, 'Defending Indigenous Peoples' Heritage' (Feb 2000) cited in *Composite Study on the Protection of Traditional Knowledge*, 28 April 2003 (WIPO/GRTKF/IC/5/8), 42 at [105].

such as the Parma Ham Consortium set their own internal rules that dictate when and how the name of a product can be used, so too the use of geographical indicators to protect traditional knowledge means that indigenous customary law can be used to determine the extent to which local knowledge is protected. In so far as geographical indications of origin ensure that it is the community rather than outsiders who set the rules that govern traditional knowledge, this accords with the principal of self-determination, while recognising that such laws must articulate with broader frameworks. It also means that Indigenous law is able to evolve and change over time.

One of the most important reasons why geographical indications of origin are thought to be particularly well suited to protect traditional knowledge is because they provide a means of recognising the connection to 'place' or 'country' that is so important for many Indigenous communities. This occurs because many of the regimes used to protect GIs are based upon the idea that the protected product is an embodiment of the location that it originates from. This manifests itself in the requirement that applicants for geographical indications of origin must show a connection between the named product and the place from where the product originates. In extreme situations, it is necessary to show that the product is 'an inimitable reflection of its geography'.[28] The idea that certain products are intrinsically shaped by the location where they are grown is reflected in the notion of *terroir*, the French concept used to describe the characteristics or attributes of a place, resulting from the land, soil, geography, climate, human and seasonal influences which contribute to the unique characteristics of agricultural products (usually wine).[29] *Terroir* is similar to the Aboriginal idea of connection to place in so far as it emphasises the relationship between human cultures, land and environment,[30] as well as the spiritual value of that relationship.[31] For some Indigenous communities, the connection to place arises because 'the land is the source of value and meaning, of rights and obligations . . . The landscape is the source of

---

[28] Rovamo (n 6) 8.
[29] *Terroir* refers to an 'area or terrain, usually small, whose soil and microclimate impart distinctive qualities to food products', see Barham (n 1) 130.
[30] Downes (n 2) 269.
[31] Barham (n 1) 131 ('Beyond the measurable ecosystem there is an additional dimension – the spiritual aspect that recognises the joys, the heartbreaks, the pride, the sweat, and the frustrations of history'), citing James E Wilson, *Terroir: The Role of Geology, Climate and Culture in the Making of French Wines* (Mitchell Beazley, 1998) 55. See also Rovamo (n 6) 8.

meaning and value and the repository of history and events and can be read as a map of itself and its own creation.'[32]

Given the various points of connection between geographical indications of origin and traditional knowledge, it is not surprising that geographical indications are being looked upon as a possible legal mechanism to support and promote indigenous interests. Despite the attention given to this topic, there have been relatively few success stories to date. While it is possible to imagine geographical indications of origin being used to protect aspects of Indigenous knowledge,[33] there are a number of questions that need to be explored before we would be in a position to reach a conclusion about whether or not GIs could be used to provide effective protection for traditional knowledge. One of the most important is whether, despite the similarities, geographical indications of origin are able to accommodate traditional knowledge on its own terms or whether, as happens so often, Indigenous interests must be modified to comply with the requisite legal requirements.

One of the primary reasons why geographical indications of origin are touted as an appropriate way of protecting Indigenous traditional knowledge is because they recognise connection to place. While there are obvious parallels between geographical indications and traditional knowledge, this does not necessarily mean that the law is in a position whereby it can capture the symbiotic relationship that exists between Indigenous communities and land. Although from an Indigenous perspective a product may be inextricably linked to and shaped by country and dreaming, one question that needs to be considered is whether this can be translated into a legal framework. In a legal context how would you show,

---

[32] David Turbull, *Masons, Tricksters and Cartographers: Comparative Studies in the Sociology of Scientific and Indigenous Knowledge* (Harwood Academic Publishers, 2000) 34 ('Everywhere is sacred since all the land was created in the Dreaming by the activities of Ancestral Beings as they moved across the landscape. These journeys left dreaming tracks, knowledge of which is recreated in song, story, and ceremony. Everyone has spiritual linkage to the land by virtue of birth such that they *are* the land. Knowledge of the Dreaming tracks, or the activities that created the land of one's birth, is therefore evidence of possession of the land and by the land.').

[33] There are a number of problems that would have to be overcome before geographical indications of origin could be used as a form of protection. In Australia, for example, there would have to be a dramatic change of heart by policy makers who see geographical indications of origin as a new form of European colonisation. Given that the objects and knowledge that would be protected under these regimes may span a number of different Indigenous communities, it would also be necessary for communities to decide who has authority to negotiate and how benefits are to be shared.

for example, the necessary connection between a dreaming story, country, a particular plant that grows on that country, and the product derived from that plant? Here we encounter the type of problem that sometimes arises with *terroir*: namely, the problem of how to prove the link between product and place. In part, this arises because of the belief that 'finding a link between a product and a terroir should, in principle, be scientifically demonstrable'.[34] This scientific model of geographical indications of origin leads to the argument that as there 'is no exact knowledge of how geographical factors affect the product'[35] it is very difficult, if not impossible, to show scientifically that a particular product was shaped by the place where it originates.[36] This has led commentators to ask:

> Could a product with exactly the same characteristics be produced in a different locality? Different products reflect their environments to greater or lesser degree and four components normally included in appellation systems (variety or species, yield, production methods, processing methods) also have varying degrees of influence for each product. For most agricultural products it is very difficult to trace specific attributes (such as trace elements in the soil) of local environments through to the crop of animal that originates there and even more difficult if the product is processed.[37]

To require as a precondition for protection scientific proof that the defining characteristics of a product are attributable to its geographic origin calls into question the very idea of natural geographical indications of origin.[38] It also creates a barrier to the suggestion that geographical indications of origin could be used to protect traditional knowledge.

There are a number of possible responses to this problem.[39] One

---

[34] Barham (n 1) 136; Rovamo (n 6) 38 ('PDOs are based on the romantic core idea of *terroir*').

[35] Rovamo (n 6) 42 ('The minimum requirement is nevertheless that the particular climatic, geographical and morphological conditions which are claimed to affect the characteristics of the products are present throughout the geographical area concerned.').

[36] Ibid. ('The 'requirement of an express link between the product and its geographical origin would preclude protection for many geographical names because a lack of scientific evidence on how the different elements of the specification affect the product. Also, if modern production techniques enable almost any product to be replicated anywhere, the geographical origin cannot be necessary for the quality and characteristics of the product.')

[37] Warren Moran, 'Rural Space as Intellectual Property' (May 1993) 12(3) Political Geography 263, 267.

[38] Ibid., 266.

[39] Another relevant development, which is not discussed here, is the increased attention being given to historical links between product and place (particularly notable in relation to PGIs).

response is to denaturalise geographical indications of origin. Instead of seeing the connection between product and place as a *natural* connection, it is possible to highlight the role that humans play in shaping that natural connection.[40] This is reflected in the comment, that draws upon the opinion of Advocate General Colomer in the *Feta* litigation before the European Court of Justice, that 'the determination of the essential or exclusive link between the product and its *terroir* is not based on strict or exact science but on a global evaluation of all factors from climate to the flora and from the fauna to the people'.[41] It is also possible to highlight the role that GIs play in helping to establish and sustain the very things that they are supposed to be based upon: namely, the uniqueness of the natural connection between product and place.[42] This somewhat counterintuitive argument builds upon the fact that as the law of geographical indications provides that a protected name can only be used in relation to products from the designated territory, it helps to ensure that the named product is unique to that territory. Another possible response to the suggestion that natural geographical indications of origin are unsustainable because they are unable to be proven scientifically is to challenge the assumptions on which it is based: namely, that the connection between product and place needs to be judged according to scientific criteria. A useful starting point for doing this is to compare the process of naming geographical indicators with the scientific naming of plants.

The taxonomic process of naming a new plant variety is, as with the naming of a geographical indication, a complex and sophisticated process. Botanical taxonomy consists of three related activities: identification (referring a plant to a previously classified and named group); classification (ordering plants into groups based on perceived similarities and differences); and nomenclature (naming groups of plants according to

---

[40] In part this builds upon the idea that 'because nature does not speak for itself but rather through the "translation" that is the production process' . . . 'the legitimation process, to be effective, must be carried out not only within the territory of production but nested within multiple levels of coordination from the local to the global. In addition, the consumer's acceptance of the product must also be legitimated, which gives the word "quality" its full meaning.' See Barham (n 1) 130.

[41] Rovamo (n 6) 42. This was based on *Federal Republic of Germany and Kingdom of Denmark v Commission of the European Communities* (Joined Cases C-465/02 and C-466/02) [2005] ECR I-9115; [2006] ETMR 16 at [AG 194]–[AG196].

[42] On the constitutive role of GI regimes, which reinforce and authorise narratives that develop group identities around place-based distinction, see Matteo Ferrari, 'The Narratives of GIs' (2014) 10 International Journal of Law in Context 222 (based on a paper presented at the International Society for the History and Theory of Intellectual Property, London 2012).

rules developed for the process).[43] The classificatory scheme used by taxonomists consists of a hierarchical series of categories that operate, in effect, like a 'box-within-a box'.[44] Under this scheme, the basic hierarchy of the plant world is divided, in descending order, into Divisions, Classes, Orders, Families, Genera (genus), Species and Varieties.[45] The process of placing plants into the appropriate taxonomic category requires taxonomists to order plants into groups based on perceived similarities and differences.[46] Since the time of Linnaeus, taxonomists have focused on morphological or physical similarities and differences as the basis on which plants are distinguished and thus classified.

The key rule for naming a plant is the requirement that the name has to be validly published.[47] For this to occur, a name of a taxon must appear in Latin in a recognised scientific publication.[48] Each species must also be given a binomial name: the first word of each binomial is the name of the genus to which the species belongs (common noun) and the second word is a specific (trivial) epithet (adjective or possessive noun).[49] Thus,

---

[43] Judith Winston, *Describing Species: Practical Taxonomic Procedures for Biologists* (Columbia University Press, 1999) 9.

[44] J McNeill, 'Nomenclature of Cultivated Plants: A Historical Botanical Standpoint' (2004) 634 Acta Horticulturae 29, 31.

[45] While all of these categories are important, the species plays a special role in that it acts as 'the empirical or basic unit of classification' (C Jeffrey, *An Introduction to Plant Taxonomy* (J & A Churchill Ltd, 1968) 17).

[46] For many years taxonomic novelty was a subjective process: what was new for one collector or botanical garden was not the same for another. While private collectors and gardens still operate in this manner, one of the triumphs of the International Code was that it introduced the idea of absolute (world-wide) novelty.

[47] While a name may be validly published, it will only become the name by which a plant is known when it is accepted and used by botanists. There is no agency to enforce them. Winston (n 43) 9. If a plant has two names, the name which is valid is the earliest one to be published (after 1753). Alphonse de Candolle began to advocate for the application of the priority principle in the middle of the nineteenth century. This was adopted at the 1867 International Botanical Congress in Paris, and has remained a component of all subsequent botanical codes.

[48] *International Code for Botanical Plant Nomenclature* (2000), Art 32.1(a) ('Publication is effected, under this Code, only by distribution of printed matter (through sale, exchange, or gift) to the general public or at least to botanical institutions with libraries accessible to botanists generally. It is not effected by communication of new names at a public meeting, by the placing of names in collections or gardens open to the public, by the issue of microfilm made from manuscripts, typescripts or other unpublished material, by publication online, or by dissemination of distributable electronic media.').

[49] *International Code for Botanical Plant Nomenclature* (2000), Art 23.1 ('The name of a species is a binary combination consisting of the name of the genus

the scientific name for the Bramley apple is *Malus Domestica* 'Bramley's seedling'. The publication contains descriptive information including the botanical name of the plant, where and when the plant was discovered, and the name of the discoverer.[50] Another important principal that underpins the naming of plants is the use of the 'type method', which ensures that a botanical name is permanently attached to its nomenclatural type, which is the element on which the description validating the publication of a name is based.[51] The type acts as an artificial reference point that helps to determine how the name is to be used in the future.[52] The type is a nomenclatural device that fixes a botanical name to a particular taxon. It does this by requiring taxonomists to attach a name for a new species to a single individual representing that species, the so-called type specimen.[53] For a name to be valid, the publication must include the name of the type and the institution where the type is deposited.[54] These rules and procedures, which are accepted and applied by botanists around the world, ensure that botanical names are unique, stable and used universally. They also ensure that a botanical name only refers to one plant and that the name is unique to that 'particular plant the world over'. In so doing, taxonomy ensures that botanical names operate as generic names: they are *the name* by which a plant species is known, at least by those professionals who commonly work with plants.

There are a number of points of similarity and connection between

---

followed by a single specific epithet in the form of an adjective, a noun in the genitive, or a word in apposition, or several words, but not a phrase name of one or more descriptive nouns and associated adjectives in the ablative (see Art. 23.6(a)), nor certain other irregularly formed designations (see Art. 23.6(c)). If an epithet consists of two or more words, these are to be united or hyphenated. An epithet not so joined when originally published is not to be rejected but, when used, is to be united or hyphenated, as specified in Art. 60.9.').

[50] It will also include ecological information such as distribution, habitat preference, reproductive season, seasonal changes and so on. The description will also include the main taxonomic traits of the plant and a 'diagnosis' that highlights the 'distinctive morphological features' that distinguish the plant from taxonomically related plants. *International Code of Botanical Nomenclature* (2000), Art 32.2.

[51] Clive Stace, *Plant Taxonomy and Biosystematics* (2nd edn, Edward Arnold, 1989) 213.

[52] The type method has been described by taxonomists 'as a legal device to provide the correct name for a taxon'. Samuel Jones and Arlene Luchsinger, *Plant Systematics* (2nd edn, McGraw-Hill, 1986) 45.

[53] For the name of a species, the type is a physical specimen lodged in an herbarium or, in certain cases, a drawing of the plant.

[54] *International Code of Botanical Nomenclature (2000)*, Arts 37–38. David Gledhill, *The Names of Plants* (3rd edn, Cambridge University Press, 2002) 27.

geographical indications of origin and the names given to botanical varieties. In taxonomic terms, plant-based geographical indications not only tend to operate at the level of the variety, they also often take a form that is similar to their botanical equivalents.[55] In many cases, geographical indications will begin with the (scientific) species name of the plant in question. This is followed by an explanation of why, despite the similarities that the geographical indication shares with other 'plants' of that species, it differs from that broad class of products. For example, the PDO application for the Chinese Grapefruit (or pomelo) '*Guanxi Mi You*' begins with the generic species name – *Citrus grandis* – and its key characteristics.[56] The remainder of the application attempts to show how the named product exhibits traits that are a product of the place where it is grown that distinguish it from other plants of that species. In the case of *Guanxi Mi You*, the specificity is attributable to the unique climate and ecology of the area where the fruit is grown. As the application states, the 'hills, plains, paddy fields and eyots below 500 metres in altitude in 16 [named] towns in Pinghe County' in Fujian Province'[57] has certain geographical characteristics which ensures that the *Citrus grandis* have a 'unique sweet-and-sour flavour'.[58] In taxonomic terms, what this application is attempting to establish is, in effect, a specific new variety: '*Citrus grandis*, Guanxi Mi You', which is, in effect, a new type of legal-scientific name. One of the consequences of including botanical names in many of the applications that underpin GIs is that it means that these geographical indications of origin are inextricably tied to the taxonomic and nomenclatural processes that generate and sustain botanical names.[59] In so far as botanical names are used in GIs, it means that the legal rights are linked to the deposited type and the accompanying description that are linked to the scientific name. The specificity of the botanical name and the detailed rules and procedures that accompany it not only help to identify the named plant, they also help to stabilize that name and the associated description. They also help to define and demarcate the things to which the GI is attached

---

[55] A variety is a specific sub-set of a species that exhibits the defining characteristics of the species (which in turn exhibits the defining characteristics of the genus). At the same time, a variety also shows a new trait that is not shared by other members of the species such as early flowering, larger fruit, or better flavour.

[56] Published in [2010] OJ C 257/3, at [3.2] (24-9-2010).

[57] Ibid., at [4].

[58] Ibid., at [5.3].

[59] While the connection may not be as direct, taxonomic practices also play a role for GIs that are not based on plant materials. This is particularly the case for animal-based products that build on zoological taxonomy and nomenclature.

(and thus where and how the name can properly be used). The inclusion of a scientific (botanical) name in a GI not only creates a high degree of certainty about the types of things which the GI (legally) signifies (which is in contrast to trade mark law's looser association between the name and the things that they are connected to), it also means that plant-based GIs can be seen as types of legal-biological hybrids.

While GIs may share many points in common with botanical names, there are also a number of important differences. One of these is that while many geographical indications of origin may draw upon and embody aspects of botanical nomenclature and taxonomy, a geographical indicator is not itself a botanical or scientific name: it is a legal name (or at least a hybrid name that is both legal and scientific).[60] Another difference between botanical names and GIs is that the legal process is more administrative and less consensual than its scientific equivalent: decisions about the validity of a GI are not made by peers but by way of administrative and legal mechanisms. A further difference is that while botanical names are universally accepted and used,[61] GIs are limited to the jurisdictions in which they are recognised. Another important difference relates to the criteria that are used to differentiate and individualise the plant or product in question. While with a botanical variety, the cause (or 'author') of the distinguishing trait may be known (it may be the result, for example, of cross breeding), this does not play a role in establishing the new name. This is in contrast to GIs where the specific traits that are used to individualise the species must be traced to the nominated geographical area.

Besides the ones mentioned above, another important difference between geographical indications of origin and botanical names relates to the nature of the name and what it represents. As we indicated, once a botanical name is accepted by the scientific community it becomes unique to that 'particular plant the world over'. While geographical indicators are also unique, there are subtle and important differences in what this means. The reason for this is that naturalised GIs not only provide the product (variety) with the traits that allow it to be differentiated from the class of products to which it belongs (species), they also go one step further and make the claim that the individualising traits are unique to the

---

[60] Given that geographical indications typically embody botanical names and the taxonomic practices that sustain them, it is perhaps better to think of them as hybrid names that are at once legal and scientific: the relative influence of each depending on the product in question. Other non-legal, non-scientific factors add to this hybridity.
[61] The universal use of scientific names tends, however, to be limited to experts.

geographical area named in the application.[62] While the uniqueness of a botanical name is manufactured (via the complex system of publication, description and type specimen that underpins botanical nomenclature), the uniqueness of a natural geographical indication is meant to be a result of the place where the named product originates.[63] (It is possible to rethink less-natural focused GIs in similar terms.)

The role that the uniqueness of the name plays within botany also differs from the role it plays in law. The unique and standardised nature of the botanical name ensures that the name is able to operate as common currency that underpins and sustains scientific research. Uniqueness performs a number of different functions in the legal grant of a geographical indication of origin. One of the most important is that it enables the law to work on the assumption that there cannot be a product from a different place with the same qualities as the protected product. This in turn enables it to be assumed that the use of a protected name on a product that is not from the specified location is necessarily misleading. This is reflected in the idea that the 'fundamental concept behind GIs is that specific geographical locations yield product qualities that cannot be replicated elsewhere. Because the *place* is said to be essential to the product, proponents argue that products outside a specified region cannot be permitted to use its place name in marketing and on product labels.'[64] By limiting protection to products that have traits that are unique to a particular geographical area, this also means that third parties from outside the geographical area are unable to argue that the grant of a legal right means that something of theirs has been taken away.[65]

In part, the criticism made of *terroir*, nature-based geographical indications of origin and, by inference, GI protection for Indigenous knowledge, is that it is not possible to establish the uniqueness that is assumed to underpin the grant of the legal right. The criticism here is not that geography and environment do not impact upon the products that are derived from the place in question, so much as that it is not possible to

---

[62] While the botanical variety is tied to the deposited type (that is named in the publication in which the new plant variety is christened), there is nothing inherent in the location where the type is stored.
[63] On this see Moran (n 37) 276.
[64] Raustiala and Munzer (n 18) 338.
[65] Winemakers are often 'striving to produce a wine that is special in the sense that it bears the "signature" of their style of vinification'. See Barham (n 1) 131 (a 'GI-dominated product is not simply from a place; it is said to have unusual, even unique qualities that the place alone can provide'). Raustiala and Munzer (n 18) 344 ('food and beverages that evoke the term terroir have *signature* qualities').

show that the requisite qualities are unique to that geographical area.[66] In the case of champagne, for example, there is dissension over some of the assumptions made about the influence of geography on the resulting wines.[67] This is reflected in the comment that there are many environments where the 'exclusive' product could be replicated and that *méthode champenoise* 'could be considered to be much more important than the physical environment which produces Pinot Noir and Chardonnay grapes of high acidity'.[68] Such arguments lead to the conclusion that 'many geographical indications would [now] be precluded from protection because modern techniques enable us to replicate and produce almost any product anywhere'.[69]

If we acknowledge the role that humans (and history) play in shaping the connection between product and place, geographical indications of origin quickly change and with them the rationales, arguments and what is expected of them. For example, it has been suggested that the ambiguity as to whether geographical indications refer solely to fixed natural features or also to (moveable) human skills has 'major implications for the normative justifications of GIs, as well as the questions of who can legitimately use the GI and who cannot'.[70] It also means that 'GIs function as something else than a combination of the origin and quality functions. This might also explain why modern geographical indications law seems to be moving away from the idea of *terroir* and afford[ing] protection to GIs solely on the basis of a certain reputation among consumers.'[71] One of the consequences of recognising the role that humans play in shaping the connection between product and place is that it means that even the purest of natural geographical indications of origin are built upon artificial, non-natural, legal factors. This, in turn, means that the connection between place and product is not something that can or should be judged exclusively on scientific terms.[72]

---

[66] Rovamo (n 6) 8 ('The lack of scientific evidence about what the terroir inputs actually do' means that 'it is far from clear that a product's characteristics derived from geography' are exclusive. The suggestion is that any 'one natural aspect of a region can also be found somewhere else'). See also Justin Hughes, 'Champagne, Feta, and Bourbon: The Spirited Debate over Geographical Indications' (2006) 58 Hastings Law Journal 299.

[67] Moran (n 37) 267.

[68] Ibid.

[69] Rovamo (n 6) 9.

[70] Raustiala and Munzer (n 18) 343.

[71] Rovamo (n 6) 9. See Dagne (n 2) 441.

[72] This is reflected in the gradual extension of GIs away from wine to cheese to crafts, textiles, toys and other non-agricultural products.

To accept that connection to place is not something that should be exclusively judged on scientific terms opens the way for the protection of Indigenous traditional knowledge. The problem with this, however, is that it begs the question of whether the law of geographical indications of origin, even in a modified denaturalised form, would ever be able to accommodate Indigenous traditional knowledge on its own terms. In this sense it goes to the heart of the question we posed at the outset: namely, is it realistic to expect that geographical indications of origin can be used to protect Indigenous knowledge? The problem here is not that we are operating, in the words of the Australian High Court, at 'the intersection between two normative systems',[73] so much as that we are bringing together two very different ways of thinking about the world. The problem confronting Indigenous communities hoping to use geographical indications of origin to protect traditional knowledge is not so much in terms of the substantive legal rules (although the doctrinal rules may cause some problems). Rather, the problems of accommodation arise in relation to the various techniques that are used to overcome the fundamentally local and defeasible character of GIs, as well as the shared agreements that underpin and sustain those techniques. More specifically, the problems arise because many of the practices, concepts, rules and procedures that underpin and sustain the substantive law were formulated within a framework that is at odds with Indigenous interests. While aspects of the law of geographical indications of origins are pre-modern, particularly in terms of the form that the law takes, in other respects, however, the law is very modern. This is particularly the case in relation to the non-doctrinal aspects of the law: that is, the adjectival and performative aspects of the law of geographical indications.[74]

While the rationalist tradition of evidence that underpins the law of geographical indications operates on the basis that 'evidence is something that testifies to the external real world of facts', in contrast 'evidence about traditional knowledge is itself evidence of a different way of knowing'.[75] This difference manifests itself in a number of ways.[76] While the evidential

---

[73] *Members of the Yorta Yorta Aboriginal Community v Victoria* (2002) 194 *ALR* 538, 550 (Gleeson CJ, Gummow and Hayne JJ).

[74] In most cases, these issues are not usually considered. One notable exception is in relation to the scientific model of geographical indications of origin, which effectively asks us to consider: how do you prove the connection between product and place?

[75] K Anker, 'The Truth in Painting: Cultural Artefacts as Proof of Native Title' (2005) 9(1) Law, Text, Culture 91, 92.

[76] The approach taken by the law of geographical indications of origin is at odds with the approach taken by Indigenous communities on a range of other

502   *Research handbook on intellectual property and GIs*

system that underpins intellectual property law assumes that knowledge and information are, for the most part, public goods accessible to all, in contrast 'the accessibility of knowledge in Aboriginal cultures [is often] highly selective and dependent on age, sex, and spiritual affiliation. Details of "connection to place"', which are needed to establish geographical indications of origin, 'are the very ones most likely to be highly secret'.[77] Another important difference relates to the transportability of evidence. While intellectual property law presumes that witnesses are able to give evidence away from the place where the evidence was generated and also that the place where the testimony is given is irrelevant, this may not be the case with Indigenous witnesses whose 'ability and authority to speak about certain places literally depends on them being physically present'.[78] The reason why the context where a testimony is delivered may be important for an Indigenous witness is because there is a 'metonymic association between following the law, walking the country and doing ceremony such as singing, dancing or painting the country. Evidence . . . is often given in "on country" hearings in recognition of the inability of some claimants to speak about country without it being beneath their feet.'[79] Another point of difference between the evidential scheme used by the law of GIs and the type of evidence that might be preferred by Indigenous communities relates to how connection to place is proved. Although the evidence used to prove connection to place (or the boundaries of a would-be geographical area) takes different forms depending on the GI in question, it is united by the fact that it confirms to certain unstated and shared presumptions about the type of evidence that will and will not be taken into account when deciding doctrinal issues such as proving connection to place. While scientific evidence that the distinguishing characteristics of a named product is directly (and uniquely) shaped by the geographical location where it is grown or produced, or historical evidence about a product's longstanding association with a particular location is clearly relevant, this is in marked contrast to the evidence that might be given by an Indigenous community. For example, paintings, or at least certain types of paintings, are used by Indigenous communities in Australia as a way of communicating connection to country to non-Aboriginal people and to the courts.[80]

---

matters including how geographical boundaries are drawn, and how products are categorised and named.

[77] Anker (n 75) 97.
[78] Ibid., 113.
[79] Ibid., 98.
[80] As Ngarralj Tommy May said of *Ngurrara Canvas II*, a painting he produced with around 50 artists as part of a native title claim:

This occurs because paintings make a 'normative claim about the basis for entitlement and the manner in which it can be proved'.[81] As Anker notes:

> [T]he 'tradition' of design, boundaries and Dreaming stories is the frame of reference which give a painting value as evidence. Designs originating in sand, rock, bark, and body painting embody relationships between ancestors and law, living people and places in the land, which makes them crucially relevant to what is being translated ... as property rights. In evidence, the painting illustrates the rights (such as those indicated by boundaries), the origin of those rights in a system of law (such as Dreaming stories) and facilitates the oral evidence of the witnesses.[82]

As matters stand, it is highly unlikely that custodians of traditional knowledge would be able to use the existing rules of evidence to prove connection to place in a way that would accord with Indigenous ideas about evidence and proof. The problem for Indigenous communities is that connection to place is very difficult if not impossible to prove without recourse to materials that would be inadmissible under the traditional rules of evidence.[83] One response to this situation is to argue that if Indigenous communities want to make use of geographical indications of origin, they need to comply with the existing rules and procedures. The problem with this is that it undermines one of the reasons why Indigenous communities might be attracted to geographical indications of origin in the first place, as well as the very thing that the law is supposed to be supporting: namely, the distinct and local aspects of traditional knowledge. In this situation, there is a choice as to whether the law is changed to allow traditional knowledge to be protected in a way that is sympathetic to Indigenous interests, or whether Indigenous peoples are subject 'to a deeper process of colonisation by reducing them to a singular unambiguous discourse known to the law'.[84] Rather than asking whether we should be 'contemplating changes and adaptations to the intellectual property regime to render it more

---

The painting is only for proof. Maybe the [white people] will say 'we don't believe you' ... That's why we made this painting, for evidence. We have painted our story for native title people, as proof. We want them to understand, so that they know about our painting, our country, our *ngurrara*. They are all the same thing. (As cited in Anker (n 75) 92)

[81] Anker (n 75) 93.
[82] Ibid., 100.
[83] Alex Reilly, 'The Ghost of Truganini: Use of Historical Evidence as Proof of Native Title' (2000) 28 Federal Law Review 453, 454.
[84] Alex Reilly, 'Cartography and Native Title' (2003) 79 Journal of Australian Studies 1.

neutral (from an ethnocentric perspective'),[85] it would perhaps be better to contemplate changes that would ensure that the law of geographical indications was more culturally specific and biased. We can get a sense of the type of changes that would be needed to do this if we look to native title law in Australia, which has a number of parallels with GIs: both apply to a specified geographical area and also require Indigenous communities to show a connection to place to establish the relevant legal rights.[86]

Native title was formally recognised in Australia in the 1992 High Court decision of *Mabo v Queensland (No 2)*.[87] While this decision paved the way for the legal and political recognition of Indigenous interests in land in Australia, it quickly became clear that additional changes were needed if Indigenous interests were to be adequately protected. The reason for this was that there were 'unique questions of proof stemming from the sui generis nature of native title' which demanded a 'unique approach to the treatment of evidence which accords due weight to the perspective of aboriginal people'.[88] A number of changes were made to allow Indigenous modes of evidence and proof to become part of the law of native title. One of the most important was in terms of how connection to place was proved. To this end, changes were made to allow songs, dances, stories, and paintings to be presented as evidence of customary law. Changes were also made to allow these non-traditional forms of evidence to be used to prove a community's connection to a particular geographical area.[89] There have also been changes in terms of the place where evidence is presented. Rather than asking or expecting Indigenous people to travel to urban locations to give evidence, courts hearing native title claims sometimes travel to remote locations to hear evidence in country. This has been

---

[85] Daniel Gervais, 'Traditional Knowledge: Are We Closer to the Answer(s)? The Potential Role of Geographical Indications' (2009) 15(2) ILSA Journal of International and Comparative Law 551, 552.

[86] Native title 'continues where there is a connection with the land that has been substantially maintained by a community, which acknowledges and observes, as far as practicable, laws and customs based on the traditional practices of its predecessors' (Reilly (n 83) 454).

[87] (1992) 175 CLR 1; [1992] HCA 23.

[88] Reilly (n 83) 467, citing *Delgamuukw v British Columbia* (1997) 153 DLR (4th) 193, 230. Procedures for proving Native Title have been 'modified to take account of cultural concerns of Indigenous claimants' (Anker (n 75) 97).

[89] Section 82(3) of the Native Title Act, 1993, as enacted, provided that 'The Court, in conducting proceedings, is not bound by technicalities, legal forms or rule of evidence'. This provision was amended in 1998 to state that the 'Federal Court is bound by the rules of evidence, except to the extent that the Court otherwise provides' (s 223(1) of the Native Title Act, 1993).

explained on the basis that 'for the Court to require indigenous people to travel perhaps thousands of kilometres to a place that they might not have been, to give evidence in a confined courtroom, to speak of Country in the abstract, to refer to maps without a context would ... offend the very notions of "access" and "justice"'.[90]

The experience in relation to native title law in Australia shows that it is possible for a legal system to adapt to and accommodate traditional knowledge on, or at least partially on, its own terms.[91] The key question for our purposes, however, is whether it is realistic to expect that similar changes might also be made to the law of geographical indications? While it is possible to imagine some changes being made that would make it easier for traditional knowledge to be protected, it is highly unlikely that the law of geographical indications would ever accommodate Indigenous traditional knowledge in its various and distinct forms, at least in a manner that takes account of indigenous-style evidence. It is unlikely, for example, that the European Court of Justice would decamp to the deserts of Central Australia to hear evidence in relation to a dispute about the alleged misuse of an Australian Indigenous geographical indication of origin that occurred in Europe. In the same way in which proponents of global Indigenous collecting societies would be unlikely to accept that the societies would be established anywhere but in 'global' centres such as Geneva, London or Washington (but certainly not in Weipa), so too it is unlikely that disputes in relation to an Indigenous GI would be heard in country, as happens with native title. It is equally unlikely that non-Australian courts would allow a painting to be introduced into evidence as proof that a named 'product' was intrinsically linked to a specified geographical area. The alien and somewhat unsettling nature of this form of evidence is exacerbated by the fact that Indigenous paintings introduced as evidence of connection to place do not merely map out or represent connection to place: they are not 'something which points to knowledge about

---

[90] Louise Anderson, 'The Law and the Desert: Alternative Methods of Delivering Justice' (2003) 30(1) Journal of Law and Society 120, 130.

[91] In many ways, the situation in relation to traditional knowledge and GIs is similar to the position of native title prior to the High Court decision of *Mabo* where the 'understanding of property rights as a question of measurement and administration contributed to the failure of the law to recognise Indigenous relationships to land in the first place'. The situation began to change with the High Court decision of *Mabo*, which required a 'reflection of epistemic and ontic commitments to place'. See Alex Reilly, 'Cartography, Property and the Aesthetics of Place: Mapping Native Title in Australia' in A Kenyon and P Rush (eds), *An Aesthetics of Law and Culture: Text, Images, Screens* (Emerald Publishing, 2004) 221, 229.

country and traditional law'. The reason for this is that 'the painting is the country', a 'painting is not just a fact about law, it is law'.[92]

Although Australian courts might be able to adapt to the evidential needs of Aboriginal and Torres Straight Islanders and Canadian courts might be able to accommodate the specific needs of Inuit peoples, it is not feasible to expect national courts or agencies who register geographical indications of origin to accommodate the evidential systems of the different Indigenous communities around the world who might want to make use of GI protection. The lack of ready access to knowledge and expertise about the distinct needs of Indigenous peoples in other countries would not only make this impractical, it would also mean that there is less likely to be the political will needed to suspend the traditional rules of evidence. As the number of Indigenous communities wanting to make use of GI protection increases, so too do the different models of evidence and proof and with it the chances of conflict and uncertainty. The conflicting and incommensurable epistemologies also mean that it would be very difficult to modify existing international treaties to accommodate traditional knowledge, except in very abstract and unhelpful ways. Perhaps the most important reason why the adjectival dimensions of GI laws are unlikely to be changed to accommodate traditional knowledge in a way that was sympathetic to Indigenous interests is because it would undermine the GI system. The reason for this is that geographical indications of origin are concerned with local knowledge, practices and products that are, almost by definition, diverse, complex and fragmented. At the same time, for local products to become part of a global GI system, they need to be observed, noted, recorded and placed in a format that enables them to be moved from one site to another. Over time the law has developed a number of devices, techniques, rules and procedures that allows this to occur. One of the problems with the idea of adapting the rules of evidence to accommodate traditional knowledge is that it would mean that local products placed into the global GI system would be accompanied by local systems of evidence and proof: a process which would undermine the mobility and thus the very possibility of geographical indications of origin.

The upshot of this is that traditional knowledge is only likely to be recognised and protected by geographical indications of origin to the extent that it complies with non-Indigenous modes of evidence and proof.[93] While the GI system does offer the possibility of protecting aspects of traditional knowledge, this will, it seems, inevitably be on non-Indigenous terms.

---

[92] Anker (n 75) 92.
[93] For a discussion of these issues in another context see Reilly (n 84).

This does not mean, however, that an unmodified GI system does not offer potential benefits to some Indigenous communities. It does mean, however, that the protection would not be as sympathetic to Indigenous interest as is often assumed. Given the costs involved in putting together a GI application and the relatively low level of benefits that are likely to flow from protection, it is necessary to ask whether it is worth it. Ultimately, this is a question for traditional owners to decide. Whatever decision is made, however, it seems that geographical indications do not, despite the similarities and points of connection, offer the panacea that many hope for.

18. Genericide: the death of a Geographical Indication?
Dev S. Gangjee

1. INTRODUCTION

Viewed in historical perspective, the determination of generic status has been the most controversial aspect of international Geographical Indications (GIs) protection for well over a century.[1] As signs which are used by more than one producer to begin with, GIs contain the potential seeds of their own destruction. The greater the success that a regional product achieves on the marketplace, the greater the risk that its designation will be treated as the general term for that type of product. Cheddar cheese and dijon mustard exemplify this. Generic terms can no longer communicate a specific geographical origin; they merely indicate the familial features for a product category. Generic status is therefore the antithesis of protected GI status. When this geographical link is severed, semantic vitality – the G in a GI – ebbs away. It is therefore surprising that this topic has remained relatively neglected[2] and the final chapter of this volume addresses this gap. Adopting a comparative perspective, this chapter identifies a menu of options for implementing the abstract test contained in Art 24.6 of the TRIPS Agreement, which merely states that there is no obligation to protect a GI where 'the relevant indication is identical with the term customary in common

---

[1] The history of this controversy is introduced in Section 2.2 below. See also GE Evans and M Blakeney, 'The International Protection of Geographical Indications Yesterday Today and Tomorrow' in G Westkamp (ed.), *Emerging Issues in Intellectual Property: Trade, Technology and Market Freedom – Essays in Honour of Herchel Smith* (Edward Elgar Publishing, 2007) 250, 283 (Today 'the issue of generic names has the potential to cause the greatest unease in negotiations for increased international protection'); TRIPS Council, 'Communication from Bangladesh et al.' 2 October 2001 (IP/C/W/308/Rev.1) at [18] ('One of the key reasons for advocating extension [of GI protection] is a desire to prevent more geographical indications from becoming generic.'). While the conflict between trade marks and GIs is also controversial, generic status has preceded it for some time.

[2] Notable exceptions are the research by Audier, le Goffic and Schoene, cited in this chapter.

language as the common name for such goods or services in the territory of that Member'.

It is an opportune moment to explore the possibilities within Art 24.6, because it is the object of renewed interest. After outlining genericide as a process, Section 2 reviews both historic episodes and contemporary disagreements illustrating its divisiveness during bilateral or multilateral negotiations. Case studies reveal the significant commercial stakes involved when deciding on the fate of a designation. While Art 24.6 has emerged as the international reference point, it is far too abstract to operate effectively as a test. Section 3 sets out the structural parameters that will serve as scaffolding for any such enquiry. The analysis reveals that despite the apparently factual nature of the enquiry into the meaning of a term, there are prescriptive choices to be made. The significance of a multifactor test is that it creates the spaces for normatively loaded presumptions within a supposedly empirical assessment. Specifying upon whom the burden of proof lies, clarifying whether a hierarchy exists amongst the factors and setting a threshold or tipping point for genericide are all extremely significant – yet entirely discretionary – choices available under TRIPS, which can greatly influence the ease or difficulty of establishing generic status. Section 4 turns to a consideration of the individual categories of evidence which assist decision makers (courts or registrars) when deciding whether a designation has become the 'common name' for the goods. These include: (i) direct as well as circumstantial evidence of consumer understanding; (ii) trade usage; (iii) expert opinion; (iv) market conditions; (v) classification in legislation or administrative rules; and (vi) the actions of rights holders themselves. Broadening out the enquiry, Section 5 reviews the extent to which genericide can be prevented while also asking if there is life after death. It considers the options for reviving terms which were formerly GIs but have entered the valley of genericide. Section 6 concludes.

2. CONTROVERSY AND COMMERCIAL STAKES

2.1 A Fatal Process

According to Art 22.1 of TRIPS, GIs are '*indications which identify a good as originating* in the territory of a Member, or a region or locality in that territory, where a given quality, reputation or other characteristic of the good is essentially attributable to its geographical origin' (emphasis added). The process of genericide is fatal because the sign no longer functions to indicate geographical origin. Since the sign no longer qualifies as a GI, it does not merit protection. As the Court of Justice of the European Union (CJEU) describes it:

> a geographical designation could, over time and through use, become a generic name in the sense that consumers cease to regard it as an indication of the geographical origin of the product, and come to regard it only as an indication of a certain type of product. That shift in meaning occurred for instance in the case of the designations 'Camembert' and 'Brie'.[3]

GIs are especially vulnerable because they are used by a group of regional producers to begin with.[4] An overview of the typical stages in this process is found in Advocate General Colomer's opinion in *Feta*.[5] Producers from outside the region of origin begin to use a geographical term to describe a product, because it has a favourable reputation and (often) migrants from the region of origin have adopted both production techniques and the associated terminology when they emigrate. As this geographical association becomes attenuated, others start using the name in good faith, to describe a product with certain characteristics and – especially where there is passivity on the part of the original group of producers – a conversion of meaning takes place.

The outcome is that generic terms form part of the linguistic public domain, whereupon any attempt to claim exclusive rights over such terms is seen as an unjustified form of protectionism.[6] In Advocate General Colomer's words, they 'form part of the general cultural and gastronomic stock and may, in principle, be used by any producer'.[7] Attempts to limit the use of such terms to specific producer groups would be viewed as a restriction on commercial speech and would impede

---

[3] *Commission of the European Communities v Federal Republic of Germany* (C-132/05) [2008] ECR I-957; [2008] ETMR 32 at [36].

[4] Latha R Nair, 'Swiss Watch International, Inc. v. Federation of the Swiss Watch Industry: The TTAB's Unwitting Message for Geographical Indications?' (2012) 102 TMR 944, 954.

[5] *Federal Republic of Germany and Kingdom of Denmark v Commission of the European Communities* (Joined Cases C-465/02 and C-466/02) [2005] ECR I-9115; [2006] ETMR 16 (AG) at [134]–[135] (hereafter *Feta 2005*).

[6] Caroline Le Goffic, *La Protection des Indications Géographiques* (LexisNexis Litec 2010) 408–409; D Rangnekar and S Kumar, 'Another Look at Basmati: Genericity and the Problems of a Transborder Geographical Indication' (2010) 13 JWIP 202, 211.

[7] *Canadane Cheese Trading AMBA and Adelfi G. Kouri Anonymos Emoriki Kai Viomichaniki Etaireia v Hellenic Republic* (C-317/95) [1997] ECR I-4681 at [AG28].

the signalling of competitive substitutability in the marketplace.[8] The theoretical basis for considering such terms to be available to producers from outside of the eponymous region is therefore clear. However, since genericide involves a dynamic process, what is less clear is the point at which one may definitively conclude that this semantic shift has occurred. Additionally, one country's GI may be another's generic term, which makes this a controversial determination amidst channels of international trade. South Africa's grievance, aired at the TRIPS Council, exemplifies this: it 'was a producer and exporter of "feta", a generic term in its territory. Currently the [EU], which had been its main export market until recently, were now blocking such exports and sending South African producers threatening letters.'[9] Arguments concerning repackaging costs are often encountered in such situations, where traders are deprived of the use of terminology they believe to be in the public domain.[10] A related contention is that generic use serves as 'free advertising' for the original regional speciality and should be permitted.[11] Conversely, if generic status 'stems from a government or administrative decision which fails to reflect linguistic usage, generic names will become mere bargaining chips' in international negotiations.[12] In the past, national governments have engaged in strategic brinkmanship, by designating certain terms to be generic in official lists. The fluidity of meaning across space and time, coupled with commercial significance, makes this an especially contentious topic.

---

[8] S Damer, 'Not Confused? Don't Be Troubled: Meeting the First Amendment Attack on Protection of "Generic" Foreign Geographical Indications' (2009) 30 Cardozo L Rev 2257.

[9] TRIPS Council, 'Special Session, Minutes of the Meeting on 16–17 March 2006', 19 May 2006 (TN/IP/M/16) at [100].

[10] See the statement of Michael Pellegrino, Vice-President, Kraft Cheese Division in *Hearings before the Committee on Agriculture, House of Representatives on the Status of the World Trade Organization Negotiations on Agriculture* (108–5) 108th Congress (2003) 325–326 (discussing Kraft having to avoid using Parmesan on non-Italian cheese in the EU).

[11] FG Zacher, 'Pass the Parmesan: Geographic Indications in the United States and the European Union – Can There Be Compromise?' (2005) 19 Emory International Law Review 427, 434 (for the argument that Kraft promoted parmesan in the US at great expense).

[12] J Audier, *TRIPS Agreement – Geographical Indications* (EC Office for Official Publications, 2000) 39 at FN40.

2.2 Controversy and Commercial Stakes

Generic use has been controversial for well over a century in international intellectual property negotiations. In the deliberations leading up to the Paris Convention,[13] it was raised as a concern when delegates were considering a rule to prohibit the use of false indications of source. On 10 November 1880, the Swedish delegate (with considerable foresight) acknowledged that delimiting the scope of protection for geographical signs was a 'delicate' matter,[14] since it needed to be balanced against the interest in preserving generic use. Referring to a dispute between Sweden and Britain concerning the use of 'Lancashire', Sweden considered it to be a generic description for metal manufactured according to a specific process, while Britain disputed this. The proposed solution was to include 'Sweden' on the metal after 'Lancashire', to clarify that it was used in the generic and not the geographical sense. The Norwegian delegate considered Champagne to be a generic term for a process of manufacture, similar to eau de Cologne,[15] while the President for the session clarified that the prohibition in draft Art 6 (ultimately Art 10) was specifically directed at false indications. It did not prevent the use of generic expressions such as Russian leather, velvet from Utrecht or eau de Cologne, which were general descriptions in the public domain.[16] Such generic usage would not qualify as a false indication of source and remained permissible. This early episode succinctly captures two enduring aspects of the genericide debate:

(i) While generic status is supposed to reflect consumer understanding of a term (the 'common name'), it is usually opposing groups of producers and their representatives who make the case on either side; and
(ii) Since rights to GIs are territorial,[17] disagreement is particularly acute in the context of international trade flows. For instance, a designation may be protected in the country of origin ('home country'), be considered generic in another country where broadly comparable products

---

[13] The Paris Convention for the Protection of Industrial Property, 20 March 1883 as revised at Stockholm on 14 July 1967, 828 UNTS 305 (1972).

[14] *Actes de la Conférence Internationale pour la Protection de la Propriété Industrielle* (Ministère des Affaires Etrangères, Impr. Nationale, Paris 1880) 85 ('Il considère la question relative à l'indication de provenance comme très délicate').

[15] Ibid., 86.

[16] Ibid., 88.

[17] WIPO Secretariat, 'Geographical Indications and the Territoriality Principle' 1 October 2002 (SCT/9/5) (GIs are established and protected on the basis of the laws applicable in a given territory, which is a limitation when the product may be sold across several territories).

are produced ('competitor country') and producers from both may compete for market share in a third country ('target market').

These two aspects reappeared in the preparatory discussions laying the groundwork for the other relevant WIPO treaties – the Madrid Agreement of 1891 and the Lisbon Agreement of 1958[18] – where generic status remained divisive.[19] The basis for such controversies is the commercial significance of such a determination. In previous decades this led to an often-acrimonious debate about the 'sins of the past'; the argument was that those using a term generically were continuing to benefit from past misappropriations and misleading uses (Californian Champagne was simply not Champagne).[20] The counter-argument emphasised that waves of (historically European) emigration to New World countries had led to emigrant producers very reasonably using terminology to describe the products and processes they were seeking to emulate in their new homes. No one was misled by such uses.[21] Entire sectors of national economies, such as those relating to viticulture or dairy production, therefore rely on the protected (or generic) status of a given term for marketing and market access purposes. For that reason the international political economy dimensions of such legal determinations are never far from the surface and the following three case studies reveal the range of commercial and political pressures involved in such disputes.

The celebrated decisions in *Spanish Champagne*[22] laid the foundations for the principal common law approach to protecting GIs in the early

---

[18] See respectively, Madrid Agreement for the Repression of False or Deceptive Indications of Source on Goods, 14 April 1891, 828 UNTS 389 (1972); Lisbon Agreement for the Protection of Appellations of Origin and their International Registration, 31 October 1958, 923 UNTS 205 (1974). See also Chapter 5 by Geuze in this volume.

[19] D Gangjee, *Relocating the Law of Geographical Indications* (CUP, 2012) 68–73 (Madrid), 152–157 (Lisbon).

[20] Ibid., 248–250.

[21] LA Lindquist, 'Champagne or Champagne? An Examination of US Failure to Comply with the Geographical Provision of the TRIPS Agreement' (1999) 27 Georgia Journal of International and Comparative Law 309, 313 ('In the 1800s, the United States experienced a huge influx of immigrants from Europe . . . Many of these immigrants brought their wine-making skills and vine cuttings with them'); J Hughes, 'Champagne, Feta, and Bourbon – The Spirited Debate about Geographical Indications' (2006) 58 Hastings Law Journal 299, 353 ('Broad-based progress requires both innovation and imitation. Imitation of technology (in the broadest sense) is often accompanied by imitation of the relevant terminology.').

[22] This encapsulates two related decisions: *J Bollinger v Costa Brava Wine Co Ltd* [1960] Ch 262 (For the preliminary points of law to decide whether an action

1960s. The UK high court concluded that using 'Champagne' – or even 'Spanish Champagne' – on bottles of sparkling wine produced in Spain was likely to mislead British consumers and was therefore impermissible under the tort of passing off. However, the civil trial was preceded by an unsuccessful criminal prosecution under the Merchandise Marks Acts 1883 to 1957. Here the defence had succeeded in convincing a jury that Champagne indicated a class or type of sparkling wine and therefore Spanish Champagne was not a false trade description. A study of declassified archival records reveals that in the gap between the failed prosecution and the favourable passing off decision, considerable pressure was brought to bear on the British government by those who wished to protect Champagne as well as those who would benefit from its free availability as a generic term.[23] As for the latter, producers of sparkling wine in Spain were heartened by the initial outcome but so was the Australian wine industry, while there are references to the potential for increased imports from South Africa. Meanwhile, French Champagne producers were understandably dismayed by this acquittal, but so were those of Spanish Sherry, who apprehended a similar fate. A little known consequence – but one which tellingly illustrates the ripple effects of generic status – is that Scotch whisky producers were implicated in this dispute. They were targeted as a potential British pressure point when France threatened retaliatory action, by suggesting that Scotch whisky might be similarly designated generic in that jurisdiction. For a brief interlude, Scotch was on the rocks.

The second illustration is more recent and showcases the divisive nature of generic status, even within a grouping of nations perceived to be like-minded in their support for GIs. Genericide continues to be controversial within the European Union (EU). A largely forgotten trace of this is the unsuccessful attempt by the European Commission to draw up a list of generic terms under Art 3(3) of Regulation 2081/92.[24] In 1996 the Commission presented a proposal for a Council decision, which was never adopted since the requisite majority was

---

could be maintained in principle); *J Bollinger v Costa Brava Wine Co Ltd* [1961] 1 WLR 277; [1961] RPC 116 (At trial).

[23] The records include declassified British Board of Trade files as well as contemporaneous newspaper reports. See D Gangjee, '*Spanish Champagne:* An Unfair Competition Approach to GI Protection' in Rochelle Cooper Dreyfuss and Jane C Ginsburg (eds), *Intellectual Property at the Edge* (CUP, 2014) 105.

[24] Council Regulation 2081/92 of July 14, 1992 on the Protection of Geographical Indications and Designations of Origin for Agricultural Products and Foodstuffs [1992] OJ L208/1.

not attained.[25] The Proposal's introductory words of caution bear repeating:

> [The] issue of generic names is a sensitive one and has always met with strong reactions. When a geographical name is registered, its use is restricted to enterprises in the area concerned and forbidden to all others. It follows from these arrangements that declaring a name to be generic has very important consequences, particularly of an economic nature, which affect the interests of private individuals. Great caution should accordingly be exercised in this area and [the provision determining generic status] should be applied in an unbiased and objective fashion.[26]

During this process, the lists submitted by individual EU Members substantially differed, making it difficult to inductively establish the criteria for inclusion within such a pan-European list. The Commission eventually included only six cheese names within the proposal: Brie, Camembert, Cheddar, Edam, Emmentaler and Gouda. It adopted the following criteria when creating the list: (i) the name had been put forward by at least eight Member States; (ii) whether the cheese was mentioned in the Stresa Agreement[27] as a protected or generic indication; and (iii) the name was not protected by any other international (including bilateral) agreement outside of the home country. Over the course of these deliberations, Feta cheese had proved very controversial, narrowly avoiding being considered a generic term.[28] The controversy continued unabated for over a decade, resulting in three Advocates' General Opinions and two decisions (the final one by the Grand Chamber) of the Court of Justice, before it was finally concluded that Feta was a valid Greek Protected Designation of Origin.[29] Similarly, the status of Parmesan, as a translation or variant of Parmigiano Reggiano cheese, has continued to be contentious within

---

[25] Proposal for a Council Decision to Establish an Indicative, Non-Exhaustive List of Names of Agricultural Products and Foodstuffs Considered Generic Names, as Referred to in Art. 3 Para 3 of Council Regulation (EEC) No. 2081/92, COM (96) 38 Final of 06/03/1996.
[26] Ibid., at [3].
[27] The Stresa International Convention for the Use of Appellations of Origin and Denominations of Cheeses, of 1 June, 1951, reproduced in WIPO Secretariat, 'Texts of International Instruments Concerning the Protection of Appellations of Origin and Other Indications of Source' 28 June 1974 (TAO/I/3). For an overview, see G Trotta, 'The Stresa Convention on the Uses of Names of Cheeses and the WIPO Draft Treaty on the Protection of Geographical Indications' [1977] Industrial Property 113.
[28] Commission Proposal (n 25), at [5].
[29] For a detailed consideration, see D Gangjee, 'Say Cheese! A Sharper Image of Generic Use through the Lens of Feta' [2007] EIPR 172.

the EU, with Germany repeatedly stressing that it is generic, but without any success.[30] Germany was subsequently on the receiving end of such claims, facing an assertion that Bayerisches Bier (Bavarian beer) was generic for a style of beer produced via a 'bottom fermentation' method.[31] Disputes along these lines continue to occur within the EU, although usually stopping short of a reference to the CJEU.[32] A review of the operation of the EU GI regime for agricultural products and foodstuffs revisited the question of whether a list of generic terms would provide commercial operators with clarity. However, it concluded that any such list by the Commission would always be subject to judicial review and therefore contingent. As the report summarises it, the 'existence of a list of generics does not necessarily mean fewer disputes about the generic character of a given name'.[33]

The third illustration relates to an ongoing dispute, primarily between US dairy industry interests and the EU. In recent years, the EU has been negotiating a series of bilateral agreements which contain GI provisions, either within the context of a broader Free or Preferential Trade Agreement (FTA or PTA), or as part of a GI-specific agreement.[34] These ensure protected GI status in the partner country, usually by resorting to lists of protected EU GIs contained in annexes. A term whose status was previously unclear, or perhaps even generic, is thereby stabilised and legally recognised. As a recent EU policy document emphasises:

> In the new generation of FTAs a satisfactory GI Chapter is a 'must have' for the EU ... The objective in EU negotiations is to add value compared to TRIPS basic provisions. What we call 'TRIPS+' means notably ... To establish a list of EU names to be protected directly and indefinitely in the third

---

[30] *Dante Bigi (Consorzio del Formaggio Parmigiano Reggiano, Third Party)* (C-66/00) [2002] ECR I-5917; [2003] ETMR 55 (CJEU) at [15]–[17]; *Commission v Germany* (n 3).

[31] *Bavaria NV, Bavaria Italia Srl v Bayerischer Brauerbund eV* (C-343/07) [2009] ECR I-5491; [2009] ETMR 61.

[32] One such example is the dispute between Italy and France over the goats' cheese Chevrotin. See Commission Regulation (EC) No. 1357/2005 of 18 August 2005 Supplementing the Annex to Regulation (EC) No 2400/96 as Regards the Entry of a Name in the Register of Protected Designations of Origin and Protected Geographical Indications – Chevrotin (PDO) [2005] OJ L 214/7.

[33] London Economics et al., *Evaluation of the CAP Policy on Protected Designations of Origin (PDO) and Protected Geographical Indications (PGI)* (Final Report for the European Commission, November 2008) 222.

[34] The agreements are listed at: http://ec.europa.eu/trade/policy/accessing-markets/intellectual-property/geographical-indications/

country, from the entry into force of the agreement [. . .and to] phase out prior uses of EU names.[35]

Examples abound in the wine agreements the EU has concluded with the US, Australia, Canada and South Africa.[36] This strategy involving lists of protected terms has been extended beyond wines and spirits, causing anxiety to US producers of agricultural products and foodstuffs. US dairy producers in particular are concerned that this could affect their ability to export products using labels with generic terms to the countries which have entered into such agreements with the EU. Therefore in 2010, in response to requests from their constituents, 56 members of the US Congressional Dairy Farmers Caucus sent a letter to the US Trade Representative, expressing their apprehensions regarding the EU-South Korea FTA.[37] They were

> very concerned that the implementing regulations of the EU-South Korea FTA will contain GI provisions that will greatly diminish, if not foreclose, the market opportunities available to many US cheeses and other agricultural products. Moreover, it must be noted that any such advantage gained by the EU will be magnified because it would set a precedent that could and likely would be, readily replicated in EU-negotiated FTAs in a number of other foreign markets of importance to the US dairy industry.[38]

Another emerging voice is the Consortium for Common Food Names (CCFN), 'an independent, international non-profit alliance whose goal is to work with leaders in agriculture, trade, and intellectual property rights' *inter alia* to object to 'any attempt to monopolize common (generic) names that have become part of the public domain'.[39] The specific threat is therefore restrictions applying to target markets, such as South Korea,

---

[35] DG AGRI, *Working Document on International Protection of EU Geographical Indications: Objectives, Outcomes and Challenges* (Advisory Group – International Aspect of Agriculture, Ares (2012) 669394, 06/06/2012) 8–9.

[36] These agreements are considered in Michael Handler and Bryan Mercurio, 'Intellectual Property' in Simon Lester and Bryan Mercurio (eds), *Bilateral and Regional Trade Agreements: Commentary and Analysis* (CUP, 2009) 317–322. The EU-Australia Wine Agreement is extensively considered in Chapter 10 by Stern in this volume.

[37] Free Trade Agreement between the European Union and the Republic of Korea [2011] OJ L127/6.

[38] Congressional Dairy Farmers Caucus, Letter addressed to Ambassador Ron Kirk, US Trade Representative, 27 September 2010; cited in C Viju, M Yeung and WA Kerr, 'Geographical Indications, Conflicted Preferential Agreements, and Market Access' (2013) 16(2) Journal of International Economic Law 409, 413.

[39] See: http://www.commonfoodnames.com

Canada or Singapore, which have all recently concluded trade agreements with the EU. In 2014, CCFN wrote to the US Senate expressing concerns directed at the EU's negotiating position within the framework of the Trans-Atlantic Trade and Investment Partnership (TTIP). It urged the Agriculture Secretary and US Trade Representative to reject any proposals which 'would in any way restrict the ability of US producers to use common meat names, such as bologna or black forest ham'.[40] The same organisation has also called on the US to counter EU strategies relating to cheese names.[41] They specifically mention the recent EU-Canada agreement, which would make the use of 'feta' produced in the US but sold in Canada subject to restrictions.[42] As *The Economist* wryly observes, 'American cheesemakers are unwilling to accept this as a feta accompli'.[43] Reacting to these potential market access impediments, the US House Ways and Means Committee Chairman was 'particularly concerned by European restrictions on the use of generic food names, which the EU improperly designates as geographical indications. This threatens the US dairy industry and cannot be tolerated.'[44]

2.3 The Hollowness of Art 24.6

As these case studies demonstrate, where the commercial stakes are significant, producers tend to speak in the names of consumers – often consumers in distant markets. While the EU is exporting protected status to third countries via the lists in bilateral agreements, the US dairy industry is attempting to do the same for generic status. Over the course of these disputes, producer interests often take priority whereas the genericide test in Art 24.6 refers to 'common' (that is general public or relevant consumer)

---

[40] Letter to Agriculture Secretary Tom Vilsack and US Trade Representative Michael Froman, 4 April 2014, at: http://www.commonfoodnames.com/current-activities/

[41] Letter to Agriculture Secretary Tom Vilsack and US Trade Representative Michael Froman, 11 Mar 2014. at: http://www.commonfoodnames.com/current-activities/.

[42] Use of feta would have to be qualified by a term such as 'style' or 'type' *and* the true origin (here, the US) would have to be visible. See Art 7.6 of Chapter 22 of the EU-Canada Comprehensive Trade and Economic Agreement (CETA) (2014). Available at: http://ec.europa.eu/trade/policy/in-focus/ceta/

[43] 'Food Names: Stressed are the Cheesemakers', *The Economist*, 19 July 2014. I am grateful to Graeme Dinwoodie and Brian Havel for this reference.

[44] Opening Statement of Devin Nunes, Hearing on Advancing the US Trade Agenda: Benefits of Expanding US Agriculture Trade and Eliminating Barriers to US Exports (11 Jun 2014), at: http://waysandmeans.house.gov

understanding and usage within the relevant jurisdiction. This underlines the need for greater clarity about the nature of the genericide enquiry. The level of abstraction contained in Art 24.6 is understandable – TRIPS establishes the general principle – yet it is unhelpful from a practical perspective. What threshold should be satisfied and which categories of evidence should be considered before concluding that 'the term [is] customary in common language as the common name for such goods'? It has been observed at the TRIPS Council that 'since there were no rules on how one could make such [a] determination, it would be difficult for the relevant authority to deal with the conflict of rights between geographical indications and generic terms'.[45]

Therefore while Art 24.6 is the default reference point for both national GI legislation and international deliberations, it requires further elaboration. According to a recent study:

> Of the 167 countries that protect GIs as a form of intellectual property, 111 (including the EU 27) have specific or *sui generis* systems of GI laws in place. There are 56 countries using a trademark system, rather than or in addition to specific GI protection laws. These countries utilize certification marks, collective marks or [regular] trade marks to protect GIs.[46]

For those countries that have opted to protect GIs within the trade mark system, national trade mark regimes usually have a relatively well established test for determining whether a mark has become generic.[47] By contrast, the majority of countries which have opted for *sui generis* protection would have designed their test for generic status with Art 24.6 as the template. Yet many of these countries have not specified any criteria to be considered when applying the test.[48]

Looking beyond national horizons, in international negotiations the

---

[45] TRIPS Council, 'Minutes of the Meeting on 17–19 September 2002' 8 November 2002 (IP/C/M/37/Add.1) at [139].

[46] D Giovannucci, T Josling, W Kerr, B O'Connor and M Yeung, *Guide to Geographical Indications: Linking Products and their Origins* (International Trade Centre, Geneva, 2009) 14.

[47] This is evident from a comprehensive WIPO survey of national trade mark laws. WIPO Secretariat, 'Grounds for Refusal of all Types of Marks' 30 Aug 2010 (WIPO/STrad/INF/5), at [35] ('Trade marks that consist exclusively of signs or indications which have become customary in the current language or in the bona fide and established practices of the trade are regularly excluded from registration.').

[48] See generally O'Connor & Co, *Geographical indications and TRIPs: 10 Years Later . . . Part II – Protection of Geographical Indications in 160 Countries around the World* (Report commissioned for EC (DG Trade) 2007).

TRIPS approach to genericide has become further entrenched. At the WTO, talks have been taking place concerning the establishment of a multilateral register for wine and spirit GIs. A draft composite text for establishing this register eventually emerged in 2011.[49] Since generic status is both the basis for an objection to registration as well as for subsequent invalidation, the need to substantiate any such assertion has been emphasised over the course of these negotiations.[50] What is missing is any further details on how such assertions may be successfully substantiated. At the time of writing, negotiations have recently concluded to revise the Lisbon Agreement, in an attempt to update its provisions and expand its membership.[51] Here a combined reading of Arts 11(2) and 12 indicates that there is no obligation to protect a term which has already become generic in a Contracting Party, prior to an application for its registration in the International Register operated under the Lisbon Agreement. An Agreed Statement as a footnote to Art 12 once again adopts the TRIPS language of 'a term customary in common language as the common name of a good or service'.[52]

Besides these multilateral initiatives, GI protection figures prominently in the ongoing negotiations for the US-EU Transatlantic Trade and Investment Partnership (TTIP) free trade agreement.[53] The question of generic status is also being considered within the context of the opaquely negotiated and controversial Trans-Pacific Partnership Agreement (TPP).[54] The TPP text is interesting because it attempts to

---

[49] The text is contained in (JOB/IP/3/Rev.1) of 20 April 2011, which is attached to the TRIPs Council, 'Report by the Chairman to the Trade Negotiations Committee' 21 April 2011 (TN/IP/21). For details, see Chapter 8 by Martín in this volume.

[50] Communication from Albania et al., 'Draft Modalities for TRIPs Related Issues' 19 July 2008 (TN/C/W/52) Annex, at [2] ('In the framework of these procedures, [when consulting the international register] domestic authorities shall consider assertions on the genericness exception laid down in TRIPS Article 24.6 only if these are substantiated.').

[51] For details, see Chapter 5 by Geuze in this volume.

[52] Geneva Act of the Lisbon Agreement on Appellations of Origin and Geographical Indications and Regulations under the Geneva Act of the Lisbon Agreement on Appellations of Origin and Geographical Indications, 20 May 2015 (LI/DC/19).

[53] See EU, *Report of the Seventh Round of Negotiations* (23 Oct 2014); available at: http://ec.europa.eu/trade/policy/in-focus/ttip/

[54] I have relied on the WikiLeaks Release of Secret Trans-Pacific Partnership Agreement (TPP) – Second Release, Intellectual Property Chapter for All 12 Nations with Negotiating Positions (16 May 2014 consolidated bracketed negotiating text), available at: https://wikileaks.org/tpp-ip2/. See also, Kim Weatherall,

flesh out the Art 24.6 test by providing guidelines for determining generic status. According to draft Art QQ.D.8, factors relevant to determining consumer understanding within a given jurisdiction include references to the term being used for a type of product in dictionaries, newspapers and relevant websites; how the product referenced by the term is marketed and used in trade in that territory; and whether the term is used in relevant international standards recognised by all the Parties to refer to a type or class of product. Cumulatively, this suggests that genericide remains a live issue, with considerable commercial significance. The following sections develop a more detailed operational framework for registrars, litigants and adjudicators attempting to resolve genericide disputes in practice.

3. STRUCTURAL FEATURES AND THRESHOLDS

In 'accordance with the principle of territoriality, protection of [GIs] is governed by the law of the country where protection is sought . . . and not by that of the country of origin'.[55] In keeping with this principle of territoriality, 'the criteria for the generic character . . . and the rules of proof . . . must be found in the laws of the country where the [GI] is claimed to be generic'.[56] Since the test is jurisdiction-specific, when it comes to setting the structural parameters for determining genericide there are four important decisions to be made.

First, is the test to be applied within the context of a *sui generis* GI protection regime or within the registered trade mark system? The former allows for greater design freedom, whereas a trade mark system usually has a fixed, stable test for establishing generic status already in place. There is less room to manoeuvre and fewer opportunities for tailoring elements of the test. By contrast, those countries who have opted for *sui generis* GI protection regimes retain the flexibility to craft an approach to genericide that is GI-specific. Therefore *sui generis* legislation may specify evidentiary factors which favour producers in the region of origin, or stipulate that once a GI is registered, it is deemed to be incapable of

---

'TPP – Section-by-section Analysis of Some Provisions People Aren't Talking About' (May 2014 Leaked Draft)' (29 Oct 2014); available at: http://papers.ssrn.com/sol3/papers.cfm?abstract_id=2516058

[55] *Exportur SA v LOR SA and Confiserie du Tech SA* (C-3/91) [1992] ECR I-5529, at [12].

[56] J Audier, 'Generic and Semi-Generic Denominations: Determination Criteria and Methods to Reduce their Effects' (2000) 22 AIDV Bulletin 29, 30.

subsequently becoming generic.[57] This will be further illustrated when we consider the question of the threshold to be satisfied later in this section.

Second, who bears the burden of proving that the contested expression is generic? This is an extremely significant question because proving genericide usually calls for extensive factual evidence. Collating this evidence is expensive and time consuming, especially where it calls for reliably robust consumer surveys.[58] According to widely accepted principles of civil procedure and evidence, it is up to the party making assertions or claims to prove them.[59] This will vary depending upon both the forum and stage at which a genericide challenge is raised in the life cycle of a protected GI. The burden of proof question is relevant: (i) during an application for registration or legal recognition in some other way; and (ii) where a protected GI is challenged.[60] Each of these will be considered in turn.

For registration-based GI protection,[61] the burden should rest with the authority that is processing the initial application for registration. In such situations, the applicant (often a producer organisation) will have initially submitted evidence that the term is a GI, so it is up to the evaluating authority to substantiate any objection that this is not the case because the term is generic.[62] This was the conclusion in a recent German dispute involving an application for Bavarian Blockmalz, a Bavarian hard candy made of sugar caramel and malt extract.[63] The application was challenged – initially successfully – on the basis of longstanding use by producers outside the Länder of Bavaria. On appeal, the case was remanded because external trade usage was insufficiently dispositive evidence to determine consumer perception of the term, which remains the ultimate benchmark. The GI

---

[57] For a thoughtful comparison of the EU and US approaches along these lines, see Caroline Le Goffic (n 6) 426–437. See also Section 5.1 below.

[58] See generally, V Schoene, 'Federal Patent Court (Bundespatentgericht – BPatG) Amends Case Law on "Generic Term" in EU Regulation 1151/2012' (2014) 9 JIPLP 765.

[59] Audier (n 56) 33–34.

[60] As identified in the context of establishing an international wine and spirits GI register managed by the WTO. See Report by the Chairman (TN/IP/21) (n 49).

[61] The features of the EU's *sui generis* registration-based system are reviewed in D Gangjee, 'Proving Provenance? Geographical Indications Certification and its Ambiguities' (2015) *World Development* [forthcoming].

[62] However, the examining authority may prefer indirectly advancing a genericide argument, by stating that the applicant has not satisfied the requirement that a link between product and place exists in the eyes of the consumers. Since it is up to the applicant to satisfy the definition of a GI, it retains the burden of proof when confronted by this argument.

[63] *Bayrisch Blockmalz* [2014] GRUR 677 (Bundespatentgericht (BPatG), 21 Nov 2013). For detailed analysis, see Schoene (n 58).

registrar – the German Patent and Trade Mark Office in this case – was asked to re-examine the application with the benefit of a broader range of evidence, including conducting a consumer survey at its own expense if this was deemed necessary. Similarly, any party opposing the registration of a GI on the basis that it is generic will bear the burden of proving this. Amongst the various grounds of opposition in the EU, Art 10(1)(d) of Regulation 1151/2012[64] requires that a 'reasoned statement of opposition' will be admissible if it 'gives details from which it can be concluded that the name for which registration is requested is a generic term'.

Once a GI is registered, the onus is on the party seeking to challenge this legally protected status by alleging generic use. This challenge can occur defensively, in the form of a counterclaim to infringement allegations by GI rights holders. In the EU context, Germany did not satisfactorily discharge this onus when claiming that Parmesan was generic[65] and neither did the owners of the trade mark 'Bavaria' for beer, who sought to challenge the protected status of Bayerisches Bier.[66] A similar burden of proof requirement will operate where GIs are protected as trade marks (usually collective or certification trade marks) and the registration is challenged. This is reinforced by the legal presumption in many legal systems that a registered trade mark is valid. Mere unsubstantiated allegations of generic status, or an over-reliance on dictionaries or other textual sources alone, will be given short shrift by adjudicators.[67]

Third is perhaps the single most important structural question: what is the tipping point for generic status? Put differently, what is the threshold for deciding when this status has been achieved and the meaning of a sign has changed? In the US, where this test has developed in the context of

---

[64] Regulation (EU) No. 1151/2012 of the European Parliament and of the Council of 21 November 2012 on Quality Schemes for Agricultural Products and Foodstuffs [2012] OJ L343/1.

[65] *Commission v Germany* (n 3) at [AG68] ('Germany, which raised the generic nature of the term "Parmesan" as a defence in the present proceedings, has failed to produce, even in respect of Germany, evidence to substantiate to any great extent its argument that the name "Parmesan" has become generic.'); (CJEU) at [52] ('It is for the Federal Republic of Germany to prove that [genericide] argument to be well founded.').

[66] *Bavaria v Bayerischer Brauerbund* (n 31) at [108] (generic status is not substantiated 'by the mere fact of the presence on the market of trade marks and labels of commercial companies including the term "Bayerisches" or translations of it as synonyms for the old Bavarian bottom-fermentation brewing method').

[67] In the US context, see *Community of Roquefort v William Faehndrich* 303 F2d 494 (1962); *Tea Board of India v The Republic of Tea* (2006) 80 USPQ 2d 1881 (TTAB).

registered trade mark law, it is satisfied where for the majority of consumers the 'primary significance' of the sign is to describe a class or category of products.[68] This majoritarian approach – it could be 60 per cent of the relevant public who consider a term to be generic – can be contrasted with that adopted in *Feta*, where generic status is achieved 'only when there is in the relevant territory no significant part of the public concerned that still considers the indication as a geographical indication'.[69] This is a high threshold, similar to the established standard under German unfair competition law[70] and designed to prevent an easy slide into genericide. The interests of those who continue to treat the sign as a GI seem to be prioritised over those who don't. This approach is once again adopted in the *Bavaria* dispute. The CJEU reasoned that since the purpose of GI protection is:

> to prevent the improper use of a name by third parties seeking to profit from the reputation which it has acquired and, moreover, to prevent the disappearance of that reputation as a result of popularisation through general use ... a name becomes generic only if the direct link between, on the one hand, the geographical origin of the product and, on the other hand, a specific quality of that product, its reputation or another characteristic of the product, attributable to that origin, *has disappeared*, and that the name does no more than describe a style or type of product. (emphasis added)[71]

Therefore in the EU, even a significant minority of the relevant public who continue to recognise the geographical significance of the term will be sufficient to sustain it as a protected GI. As Schoene puts it, for 'the purposes of Regulation 1151/1012 one may therefore uphold traditional [German] principles ... according to which a geographical term only becomes generic when merely negligible parts of consumer and trade

---

[68] Articulated in the context of cancellation of a registered trade mark on the basis of genericide, the test is found in the Lanham Act § 14(3), 15 USCA § 1064(3) ('The primary significance of the registered mark to the relevant public ... shall be the test for determining whether the registered mark has become the generic name of goods or services on or in connection with which it has been used.'). See also JT McCarthy, *McCarthy on Trade Marks and Unfair Competition*, 4th edn (Thomson West, June 2014 update) at §12:6.

[69] Recital 23 of Commission Regulation (EC) No 1829/2002 of 14 October 2002 Amending the Annex to Regulation (EC) No 1107/96 with Regard to the Name Feta [2002] OJ L277/10.

[70] H Harte-Bavendamm 'Ende der geographischen Herkunftsbezeichnungen? "Brüsseler Spitzen" gegen den ergänzenden nationalen Rechtsschutz' [1996] GRUR 717, 718.

[71] *Bavaria v Bayerischer Brauerbund* (n 31) (CJEU) at [106]–[107].

circles understand it in a geographical sense'.[72] In *sui generis* GI protection systems it is therefore possible to set this threshold in a manner which makes it considerably more difficult to successfully prove genericide. This aspect highlights the normative choices to be made when designing such a test.

Fourth, how are the categories of evidence to be approached? Here there are at least two relevant considerations: (i) should the categories be open-ended or operate as a closed list?; and (ii) ought there to be any hierarchy established amongst the various categories? While the individual categories are considered in Section 4 below, this general question relates to the approach to be taken when assessing them. For instance, the present position in the EU is to have an open-ended list of factors, while specifically indicating a few in the relevant legislation. These serve as a stable core and starting point. Thus, according to Art 41(2) of Regulation 1151/2012:

> To establish whether or not a term has become generic, account shall be taken of all relevant factors, in particular:
> (a) the existing situation in areas of consumption;
> (b) the relevant national or Union legal acts.

A similar test operates for wines in the EU.[73] Turning to the question of a hierarchy, in *Feta* the CJEU favoured an even-handed approach when considering the factors, confirming that no one factor should be considered dispositive.[74] By focusing on the perception of the term in Greece and failing to adequately appreciate the extent of external feta production elsewhere in the EU, the Commission had not appropriately considered all the factors required by Regulation 2081/92 when rejecting the genericide challenge. Subsequently, in an apparent inversion of the *Feta* reasoning, when applying the CJEU's guidance the German Federal Patent Court had prioritised market factors such as the extent of external production beyond the region of origin.[75] The (over)-

---

[72] Schoene (n 58) 766.
[73] See Art 118k of Council Regulation (EC) No 491/2009 of 25 May 2009 Amending Regulation (EC) No. 1234/2007 Establishing a Common Organisation of Agricultural Markets and on Specific Provisions for Certain Agricultural Products [2009] OJ L 154/1.
[74] *Kingdom of Denmark, Federal Republic of Germany and French Republic v Commission of the European Communities* (Joined cases C-289/96, C-293/96 and C-299/96) [1999] ECR I-1541 (hereafter *Feta 1999*).
[75] Münchner Weißwurst [2009] Wettbewerb in Recht und Praxis 472 (BPatG, 17 Feb 2009); Thüringer Klöße, 30W(pat) 78/06 (BPatG, 2 Oct 2009).

emphasis on this factor was rectified in the recent Bavarian Blockmalz decision, such that it now has to be balanced against all the other evidence relating to consumer perception.[76] Overall, the argument in favour of an open-ended test, without any hierarchy prioritising certain factors, is persuasive: 'The relative importance of a factor in relation to others is not decisive, because what matters is the overall result. The combination of factors determines whether or not [a GI] is generic as claimed.'[77] What decision makers should look for is a cumulative steer; whether the preponderance of evidence aligns to support a conclusion one way or the other.

4. CATEGORIES OF EVIDENCE

When considering categories of evidence, it is desirable that these categories should directly relate to the overarching enquiry.[78] Under Art 24.6, the aim is to establish whether 'the relevant indication is identical with the term customary in common language as the common name for such goods or services in the territory of that Member'. In the equivalent trade mark context, when deciding on generic status some courts 'have focused on the efforts of mark owners in defending their rights, as opposed to how a term is understood by the consuming public'.[79] They have misguidedly adopted an acquiescence-focused enquiry, such as the one required by Art 24.7 of TRIPS,[80] rather than one which inductively pieces together consumer understanding. With this general caveat in mind, it is helpful to enumerate the specific categories which have been relied on by decision makers in the past.

---

[76] *Bayrisch Blockmalz* (n 63).

[77] Audier (n 56) 30.

[78] B Beebe, 'An Empirical Study of the Multifactor Tests for Trademark Infringement' (2006) 95 California Law Review 1581, 1645. (For a cautionary tale of how factors could end up drifting away from the original test; in this particular case when analysing the likelihood of confusion test in trade mark law, Beebe demonstrates that certain 'factors as well as their overall design often distract from their ultimate purpose: to estimate what is actually occurring or will occur in the marketplace'.)

[79] A Pickett, 'The Death of Genericide? A Call For a Return to the Text of the Lanham Act' (2007) 9 Tulane Journal of Technology and Intellectual Property 329, 331.

[80] The provision essentially states that there is a five-year time limit on requests to prohibit the use of, or cancel, registered trade marks that conflict with GIs as per Arts 22(3), 23(2) or 24(5).

4.1 Consumer Understanding

The most all-embracing of the categories, consumer understanding usually covers direct and indirect (or circumstantial) evidence of common use and understanding. Before turning to each of these sub-categories, two threshold questions which need to be addressed are: (1) who is the 'intended audience' for the sign?; and (2) what are the characteristics attributed to this audience (for example sophistication versus imprecision)?

The references to 'common language' and 'common name' in Art 24.6 suggest that the target audience should be the average consumer for that product and where the product category is one that is widely consumed (for example cheese or tea within a given jurisdiction), this audience is potentially synonymous with the general public. Historically, the audience has included some combination of the general public, consumers of that product, traders and other market participants in the relevant product sector and expert opinion.[81] The sub-categories of traders and experts are considered in greater detail below. This general question was considered by the Commission in *Feta*, when deciding whether the 'relevant public' understands the term to be commonly used for a type of white cheese in brine. Identifying the appropriate group to which the test is applied is an important preliminary stage and the Scientific Committee,[82] which advised the European Commission during the registration process, reasoned in the following manner:

> As regards the public concerned, the assessment will depend on the kind of product and the public to which that product is addressed. In the present case, the product being a cheese for consumption primarily by the final consumer (but also by commercial purchasers, such as restaurants, food factories, etc.), the general public is relevant. Therefore, it is with regard to the general public that the designation or indication in question must have lost its original geographical meaning.[83]

When it comes to determining the perceptiveness, attention to detail or background knowledge of this relevant public, care should be taken to rely upon empirical evidence as far as possible. As Lorvellec rhetorically

---

[81] Audier (n 56) 32.
[82] A committee of experts established to assist the Commission with all technical matters relating to registration of GIs, including determining generic status. See Commission Decision of 21 December 1992 Setting Up a Scientific Committee for Designations of Origin, Geographical Indications and Certificates of Specific Character (93/53/EEC) [1993] OJ L13/16.
[83] Recital 23 of Commission Regulation (EC) No 1829/2002 (n 69).

asks: 'Should the law, however, search for the least educated consumer from the bottom of the deepest well of ignorance to determine whether a sign deserves protection?'[84] This is all the more apposite where levels of consumer knowledge and sophistication are increasing over time for the product in question. Once again wine is the paradigmatic example.[85]

Direct evidence usually takes the form of consumer polls or surveys, either of the general public or consumers of that product. When the application for Feta proved controversial, the Commission resorted to a Eurobarometer survey of 12,800 nationals across the (then) 12 Member States. On the basis of this survey it concluded that the cheese continued to connote a Greek origin for most of those who knew of it.[86] However, the probative value of surveys is often controversial, since their design, implementation and analysis must be rigorously monitored to ensure that the results are not skewed in favour of the party commissioning the survey.[87] An example of a flawed genericide survey is found in *Darjeeling*, where a US trade mark applicant wishing to use the term attempted to cancel Darjeeling as a registered US certification trade mark. Here the genericide survey was challenged both on the basis of its open-ended, unhelpful questions – for example, 'What is Darjeeling tea?' – as well as the opaque manner in which the respondents' answers were characterised and grouped.[88] It failed to either show or support an inference that the primary significance of Darjeeling was generic for US tea drinkers.

Indirect evidence of consumer understanding is usually found in textual sources. As was indicated above, draft Art QQ.D.8 of the TPP Agreement

---

[84] L Lorvellec, 'You've Got to Fight for Your Right to Party: A Response to Professor Jim Chen' (1996) 5 Minnesota Journal of Global Trade 65, 72.

[85] See e.g. BWA Ben Dewald, 'The Role of the Sommeliers and their Influence on US Restaurant Wine Sales' (2008) 20(2) International Journal of Wine Business Research 111, 120 ('The globalization of the wine industry and the increasing sophistication and knowledge of the consumer requires that sommeliers stay current on a rapidly changing business'); J Bruwer and G Wood, 'The Australian Online Wine-Buying Consumer: Motivational and Behavioural Perspectives' (2005) 16(3) Journal of Wine Research 193 (the study found that online wine buyers were mostly well-educated, high-income males in the 35- to 44-year-old age group).

[86] See *Feta 1999* (n 74) at [36]–[38].

[87] For similar concerns relating to methodology and implementation expressed in the context of trade mark law, see A Niedermann, 'Surveys as Evidence in Proceedings before OHIM' (2006) 37 IIC 260; Special Report, 'The Value and Treatment of Survey Evidence in Different Jurisdictions' (2010) 100 Trade Mark Reporter 1373; Shari Diamond and David Franklyn, 'Trade Mark Surveys: An Undulating Path' (2014) 92 Texas Law Review 2029.

[88] *Tea Board of India v The Republic of Tea* (n 67).

refers to whether the term is used for a type of product in dictionaries, newspapers and relevant websites. In *Feta*, the Commission's decision to re-register the term as a protected designation was in part based on the meaning conveyed by '[g]eneral reference works such as dictionaries and encyclopaedias, and specialised publications sent in by the Member States'.[89] While most of these sources referred to a cheese made from ewes' and goats' milk, several additionally referred to the cheese being of Greek origin. As McCarthy points out in the comparable context of trade mark genericide, while 'not determinative, dictionary definitions are relevant and sometimes persuasive in determining public usage. This is based upon the assumption that dictionary definitions "usually reflect the public's perception of a word's meaning and its contemporary usage".'[90] In *Fontina*, the US Trade Mark Trial and Appeal Board (TTAB) affirmed that textual sources referred to a type of cheese with a certain hardness, texture and flavour characteristics, regardless of origin, since they used lower-case letters for the term.[91] Furthermore, in the prevalent semiotic ecosystem of the internet, there is an understandable reliance on websites and online resources. Recent trade mark scholarship suggests that, given the way search engines work (predicting what online consumers associate with a search term), the highest ranked search results should have probative value – is the term being used generally for a product class on those websites?[92] Searches can be tailored to produce country specific results, be attempted across a range of different search engines and conducted at different points of time to show that the meaning (generic or origin specific) of that term is stable. However, at least one court has concluded that the frequency with which a term appears on the internet is not, of itself, capable of establishing the generic nature of a GI.[93]

Two final aspects relating to consumer understanding are also worth considering. First, is there a difference between informal or casual and commercial usage? This distinction has been considered in the trade mark context:

> Buyers or users of a product may sometimes use a trade mark in a generic sense in casual conversation even though when questioned, those persons are fully aware of the trademark significance of the term. For example, persons may use

---

[89] Recital 21 of Commission Regulation (EC) No 1829/2002 (n 69).
[90] McCarthy (n 68) § 12:13.
[91] *In re Cooperativa Produttori Latte E Fontina Valle D'Aosta*, 230 USPQ 131 (TTAB 1986).
[92] Lisa Larrimore Ouellette, 'The Google Shortcut to Trade Mark Law' (2014) 102 California Law Review 351, 397–398.
[93] *Consorzio per la tutela del formaggio Grana Padano v OHIM, Biraghi* (T- 291/03) [2007] ECR II-3081 at [71].

XEROX or KLEENEX in a generic sense when asking someone to 'Make me a Xerox of this' or asking a friend for 'a Kleenex,' even though when going to purchase a photocopy machine or a box of tissues, they know that XEROX and KLEENEX identify the commercial source of those products.[94]

This dichotomy was also acknowledged when the UK high court was deciding whether Sherry was generic for a type of fortified wine: 'In the first place even people who are knowledgeable about wine and would expect if they asked in a bar for a glass of dry sherry to be given a wine from Spain may on other occasions use the word "sherry" to include "sherry type" wines.'[95] Therefore, generic usage in a context which is not commercially significant should be bracketed and discounted. Second, what if there is a split in consumer understanding, with some considering the term to be generic while others rely on its geographical origin-indicating ability? A straightforward resolution to this issue would be to approach it via the threshold or tipping point for genericide, considered in Section 3 above. If 30 per cent of those surveyed – a significant minority – still consider a term to have geographical significance, this should be sufficient to prevent genericide in the EU. Alternatively, a US-style majoritarian approach could be adopted, where any view which is supported by over 50 per cent of the target audience is dispositive. There have been other approaches to this issue in the past as well. One has been to fashion injunctive relief in accordance with the split in perception. Thus where medical professionals and the trade considered Aspirin to be a trade mark but the general public considered it to be generic, the permissible use of the term was conditioned in consonance with this. Aspirin could be used generically in direct sales to the public but the professional generic designation – acetyl salicylic acid – would have to be used in sales to professional customers.[96] An alternative approach would be to consider such terms as semi-generic,[97] allowing suitably qualified use to

---

[94] McCarthy (n 68) § 12:8. See also, RH Folsom and LR Teply, 'Trade Marked Generic Words' (1980) 89 Yale Law Journal 1323; I Simonson, 'An Empirical Investigation of the Meaning and Measurement of "Genericness"' (1994) 84 TMR 199.

[95] *Vine Products Ltd v Mackenzie & Co Ltd (No.3)* [1967] FSR 402, 423 (Ch D).

[96] *Bayer Co. v United Drug Co.*, 272 F. 505 (DNY 1921).

[97] For details of the US approach to semi-generic wine designations, see PM Brody, '"Semi-Generic" Geographical Wine Designations: Did Congress Trip over TRIPs?' (1999) 89 TMR 979; LA Zahn, 'Australia Corked Its Champagne and So Should We: Enforcing Stricter Protections for Semi-Generic Wines in the United States' (2012) 21 Transnational Law & Contemporary Problems 477.

be made of them, such as indicating the actual origin of the product (for example Californian Chablis).

4.2 Trade Opinion

There are two issues to be kept in mind when relying on opinions from traders in the relevant product sector as well as related commercial operators such as retailers. The first is the extent to which trade opinion is reliable, since there is always the possibility of a bias in either direction. If the traders who are alleging generic status are based outside the region of origin, it is in their own interest to make such a claim. Where Halloumi was registered as a US certification trade mark for Cypriot cheese and its cancellation was subsequently sought on the basis of generic use, the cancellation petitioner's own export manager appeared as a witness.[98] However, his evidence related to the use of Halloumi outside the US and was therefore considered irrelevant for determining its meaning in that market. Conversely, claims by 'friendly' traders that the term still retains geographical significance should be treated with similar scepticism.[99] The second issue concerns a divergence between trade opinion and consumer or general public perception. An example is found in a decision concerning an application to register Stilton as a UK certification trade mark, where the concern was that the sign may be descriptive of a type of cheese (that is generic).[100] While rebutting the generic use argument, the trade gave evidence that cheese made outside of the geographical boundaries specified in the certification mark application had not been sold on the market as Stilton. While many consumers may have thought that the name referred to a type of cheese, trade and gastronomic experts felt that the name reflected not only qualities and physical characteristics but also a certain method of manufacture and a particular region where the cheese was produced. In such situations, traders may be considered to have domain-specific expertise and their usage may be considered more accurate or knowledgeable. However, the more convincing view is that the Art 24.6 enquiry is concerned with common usage and prioritises consumer (or general public) perception of the sign. Trade opinion is then subsumed within this

---

[98] *Danish Dairy Board, Inc. v The Ministry of Commerce and Industry of the Republic of Cyprus*, Cancellation No. 19,815, 11 Aug 1999, TTAB (Non precedential) (Unreported).

[99] The potential for bias is also acknowledged in the related trade mark context. See McCarthy (n 68) § 12:13.

[100] *'Stilton' Trade Mark* [1967] FSR 15, 19–20 (Ch D).

broader enquiry such that it is relevant only to the extent that it shapes broader consumer or public opinion.

Precisely this issue was addressed by the CJEU in *Kornspitz*, which centred on the question of whether a trade mark had become generic and was liable to be revoked.[101] While the Kornspitz trade mark (used for a pre-prepared baking mix to produce bread rolls) had arguably become the generic name for a particular type of bread roll for consumers, the same was not true for the bakers who purchased the baking mix and subsequently sold the rolls. The Advocate General refers to the possibility that 'the way in which [commercial] intermediaries understand the term is relevant only if it influences the end consumer's decision to purchase'.[102] The CJEU initially confirmed that:

> whether a trade mark has become the common name in the trade for a product or service in respect of which it is registered must be assessed not only in the light of the perception of consumers or end users but also, depending on the features of the market concerned, in the light of the perception of those in the trade, such as sellers.[103]

Yet while trade evidence remained relevant, given the nature of the test, 'in general, the perception of consumers or end users will play a decisive role'.[104] Here the facts suggested that traders' perceptions did not influence consumer perceptions – the bread rolls were usually sold 'unmediated' to consumers.[105] Therefore, trade opinion could not act as a brake to halt genericide, in circumstances where end-consumers otherwise used the term generically.

4.3 Expert Opinion

Expert opinion has been mentioned as a resource in previous attempts to identify criteria for determining generic status. WIPO's Model Law for Developing Countries states in s 5(c) that generic names would be excluded from protection, 'a name being understood as generic from the time when it is generally considered as such by experts on the subject and by the general public'.[106] The expertise contemplated appears to

---

[101] *Backaldrin Österreich The Kornspitz Company GmbH v Pfahnl Backmittel GmbH* (C-409/12) [2014] ETMR 30.
[102] Ibid., at [AG21].
[103] Ibid., at [28].
[104] Ibid., at [29].
[105] Ibid., at [23]–[26].
[106] WIPO 'Draft of the Model Law for Developing Countries on Appellations of Origin and Indications of Source' (TAO/I/INF.1) 30 Oct 1974.

encompass not only linguistic expertise directed at analysing the meaning contained in textual sources or qualitative research expertise to interpret survey results, but also trade experts (such as food critics or sommeliers) who could testify, based on their own experience, as to consumer or retail sector perceptions of the term.[107]

4.4 Market Conditions

Market conditions provide the context for shaping consumer understanding. Thus where Halloumi was applied for as a certification trade mark in Canada to indicate a sheep or goats' milk cheese of Cypriot origin, it was successfully opposed on the basis of market evidence. It was established that: (1) it was possible to procure from various Canadian sources cheese with packaging bearing the name Halloom, Hallouh or Haloumi; (2) a number of Canadian producers had sold substantial quantities of cheese designated as Halloom in Canada; and (3) several industry stakeholders testified that the term 'halloumi' designated a type of cheese and was used generically.[108] While market evidence can therefore provide a sense of the labelling consumers are habitually exposed to and patterns of purchasing behaviour, it should not become an all-too-convenient shortcut or proxy for consumer understanding. It constitutes one strand of the multifactor analysis. With this in mind, the following overlapping sub-categories have been identified as having probative value:[109]

- the method of distribution of the product, which is relevant for determining the extent to which commercial intermediaries play a role in shaping public perception;
- the extent to which the GI product specification has been respected in the 'home' country, or alternatively whether unregulated usage in this country has undermined any basis for international protection;
- the ratio of goods on the market which satisfy the product specification, including the origin requirement, to those which use the term but do not satisfy these conditions;

---

[107] Audier (n 56) 34.
[108] *Ministry of Commerce & Industry of the Republic of Cyprus v Les Producteurs Laitiers du Canada et al.*, 2010 FC 719; Affirmed on appeal: *Ministry of Commerce & Industry of the Republic of Cyprus v Les Producteurs Laitiers du Canada et al.* 2011 FCA 201.
[109] See generally, Audier (n 56) 32; *Grana Padano v OHIM* (n 93) at [65]–[66]; Recitals 11 to 21 of Regulation 1829/2002 (n 69).

- the absolute quantities of goods which use the term as a GI and those which use it generically in a defined market;
- the market shares of goods which have GI usage on the label compared to goods using the term generically.
- statistics relating to both production and consumption of the GI-labelled and generic-labelled products;
- publicity materials relating to the product, including the packaging and labelling used.

This final sub-category relating to labelling deserves closer consideration, since it has proved influential in the EU. It informed the Commission's conclusions when it recognised Feta as a protected designation once again in 2002. Whilst gauging the semantic content of Feta across the EU, the Commission reasoned that because non-Greek cheese labelling (that is on the supposedly generic product) referred to Greek images and iconography, this suggested an ongoing association of Greek origin which still mattered to consumers.

> According to the information sent by the Member States, those cheeses actually bearing the name 'Feta' on Community territory generally make explicit or implicit reference to Greek territory, culture or tradition, even when produced in Member States other than Greece, by adding text or drawings with a marked Greek connotation. The link between the name 'Feta' and Greece is thus deliberately suggested and sought as part of a sales strategy that capitalises on the reputation of the original product, and this creates a real risk of consumer confusion.[110]

This reasoning was subsequently endorsed by the CJEU in its final *Feta* decision.

> The evidence adduced to the court also shows that, in Member States other than Greece, feta is commonly marketed with labels referring to Greek cultural traditions and civilisation. It is legitimate to infer therefrom that consumers in those Member States perceive feta as a cheese associated with the Hellenic Republic, even if in reality it has been produced in another Member State.[111]

This was also reiterated in the *Parmesan* decision.

> According to the documents in the case, in Germany, certain producers of cheese called 'Parmesan' market that product with labels referring to Italian cultural traditions and landscapes. It is legitimate to infer from this that

---

[110] Recital 20 of Commission Regulation (EC) No 1829/2002 (n 69).
[111] *Feta 2005* (n 5) at [87].

consumers in that Member State perceive 'Parmesan' cheese as a cheese associated with Italy, even if in reality it was produced in another Member State.[112]

However the conflation of reasoning when making such inferences is problematic. A significant distinction, which seems to have been conveniently overlooked, is whether the labelling is: (i) misleading as to the origin of the product – falsely suggesting Greek or Italian origin for the products described above; or (ii) merely allusive or evocative, implying that while the product may have historically originated in a certain region or cultural context, the specific product being considered by the consumer is not produced there. After all, there is a difference between a sign which says 'Chinese Restaurant' (often accompanied by references to Chinese cultural imagery and characters) and the label 'Made in China'. The former is merely an allusive practice found throughout the world, based on the historic origins of the style of cuisine, whereas the latter communicates present-day origin more definitively. The former is generic usage while the latter is not. The Commission and subsequently the CJEU appear to have collapsed the distinction between the two.

4.5 Status in Legislation or Official Classifications

The status of a term as it appears in legislation is sometimes considered a useful marker when determining its meaning.[113] In the EU, the test for determining generic status in Art 41(2) of Regulation 1151/2012 specifically refers to 'the relevant national or Union legal acts'. In *Feta*, this category of evidence was considered in some detail. Across the EU membership, generic use was permitted by national marketing rules for cheese in Denmark from 1963, in the Netherlands from 1981 and in Germany from 1985. At the community level, feta was treated as a commonly used term for a category of cheese in harmonised customs legislation and for obtaining export refunds for milk products.[114] However, the importance to be given to such legislation depends on its specific relationship to the product in question, or the extent to which it shapes (or reflects) consumer understanding of the term. Therefore, the Greek national rules regulating

---

[112] *Commission v Germany* (n 3) at [55].
[113] See e.g. Melinda Wallman 'Recent Developments in Australian Intellectual Property Law: The Protection of Geographical Indications of Wines' (1994) 5 Australian Intellectual Property Journal 113, 116 (considering in detail how the various Australian Federal and State Food Acts recognised terms like Champagne and Chablis as descriptive of type rather than distinctive of geographical origin).
[114] *Feta 1999* (n 74) at [59].

the use of 'Feta' were relevant because they 'embod[ied] in legislation the traditional use of the name "Feta" in Greece over the centuries'.[115] By contrast, Community rules on customs duties and export refunds were of little consequence as they 'reflect an approach specific to customs matters and [were] not in any way intended to govern industrial property rights'.[116] Thus, national or Community legislation was only occasionally a helpful guide when evaluating public perception.[117] Similarly, the other Members' national legislation merely permitted the use of Feta, and this was not sufficiently probative when establishing consumer perceptions of the term.[118]

Nevertheless, the extent to which legislation or administrative regulations permit the use of a generic term is significant where it encourages producers outside of the region of origin to adopt it in their labelling. Widespread generic use by such traders will then go on to influence public perception. In other situations, the status in legislation or administrative rules presumably rests on an initial (often bureaucratic) decision that the term is generic. In the US, regulations administered by the Food and Drug Administration (FDA) have been referred to in disputes over generic status. In an enquiry into the status of Roquefort, it was unsuccessfully argued that FDA standards of product identity appeared to refer to Roquefort as the alternative to (and therefore synonymous with) generic sheep's milk cheese containing blue mould. These standards of identity did not help establish consumer understanding of the term.[119] Such standards could also be the basis for indirectly concluding that a term is generic. The Federal Trade Commission (FTC) was approached for a ruling on whether the use of basmati and jasmine on varieties of rice which did not originate in the Indian sub-continent or Thailand respectively was misleading labelling and should be prohibited.[120] The FTC denied this petition essentially on the basis that these terms were generic. There would be no significant consumer injury and the use was not sufficiently misleading as to origin or quality, since US Department of Agriculture regulations

---

[115] Ibid., at [71].
[116] Ibid.
[117] Ibid., at [72].
[118] *Feta 2005* (n 5) at [88], [91]–[92], [98].
[119] *Community of Roquefort v William Faehndrich, Inc* 198 F.Supp. 291 (DC, SDNY, 1961); On appeal *Community of Roquefort v William Faehndrich, Inc* 303 F.2d 494, 133 USPQ 633 (2nd Cir., 1962).
[120] Commission Denial of Petition for Rulemaking Procedure: The Advertising and Marketing of Basmati and Jasmine Rice, 15 May 2001 (FTC File No. P014506); At: http://www.ftc.gov/opa/2001/05/fyi0131.htm

referred to basmati and jasmine as examples of 'aromatic rough rice' and were not limited to rice grown in any particular country.[121]

At other times, legislation will directly stipulate that certain terms are generic or semi-generic. In the US Code of Federal Regulations, this has been done for certain types of wine. Semi-generics can be used for wine produced outside of the eponymous region, provided the actual place of origin is also specified (for example Californian Chablis).[122] This classification is the result of a bureaucratic decision-making process and the basis or reasoning for such determinations has never been made clear.[123] Making such an explanation publicly available is important because permitting semi-generic usage results in concerns that such marketing practices would lead to full-blown genericide over time.[124] What is more curious – and seems to have escaped comment up to now – is the official 'freezing' of generic meaning which occurs through such legislative strategies. While the US routinely accuses the EU of trying to reclaim generic terms through the introduction of lists of protected terms in FTAs, its own approach to 'preserving' generic status in this manner is equally vulnerable as an artificial imposition. Allusions to this artificiality are found in a decision by the US Federal Circuit, which concluded that Chablis was generic for a type of wine. The court rejected the argument that the regulations specifying that Chablis was semi-generic were dispositive in any way, since they did 'not establish how the term is understood by American consumers of wine'.[125]

A final resource worth mentioning is international standards for product classifications such as those set out by the Codex Alimentarius Commission, an intergovernmental organisation created by the Food and Agriculture Organization (FAO) and the World Health Organization (WHO).[126] Codex committees create commodity standards, which are

---

[121] See Grain Inspection, Packers and Stockyards Administration (GIPSA), *United States Standards for Rice* (US Department of Agriculture, 2009), at: http://www.gipsa.usda.gov/fgis/standards/ricestandards.pdf

[122] See US Code of Federal Regulations, Title 27 § 4.24 ((2) 'Examples of semi-generic names which are also type designations for grape wines are Angelica, Burgundy, Claret, Chablis, Champagne, Chianti, Malaga, Marsala, Madeira, Moselle, Port, Rhine Wine (syn. Hock), Sauterne, Haut Sauterne, Sherry, Tokay').

[123] The background to this legislative provision, against the backdrop of the signing of the TRIPS Agreement, is provided in Brody (n 97).

[124] Zahn (n 97) 483.

[125] *L'Institut National Des Appellations D'Origine v Vintners International Co.*, 958 F.2d 1574, 1581 (Fed. Cir. 1992).

[126] See generally, M Echols, *Geographical Indications for Food Products: International Legal and Regulatory Perspectives* (Kluwer, 2008) 181–199.

non-mandatory but serve as an international reference point. Therefore, the relevant standard for cheddar cheese contains a description (a ripened hard cheese of a certain colour ranging from ivory to orange; the ripening requirements); its essential composition and quality factors (raw materials; permitted ingredients); permissible additives, and so on.[127] It is evident from these benchmarks that cheddar can be used for any cheese which possesses these features, regardless of its geographical origin. In the recent past, this technical forum became the unlikely venue for a fascinating genericide gambit. A proposal was submitted to establish an international generic standard for Parmesan cheese, which was strongly resisted by Italy and the EU.[128] While the EU resisted the international standard, it recognised the right of individual countries to treat Parmesan as a generic term. A related aspect of the controversy was whether the Codex ought to pay heed to intellectual property norms when defining a standard, resulting in a difference of opinion between the WIPO observers and the FAO as well as the WHO's legal positions. Due to the impasse, the proposal has been deferred indefinitely but such a standard would considerably undermine the value of Italian Parmigiano Reggiano as a PDO under EU law.[129] This incident provides the background context for the TPP's inclusion of 'relevant international standards recognized by all the Parties to refer to a type or class of product' within the genericide determination criteria (see Section 2 above).

4.6 The Action (or Inaction) of Rights Holders

There are two related lines of enquiry to pursue in this regard: whether the rights holders were vigilant and proactive in protecting the GI against unauthorised use by third parties, as well as whether there were effective remedies available to rights holders as a condition precedent. Advocate General Colomer recognised this in *Feta*, observing that:

> Another factor which has a bearing on the process is the passivity of interested parties. The name will decline if no action is taken . . . in response to its misuse, whereas it will be strengthened if there is an appropriate reaction. However, it must be recalled that protective measures can be weakened by a shortage of

---

[127] Codex Standard 263-1966; Available from: www.codexalimentarius.org.
[128] Codex Alimentarius Commission, *Report of the Twenty-Seventh Session* (Geneva, 28 June–3 July 2004) at Appendix X; Codex Alimentarius Commission, *Report of the Twenty-Eighth Session* (Rome, 4–9 July 2005) at [167]–[176].
[129] Dossier No. IT/PDO/0117/0016; Registered 06/09/2003. Available at: http://ec.europa.eu/agriculture/quality/door/list.html

legal provisions, which did not exist at all until very recently, and by apathy on the part of national courts.[130]

Where remedies do exist but no action is taken, this greatly increases the likelihood of genericide. Thus Sherry was found to be essentially semi-generic in the UK, due to the acquiescence of sherry producers to the use of 'British Sherry' for many decades. While Sherry *per se* retained its geographical associations with the Jerez district of Spain, the qualified use was permissible due to inaction and the loss of a right to complain of passing off.[131]

Passivity or inaction has an established track record as a factor, being relied on by French courts as early as 1926 when deciding that Camembert had become generic and was not restricted to cheese produced in certain parts of Normandy alone.[132] Factors which influenced the trial and appeal courts included: the apparent absence of a *terroir* anchor to the geographical influences of the region; a Paris Congress in 1909 defining the cheese with reference to milk type and fat composition but not to any specified region of origin; lists of producers beyond Normandy who produced a cheese under that name from 1856 to 1909; and the availability of penal sanctions and civil remedies which were not utilised to inhibit such external uses. Historical research suggests that the inaction may have also been due to the fact that production techniques were initially associated with a single family and the equivalent of trade secret or know-how protection was preferred. Given the historic patterns of the division of labour, women had a prominent role in the farmhouse-based production of this cheese and the manufacturing process gradually spread via a combination of trade secret 'misappropriation', the hiring of former female employees at attractive wages to learn their secrets and, in some cases, the outright seduction of those with the requisite savoir faire.[133]

---

[130] *Feta 2005* (n 5) at [AG135]. (The official version in Spanish refers to 'interesados' i.e. interested parties or stakeholders, while the English version mistakenly refers to 'consumers'.)

[131] *Vine Products* (n 95). (It must be noted that the remedial option – in this case the extended passing off action recognised in *Spanish Champagne* – had been acknowledged only two years prior to the Sherry litigation. Sherry producers thought they did not have a viable remedy prior to *Spanish Champagne*, which makes the outcome perhaps unduly formalistic.)

[132] *Syndicat du Véritable Camembert de Normandie v Laiterie Coopérative de Ligueil et Autres Syndicats Intervenants*, Cour d'appel d'Orléans, 20 January 1926; Noted in [1926] La Propriété Industrielle 172–173.

[133] For an entertaining account, see P Boisard, *Camembert: A National Myth* (Berkeley: University of California Press, 2003).

Three other aspects relating to right holder conduct are also worth considering. First, how much action ought to be taken to prevent genericide? In *Darjeeling* the party alleging genericide claimed that the Tea Board of India had failed to police its certification trade mark, leading to unlicensed and unregulated third party usage. The TTAB's response to this 'loss of control' argument provides a helpful steer:

> Even if control is not maintained and misuse occurs, it must be shown that *the misuse was of such significance* to permit an inference that the mark is generic ... We have no information about *the nature or extent of the past misuse* of DARJEELING, let alone whether the misuse was of such extent and duration that we could presume that DARJEELING has lost all significance as a mark. (emphasis added)[134]

By contrast, the registration of the certification trade mark and associated licensing activity demonstrated that the Tea Board was not passive. Therefore, the focus should be on whether rights holders have made genuine attempts to enforce their rights, rather than evaluating whether the attempts have been entirely successful in preventing unauthorised use. Also – crucially – the effect of this passivity or inaction is the more important issue. Has it led to an overall environment of generic use and understanding? Mere passivity by itself is therefore not sufficient. Second, evidence of the rights holders own misuse and promotion of the term in a generic manner will always be relevant. Third, in the context of international protection, it is worth remembering that a 'term that has become generic is often the result of a country's inability to enforce a geographic indication because of the absence of an effective international treaty'.[135] The Advocate General's opinion in the final *Feta* decision highlights this dimension of helplessness, especially in the context of third party use across jurisdictional boundaries. Giving undue importance to the inaction theory would

> fail to take account of the fact that many undertakings were unable to enforce their legal rights prior to the entry into force of the legislation concerned, as a result of which they were placed at a serious disadvantage because a process of generalisation, resulting merely from infringement of the name by unauthorised third parties, was permitted to take place.[136]

---

[134] *Tea Board of India v Republic of Tea* (n 67) 1886.
[135] RW Benson, 'Toward a New Treaty for the Protection of Geographical Indications' [1978] Industrial Property 127, 129.
[136] *Feta 2005* (n 5) at [AG156].

5. BACK FROM THE DEAD?

The categories of evidence identified above will help in deciding whether the process of genericide has advanced to the point where a GI no longer merits protection. Nevertheless genericide is a process and processes may be reversible. So what are the options for reclaiming some degree of protection for terms which have been found to be generic?

5.1 Preventive Measures: The Freezing of Meaning

Before considering the options for reclaiming terms it is worth reviewing some of the distinctive features of *sui generis* GI protection regimes that are designed as preventive measures. We have considered the potential for both a higher burden of proof and threshold or 'tipping point' standard in Section 3 above. In addition, such regimes often expressly prohibit the qualified or semi-generic use of protected signs within the infringement provisions, which helps to preserve origin-specific meaning.[137] Art 23.1 of TRIPS is a prominent example, where wines which are not from the eponymous regions cannot use those GIs 'even where the true origin of the goods is indicated or the geographical indication is used in translation or accompanied by expressions such as "kind", "type", "style", "imitation" or the like'. Spanish Champagne or Champagne-style wine are therefore prohibited. Besides these, one additional (and controversial) feature is the 'freezing' of meaning for protected terms, so that they are subsequently shielded from genericide.

An example of such a provision is found in Art 6 of the Lisbon Agreement of 1958: 'An appellation which has been granted protection in one of the countries of the Special Union ... cannot, in that country, be deemed to have become generic, as long as it is protected as an appellation of origin in the country of origin.' Similarly, Art 12 of the Geneva Act of the revised Lisbon Agreement (2015) states that once recognised, protected terms 'cannot be considered to have become generic in a Contracting Party'. Thus once an appellation is accepted for registration and there are no objections during the initial windows of opportunity, the appellation is subsequently insulated against genericide across the other 27 contracting parties without the rights holder having to take further action. A similar shielding provision is found in the Madrid Agreement of

---

[137] See Le Goffic (n 6) 416–418.

1891 for products of the vine;[138] in French national law;[139] in Regulation 1151/2012;[140] and in other EU Regulations relating to wine.[141] Endorsing such provisions, Professor Audier has referred to them as a legislative vaccination for the 'virus' of generic use.[142] Others have expressed concerns that such provisions interfere with the organic evolution of language. During the recent revisions to the Lisbon Agreement, when debating the proposed successor to Art 6, the US delegation was concerned 'that the artificial constraint against genericism undermined trade mark system principles'.[143] The Australian delegation emphasised that:

> the issue of whether a term had become generic in a particular member State had to be a matter for the national law and circumstances of that member State, independently of what happened in the country of origin of the geographical indication ... no other intellectual property right operated perpetually or extraterritorially without an opportunity for review at the national level.[144]

The manner in which Art 6 departs from other general principles in IP law has also been raised, such as use being required within a jurisdiction to maintain rights and the onus of enforcement being on the rights holder.[145] A similar debate took place when considering Art 13(3) of Regulation 2081/92, which established that protected names 'may not become generic'. The Economic and Social Committee opined that such

---

[138] Art 4 states: 'The courts of each country shall decide what appellations, on account of their generic character, do not fall within the provisions of this Agreement, regional appellations concerning the source of products of the vine being, however, excluded from the reservation specified by this Article.'

[139] See Art 10 of the *Loi du 6 Mai 1919 Relative à la Protection des Appellations d'Origine*, 8 Mai 1919, Journal Officiel 4726 ('Les appellations d'origine des produits vinicoles ne pourront jamais être considérées comme présentant un caractère générique et tombées dans le domaine public'); Art L.643-1 of the *Code Rural et de la Pêche Maritime* ('L'appellation d'origine ne peut jamais être considérée comme présentant un caractère générique et tomber dans le domaine public').

[140] See Art 13(2) ('Protected designations of origin and protected geographical indications shall not become generic.').

[141] See e.g. Art 45(3) of Council Regulation (EC) No 479/2008 of 29 April 2008 on the Common Organisation of the Market in Wine [2006] OJ L148 ('Protected designations of origin or protected geographical indications shall not become generic in the Community...').

[142] J Audier, 'Indications Géographiques: Le Virus "Générique"' (2003) 8 Revue Propriétés Intellectuelles 252.

[143] WIPO 'Report of the Lisbon Working Group' 12 June 2014 (LI/WG/DEV/8/7) at [25].

[144] Ibid., at [30].

[145] WIPO, 'Results of the Survey of the Lisbon System', 18 June 2010 (LI/WG/DEV/2/2) 26–28.

robust protection was needed since prestigious names which have been 'established through the care and hard work of certain producers must not be allowed to become generic designations simply because they are not properly protected'.[146] Yet the Committee for Consumer Protection saw this as an 'edict from the Thought Police, attempting to control the evolution of language' by legislation.[147]

Of greater legal significance is the manner in which such a provision is to be interpreted. In so far as Art 6 of the Lisbon Agreement is concerned, 'some delegations would consider such a provision to establish a rebuttable presumption, while others consider it to establish an absolute ban'.[148] The drafting history of Art 6 suggests it was intended as an absolute ban, but certain national courts of Lisbon members (for example the Italian Supreme Court of Cassation) have held that it only establishes a rebuttable presumption that the term is non-generic.[149] The Supreme Court's reasoning is questionable because it would deprive Art 6 of any meaningful content – as we have seen above, any party claiming that an appellation is generic *already has to satisfy this burden* in the normal course of litigation. Looking beyond the specifics of Lisbon, one study queries whether 'cannot ... be deemed to have become generic' operates as a bright line rule at all, raising the possibility that generally applicable defences such as acquiescence (based on sufficiently long-standing and uninterrupted use by external traders) would continue to operate, permitting certain individual traders to make commercial use of such terms.[150] Therefore, while such a rule does exist, its operational impact has not yet been adequately explored.

---

[146] Opinion on the proposal for a Council Regulation (EEC) on the Protection of Geographical Indications and Designations of Origin for Agricultural Products and Foodstuffs OJ C 269, 14/10/1991 p 62 at [1.7].

[147] Ken Collins (Chairman), Opinion of the Committee on the Environment, Public Health and Consumer Protection annexed to the Report of the Committee on Agriculture, Fisheries and Rural Development on the Commission Proposals for Council Regulations (SEC(90) 2415 final) and (SEC(90) 2414) (30 October 1991; Session Document A3-0283/91) 28, 29.

[148] WIPO, 'Report Adopted by the Working Group' 16 November 2012 (LI/WG/DEV/5/7) at [222].

[149] *Pilsen Urquell v Industrie Poretti SpA* [1998] ETMR 168 (whether the Czech appellation 'Plzen' and its translation 'Pilsener' were generic in Italy). For a more detailed critique, see Gangjee (n 19) 153–154; Chapter 14 by Heath in this volume.

[150] M Ficsor, 'Challenges to the Lisbon System' 31 October 2008 (WIPO/GEO/LIS/08/4) at [30].

5.2 Reclaiming Generic Terms

5.2.1 Negotiated reclamation

Especially where it relates to international GI protection, in recent decades the bilateral reclamation of designations has proved to be the EU's preferred strategy, since multilateral options are presently limited. When the EU has attempted to unilaterally impose its views in the past, this has met with vocal opposition. This is best exemplified by the resistance to its so-called 'claw back' list of 41 terms, which was proposed in the context of WTO agricultural negotiations. The EU was seeking to re-establish internationally protected GI status for terms considered to be generic in some jurisdictions and the list included Roquefort cheese, Parma ham, Rioja wine and Feta cheese.[151] The list proved unpalatable but bilateral negotiations have been far more effective. This 'contractual' approach to GI protection or reclamation has a long history and is by now well established:[152] three such examples are the protection of Port and Madeira in the UK;[153] wine-specific provisions being introduced into the Treaty of Versailles to ensure protection in Germany;[154] and the network of bilateral European agreements including the influential Franco-German one, which preceded the harmonised EU regimes.[155] Most recently, such negotiated reclamation has been visible in wine agreements between the EU and new world wine producers such as Australia, South Africa and the US,[156] in return for improved market access in the EU. While this generates

---

[151] EC Press Release 'WTO Talks: EU Steps Up Bid for Better Protection for Regional Quality Products' (IP/03/1178) 28 August 2003. Available at: http://europa.eu/rapid/press-release_IP-03-1178_en.htm. For representative critiques, see M Handler, 'The EU's Geographical Indications Agenda and its Potential Impact on Australia' (2004) 15 Australian Intellectual Property Journal 173; B Goebel, 'Geographical Indications and Trade Marks – The Road from Doha' (2003) 93 TMR 964, 991–994.

[152] See A Conrad, 'The Protection of Geographical Indications in the TRIPs Agreement' [1996] TMR 11, 27–28; Chapter 14 by Heath in this volume.

[153] See Section 1 of the Anglo-Portuguese Commercial Treaty Act 1914.

[154] See Arts 274 and 275 of the Treaty of Versailles, 28 June 1919.

[155] The structure of these agreements is discussed in R Plaisant, 'The Revision of the International Treaty Provisions Dealing with Appellations of Origin and Indications of Source' [1980] Industrial Property 182.

[156] See respectively, Agreement between the European Community and Australia on Trade in Wine [1994] OJ L 86/94, superseded by the Agreement between the European Community and Australia on Trade in Wine [2009] OJ L28/3; Agreement between the European Community and the Republic of South Africa on Trade in Wines [2002] OJ L28/4; Agreement between the European Community and the Republic of South Africa on Trade in Spirits [2002]

initial resentment at the re-monopolisation of formerly generic names,[157] Taubman identifies such a 'fixed outcomes' approach, as opposed to a 'fixed rules' one, to be more in keeping with the general tenor of international trade negotiations. While consistency with established principles is sacrificed (a harmonised, rule-based approach to determining generic status), it could lead to improvised, pragmatic and equitable outcomes in individual cases.[158] The extent of the reclamation is also varied, as evinced by Art 7.6 of the EU-Canada CETA.[159]

- The use of certain terms in a semi-generic manner is permitted on an ongoing basis, provided the true place of origin is also indicated.
- Some terms are no longer prospectively generic, but existing producers using them generically are 'grandfathered', provided they have been doing so: (i) prior to a designated date (in this case, 18 Oct 2013); and (ii) for a sufficiently long qualifying period (in this case, either five or ten years depending on the term in question).
- For those producers who have been using the term generically but do not qualify under the grandfathering provisions, a phase out or transition period is envisaged.

5.2.2 Reinventing a narrower GI

We have seen above that French courts declared Camembert to be generic in 1926.[160] Nevertheless 'Camembert de Normandie' is a protected designation in France as well as the EU today.[161] The revised product specification attempts to explain why the narrower PDO deserves recognition, despite 'camembert' being generic:

> It became truly well-known at the start of the 20th century and during the First World War, when cheese-makers endeavoured to provide camemberts to the French army and meet growing national demand. Although, as result of this demand, the number of products of diverse origin described as 'camembert' increased considerably, only the 'Camembert de Normandie' has finally been

---

OJ L28/113; Agreement between the European Community and the United States of America on Trade in Wine [2006] OJ L87/2. See also the references in Handler and Mercurio (n 36).

[157] Andries van der Merwe, 'Geographical Indication Protection in South Africa with Particular Reference to Wines and to the EU' (2009) 10(1) Estey Centre Journal of International Law and Trade Policy 186.
[158] See Chapter 9 by Taubman in this volume.
[159] See (n 42).
[160] *Syndicat du Véritable Camembert de Normandie* (n 132).
[161] Dossier No. FR/PDO/0017/0112), registered on 21 June 1996.

awarded recognition in France with a protected designation of origin, attesting to its links with Normandy and with technical methods rooted in tradition: raw milk, split moulding, and being left to drain naturally.[162]

West Country Farmhouse Cheddar represents a similar attempt at obtaining a narrower GI,[163] as does Noord-Hollandse Edammer.[164] However, it is evident that the narrower registration in such cases cannot be used to prevent the use of the broader generic term. Therefore, for Camembert de Normandie, any claims to exclusivity over camembert *per se* are disclaimed.[165]

5.2.3 Winning hearts and minds (again)

The final possibility is to reverse public perception and re-establish origin-specific salience for a generic term. For obvious reasons, this is the most difficult option. The starting point is the premise that 'the problem of generic nature is viewed in dynamic perspective'.[166] It has been pointed out that, with sufficient effort and over time, formerly generic terms may become 're-localised'.[167] This seems to have occurred in the case of Bayerisches Bier:

> As regards, more specifically, the argument . . . that the indication at issue was used historically to designate – since the 19th century – a particular method of production, based on bottom fermentation, originating in Bavaria but [which] has since spread throughout Europe, it should be noted . . . that even if that may have been the case, just as a name originally linking products to a certain region may become generic over time, it is possible for formerly generic terms to be used again in the sense of a geographical indication of a product, as was, according to the Commission and the Council, the case with 'Bayerisches Bier' after 1940.[168]

Champagne may be well on its way to finally achieving this in the US market. Professor Hughes refers to previous advertising campaigns

---

[162] Publication of an Amendment Application Pursuant to Article 50(2)(a) of Regulation (EU) No 1151/2012 of the European Parliament and of the Council on Quality Schemes for Agricultural Products and Foodstuffs [2013] OJ C140/20.
[163] Dossier No. UK/PDO/0017/0279.
[164] Dossier No. NL/PDO/0017/0315.
[165] See FN7 in the Annex to Council Regulation 1107/96 [1996] OJ L148/1 ('Protection of the name "Camembert" is not sought.').
[166] *Feta 1999* (n 74) at [AG10].
[167] Audier (n 56) 39; Lindquist (n 21) 334–335; G Schricker, 'Protection of Indications of Source, Appellations of Origin and other Geographic Designations in the Federal Republic of Germany' [1983] IIC 307, 316.
[168] *Bavaria v Bayerischer Brauerbund* (n 31) at [AG117].

'launched in English-language publications to convince high-end consumers that "Champagne" designates only sparkling wines from the Champagne district'.[169] The campaign continues with an online presence, which announces that Champagne sales have been increasing since the economic downturn in 2009, with over 17.8 million bottles being shipped from France to the US in 2013.[170] Meanwhile, Art 6 of the EU-US Wine Agreement is intended to prospectively reduce the generic use of Champagne, although existing users are grandfathered.[171] More intriguingly, the Comité Interprofessionel du Vin de Champagne (CIVC) recently asserted common law (that is non-registration based) certification trade marks rights to oppose the registration of Champarty as a trade mark in the US.[172] While the TTAB rejected the opposition on the basis that the marks were so different that consumer confusion was unlikely, it did not rule out the possibility that rights to such a certification mark existed.

6. CONCLUSION

This chapter provides a legal framework for deciding when a GI has crossed over the line into generic usage. Genericide is a process and one that is highly controversial. Since Art 24.6 has emerged as the international reference point, it is in need of further elaboration. The question of whether a designation has become the 'common name' for a type of product must receive an empirically informed answer. However, when identifying and evaluating the categories of evidence, there is space for normative inflections to be introduced. Therefore Section 3 set out the four main structural issues to be addressed (which regime to opt for; who bears the burden of proof; what is the threshold or tipping point for genericide; and how the factors should interact). Section 4 proceeded to analyse the categories of evidence, drawing on comparative experiences with such categories in operation. Here the perception of the target audience (consumers or

---

[169] Hughes (n 21) 377–380. (Hughes also draws parallels with the successful reclamation of the trade marks Singer for sewing machines and Goodyear for rubber products in the US.)

[170] See 'Champagne Shipments in US'; Available at: http://www.champagne.us/

[171] See B Rose, 'No More Whining about Geographical Indications: Assessing the 2005 Agreement between the United States and the European Community on the Trade in Wine' (2007) 29 Houston Journal of International Law 731, 760–761.

[172] *Comité Interprofessionel du Vin de Champagne and Institut National de l'Origine et de la Qualité v Shlomo David Jehonadav*, Opposition No. 91195709, 8 March 2013, TTAB (Unreported).

the general public, as informed by trade and expert opinion) must be gauged against the contextual backdrop of market conditions, legislative or bureaucratic classifications of the term and the actions of the rights holder. Section 5 concluded with a review of the options for preventing generic use in *sui generis* GI regimes, while also suggesting the avenues for reviving terms that were formerly GIs but have been declared generic. The comparative analysis is offered as a practical resource for decision makers seeking to make such determinations, but it also serves as a reminder that the overarching enquiry should remain focused on consumer (or general public) perception of the term. Traders often speak in the name of consumers and are better placed to provide information relating to certain categories of evidence. The test should ultimately seek to establish how the audience actually perceives a designation, rather than how traders on either side of the divide would wish them to perceive it.

Index

Abélès, Marc 90
Abricot Luizet du Valais, Switzerland 449
Addor, F 125, 158, 184, 220, 234, 293, 325
administrative issues
　approval procedure for product labels, WIPO-administered treaties 108–9, 113–14, 115
　China, GI protection 336, 350, 355, 357–8
　TRIPS Agreement, protection of GIs 139–40, 143
　see also registration
Adolphe, Cindy 85
Africa, *terroir* definition 77–8
Agdomar, M 168
Agier, Michel 91
Agneau du Quercy (sheep meat), France 449
agriculture
　diversity in agricultural production, call for 172–3
　environmental conditions, traditional practice and sense of place 81, 85
　farmers' markets 89–90
　farming and land ownership schemes as obstacles 81–2
　handicraft goods and foodstuffs 311–14, 321–2
　income and support to local employment, effects on 453–5
Akerlof, George 443, 482
Alaska mineral water 385
Allaire, G 306
Allen, Max 484
Alphandéry, Pierre 90
Alsace geranium, France 310
'ambush marketing' legislation 171
　see also legislation; markets; TRIPS Agreement

An, Qinghu 329, 355
Anderson, K 251
Anheuser-Busch (AB) company (US) 397–8, 401, 402–3, 405–6, 412, 425, 431, 437, 438
　see also Budweiser cases and GIs v trade marks
Anker, K 501, 502, 503, 504, 506
Anthony, Dan 307
Appadurai, Arjun 91
appeal process, Australia, wine GI legislation 272, 274–7
Appellation d'Origine Contrôlée (AOC) labelling
　early collective wine branding *see under* France, collective wine branding (19th–20th centuries)
　and *terroir* concept *see* global labelling challenge, AOC labelling and *terroir* concept
Appellations of Origin
　Lisbon Agreement *see* WIPO-administered treaties, Lisbon Agreement for the Protection of Appellations of Origin
　terroir and sense of place 79–80
　WIPO definition 100–102
Aranmula Metal Mirror, India 316
Armagnac, France 17, 27
Audier, J 297, 511, 521, 522, 533, 542, 546
Augé, Marc 90
Austin, G 163
Australia
　EU bilateral trade agreement 178–9, 366
　generic product descriptions, use of 154–5, 170
　market considerations 153–4
　Victorian Champagne Company 154–5
　wine recognition 42

549

Australia, cases
   Anheuser-Busch v Budejovicky Budvar 397, 400, 401
   Beringer Blass Wine Estates v Geographical Indications Committee 277
   Comité Interprofessionel du Vin de Champagne (CIVC) v NL Burton (Freixenet) 252–6
   Comité Interprofessionnel des Vins de Côtes de Provence v Stuart Alexander Bryce (La Provence wine) 264–6, 288
   Comité Interprofessionnel du Vin de Champagne v Powell 178
   Mabo v Queensland (No 2) 504, 505
   Thomson v B Seppelt and Great Western trade mark 247–9, 281–3
   Yarra Valley Dairy v Lemnos Foods 155
   Yorta Yorta Aboriginal Community v Victoria 501
Australia, Indigenous knowledge and GIs of origin 484–507
   botanical names and GIs, differences between 498–9
   botanical taxonomy issues 494–8
   connection to 'place' or 'country', importance of, and notion of *terroir* 491–4, 499–504, 506
   evidence, transportability of 502–3, 504–6
   GI of origin definition 486–7, 488
   group rights 490
   Indigenous, criteria associated with 486
   and Indigenous law 490–91
   legal mechanism to support and promote indigenous interests 492, 501–7
   native title, formal recognition 504–6
   natural geographical indications of origin 487
   outsider status of GIs 488–90
   paintings as communication method 502–3, 505–6
   protection period, lack of limitation 490
   quality and authenticity issues 490
   'traditional knowledge' definition 485–6, 488, 489–90
Australia, wine GI legislation 245–91
   appeal process 272, 274–7
   application fee as registration deterrent 289
   Australian Grape & Wine Authority (AGWA) 251, 270, 286–7, 289
   Australian Wine and Brandy Corporation (AWBC) Act 251–2, 260, 263–73, 275–6, 280, 282–3, 286–8
   Australian Wine and Brandy Corporation (AWBC) Act, trade mark v GI disputes 278–9
   Beaujolais litigation 256–7, 258, 261, 262
   Champagne, meaning to Australian purchasers 255–6
   Competition and Consumer Law 252, 254
   Coonawarra litigation 207, 249–50, 268–9, 273–7
   EU GI, use in non-infringing manner 280–81
   EU-Australia Wine Treaty 249, 258–63, 264–5, 286–9, 517
   European protection attempts 252–8
   export growth 260
   fee as registration deterrent, Australia 289
   'FEET FIRST' trade mark and use of German GI 'First' 280–81
   'first in time, first in right' approach 279
   foreign GIs, protection for 264–6
   free trade agreements 246–7
   generic names, lack of protection for 253–4, 258, 259–61
   generic names, phase-out tranches 261–2
   GI criteria problems 268–9
   GI definition 266–9
   GI protection, narrowing of (2010) 287–9
   GIC (Geographical Indications Committee) and geographical boundaries 271–2, 273–7, 278–9, 284–5
   Grape and Wine Authority Act 245

Great Western trade mark 247–9, 281–3
history 245–63
injunctions 269
Label Integrity Program 252
misleading descriptions 263–4
misuse of GI 256–8, 269–71, 280–83, 287–8
regional hierarchy 269, 275–6
Register of Geographical Indications 272–3, 289
registered GIs, protection for 263–4, 272–3, 289
Registrar of Trade Marks 278–9, 283, 284–5
rights of interested persons 269–71
Rothbury trade mark dispute 284–5
Spanish Champagne and generic problem 252–6
trade mark protection, earliest case 247–9
trade mark v GI disputes 278–9
Trade Marks Act 249, 280
Trade Practices Act (TPA) and passing-off action 252–6, 288
traditional expressions, protection for 262–3, 289
US–Australia Free Trade Agreement 278–85
value of wine GI 249–50
Wine Australia Corporation Act 267–8, 269, 286–7
wine production history 251
Winemakers Federation of Australia (WFA) 260
Ayu, M 245

Babcock, B 161
Baeumer, Ludwig 365, 375
Bagal, M 49
Baham, Elizabeth 89
Bailey, Robert 68
Balganesh, S 157
Barham, Elizabeth 20, 46–71, 87, 89, 126, 134, 135, 200, 204, 207, 292, 297, 484, 488, 491, 493, 494, 499
Barjolle, Dominique 53, 88, 238, 239, 240, 440–62, 466, 479
Barnes, D 169
Bashaw, B 339, 345, 347, 349, 357

Basmati rice, India 213–14, 219–20, 536–7
Bassett, Thomas 78
Bäumer, L 430–31
Bavarian Bier, Germany 380, 381, 388–9, 515–16, 523, 524, 546
Bavarian Blockmalz candy, Germany 522–3
Beaujolais litigation, Australia 256–7, 258, 261, 262
Beebe, Barton 168, 171, 182, 526
beer
    Bavarian Bier, Germany 380, 381, 388–9, 515–16, 523, 524, 546
    Budweiser *see* Budweiser cases and GIs v trade marks
    Newcastle Brown Ale, UK 478
    Plzen, Czech Republic 422, 543
    Sapporo, Japan 362
    see also wines and spirits
Beeston, J 251
Beier, F 430
Beiter, K 403
Belletti, Giovanni 458
Benson, R 540
Bently, L 217
Bérard, Laurence 47, 59, 61, 67, 69, 70, 72–91, 135, 200, 207, 301, 318, 319
Beresford, Lynne 328
Bergamote de Nancy sweets, France 309
Bergeron, L 362
Bertrand, George 77
Bessière, J 65
Bhagwati, Jagdish 204, 237
Biagioli, Mario 489
Biénabe, E 311
Bienaymé, M 313
bilateral trade agreements 178–82, 234–7, 246–7, 356, 402–4, 409, 415–17, 435
    see also trade
Bingen, Jim 89
Blakeney, M 175, 183, 325, 508
blending ('mélanges') practice, France 21, 23, 24, 36
Blowen, S 59
Blue de Bresse cheese, France 478
Bodenhausen, G 99, 100, 110, 232

Boeren Leidse met Sleutels cheese,
    Netherlands 449, 452, 454, 455,
    457
Boisard, P 539
Boisvert, Valérie 85
Boltanski, L 54, 55, 459
Bonnet, C 161, 480
Bons Bois spirit, France 115
Bordeaux wine 17, 18, 19–20, 21, 22,
    23, 26, 27, 31, 35, 38, 261
botanical taxonomy issues, Australia
    494–9
Bourgeon, J 16
Bowen, Sarah 47, 49, 60, 125, 126
Bramley, C 14–15, 159, 484, 486, 489
branding
    advantages, India, Feni liqueur 480,
        482, 483
    'brand entrepreneurs' and incentive
        preservation concerns 165
    see also labelling
Bresse poultry production, France 81
Brie cheese, France 515
Brody, P 134, 151, 530
Broude, T 173, 220
Brunet, Roger 75–6
Bruwer, J 528
Budweiser beer, US 51–2
Budweiser cases and GIs v trade marks
    396–439
    Anheuser-Busch (AB) company
        (US) 397–8, 401, 402–3, 405–6,
        412, 425, 431, 437, 438
    bilateral treaties on GIs 402–4, 409,
        415–17, 435
    Budweiser Budvar (BB) brewery
        398–403, 405–6, 410–12, 439
    conflicting prior trade marks 406
    and Czech Republic Accession
        Treaty 409, 417–18
    defence strategies 408–12
    'descriptive use' defence 408
    dilution of indications in bilateral
        treaties 416
    'honest practices' issue 412
    lawsuits 400–404
    Madrid Agreement 414, 415, 421,
        426, 429
    misleading indications 412–14,
        416–17, 418
    Paris Convention and GIs 412–13
    registrability issue 405–8, 409,
        416–17
    right to use as defence strategy
        408–10
    trade mark and trade name issues
        401, 404–12
    translations of indications 399,
        410–11, 421
    Versailles Peace Treaty as
        multilateral agreement 415
    WTO Panel Report 408–9
Budweiser cases and GIs v trade
    marks, Lisbon Agreement for the
    Protection of Appellations of
    Origin 399–402, 406, 407, 409,
    410, 414, 418–39
    community law relationship 435–6
    generic appellations of origin 421–2,
        429
    history of 418–19
    implementation regulations 435
    indications eligible for protection
        425–6
    infringement procedures 429–30
    judicial review 429–35
    Lisbon Register 422–4
    national courts 424–35
    prior trade mark rights 420–21
    products eligible for protection 424
    protection scope 426
    registration procedure 419–21
    scope and contents 421–2
    specific Budweiser issues 422–36
    third party protection 427–9
Budweiser litigation
    Austria 403–4, 416, 417, 437
    Bulgaria 401, 430
    Denmark 411
    France 406
    Hungary 401, 426, 430, 438
    Israel 401, 430, 431–5, 438
    Italy 402, 406, 409, 422, 430, 431,
        437, 438
    Portugal 402–3, 422, 430–31, 438
    Spain 402, 431
    Switzerland 403
Burgundy wine, France 23, 28, 38, 208
Burrell, R 154, 160, 161, 165, 169
Buttel, F 54

Café Chiapas, Mexico 115
cahier de charges (certification requirements) 70
  see also global labelling challenge, AOC labelling and *terroir* concept
Calboli, I 162
Calisson d'Aix sweets, France 310
Camembert cheese, France 82, 480, 515, 539, 545–6
Campbell, H 70
Canada
  Delgamuukw v British Columbia 504
  EU bilateral trade agreement 179, 237
  Halloom cheese 533
Cantal cheese, France 449, 452–3, 454, 455, 457
Caribbean countries, EU Economic Partnership Agreement 180
Carmody, C 225
Carrier, J 56
Carter, T 236
Carty, H 165
Casabianca, F 69, 80
CCFN (Consortium for Common Food Names) 517–18
ceramics
  Moustiers earthenware, France 307
  Porcelaine de Limoges and Porcelaine de Nevers, France 307
  Poterie de Valauris, France 314
certification, *cahier de charges* (certification requirements) 70
certification marks
  China 328–9, 330, 333–4, 335–6, 347–50, 353, 354, 355, 356–8
  TRIPS Agreement, protection of GIs 136, 139
  WIPO-administered treaties 107–8
  see also collective marks
Chablis wine 43, 134, 537
  Chablis with a Twist, US 207, 211–12, 233
Chamberlin, E 442
Champagne, France 134, 546–7
  advertising campaign, US 546–7
  collective wine branding (19th–20th centuries) 17, 20, 23, 24, 27, 29
  meaning to Australian purchasers 255–6
  Spanish Champagne 160, 252–6, 513–14
  Victorian Champagne Company, Australia 154–5
Chandola, H 216
Chappuis, J 445
Chardonnay wine 134, 500
Chartier, Roger 75
Châtaigne d'Ardèche (chestnuts), France 311–12
Cheddar cheese, UK 449, 452, 454, 455, 457, 515, 538, 546
cheese
  Blue de Bresse, France 478
  Boeren Leidse met Sleutels, Netherlands 449, 452, 454, 455, 457
  Brie, France 515
  Camembert, France 82, 480, 515, 539, 545–6
  Cantal, France 449, 452–3, 454, 455, 457
  Cheddar, UK 449, 452, 454, 455, 457, 515, 538, 546
  Chevrotin, France 516
  Comté, France 449, 452–5, 457
  Edamer, Netherlands 449, 452, 453, 454, 455, 457, 515, 546
  Emmental, Switzerland 515
  Feta *see* Feta cheese
  Fontina, Italy 449, 452–5, 457, 529
  Gorgonzola, Italy 387–8
  Gouda, Netherlands 515
  Gruyère, Switzerland 449, 452, 453–5, 457
  Halloumi, Cyprus 155, 531, 533
  kanterkaas, Netherlands 147, 161
  L'Etivaz, Switzerland 445
  Parmesan, Italy 112, 138, 449, 452, 453–5, 457, 515–16, 523, 538
  Reblochon, France 115
  Roquefort, France 112, 235, 313, 523, 536, 544
  Stilton, UK 113, 531
Chen, J 153
Chen, Shiping 357
Chevrotin cheese, France 516
Chiappetta, V 165

Chiemsee trade mark, Germany 384–5, 405–6, 411
Chile
 EU bilateral trade agreement 179
 terroir use 86
 wine recognition 42
China
 Great China Wall trade mark 385
 Guanxi Sweet Shaddock (pomelo) trade mark 356, 497
 Hong Kong, China Proposal, WTO multilateral register negotiations 187, 189–90, 199–200
 Jinhua Ham 347, 348–9, 351–2
 Shanxi Laochencu vinegar 353
 Xianglian lotus seeds 350
China, GI protection 327–58
 administrative appeal and judicial review 350, 357–8
 administrative forces for industry and commerce (AICs) 336, 355, 357
 Anti-Unfair Competition Law 345
 bilateral trade agreements, scarcity of 356
 cancellation of ordinary trade marks containing GIs, call for 349
 certification and collective marks 328–9, 330, 333–4, 335–6, 347–50, 353, 354, 355, 356–8
 challenges 346–52
 coexistence as conflict solution 349
 Consumer Rights and Interests Protection Law 345
 control and supervision of GI use 335, 340, 344
 cost factors 353
 enforcement 335–7, 340, 344, 354–5
 exclusive rights, obtaining and enforcement 354–5
 fair use exemption of geographical name under trade mark system 334, 352
 'first in time/ first in right' principle 348, 350
 foreign applicants 332, 340–41, 344, 357
 French Cognac application 341
 General Administration of Quality Supervision, Inspection and Quarantine (AQSIQ) 336–41, 346, 351–2, 354, 355–6, 357
 GI definition in Trade Mark Law 331
 history of 327–8
 infringement cases 336, 340, 355
 international protection 356–7
 legal considerations 329–30, 355–8
 Lisbon Agreement 356
 Madrid international registration system 356–7
 Ministry of Agriculture (MOA) supervision 336–7, 341–4, 346, 351, 354, 355–6, 357
 misleading trade marks 347, 350, 354, 355
 producer application to use a GI product 339, 343
 Product Quality Law 346
 Provisions on the Protection of Geographical Indication Products (PPGIP), scope of GI products 337–8, 340, 355–6, 357
 registration numbers 336
 registration requirements 331–2, 338–9, 342–3, 347
 State Administration for Industry and Commerce (SAIC) 327, 329, 330, 331, 332–3, 334, 346, 355
 sui generis regime and trade mark regime comparison 351–2, 353–8
 sui generis systems 336–44, 346–7
 Trade Mark Law 113, 328, 329–31, 333, 334, 335–6, 347–50, 351–2, 355–6, 357, 497
 Trade Mark Law, GIs included in 329–30
 trade mark law ranks higher than AQSIQ and MOA administrative regimes 355–6
 trade mark regime, conflict between GIs and trade marks and proposed solutions 347–50
 Trade Mark Review and Adjudication Board (TRAB) 350, 357
 TRIPS compliance 332, 356, 357

wines and spirits, special provisions 332–3
Zhangqiu Scallion trade mark 356
Zhejiang Food v Shanghai Taikang Food 351
Cholet textiles, France 316
Christiansen, A 239
Clarkin, M 154, 155
classification
  Ethiopia, coffee-growing 85–6
  grands crus 38–9
  India, Feni liqueur, reclassification plans 471–2
  international standards for product classifications, generic status and genericide 537–8
  term status in official classifications as evidence 535–8
  see also geographical indications; trade marks
co-existence provisions 366, 368–70, 371–2, 381–2, 387, 391–3
  China 349
  see also conflicts between trade marks and GIs, resolving
Coase, R 445
Cocks, Sam 361, 383
Coelho, A 41
Coestier, B 16
coffee
  Café Chiapas, Mexico 115
  classification and grading system, Ethiopia 85–6
  Monsooned Malabar, India 313–14
  see also teas
Cognac, France 27, 341
collective dimension of GIs, handicrafts and link to origin in culture 318–19
collective marks
  EU 112–13
  TRIPS Agreement 136, 139
  WIPO-administered treaties 107, 108, 113
  see also certification marks
collective organisations, and coordination of production and sales in rural development 445–6, 447, 449–51, 456, 458, 460–61

collective wine branding *see* France, collective wine branding (19th–20th centuries)
Colombia
  Denomination of Origin 325
  woven handicrafts 307, 315
colonial legacy, India 471–2
commercial performances directly linked to organisational performance 448–58
  see also intellectual property tools for rural development objectives
competition *see* unfair competition
Comté cheese, France 449, 452–5, 457
conflicts between trade marks and GIs, resolving 361–95
  Alaska mineral waters trade mark 385
  appropriate reference date for priority and protection of GIs prior to their registration 380–81
  bad faith question 387–8
  Bavaria v Bayerisches Bier case 380, 381, 388–9, 515–16, 523, 546
  Chiemsee trade mark 384–5
  co-existence provisions 366, 368–70, 371–2, 381–2, 387, 391–3
  EU Regulation No 1151/2012 on quality schemes for agricultural products and foodstuffs 379–82
  EU Wine Regulation 366, 377–9
  European Court of Human Rights (ECHR), *Anheuser-Busch v Portugal* (Budweiser trade mark) 367, 372–4, 382, 390–91, 392
  exclusivity to prior well-known marks 366, 368–71, 376, 378
  fair use of descriptive terms 369, 370, 392–3
  Fuerteventura trade mark 385
  future direction 393–5
  Geneva Act 395
  geographically descriptive names as trade marks 384–5
  GI protection systems 364–5
  Gorgonzola/Cambozola case 387–8
  Great China Wall trade mark 385

internal trade mark law defense 370–71
Kerry Spring mineral water 392, 411
Lisbon Agreement on appellations of origin 365–6, 374–7, 394–5
misleading trade marks 371, 383–5, 386, 389–90
monopoly avoidance 382–93
multilateral register proposal for GIs 393–4
national level protection 380–82, 387–8, 390–91, 394
principles 363–7
priority, exclusivity and territoriality (PET) protection 363–4, 366–7, 371–82, 386–8
protection of GIs against use in translation 372–3
refusal of an international registration 375–6
refusal of trade mark application because of GI 380
refusal of trade mark because of conflict with EU Community GI 386–90
refusal of trade mark because of conflict with national GI 390–91
refusal of trade mark as geographically descriptive 383–5
registered trade marks as 'possessions' 373–4
reputation of trade mark 371, 378–9, 386, 387, 389–90
Sapporo beer 362
sui generis protection systems 365, 366–7, 369
superiority-based rule 365–6, 367, 371–2
third-party use and phase-out period 365, 366, 374–5
Torres Vedras wine 378–9
trade mark application an object of private property 372–4
TRIPS Agreement effects 133, 367, 369–70, 371–2, 375, 377, 378, 379, 381–2, 383, 387
WTO Doha Round negotiations 393–4

WTO Panel Report 367–72, 379, 387, 394, 408–9
Conlinsk, J 16
connection to 'place' or 'country', importance of 491–4, 499–504, 506
see also terroir
Conrad, Albrecht 131, 544
Consortium for Common Food Names (CCFN) 517–18
consumer protection 31–2, 33, 159–63
  China Consumer Rights and Interests Protection Law 345
  consumer deception test 223–5
  consumer understanding effects, generic status 512, 524–5, 527–31, 546–7
Coombe, Rosemary 489
Coonawarra wine litigation, Australia 207, 249–50, 268–9, 273–7
Coquet, L 26
Cornish, W 147
Correa, C 175
Corte-Real, Antonio 397, 431
cost factors
  application fee as registration deterrent, Australia 289
  China 353
  products differentiated by geographical origin, costs and benefits comparison 444–6
  TRIPS Agreement, GI extension rethink and EU policies 158, 161–2
  WTO multilateral register negotiations 197–8
  see also economic factors
counterfeits, marginal counterfeits, emergence of 452–3
  see also intellectual property tools for rural development objectives
Covarrubia, P 180
Craven, E 181
Critz, J 42
Croce, E 65
culture, and handicrafts *see* handicrafts and link to origin in culture
customers *see* consumer protection
Cyprus, Halloumi cheese 155, 531, 533

Czech Republic
  Budweiser *see* Budweiser cases and GIs v trade marks
  Plzen beer 422, 543

Daes, E 490
Dagne, Teshager Worku 484
Damer, S 511
Darby, M 443
Darjeeling tea, India 312, 523, 528, 540
Das, K 125, 221
Davis, J 217
De Sainte Marie, C 69
De Saussure, F 210
De Zwart, M 155
Dedeire, Marc 78
Deffontaines, Jean-Pierre 79
definitions
  'appellation of origin', WIPO-administered treaties 100–102
  French *terroir* definition debate 76–8, 79–80, 84–5
  GI definition, Australia 266–9
  GI definition in Trade Mark Law, China 331
  GI definition, TRIPS 46–7, 48, 188, 190, 191, 193, 196–8
  GI definition, WIPO-administered treaties 102–4
  GI of origin definition, Australia 486–7, 488
  'identification' and 'indication' of a product, differences between (TRIPS definition) 208–10, 213, 223–4, 236–7
  'indication of source', WIPO-administered treaties 99–101
  'local traditions', France 25
  'origin' of wine, uncertain legal definitions, France 21
  terroir, Africa 77–8
  'traditional knowledge', Australia 485–6, 488, 489–90
Delfosse, Claire 47, 59, 76
DeLind, Laura 89
Demaret, P 146
Demossier, Marion 60, 79
Dentelle du Puy (lace), France 304
Desai, A 472

'descriptive use' defence 408
  see also Budweiser cases and GIs v trade marks
developing countries
  and domestic protection 218–20, 221–3
  production systems 85–6
  WTO multilateral register negotiations 195–6
  WTO TRIPS Agreement, protection of GIs 124–5, 140, 174–7
  see also individual countries
Devlétian, A 430
Dewald, B 528
Diamond, Shari 528
Dibden, J 154
differentiation, products
  differentiated by geographical origin 444–6
dilution of famous trade marks 43–4, 108, 163–8
  see also trade marks
dilution of indications in bilateral treaties 416
Dion, Roger 77
discrimination on basis of subject matter categories, lack of justification for 316–17
  see also handicrafts and link to origin in culture
diversification practices and artisanal product 453, 455
diversity in agricultural production, call for 172–3
Doha Round *see* WTO Doha Round
Dombes carp farming, France 81–2
Downes, David 484, 489, 490, 491
Drahos, Peter 215
Duby, G 23

Echols, M 537
ECHR (European Court of Human Rights), *Anheuser-Busch v Portugal* (Budweiser trade mark) 367, 372–4, 382, 390–91, 392
eco-labels 53
economic factors
  collective wine branding, France (19th–20th centuries) 23, 36–7

economic advantages for producers in international markets, securing 152–3, 176
economic performance, costs and benefits comparison 444–5
traditional economic arguments, justification through 159–63
see also cost factors; value
Economides, N 159
Edamer cheese, Netherlands 449, 452, 453, 454, 455, 457, 515, 546
Edmond, G 273
Emmental cheese, Switzerland 515
Enneking, U 481
entry deterrence and market concentration, India, Feni liqueur 478–9
environmental factors
  agriculture, environmental conditions and traditional practice 81, 85
  eco-labels 53
  environmental significance of GIs 126
  handicrafts, and processing under special environmental factors 313–14
  sustainability 66–8, 136, 173
Escudero, S 430, 486
Ethiopia, coffee-growing classification and grading system 85–6
EU
  Alaska mineral water 385
  Australia bilateral trade agreement 178–9, 366
  Australia, wine GI legislation protection attempts 252–8
  bilateral trade agreement effects 516–18, 544–5
  Budweiser cases *see* Budweiser cases and GIs v trade marks
  Canada bilateral trade agreement 179, 237
  Chile bilateral trade agreement 179
  collective, guarantee or certification marks, use of 112–13
  commercial performances directly linked to organisational performance 448–58
  Community Trade Mark Regulation (CTMR) 113, 384, 387
  Economic Partnership Agreement, Caribbean countries 180
  EU-Australia Wine Treaty 249, 258–63, 264–5, 286–9, 517
  EU-Korea Free Trade Agreement 180, 517–18
  exclusivity to prior well-known marks 366
  French rural policy, influence on EU policies 152–4
  GI, Australian use in non-infringing manner 280–81
  GI extension *see* WTO TRIPS Agreement, GI extension rethink and EU policies
  Great China Wall trade mark 385
  handicrafts and link to origin in culture 293–5, 322–5
  labels of origin, administration of 52–3
  legislation on geographical indications protection, French influence on 41, 44
  list of generic terms under Regulation 2081/92.24 (agricultural products and foodstuffs) 514–16
  place names, pre-existing European place names, use of, global labelling challenge 50–51
  Protected Designation of Origin (PDO) 83–4, 299–300, 315, 324–5, 448–58
  Protected Geographical Indication (PGI) 83–4, 300–301, 307, 308, 309–10, 324–5, 448–58
  refusal of trade mark because of conflict with a Community GI 386–90
  Regulation 1151/2012 on Quality Schemes for Agricultural Products and Foodstuffs 83–4, 379–82
  sui generis protection systems 112–13
  Switzerland bilateral trade agreement 179
  Trade Marks Directive 169, 368, 383, 388, 390, 391–2, 408, 410, 411

Transatlantic Trade and Investment
    Partnership (TTIP) 520
  US bilateral trade agreement 179,
    180–81, 236, 516–17, 518
  'Wine Pact' with US 134–5, 517, 547
  Wine Regulation 366, 377–9
  and WTO multilateral register
    negotiations 187–8, 190–92,
    194, 195, 197–8, 199–200, 201
  see also individual countries
EU, cases
  Anheuser-Busch v Budějovický
    Budvar 363
  Anheuser-Busch v Rudolf Ammersin
    379, 397
  Bavaria v Bayerisches Bier 380, 381,
    388–9, 409, 515–16, 524
  Budejovicky Budvar narodni podnik
    v OHIM 211, 390, 400, 407–8,
    436
  Budejovicky Budvar v Rudolf
    Ammersin 379, 397, 403, 404,
    409, 416, 436
  Bureau national interprofessionnel du
    Cognac v Gust Ranin Oy 386
  Canadane Cheese Trading v Hellenic
    Republic 155, 208, 510
  Carl Kühne v Jütro Konservenfabrik
    301
  Chiemsee 384–5, 405–6, 411
  Commission v Germany (on
    'Parmesan') 147, 534–5
  Consorzio del Prosciutto di Parma v
    Asda Stores Ltd (Parma ham)
    229–30
  Diageo v Intercontinental Brands 253
  Exportur SA v LOR SA and
    Confiserie du Tech 300
  Germany and Denmark v Commission
    (on 'Feta') 147, 207–8, 223, 381,
    494, 510–11, 524–9, 534, 535–6,
    538–9, 540, 546
  Gerri/Kerry 392
  Gorgonzola/Cambozola 387–8
  Grana Padano v OHIM 529, 533
  Koninklijke Philips Electronics v
    Lucheng Meijing Industrial
    Company 363
  Kornspitz 532
  L'Oréal v Bellure 169

  Nokia Corporation v HMRC 363
  European Court of Human Rights
    (ECHR), Anheuser-Busch v
    Portugal (Budweiser trade mark)
    367, 372–4, 382, 390–91, 392
  Evans, G 175, 301, 325, 508
  Evans, L 251
  evidence
    contested expression is generic
      522–3, 525–40
    transportability of, Australia,
      Indigenous knowledge 502–3,
      504–6
  exclusive rights, obtaining and
    enforcement, China, GI protection
    354–5
  exclusivity to prior well-known marks
    366, 368–71, 376, 378
    see also prior trade mark rights
  expert opinion as evidence 532–3
    see also generic status and genericide
  export growth
    Australian wine 260
    French AOC export revenue rise 65

Fabre, Daniel 73
fabrics
  Cholet textiles, France 316
  Dentelle du Puy (lace), France 304
  Harris Tweed certification, UK 307
  Kancheepuram Silk, India 304, 306,
    317–18
  Kashmir Pashmina, India 315
  Mysore Silk, India 306–7
  Pipli Appliqué Work, India 318
  Pochampally Ikat and Orissa Ikat
    (tie-dye), India 305, 320
  Shetland Wool, UK 315
  woven handicrafts, Colombia 307,
    315
'fair trade' rules 202–42
  bilateral trade negotiations 234–7
  consumer deception test 223–5
  controversies over legitimate use of
    GIs, long-running 217–18
  developing countries and domestic
    protection 218–20, 221–3
  Doha Round of trade negotiations
    215
  GATT on origin marking 227–8, 236

GATT 'trade-in-goods' paradigm 219
generic and semi-generic terms 211, 212–13, 222, 223, 236
'identification' and 'indication' of a product, differences between (TRIPS definition) 208–10, 213, 223–4, 236–7
identity preservation 218–20
international trade negotiation methodologies 204–6
IP protection as policy objective 225–7
judicial analysis of function of trade marks and GIs 211–13
language use and geographical terms 207–10, 213–16, 221–2, 232–3
legal history of GIs 216–18
legitimacy negotiations 217–18, 224, 225–7, 228, 230–31, 232–4, 236, 237–41
legitimacy negotiations, case-by-case basis (rule of reason) and *per se* choice 238–9, 240–41
Lisbon Agreement 232–3
locality and global trade 206–8
Madrid agreement on false indications of origin 232–3
misleading use of GIs, protection against 223–4, 229–33, 236
Paris Convention legacy 216, 218, 224–5, 230–32
policy objectives of GI protection 216–17
public domain intrusions 203–4, 211–12
traditional production 203, 219
TRIPS, collective management of 221–5
TRIPS and extension to products other than wines and spirits 223–4
TRIPS 'fix-rule' approach 240–41
TRIPS 'IP-is-trade' paradigm 219, 240
TRIPS and language usage 213–16, 221–2
TRIPS non-compliance 225–6, 227–8, 240

TRIPS standards on GI protection 208–10, 213, 215–16, 218, 233–4, 236, 239–40
WTO Uruguay Round trade negotiations 128–9, 148–9, 156, 183–6, 218
fair use of descriptive terms 138, 352, 369, 370, 392–3
false indications
 Madrid agreement 232–3
 see also misleading indications
false labelling prevention, France, collective wine branding 26
farming *see* agriculture
Farrer, K 155
Faure, E 430
Faure, M 60
'FEET FIRST' trade mark, Australia 280–81
Feni liqueur, India *see* India, Feni liqueur and welfare enhancement
Ferguson, James 91
Ferrari, Matteo 494
Feta cheese 449, 452, 453–5, 457, 515, 544
 Germany and Denmark v Commission 147, 207–8, 223, 381, 494, 510–11, 524–9, 534, 535–6, 538–9, 540, 546
Ficsor, Mihály 375–6, 543
Fine, Ben 47
'first in time, first in right' approach 128, 136–9, 279, 348, 350
FitzSimmons, M 47–8, 61
Folsom, R 530
Fonte, M 47
Fontina cheese, Italy 449, 452–5, 457, 529
foreign GIs, protection for
 Australia 264–6
 China 332, 340–41, 344, 357
 see also international trade effects
Fowler, P 157
France
 Agneau du Quercy (sheep meat) 449
 Alsace geranium 310
 AMAP (Association for the Preservation of Peasant Agriculture) 90

AOC labelling administrative process
    see under global labelling
    challenge, AOC labelling and
    terroir concept
Armagnac 17, 27
Blue de Bresse cheese 478
Bons Bois spirit 115
Bordeaux wine 17, 18, 19–20, 21, 22,
    23, 26, 27, 31, 35, 38, 261
Bresse poultry production 81
Brie cheese 515
Burgundy wine 23, 28, 38, 208
Calisson d'Aix sweets 310
Camembert cheese 82, 480, 515, 539,
    545–6
Cantal cheese 449, 452–3, 454, 455,
    457
Chablis *see* Chablis
Champagne *see* Champagne
Chardonnay wine 134, 500
Châtaigne d'Ardèche (chestnuts)
    311–12
Chevrotin cheese 516
Cholet textiles 316
Cognac 27, 341
Comté cheese 449, 452–5, 457
Dentelle du Puy (lace) 304
Dombes carp farming 81–2
GI protection system 152–3
Gironde wine production 21–2, 23,
    24, 26, 27, 31, 35
GIs for non-agricultural products
    295, 297–8
handicrafts *see* handicrafts and link
    to origin in culture, France
Institut National des Appellations
    d'Origine et la Qualité
    (INAO) 63–5, 66, 67, 68, 135–6,
    321–2
*Institut National des Appellations
    d'Origine v Yves Saint Laurent*
    211
La Provence wine 264–6, 288
Lactalis dairy cooperative 82
Merville potatoes 449
Morteau sausage 314
Moustiers earthenware 307
Nyons olive oil 115, 312–13, 449
Pâtes d'Alsace 309–10
Pierre de Bourgogne 322

Porcelaine de Limoges and
    Porcelaine de Nevers 307
Poterie de Valauris 314
Reblochon cheese 115
Roquefort cheese 112, 235, 313, 523,
    536, 544
rural development benefits 65–6
rural policy and GI regulation,
    influence on EU policies 152–4
Saint Joseph wine 67–8
Saint-Émilion wine 22, 261
Sancerre wine 261
*Société Empresa del Tabaco
    Cubatabaco v Aramis* 211
sui generis system creation 105–6
terroir logic 1–2
terroir and sense of place *see terroir*
    and sense of place
France, collective wine branding
    (19th–20th centuries) 13–45
    Appellation d'Origine Contrôlée
        (AOC) labels 13–14, 25, 37–9,
        105–6, 135–6, 231, 268–9
    Appellation d'Origine Contrôlée
        (AOC) labels, Decree of 30, 52
    Appellation d'Origine Contrôlée
        (AOC) labels, 'fair' customs 40
    Appellation d'Origine Contrôlée
        (AOC) labels, production
        technique control 38
    appellation d'origine and definition
        of quality 32–5
    appellations and 'established local
        customs' 27
    atmospheric conditions, influence of
        31, 36
    collective appellations 39–41, 44
    consumer protection considerations
        31–2, 33
    contemporary trends 41–5
    economic information, circulation
        of 36–7
    European legislation on geographical
        indications protection, influence
        on 41, 44
    false labelling prevention 26
    fraud concerns 28–9, 32
    freedom of association law (1887) 21
    generic names and collective
        designations 15–17

geographical delimitation effects 27–30, 33
global registry, call for 48
grands crus classification 38–9
identification of place names with territory 38
institutional framework protection 40
intellectual property protection in 19th century France 15–39
judicial construction of wine market (1908–1914) 25–30
judicial interpretations on economic behaviours 22–3
'local traditions' definition 25
'mélanges' (blending) practice 21, 23, 24, 36
name indicating place of vintage 17–18
'origin' of wine, uncertain legal definitions 21
place of manufacture, understanding of 20
premier crus wine selling 'by subscription' to consortium of traders (1800s) 19
price drop, taxation, and productivity restrictions (1931) 37–8
product definitions, disagreement over 24–5
production and demand, and changing economic environment 23
public quality signs, controversy over 16
quality considerations 16, 21, 22, 28–34, 38, 44
regional boundaries of appellation regions, local commissions to fix 26–7
regional name protection 20, 22
state authorities' attitude 30–37
'their composition and their substantial qualities', removal from bill on appellations of origin 30–31
total area in vineyards with no right to an appellation 41–2
trade and competition regulation (1830s) 18–19
trade mark and the geographical denomination, shift of emphasis to 36
trade mark protection comparison 15–16
trade marks and brands law (1857) 20
US comparison 42–4
VDQS (*vin délimité de qualité supérieure*) status 42
wine traders and retailers, growth of 23–4
Frankel, S 489
Franklyn, David 169, 528
free trade agreements *see* bilateral trade agreements
free-riding concerns 158, 168–70, 197
fruit and vegetables
  Abricot Luizet du Valais, Switzerland 449
  Guanxi Sweet Shaddock (pomelo) trade mark, China 356, 497
  Jersey Potatoes, UK 449
  Merville potatoes, France 449
  Opperdoeze ronde potatoes, Netherlands 449
  Zagora apples, Greece 449
Fuerteventura trade mark 385
Furger, Alex 216
future directions
  conflicts between trade marks and GIs, resolving 393–5
  India, Feni liqueur and welfare enhancement 482–3
  WTO multilateral register negotiations 186–7, 195–200

Gangjee, Dev S 1–9, 15, 53, 57, 83, 130, 133, 139, 147, 152, 155, 162, 165, 169, 170–71, 172, 176, 192, 206, 207, 214, 220, 230, 254, 294, 295, 298, 309, 319, 348, 361, 363, 372, 413, 451, 464, 487, 508–48
Gasnier, Thierry 73, 74, 84
GATT *see* WTO GATT
Gaytán, M 47
Geiger, Christophe 143, 324, 326

generic status
  appellations of origin, Budweiser cases 421–2, 429
  'fair trade' rules 211, 212–13, 222, 223, 236
  Food and Drug Administration (FDA) and generic status, US 536–7
  France, collective wine branding (19th–20th centuries) 15–17
  generic names, phase-out tranches, Australia 261–2
  generic product descriptions, use of, Australia 154–5, 170
  'generic quality' of raw material 306
  genericide rules, TRIPS Agreement, protection of GIs 139
  names, lack of protection for, Australia 253–4, 258, 259–61
  product descriptions, TRIPS Agreement 154–5, 162–3, 165–8
  WTO multilateral register negotiations *see* WTO multilateral register negotiations and international protection of GIs
generic status and genericide 508–48
  Consortium for Common Food Names (CCFN) 517–18
  consumer understanding effects 512, 524–5, 527–31, 546–7
  controversy and commercial stakes 512–18
  EU bilateral trade agreements, effects of 516–18, 544–5
  EU list of generic terms under Art 3(3) of Regulation 2081/92.24 (agricultural products and foodstuffs) 514–16
  evidence that contested expression is generic 522–3, 525–40
  expert opinion as evidence 532–3
  'freezing' of meaning for protected terms 541–3
  generic terms as part of linguistic public domain 510–11, 519, 524–5, 527–31
  genericide as fatal process 509–11
  GI registration challenges 523
  GIs as indicators of territory 509
  informal or commercial usage, consumer understanding of 529–30
  international standards for product classifications 537–8
  international trade effects 512–13
  labelling factors 534–5, 536
  Lisbon Agreement 513, 541, 542–3
  Madrid Agreement 513, 541–2
  market conditions as evidence 533–5
  misappropriations and misleading uses, claims of 513
  narrower GI, reinvention of 545–6
  negotiated reclamation of generic terms 544–5
  Paris Convention 512
  protection reclamation possibilities 541–7
  reclaiming generic terms 544–7
  rights holders' action as evidence 538–40
  semi-generic terms *see* semi-generic terms
  sherry as generic term, UK 530–31, 539
  Stresa Agreement 515
  structural features 521–6
  sui generis principle protection regime 521–3, 525, 541
  term status in legislation or official classifications as evidence 535–8
  territoriality principle 521–2
  thresholds for generic status 523–5, 541
  trade mark protection regime 519, 521, 523–4, 529–30, 532, 540
  trade opinions as evidence 531–2
  Trans-Pacific Partnership Agreement (TPP) 520–21, 528–9, 538
  TRIPS Agreement Art. 24.6 508–9, 518–21, 526, 531–2
  US-EU Transatlantic Trade and Investment Partnership (TTIP) free trade agreement 520
Geneva Act *see under* WIPO-administered treaties, Lisbon Agreement
geographical area, relationship with, WIPO-administered treaties 111–13

geographical indications
  China *see* China, GI protection
  collective dimension, handicrafts and link to origin in culture 318–19
  definition, Australia, wine GI legislation 266–9
  definition, WIPO 102–4
  'fair trade' rules 208–10, 213, 215–18, 233–4, 236, 239–40
  generic status and genericide 523
  Geographical Indications Committee (GIC), Australia 271–2, 273–7, 278–9, 284–5
  global labelling challenge 52–7
  India, Feni liqueur 463–4, 477, 478–80, 482–3
  India Geographical Indications of Goods (Registration and Protection) Act 302–3
  'indication of source' definition 99–100
  and Indigenous knowledge *see* Australia, Indigenous knowledge and GIs of origin
  non-agricultural and non-foodstuff products 293–4, 295, 297–8
  protection and TRIPS *see* WTO TRIPS Agreement, protection of GIs
  and trade marks, resolving conflicts *see* conflicts between trade marks and GIs, resolving
  TRIPS extension rethink *see* WTO TRIPS Agreement, GI extension rethink and EU policies
  see also classification; trade marks
George, Pierre 77
Germany
  Bavarian Blockmalz candy 522–3
  Bayerisches Bier (Bavarian beer) 380, 381, 388–9, 515–16, 523, 524, 546
  Chiemsee trade mark 384–5, 405–6, 411
  'FEET FIRST' trade mark, Australia 280–81
  Kerry Spring mineral water 392, 411
Gervais, Daniel 46, 103, 123–45, 146, 183, 201, 233, 326, 375, 376, 504
Gettler, Leon 236
Geuze, Matthijs 95–122, 185, 186, 190, 215, 296, 326, 419, 466, 520
Gevers, Florent 129, 138, 361
GI *see* geographical indications
Gibson, C 175
Gilg, A 48
Giovannucci, Daniele 48, 87, 172, 328, 334, 354, 443, 519
Giraud, G 309
Gironde wine production, France 21–2, 23, 24, 26, 27, 31, 35
global labelling challenge 46–71
  eco-labels 53
  embeddedness of products 54, 56
  GIs and conventions of place 52–7
  global label of origin types, variations in 49, 53–4, 56
  pre-existing European place names, use of 50–51
  process feature, significance of 53
  'quality' foods 54
  social constraints and conventions theory 54–5
  TRIPS definition of geographical indications 46–7, 48
  US opposition to GIs and global registry 50–52
  WIPO on GIs as intellectual property 53
  see also multilateral register
global labelling challenge, AOC labelling and *terroir* concept 57–70
  AOC delineation and ecological analysis 68, 69
  cahier de charges (certification requirements) 70
  cultural re-evaluation 60–61
  French administrative process 62–6
  French administrative process, Institut National des Appellations d'Origine et la Qualité (INAO) 63–5, 66, 67, 68, 135–6, 321–2
  French administrative process, producers' union 62–3
  French AOC export revenue rise 65

French rural development benefits 65–6
human factors (*savoir faire*) 68–70
natural factors as determinants, and environmental sustainability 66–8, 136
patrimonialization 59, 60, 62
Saint Joseph wine and environmental gain 67–8
'shared corporeality' of agro-food practice 61
taste factors 69–70
terroir interpretation 57–62
xenophobic interpretation of *terroir* 61
globalisation versus localisation debate 90–91
see also terroir and sense of place
Goebel, Burkhart 46, 136, 162, 197, 279, 361–95, 403, 408, 544
Goldberg, S 170, 235
Goodman, David 47–8, 54, 61, 70, 126
Gopalakrishnan, N 318–19
Gorgonzola cheese, Italy 387–8
Gouda cheese, Netherlands 515
grands crus classification 38–9
Grazioli, A 125, 158, 184, 220, 234, 293, 325
Great China Wall trade mark, EU 385
Great Western trade mark, Australia 247–9, 281–3
Greece
  Feta cheese *see* Feta cheese
  Peza Olive Oil 449
  Zagora apples 449
Groeschl, Manuela 46, 136, 162, 279, 361–95, 403, 408
Gruyère cheese, Switzerland 449, 452, 453–5, 457
Guanxi Sweet Shaddock (pomelo) trade mark, China 356, 497
Gupta, Akhil 91
Guy, K 23

Haight Farley, C 164
Halloumi cheese, Cyprus 155, 531, 533
Hamilton, M 240
handicrafts and link to origin in culture 292–326

agricultural goods and foodstuffs 311–14, 321–2
agricultural raw products obtained from a local variety 311–12
characterisation of handicraft goods 303–7
collective dimension of GIs 318–19
Colombia *see* Colombia
concepts underlying GIs 297–303
discrimination on basis of subject matter categories, lack of justification for 316–17
EU, implications for 322–5
EU interpretation and *sui generis* regime 293–5
EU, Protected Designation of Origin (PDO) 299–300, 315, 324–5
EU, Protected Geographical Indication (PGI) 300–301, 307, 308, 309–10, 324–5
foodstuffs 308–10
'generic quality' of raw material 306
geographical origin definition 318–19, 324
'geographical origin' interpretation 302–3
human factors 303–17
human factors, validity of GIs based on 317–20
international recognition of TRIPS distinction between wines and spirit and other goods 325–6
link requirement, evaluation consistency 316–20
link strength 320
Lisbon Agreement on appellations of origin 298–9, 303, 307, 314, 326
natural and human factors, combination of 311–17, 319
non-agricultural goods 314–16, 322–3
processed products obtained from local raw material 312–13
processing, natural factors influencing 316
processing under special environmental factors 313–14
production methods, sophisticated 304–5

proof of historically verifiable
    human factors 319
  raw material, absence of specific
    origin requirement 306-7,
    308-10
  raw material, natural factors as
    source 314-15, 319
  traditional designs and drawings
    304-5, 319-20
  TRIPS and GI protection 301-3
  UK *see* UK
  'uniqueness' issues 305, 307
  Vietnam, Hué hat 315
handicrafts and link to origin in
    culture, France
  Alsace geranium 310
  Bergamote de Nancy sweets 309
  Calisson d'Aix sweets 310
  Châtaigne d'Ardèche 311-12
  Cholet textiles 316
  Dentelle du Puy (lace) appellation of
    origin 304
  implications for 320-22
  Morteau sausage 314
  Moustiers earthenware 307
  Nyons olive oil 312-13
  Pâtes d'Alsace 309-10
  Pierre de Bourgogne 322
  Porcelaine de Limoges and
    Porcelaine de Nevers 307
  Poterie de Valauris 314
  Roquefort cheese 112, 235, 313, 523,
    536, 544
  terroir concept 297
handicrafts and link to origin in
    culture, India
  Aranmula Metal Mirror 316
  Darjeeling Tea 312, 523, 528, 540
  Feni liqueur *see* India, Feni liqueur
    and welfare enhancement
  Kancheepuram Silk 304, 306,
    317-18
  Kashmir Pashmina 315
  Konark Stone Carving 306
  Kondapalli Bommalu figurines 304,
    306
  Monsooned Malabar coffee 313-14
  Mysore Silk 306-7
  Navara Rice 311
  Pipli Appliqué Work 318

Pochampally Ikat and Orissa Ikat
    (tie-dye) 305, 320
Handler, Michael 44, 48, 108, 124, 141,
    146-82, 223, 236, 325, 393, 466,
    517, 544, 545
Harris Tweed certification trademark,
    UK 307
Harte-Bavendamm, Henning 138, 361,
    524
Heath, Christopher 51, 133, 217,
    396-439, 543
Hénin, Stéphane 75
Hinrichs, C 54
Hong Kong, China Proposal, WTO
    multilateral register negotiations
    187, 189-90, 199-200
Hué hat, Vietnam 315
Hughes, Justin 150, 162, 175-6, 224,
    361, 372, 383, 463, 500, 513, 546-7
human factors
  global labelling challenge 68-70
  handicrafts and link to origin in
    culture 303-20

'identification' and 'indication' of
    a product, differences between
    208-10, 213, 223-4, 236-7
  see also 'fair trade' rules
Ilbery, B 484
IMFL ('Indian Made Foreign
    Liquors') 464-5, 473, 476-7, 481
imitation concerns 161-2
INAO, (Institut National des
    Appellations d'Origine et la
    Qualité), France 63-5, 66, 67, 68,
    135-6, 321-2
India
  Basmati rice 213-14, 219-20, 536-7
  Darjeeling tea 312, 523, 528, 540
  Geographical Indications of Goods
    (Registration and Protection)
    Act 302-3
  GIs for non-agricultural and non-
    foodstuff products 293-4
  handicrafts *see* handicrafts and link
    to origin in culture, India
  Monsooned Malabar coffee 313-14
  Navara Rice 311
  Proof of Origin (Historical Records)
    principle 319

tea and rice GIs 125, 214
India, Feni liqueur and welfare enhancement 308, 463–83
  alcoholic strength 469, 470
  branding advantages 480, 482, 483
  cashew varieties and production drop 473–4
  colonial legacy 471–2
  demand-side factors and production decline 476–7
  distillers as distributors 482
  entry deterrence and market concentration 478–9
  Feni evolution, specification and cultural practices 467–71
  Feni production decline 464–5, 473–7
  future scenarios, possible 482–3
  GI registration 463–4, 477, 478–9, 482–3
  GI specifications 465–7
  Goa economic development 464
  'Indian Made Foreign Liquors' (IMFL) 464–5, 473, 476–7, 481
  institutional issues and market segmentation 472–3, 478, 479
  market size and concentration 480–82
  monopoly rents issue 477, 480–81
  political economy and production drop 474–6
  product differentiation and GIs 479–80
  reclassification plans 471–2
  reputation, quality and price margins 480–81
  sale restrictions 471–2
  taxation and under-reporting of output 474–6
'Indian Made Foreign Liquors' (IMFL) 464–5, 473, 476–7, 481
indications *see* geographical indications
Indigenous knowledge, Australia *see* Australia, Indigenous knowledge and GIs of origin
information
  economic information, circulation of, France 36–7
  market failure and information asymmetry links 443
  WTO database proposal 193, 196
infringement procedures 336, 340, 355, 429–30
injunctions, Australia 269
Institut National des Appellations d'Origine et la Qualité (INAO), France 63–5, 66, 67, 68, 135–6, 321–2
institutional issues
  France, collective wine branding (19th–20th centuries) 40
  India 472–3, 478, 479
intellectual property protection
  19th century France 15–39
  'fair trade' rules 225–7
  World Intellectual Property Organization *see* WIPO
intellectual property tools for rural development objectives 440–62
  agricultural income and support to local employment, effects on 453–5
  collective organisations, and coordination of production and sales 445–6, 447, 449–51, 456, 458, 460–61
  commercial performances directly linked to organisational performance (EU PDO-PGI research programme) 448–58
  diversification practices and artisanal product 453, 455
  economic performance, costs and benefits comparison 444–5
  economic value linked to qualities and added value, distinction between 441–3, 459
  legal protection of commercial value through intellectual property rights 451–3
  marginal counterfeits, emergence of 452–3
  market failure and information asymmetry links 443
  market function 441
  market as place where economic value is generated 441–4, 446, 458, 459–60

name protection, importance of 452–3
organisation function 441
organisation as instrumental in retaining economic value 444–6
organisational performance
  as determining factor for commercial performance 449–51
  product differentiation and economic value creation 442, 459–60
  production volumes in collective organisations, regulation of 446, 450
  products differentiated by geographical origin, costs and benefits comparison 444–6
  protection of local landscapes 456
  quality products, effects of increasing demand for 443–4
  territorial performance 456–8, 459–61
  territory as beneficiary of impacts 447–8
  territory function 441
international markets *see* markets
international protection
  China, GI protection 356–7
  multilateral register negotiations *see* WTO multilateral register negotiations and international protection of GIs
  refusal of international registration 375–6
  WIPO registration system review 116–21
international standards *see* Madrid Agreement
see also WIPO-administered treaties, Madrid Agreement; WTO TRIPS Agreement
international trade effects
  'fair trade' rules 204–6
  generic status and genericide 512–13
  see also foreign GIs, protection for
Italy
  Fontina cheese 449, 452–5, 457, 529
  Gorgonzola cheese 387–8
  Parma ham 115, 449, 544

Parmesan cheese 112, 138, 449, 452, 453–5, 457, 515–16, 523, 538
Pilsen Urquell v Industrie Poretti 543

Jackson, John 239
Jacobsen, Rowan 47
Jacoby, J 164
Jamón de Terruel (ham), Spain 449
Janke, Terri 484–5
Japan
  customs duties, taxes and labelling practices on imported wines 228
  Sapporo beer 362
Jenkins, T 467
Jersey Potatoes, UK 449
Jing Dai 337
Jinhua Ham, China 347, 348–9, 351–2
Jordan, R 251
Josling, Tim 149, 173, 361
judicial review 22–3, 25–30, 211–13, 350, 357–8, 429–35
see also cases under individual countries; legislation

Kancheepuram Silk, India 304, 306, 317–18
kanterkaas cheese, Netherlands 147, 161
Karni, E 443
Kashmir Pashmina, India 315
Keon, J 233
Kerber, W 239
Kerr, W 176
Kerry Spring mineral water, Germany 392, 411
Kiesel, K 479–80
Kireeva, Irina 328–9, 337, 344, 351
Kirsten, J 15, 159, 484, 486, 489
Klieger, R 164
Knaak, Roland 151, 381–2, 403, 417–18, 430
Kneafsey, M 484
Knoll, A 207
Koenig, M 323
Konark Stone Carving, India 306
Kondapalli Bommalu figurines, India 304, 306
Kongolo, T 325
Kumar, Arun 475

Kumar, S 510
Kur, Annette 176, 361, 383

La Provence wine, France 264–6, 288
labelling
　administrative approval procedure for product labels 108–9, 113–14, 115
　Appellation d'Origine Contrôlée (AOC) labels *see under* France, collective wine branding (19th–20th centuries)
　Australia Label Integrity Program 252
　false labelling prevention, France, collective wine branding (19th–20th centuries) 26
　generic status and genericide 534–5, 536
　global *see* global labelling challenge
　labels of origin, EU administration of 52–3
　US Bureau of Alcohol, Tobacco and Firearms (ATF) wine-labelling regulatory scheme 42–3
　US certificate of label approval (COLA) 135
　see also branding
Lachiver, M 19, 21, 22, 23, 26
Lackert, Clark 138
Lacour, L 17, 20
Lactalis dairy cooperative, France 82
Ladas, S 208, 232, 413, 414, 415, 428, 435
LaFrance, M 164
Laing, S 206
Lamont, M 59
Landes, W 15, 16, 159
Lang, A 157, 168
language
　botanical taxonomy issues, Australia 494–9
　generic terms as part of linguistic public domain 510–11, 519, 524–5, 527–31
　use, 'fair trade' rules 207–10, 213–16, 221–2, 232–3
　see also translation
Larrimore Ouellette, Lisa 529
Le Goffic, Caroline 13, 510, 522, 541

legislation
　'ambush marketing' 171
　China 329–30, 355–8
　'fair trade' rules 216–18
　Indigenous law, Australia 490–91, 492, 501–7
　legal effects of registration 192–3, 194, 196–8
　legal protection of commercial value through intellectual property rights 451–3
　term status in legislation 535–8
　wine GI legislation, Australia *see* Australia, wine GI legislation
　see also cases under individual countries; judicial review
legitimacy negotiations, 'fair trade' rules 217–18, 224, 225–7, 228, 230–31, 232–4, 236, 237–41
Lenclud, Gérard 86
Leopold, E 47
Letablier, M 80
L'Etivaz cheese, Switzerland 445
Li, Xiaoxia 358
Liepins, R 70
Lindquist, L 184, 513, 546
Linnemer, L 16
Lisbon Agreement *see* WIPO-administered treaties, Lisbon Agreement
Lister, C 156
localisation versus globalisation debate 90–91
　see also terroir and sense of place
locally-produced foods and sustainability 173
Long, A 47
Long, C 164
Longdin, L 171–2
Lorvellec, L 527–8
Loureiro, M 176
Lyson, T 54

Maby, Jacques 78
McBride, W 176–7
McCarthy, J 126–7, 524, 529, 530, 531
McCluskey, J 176
Madrid Agreement *see* WIPO-administered treaties, Madrid Agreement

Mahé, L 207
Maher, M 42, 43, 466
Mandal, R 472, 473
Marchenay, P 47, 59, 61, 67, 70, 81–2, 301, 318, 319
Marescotti, A 55
Marie-Vivien, Delphine 38, 62, 65, 85, 135, 219, 292–326, 365
markets
　'ambush marketing' legislation 171
　economic advantages in international 152–4, 176
　efficiency and trade marks 159–60
　entry deterrence and market concentration, India 478–9
　market conditions as generic status evidence 533–5
　as place where economic value is generated 441–4, 446, 458, 459–60
　segmentation, India 472–3, 478, 479
Markham, D 19
Marsden, T 47
Martin, B 456
Martín, José Manuel Cortés 48, 51, 183–201, 223, 393, 466
Martino, T 165
Mather, C 181
meat
　Agneau du Quercy (sheep meat), France 449
　Bresse poultry production, France 81
　Jamón de Terruel (ham), Spain 449
　Jinhua Ham, China 347, 348–9, 351–2
　Melton Mowbray Pork Pie, UK 308–9
　Morteau sausage, France 314
　Parma ham, Italy 115, 449, 544
　Scottish Lamb, UK 449
　Ternasco de Aragon (lamb meat), Spain 449
'mélanges' (blending) practice, France 21, 23, 24, 36
Melton Mowbray Pork Pie, UK 308–9
Menival, David 135–6
Mercurio, Bryan 179, 181, 517, 545
Merville potatoes, France 449

Mexico
　Café Chiapas 115
　EU bilateral trade agreement 179
Meyer, Mireille 74
Michelet, Jules 74
Miller, D 56
mineral water
　Alaska 385
　Kerry Spring, Germany 392, 411
minimalist approach, WTO multilateral register negotiations 187, 188–9, 192–4, 195–7, 199
misleading indications
　Australia, wine GI legislation 256–8, 263–4, 269–71, 280–83, 287–8
　Budweiser cases and GIs v trade marks 412–14, 416–17, 418
　'fair trade' rules 223–4, 229–33, 236
　see also false indications
misleading trade marks
　China 347, 350, 354, 355
　conflicts between trade marks and GIs, resolving 371, 383–5, 386, 389–90
misrepresentation and misappropriation 158, 159–63, 168–70, 513
monopoly avoidance 382–93
monopoly rents issue, India 477, 480–81
Monsooned Malabar coffee, India 313–14
Moran, Warren 152, 493, 499, 500
Morlat, R 76–7
Morrin, M 164
Morteau sausage, France 314
Moschini, GianCarlo 478, 480, 482
Moustiers earthenware, France 307
Mukhopadhyay, Pranab 49, 463–83
multilateral register
　global registry, call for 48, 520
　negotiations see WTO multilateral register negotiations and international protection of GIs
　proposal 393–4
　see also global labelling challenge
Munzer, Stephen 159, 463, 466, 489, 490, 499, 500
Murdoch, J 47
Mysore Silk, India 306–7

Nair, Latha 510
national treatment
  conflicts between trade marks and GIs, resolving 380–82, 387–8, 390–91, 394
  developing countries and domestic protection 218–20, 221–3
  domestic legal systems and WTO multilateral register negotiations 191–2
  WIPO-administered treaties 100–101, 102, 103–11
  see also individual countries
native title, formal recognition, Australia 504–6
  see also Australia, Indigenous knowledge and GIs of origin
Navara Rice, India 311
Nelson, P 443
Netanel, N 151
Netherlands
  Boeren Leidse met Sleutels cheese 449, 452, 454, 455, 457
  Edamer cheese 449, 452, 453, 454, 455, 457, 515, 546
  Gouda cheese 515
  kanterkaas cheese 147, 161
  Opperdoeze ronde potatoes 449
New Zealand, *Wineworths Group Ltd v Comité Interprofessionel du Vin de Champagne* 160
Newcastle Brown Ale, UK 478
Nicolas, F 80
Niedermann, A 528
Nigh, T 68
notification systems *see* registration
Nuckton, C 42
Nyons olive oil, France 115, 312–13, 449

O'Connor, Bernard 293, 358, 415, 486, 519
Okediji, Ruth 146, 361, 383, 390
olive oil
  Nyons, France 115, 312–13, 449
  Peza, Greece 449
Olson, Mancur 463
Olszak, Norbert 297, 310, 365
Opperdoeze ronde potatoes, Netherlands 449

organisation function, intellectual property tools for rural development objectives 441, 444–6, 449–51
origin, and Indigenous knowledge *see* Australia, Indigenous knowledge and GIs of origin
Osmond, R 251
Ostertag, M 231

paintings as communication method, Australia, Indigenous knowledge 502–3, 505–6
Panizzon, Marion 485
Papadopoulos, A 47
Paris Convention *see* WIPO-administered treaties, Paris Convention
Parma ham, Italy 115, 449, 544
Parmesan cheese, Italy 112, 138, 449, 452, 453–5, 457, 515–16, 523, 538
passing off action 160–61, 252–6, 288
Pâtes d'Alsace, France 309–10
patrimonialization, global labelling challenge 59, 60, 62
PDO (Protected Designation of Origin), EU 83–4, 299–300, 315, 324–5, 448–58
Pecqueur, B 447, 458
Peer, S 59
Pélisser, Paul 77, 78
Pellegrino, Michael 511
Perri, G 65
Perrot, A 16
PET (priority, exclusivity and territoriality) protection 363–4, 366–7, 371–82, 386–8
Peza Olive Oil, Greece 449
PGI (Protected Geographical Indication), EU 83–4, 300–301, 307, 308, 309–10, 324–5, 448–58
phase-out period, and third-party use 365, 366, 374–5
phylloxera root louse 14, 23, 39–40
Pickett, A 526
Pierre de Bourgogne, France 322
Pipli Appliqué Work, India 318
Pires de Carvalho, Nuno 381

place
  conventions of, global labelling challenge 52–7
  of manufacture, understanding of 20
  names, pre-existing European 50–51
  and *terroir see* terroir
Plaisant, M 430
Plaisant, R 544
Plzen beer, Czech Republic 422, 543
Png, I 16
Pochampally Ikat and Orissa Ikat (tie-dye), India 305, 320
Polanyi, K 54
political economy and production drop, India, Feni liqueur 474–6
Pollan, M 172
Poncet, Yveline 85, 86
Porcelaine de Limoges and Porcelaine de Nevers, France 307
port wine 237, 544
Porter, M 442
Portugal, Torres Vedras wine 378–9
Posner, R 15, 16, 159
Postel-Vinay, G 35
Poterie de Valauris, France 314
Prabhudesai, S 465, 481
Prat, Andrea 463
premier crus wine selling 'by subscription' 19
prior trade mark rights
  Budweiser cases 420–21
  exclusivity 366, 368–71, 376, 378
  WIPO-administered treaties 118, 121
  see also trade marks
priority, exclusivity and territoriality (PET) protection 363–4, 366–7, 371–82, 386–8
processed products, handicrafts and link to origin in culture 312–14, 316
producer application to use a GI product, China 339, 343
product classification *see* classification
product definitions, disagreement over, France, collective wine branding (19th–20th centuries) 24–5
product differentiation
  and economic value creation 442, 459–60

India, Feni liqueur and welfare enhancement 479–80
product reputation *see* reputation
production methods
  developing economies 85–6
  sophisticated, handicrafts and link to origin in culture 304–5
production technique control, France, collective wine branding (19th–20th centuries) 38
production volumes in collective organisations, regulation of 446, 450
Profeta, A 324
Proof of Origin (Historical Records) principle, India 319
Protected Designation of Origin (PDO), EU 83–4, 299–300, 315, 324–5, 448–58
Protected Geographical Indication (PGI), EU 83–4, 300–301, 307, 308, 309–10, 324–5, 448–58
protection reclamation possibilities 541–7
  see also generic status and genericide
public domain intrusions, 'fair trade' rules 203–4, 211–12

quality issues
  Australia, Indigenous knowledge 490
  economic value linked to qualities and added value, distinction between 441–3, 459
  EU quality schemes for agricultural products and foodstuffs 379–82
  France, collective wine branding (19th–20th centuries) 16, 21, 22, 28–34, 38, 44
  and imitation concerns 161–2
  INAO (Institut National des Appellations d'Origine et la Qualité), France 63–5, 66, 67, 68, 135–6, 321–2
  increasing demand for products, effects of 443–4
Product Quality Law, China 346
public quality signs, controversy over, France 16
'quality' foods, global labelling challenge 54

reputation, quality and price
   margins, India, Feni liqueur
   480–81
TRIPS Agreement, GI extension
   rethink and EU policies 161–2,
   171–4, 176
VDQS (*vin délimité de qualité
   supérieure*) status, France 42
WIPO-administered treaties 100,
   101, 102–3, 104, 106, 111, 113,
   120

Rafael, R 480
Rangnekar, Dwijen 49, 308, 463–83,
   488, 510
Rastoin, J 41
Raustiala, Kal 159, 463, 466, 489, 490,
   499, 500
Rautenberg, M 59
raw materials, handicrafts and link to
   origin in culture 306–7, 308–10,
   312–13, 314–15, 319
Rawls, J 206
Ray, C 61
Raynaud, E 446
Reblochon cheese, France 115
refusal of trade mark application *see
   under* conflicts between trade
   marks and GIs, resolving
regional hierarchy, Australia, wine GI
   legislation 269, 275–6
regional name protection, France,
   collective wine branding
   (19th–20th centuries) 20, 22, 26–7
registration
   application fee as registration
      deterrent, Australia 289
   of certification mark, US 113
   challenges, generic status and
      genericide 523
   Geographical Indications of Goods
      (Registration and Protection)
      Act, India 302–3
   global registry, call for 48
   Lisbon register *see* WIPO-
      administered treaties, Lisbon
      Agreement
   Madrid Agreement *see* WIPO-
      administered treaties, Madrid
      Agreement

multilateral register for wine and
   spirit GIs, call for 520
numbers, China 336
protection, Australia 263–4, 272–3,
   289
refusal of an international
   registration 375–6
requirements, China 331–2, 338–9,
   342–3, 347
and trade mark law 106–8
trade marks as 'possessions' 373–4
TRIPS Agreement, notification and
   registration system 139–44
WIPO international registration
   system review 116–21
see also administrative issues
Reichman, J 149
Reilly, Alex 503, 504, 505, 506
Reitman, D 16
Rendu, A 17, 20
Renting, H 447
reputation factors
   conflicts between trade marks and
      GIs, resolving 371, 378–9, 386,
      387, 389–90
   reputation, quality and price
      margins, India 480–81
   TRIPS Agreement, GI extension
      rethink and EU policies 159,
      160, 167–70, 174–5, 176
   WIPO-administered treaties 101–3,
      104, 110, 111, 113, 120
Resinek, N 378
Revel, J 59
Réviron, S 445
Ribeiro de Almeida, Alberto Francisco
   126
rice
   Basmati, India 213–14, 219–20,
      536–7
   Navara, India 311
Richardson, M 165
right to use as defence strategy,
   Budweiser cases 408–10
rights holders' action as generic status
   evidence 538–40
Rioja wine, Spain 544
Robinson, D 175
Roep, D 448
Roncin, François 66, 85

Ronga, G 430
Roquefort cheese, France 112, 235, 313, 523, 536, 544
Rose, B 51, 179, 547
Rothbury wine, Australia 284–5
Roudié, Philippe 19, 21, 23, 26, 27, 28, 77
Rovamo, Oskari 486, 487, 490, 491, 493, 494, 500
Roy, R 465
rural development
　French policies 65–6, 152–4
　and intellectual property tools *see* intellectual property tools for rural development objectives
Ryan, Ó 176

Sabot, S 67
Saint Joseph wine, France 67–8
Saint-Émilion wine, France 22, 261
Sancerre wine, France 261
Sapporo beer, Japan 362
Sautter, Gilles 77, 78
Schechter, F 165, 466
Scheffer, Sandrine 66, 85
Schoene, Volker 388, 522, 524–5
Schricker, G 546
Schroeder, W 68
Scotland *see under* UK
Sekiguchi, T 480
semi-generic terms 43–4, 530–31, 537
　'fair trade' rules 211, 212–13, 222, 223, 236
　TRIPS Agreement 134–6
　see also generic status
Shanxi Laochencu vinegar, China 353
Shapiro, C 15, 443
Sherman, Brad 217, 484–507
sherry wine, Spain 160, 167, 529–31, 539
Shetland Wool, UK 315
Shi, Xinzhang 350
Simioni, M 161, 480
Singapore, *Novelty Pte Ltd v Amanresorts* 168
Singh, Ranjay 484
Singhal, Shivani 485, 486, 490
Skol, A 378
Smallwood, D 16
Smith, A 41

Soam, S 125
South Africa
　EU bilateral trade agreement 179
　wine recognition 42
South Korea, EU-Korea Free Trade Agreement 180, 517–18
Spain
　Fuerteventura trade mark 385
　Jamón de Teruel (ham) 449
　Rioja wine 544
　sherry as generic term 530–31, 539
　sherry wine 160, 167, 529–31, 539
　Spanish Champagne 160, 252–6, 513–14
　Ternasco de Aragon (lamb meat) 449
Spence, M 170
Spencer, D 223
Spennemann, C 179
Stanziani, Alessandro 13–45, 59, 76, 84, 152, 257, 365
Stern, Stephen 43, 51, 85, 235, 245–91, 488, 517
Stevenson, I 14
Stewart, T 131, 183
Stilton cheese, UK 113, 531
Stresa Agreement 515
sui generis regime
　China 336–44, 346–7, 351–2, 353–8
　conflicts between trade marks and GIs, resolving 365, 366–7, 369
　EU 112–13, 293–5
　France 105–6
　generic status and genericide 521–3, 525, 541
　TRIPS Agreement 130–31, 143
　WIPO-administered treaties 105–6, 112–14
Sunder, Madhavi 221, 485
superiority-based rule 365–6, 367, 371–2
　see also conflicts between trade marks and GIs, resolving
sustainability 66–8, 136, 173
　see also environmental factors
sweets
　Bavarian Blockmalz candy, Germany 522–3
　Bergamote de Nancy sweets, France 309

Calisson d'Aix sweets, France 310
Pâtes d'Alsace, France 309–10
Switzerland
   Abricot Luizet du Valais 449
   Chocosuisse Union des Fabricants Suisses de Chocolat v Cadbury, UK 160
   Emmental cheese 515
   EU bilateral trade agreement 179
   GIs for non-agricultural goods 294
   Gruyère cheese 449, 452, 453–5, 457
   L'Etivaz cheese 445
Sylvander, B 46, 47, 49, 54, 55, 66, 87, 238, 239, 240, 292, 444, 446, 459

taste factors 69–70
   see also global labelling challenge, AOC labelling and *terroir* concept
Taubman, Antony 46, 147, 182, 202–42, 545
taxation 37–8, 474–6
teas 125, 214
   Darjeeling, India 312, 523, 528, 540
   see also coffee
Teply, L 530
term status in legislation or official classifications 535–8
   see also generic status and genericide
Ternasco de Aragon (lamb meat), Spain 449
territorial performance, rural development objectives 441, 447–8, 456–8, 459–61
territoriality principle 521–2
terroir
   and AOC labelling *see* global labelling challenge, AOC labelling and *terroir* concept
   handicrafts and link to origin in culture 297
   xenophobic interpretation 61
terroir and sense of place 72–91
   agriculture, environmental conditions and traditional practice 81, 85
   anthropological interpretation 86–91
   Australia 491–4, 499–504, 506
   Bresse poultry production 81
   concept issues 81–2

Dombes carp farming 81–2
farmers' markets 89–90
farming and land ownership schemes as obstacles 81–2
French AOC system 79–80
French context and history 73–80, 135–6
French context, human geography and pedology 74–6, 85
French supermarket sector, influence on food-processing industry 82
French *terroir* definition debate 76–8, 79–80, 84–5
globalisation versus localisation debate 90–91
Lactalis dairy cooperative 82
local development and spatial planning tool 78–9
local production, motivation and management methods 87–90
localised production systems as '*produits de terroir*' 86–7
misunderstandings and complex social factors 80–86
Normandy Camembert AOC 82
production systems of emerging and developing economies 85–6
Protected Designation of Origin (PDO) 83–4
Protected Geographical Indication (PGI) 83–4
'sense of place' concept 86–7
'Terroir et Cultures' organisation 80, 81
translation from French, problems with 84–6
TRIPS Agreement protection of GIs 83
UNESCO 'Planète Terroirs' 80
Thailand, GIs for handicraft goods 294
Thévenod-Mottet, E 302, 325, 440, 444
Thévenot, L 54, 55, 59, 459
third party protection, Budweiser cases 427–9
third-party use and phase-out period 365, 366, 374–5
thresholds for generic status 523–5, 541
   see also generic status and genericide
Tian Furong 345, 346, 347
Torres Vedras wine, Portugal 378–9

TPP (Trans-Pacific Partnership
    Agreement) 520–21, 528–9, 538
Tracy, M 42
trade
    bilateral trade agreements 178–82,
        234–7, 356, 402–4, 409, 415–17,
        435
    'fair trade' rules *see* 'fair trade'
        rules
    international trade effects, generic
        status and genericide 512–13
    regulation (1830s), France 18–19
    trade opinions as generic status
        evidence 531–2
trade marks
    Australia 247–9, 278–9, 280, 283,
        284–5
    Budweiser cases *see* Budweiser cases
        and GIs v trade marks
    China, Trade Mark Law 113, 328,
        329–31, 333, 334, 335–6,
        347–50, 351–2, 355–6, 357,
        497
    conflict resolution *see* conflicts
        between trade marks and GIs,
        resolving
    dilution of famous trade marks
        43–4, 108, 163–8
    EU Trade Marks Directive 169, 368,
        383, 388, 390, 391–2, 408, 410,
        411
    France, collective wine branding
        (19th–20th centuries) 15–16,
        20, 36
    prior trade mark rights *see* prior
        trade mark rights
    protection regime, generic status
        and genericide 519, 521, 523–4,
        529–30, 532, 540
    TRIPS Agreement protection 131,
        133, 136–9, 159–68
    US 43–4, 52, 217
    WIPO-administered treaties 96–7,
        106–8
    see also classification; geographical
        indications
traditional designs and drawings
    304–5, 319–20
traditional economic arguments,
    justification through 159–63
traditional expressions, protection for,
    Australia 262–3, 289
'traditional knowledge' definition,
    Australia 485–6, 488, 489–90
traditional practice in agriculture 81,
    85
traditional production, 'fair trade' rules
    203, 219
Trans-Pacific Partnership Agreement
    (TPP) 520–21, 528–9, 538
Transatlantic Trade and Investment
    Partnership (TTIP) 520
translation
    from French, problems with 84–6
    of indications, Budweiser cases 399,
        410–11, 421
    protection of GIs against use in
        translation 372–3
    see also language
Tregear, A 176, 309
TRIPS Agreement *see* WTO TRIPS
    Agreement
Troplong, R 19
Trotta, G 515
Trubek, Amy 47, 88
TTIP (Transatlantic Trade and
    Investment Partnership) 520
Turbull, David 492
Tushnet, R 163

UK
    Cheddar cheese 449, 452, 454, 455,
        457, 515, 538, 546
    Jersey Potatoes 449
    Melton Mowbray Pork Pie 308–9
    Merchandise Marks Acts 514
    Newcastle Brown Ale 478
    Port and Madeira protection 544
    Scotland, Harris Tweed certification
        trademark 307
    Scotland, Native Shetland Wool 315
    Scottish Lamb 449
    sherry as generic term 530–31, 539
    Shetland Wool 315
    Stilton cheese 113, 531
    unfair competition rules 231
UK, cases
    *Bollinger v Costa Brava Wine
        Company* (Spanish Champagne)
        160, 253, 513–14

Chocosuisse Union des Fabricants Suisses de Chocolat v Cadbury 160
Fage UK v Chobani UK 418
John Walker & Sons Ltd v Henry Ost 160
Northern Foods v The Department for Environment, Food and Rural Affairs 309
Taittinger v Allbev 160
Vine Products Ltd v MacKenzie & Co Ltd (Sherry) 160, 167, 529–30, 539
Ulin, R 59
UNESCO 'Planète Terroirs' 80
unfair competition concerns
  China, Anti-Unfair Competition Law 345
  Paris Convention 109–10
  TRIPS Agreement 130–31, 139
  UK 231
  unfair business practices, laws focusing on 109–11
'uniqueness' issues 305, 307
  see also handicrafts and link to origin in culture
Urry, J 48
Uruguay Round *see* WTO Uruguay Round of negotiations
US
  American Origin Products Association 89
  Budweiser beer 51–2
  Bureau of Alcohol, Tobacco and Firearms (ATF) wine-labelling regulatory scheme 42–3
  certificate of label approval (COLA) 135
  Champagne advertising campaign 546–7
  'dilution' of famous trade marks 43–4
  EU bilateral trade agreement 179, 180–81, 236, 516–17, 518
  Food and Drug Administration (FDA) and generic status 536–7
  France, collective wine branding (19th–20th centuries) comparison 42–4
  'local food' and GIs 48

  market considerations and TRIPS Agreement 153–4
  opposition to GIs and global registry 50–52
  Pan-American industrial property conventions 231–2
  preferential trade agreements 179, 180–81
  registration of certification mark 113
  semi-generic terms, use of 43–4, 530, 537
  'Taste of Place' conference, Vermont 88–9
  Trademark Dilution Revision Act 217
  Transatlantic Trade and Investment Partnership (TTIP) 520
  US–Australia Free Trade Agreement 278–85
  'Wine Pact' with EU 134–5, 517, 547
  wine recognition 42
  wine trade mark protection 43–4, 52
US, cases
  Anheuser-Busch 397
  Bayer Co. v United Drug Co. 530
  Community of Roquefort v William Faehndrich 235, 523, 536
  Fontina 529
  Institut National des Appellations D'Origine v Vintners Int'l (Chablis with a Twist) 207, 211–12, 233
  Moseley v Victor's Secret Catalogue 163, 217
  Syncom Formulations v SAS Pharmaceuticals 211
  Tea Board of India v The Republic of Tea (Darjeeling) 523, 528, 540
  Ty v Perryman 163

Valadier, A 65
Valeschini, E 446
value
  Australian wine GI 249–50
  economic value linked to qualities 441–3, 459
  legal protection through intellectual property rights 451–3

market as place where economic
value is generated 441–4, 446,
458, 459–60
organisation as instrumental in
retaining economic value 444–6
product differentiation and
economic value creation 442,
459–60
see also cost factors; economic
factors
Van Caenegem, William 126, 152, 153
Van de Kop, Petra 49, 87
Van der Merwe, Andries 545
Van der Ploeg, J 47
Vanzetti, A 437
VDQS (*vin délimité de qualité
supérieure*) status, France 42
Versailles Peace Treaty 231, 415, 544
Vietnam, Hué hat 315
Viju, C 180, 517
Vittori, Massimo 49, 126, 158, 175
Vivas-Eugui, David 124, 161, 179, 234

Wadlow, C 231
Wallman, Melinda 535
Wallon, A 23
Wang, Min-Chiuan 125, 327
Watal, Jayashree 123
Watts, M 54
Waye, V 178
Weatherall, Kim 520–21
Weinrib, E 170
welfare enhancement *see* India, Feni
liqueur and welfare enhancement
Wilkinson, J 54
Wilkinson, Percy 245–6
Wilson, James 58, 136, 491
wines and spirits
Armagnac, France 17, 27
Australia *see* Australia, wine GI
legislation
Beaujolais litigation, Australia
256–7, 258, 261, 262
Bons Bois spirit, France 115
Bordeaux, France 17, 18, 19–20, 21,
22, 23, 26, 27, 31, 35, 38, 261
Burgundy, France 23, 28, 208
Chablis *see* Chablis
Champagne *see* Champagne
Chardonnay 134, 500

China, special provisions 332–3
Cognac, France 27, 341
collective wine branding *see* France,
collective wine branding
(19th–20th centuries)
Coonawarra litigation, Australia
207, 249–50, 268–9, 273–7
EU Wine Regulation 366, 377–9
EU-US 'Wine Pact' 134–5, 517, 547
Gironde wine production, France
21–2, 23, 24, 26, 27, 31, 35
La Provence, France 264–6, 288
multilateral register *see* multilateral
register
port wine 237, 544
Rioja, Spain 544
Rothbury trade mark dispute,
Australia 284–5
Saint Joseph, France 67–8
Saint-Émilion, France 22, 261
Sancerre, France 261
sherry, Spain 160, 167, 529–31, 539
Torres Vedras, Portugal 378–9
TRIPS Agreement protection 132–3,
139–41
US wine trade mark protection
43–4, 52
wine recognition 42
see also beer
WIPO-administered treaties 95–122
administrative approval procedure
for product labels 108–9,
113–14, 115
'appellation of origin' definition
100–102
certification and guarantee marks
107–8
collective mark use 107, 108, 113
'dilution' protection for reputed
marks 108
geographical area, relationship with
111–13
geographical indications definition
102–4
geographical signs, exemption of 107
global labelling challenge 53
goods of the same kind not
originating in the geographical
area of origin 120–21
history 95–8

'indication of source' definition 99–100
national treatment provisions 100–101, 102, 103–4, 105–6
national treatment provisions, differences in means of protection 104–11
non-geographical denominations 115, 119
prior trade marks, protection for 118, 121
product reputation factors 101–3, 104, 110, 111, 113, 120
quality factors 100, 101, 102–3, 104, 106, 111, 113, 120
Standing Committee on the Law of Trade Marks (SCT) 96–7
sui generis protection system 105–6, 112–14
trade mark law and registration 106–8
unfair business practices, laws focusing on 109–11
WIPO-administered treaties, Lisbon Agreement 95–8, 101–2, 103–4, 111, 113–16
Appellation of Origin (AO) and geographical indications, differences between 128–30
Budweiser cases *see* Budweiser cases and GIs v trade marks, Lisbon Agreement for the Protection of Appellations of Origin
China, GI protection 356
conflicts between trade marks and GIs, resolving 365–6, 374–7, 394–5
'fair trade' rules 232–3
generic status and genericide 513, 541, 542–3
Geneva Act 102, 104, 119–21, 143, 395
handicrafts and link to origin in culture 298–9, 303, 307, 314, 326
international registration system review 116–21
limited membership 114, 116
Madrid and Hague systems, parallels with 116–17, 118
register 125–6
Secretariat on Lisbon Agreement 375–6
Working Group on the Development of the Lisbon System (Appellations of Origin) 97–8, 117–19
WIPO-administered treaties, Madrid Agreement
Budweiser cases 414, 415, 421, 426, 429
China and international registration system 356–7
'fair trade' rules 232–3
generic status and genericide 513, 541–2
on Indications of Source 100
international registration system 108, 356–7
Lisbon and Hague systems, parallels with 116–17, 118
WIPO-administered treaties, Paris Convention
Budweiser cases 412–13
'fair trade' rules 216, 218, 224–5, 230–32
generic status and genericide 512
Indications of Source (IS) and Appellations of Origin (AO) definitions 99, 100–101
on misleading allegations 131, 137
on unfair competition 109–10
Wiseman, Leanne 484–507
Wiskerke, H 448
Wood, G 528
World Intellectual Property Organization *see* WIPO
World Trade Organization *see* WTO
WTO
Doha Development Agenda 133–44, 187, 192, 195–6, 215, 393–4
GATT Agreement 148–9, 156, 219, 227–8, 236
Panel Report 367–72, 379, 387, 394, 408–9
Uruguay Round negotiations 128–9, 148–9, 156, 183–6, 218
WTO multilateral register negotiations and international protection of GIs 183–201

'Built-In Agenda' for future
    negotiations 186–7
cost factors 197–8
critical appraisals 192–4
developing countries 195–6
and domestic legal systems 191–2
Draft Composite Text on
    proposed multilateral register
    195–6, 201
EU-led Proposal (comprehensive
    approach) 187–8, 190–92, 194,
    195, 197–8, 199–200, 201
free-riding concerns 197
future direction 195–200
generic status consideration 191–2
Hong Kong, China Proposal 187,
    189–90, 199–200
information database proposal 193,
    196
legal effects of registration 192–3,
    194, 196–8
minimalist approach (US-led Joint
    Proposal Group) 187, 188–9,
    192–4, 195–7, 199
participation as voluntary or
    mandatory 198–9
presumption of eligibility for
    protection, provision of 197–8
product coverage issues 199–200
Seoul Summit Document 187
submission divisions 187–8
system participation concerns
    193–4
WTO TRIPS Agreement
    collective management and 'fair
        trade' rules 221–5
    compliance, China 332, 356, 357
    conflicts between trade marks and
        GIs 133, 367, 369–70, 371–2,
        375, 377, 378, 379, 381–2, 383,
        387
    definition of GI 46–7, 48, 188, 190,
        191, 193, 196–8
    distinction between wines and spirit
        and other goods 325–6
    'fair trade' rules 208–10, 213,
        215–16, 218, 233–4, 236, 239–40
    'fix-rule' approach 240–41
    generic status and genericide 508–9,
        518–21, 526, 531–2

'identification' and 'indication' of
    a product, differences between
    208–10, 213, 223–4, 236–7
'IP-is-trade' paradigm 219, 240
and language usage 213–16, 221–2
multilateral register negotiations
    183–5
WTO TRIPS Agreement, GI extension
    rethink and EU policies 146–82,
    223–4
'ambush marketing' legislation 171
bilateral preferential trade
    negotiations, significance of
    178–82
'brand entrepreneurs' and incentive
    preservation concerns 165
costs and benefits concerns 158,
    161–2
customer protection and trade
    marks (misrepresentation-based
    standard) 159–63
dilution of famous trade marks,
    prevention of 163–8
diversity in agricultural production,
    call for 172–3
dual minimum standards of GI
    protection (Articles 22 and 23)
    146–8, 156–7, 158
economic advantages for producers
    in international markets,
    securing 152–3, 176
'extended passing off' action 160–61
free-riding concerns 158, 168–70
French rural policy and GI
    regulation, influence on EU
    policies 152–4
generic product descriptions 154–5,
    162–3, 165–8
interests and motivations of parties
    concerned 152–6
justification through traditional
    economic arguments 159–63
justifications, assessment of
    purported 156–77
locally-produced foods and
    sustainability 173
market considerations in US and
    Australia 153–4
misappropriation prevention 158,
    168–70

quality goods, fostering production and consumption 171–4, 176
quality, and imitation concerns 161–2
reputation factors 159, 160, 167–70, 174–5, 176
stand-alone arguments 157–8
trade marks and market efficiency 159–60
WTO TRIPS Agreement, protection of GIs 123–45
administrative issues 139–40, 143
collective or certification marks 136, 139
developing countries 124–5, 140, 174–7
dual-purpose denominations 134–6
emergence of 127–9
environmental significance of GIs 126
'fair use' exception to rights conferred by a trade mark 138
'first-in-time, first-in-right' approach 128, 136–9
genericide rules 139
geographical origin concept 129–30
higher protection for products other than wines and spirits 141–2
homonymous indications 131–2, 139
notification and registration system 139–44
opponents' case against GI protection 126–7
semi-generic denominations 134–6
substantive protection 130–33
sui generis system concerns 130–31, 143
terroir and sense of place 83
trade marks, protection under 131, 136–9
unfair competition concerns 130–31, 139
for wines and spirits 139–41
wines and spirits protection 132–3

xenophobic interpretation of *terroir* 61
see also global labelling challenge, AOC labelling and *terroir* concept
Xianglian lotus seeds, China 350

Yeung, M 176

Zacher, F 511
Zagora apples, Greece 449
Zahn, L 530, 537
Zalik, A 157
Zhang, Yumin 355
Zhangqiu Scallion trade mark, China 356
Zhao Xiaoping 347, 348
Zheng, Haiyan 49, 113, 327–58